Springer Series on Atomic, Optical, and Plasma Physics

Volume 82

AF774200

For further volumes:
http://www.springer.com/series/411

The Springer Series on Atomic, Optical, and Plasma Physics covers in a comprehensive manner theory and experiment in the entire field of atoms and molecules and their interaction with electromagnetic radiation. Books in the series provide a rich source of new ideas and techniques with wide applications in fields such as chemistry, materials science, astrophysics, surface science, plasma technology, advanced optics, aeronomy, and engineering. Laser physics is a particular connecting theme that has provided much of the continuing impetus for new developments in the field, such as quantum computation and Bose-Einstein condensation. The purpose of the series is to cover the gap between standard undergraduate textbooks and the research literature with emphasis on the fundamental ideas, methods, techniques, and results in the field.

Michael Bonitz · Jose Lopez
Kurt Becker · Hauke Thomsen
Editors

Complex Plasmas

Scientific Challenges and Technological Opportunities

Springer

Editors
Michael Bonitz
Institut für Theoretische Physik und Astrophysik, Lehrstuhl Statistische Physik
Christian-Albrechts-Universität zu Kiel
Kiel
Germany

Jose Lopez
Seton Hall University
South Orange, NJ
USA

Kurt Becker
Polytechnic Institute of New York University
Brooklyn, NY
USA

Hauke Thomsen
Institut für Theoretische Physik und Astrophysik, Lehrstuhl Statistische Physik
Christian-Albrechts-Universität zu Kiel
Kiel
Germany

ISSN 1615-5653 ISSN 2197-6791 (electronic)
ISBN 978-3-319-37587-8 ISBN 978-3-319-05437-7 (eBook)
DOI 10.1007/978-3-319-05437-7
Springer Cham Heidelberg New York Dordrecht London

Printed on acid-free paper

Springer is part of Springer Science+Business Media (www.springer.com)

Preface

Plasma physics is an extremely broad field covering charged particle systems with temperatures from micro-Kelvins to millions of degrees, and densities ranging from the dilute matter in interstellar space to plasmas in the interior of compact stars that are million times denser than ordinary condensed matter. *Complex plasmas* are a special type of low-temperature plasmas characterized by the presence of nano or microparticles, by chemical reactions as well as by the interaction of plasmas with solid surfaces. These additional ingredients give rise to many technological applications that have emerged only recently. At the same time, complex plasmas have many peculiar properties and are experimentally and theoretically quite complicated. Complex plasmas have strong overlap with other fields of physics, including atomic and molecular physics, condensed matter physics, and material science. They require special techniques and approaches that are often not covered in standard plasma physics textbooks or courses.

Responding to this challenge, the present book provides a collection of tutorial style review articles that give graduate students and young researchers an introduction into various aspects of complex plasmas. This book continues and extends the material of the successful book "Introduction to Complex Plasmas," edited by M. Bonitz, N. Horing, and P. Ludwig that was published by Springer in 2010. It contains new results in basic and applied research and has a particular focus on modern technological applications of complex plasmas.

Part I of this book is devoted to dusty plasmas and covers fundamental issues such as phase transitions and nonequilibrium effects in streaming plasmas. Part II is devoted to quantum effects of the electrons that become increasingly important in complex plasmas and their description with modern simulation techniques. Part III covers fundamental properties of lower temperature plasmas including experimental diagnostics and modern computer simulations. Part IV concludes the book with an overview of various modern technological applications of complex plasmas.

This book is based on tutorial lectures given at the Third Graduate Summer Institute on "Complex Plasmas" at *Seton Hall University*, South Orange, NJ (USA) from July 30 to August 8, 2012. The workshop was jointly organized by the SFB-TR24 "Fundamentals of Complex Plasmas" Greifswald/Kiel (Germany) and Seton Hall University. We gratefully acknowledge financial support by the Deutsche Forschungsgemeinschaft via SFB-TR24, the Army Research Office

(ARO), National Science Foundation (NSF), the U.S. Department of Energy (DoE), the Polytechnic Institute of New York University, and the Princeton Plasma Physics Laboratory (PPPL) and the help of many people who have made the workshop and this tutorial book possible.

Kiel, NY, USA, October 2013

Michael Bonitz
Jose Lopez
Kurt Becker
Hauke Thomsen

Contents

Part I Dusty Plasmas

Part III Low-Temperature Plasmas

Part I
Dusty Plasmas

Chapter 1
Phase Transitions in Dusty Plasmas

Hauke Thomsen, Jan Schablinski and Michael Bonitz

Abstract This chapter is devoted to phase transitions in dusty (complex) plasmas which contain micrometer-size highly charged particles. These systems allow for studying the thermodynamic many-body features on a single-particle level. We start with a brief introduction to dusty plasmas and the parameters that can be used to distinguish different phases in harmonically confined dust crystals in two and three dimensions. These finite systems behave differently than bulk systems as their structure is dominated by centric rings in 2D or spherical shells in 3D. Both Monte Carlo and (Langevin) molecular dynamics are addressed as two important simulation methods for classical interacting many body systems. In order to study phase transitions in complex plasma experiments, selective control over the dust kinetic temperature is essential. This control can be achieved by the laser manipulation method where dust grains are accelerated by randomly moving laser spots. Furthermore, parameters are discussed that allow one to detect phase transitions in finite systems in simulations and experiments: the relative inter-particle distance fluctuations and the bond angular order parameter.

H. Thomsen (✉) · M. Bonitz
Institut für Theoretische Physik und Astrophysik, Christian-Albrechts-Universität zu Kiel, Leibnizstr. 15, 24118 Kiel, Germany
e-mail: thomsen@theo-physik.uni-kiel.de

M. Bonitz
e-mail: bonitz@theo-physik.uni-kiel.de

J. Schablinski
Institut für Experimentelle und Angewandte Physik, Christian-Albrechts-Universität zu Kiel, Leibnizstr. 15, 24118 Kiel, Germany
e-mail: schablinski@physik.uni-kiel.de

M. Bonitz et al. (eds.), *Complex Plasmas*, Springer Series on Atomic, Optical, and Plasma Physics 82, DOI: 10.1007/978-3-319-05437-7_1,

1.1 Introduction

The plasma state is often referred to as the *fourth state of matter*. A *solid* body is characterized by a well-defined shape. Although the atoms are not necessarily arranged in a fixed lattice, they cannot move freely through the solid body. The *liquid* phase begins with the melting of a solid. This phase is characterized by a fixed volume but a variable shape. The atoms are not fixed to lattice positions, and they can move freely through the liquid. In contrast to a solid, in a liquid a long-range order does not exist. In the *gaseous* state, the system has variable shape as well as variable volume, and expands to the available volume. The atoms move through the volume with rare collisions, but they are still neutral as the electrons are bound to the atomic cores. Finally, the *plasma* state is characterized by a significant fraction of at least partially ionized atoms. Electrons and atomic cores are no longer strictly bound. In common dusty plasma experiments, the plasma is generated by an rf-discharge in an argon gas with a pressure of about $1 - 100\,\mathrm{Pa}$. The widely used rf-frequency is 13.56 MHz and the applied discharge powers are several watts [1, 2].

Dusty plasmas are an excellent model system to study strong correlation effects and phase transitions on an atomic level [3]. Due to the large size of the dust grains, compared to electrons, ions and most molecules, they can be traced by a standard CCD camera. Stereoscopic setups of several cameras in combination with a reconstruction algorithm allow to record the full 3D trajectories of a Yukawa ball [4]. Moreover, the typical time scale of the dust's dynamics is on the order of seconds. Therefore, filming the dust cloud with frame rates of about 100 fps is sufficient to record the full time development.

Phase transitions are a paramount feature of many-body systems. When a system parameter, in most cases the temperature, is continuously increased, the macroscopic properties dramatically change at the critical point as the system moves from one phase to another. Perhaps the most well-known examples are the *melting* of ice as a *solid-liquid* and the *boiling* of water as a *liquid-gas* phase transition. These two phase transitions, under standard conditions for pressure, were utilized by Anders Celsius to define a temperature scale in 1742 [5]. Another famous example is the phase transition of a ferromagnet, where the spontaneous magnetization vanishes above a critical temperature. Most phase transitions are investigated in the *macroscopic* (or *thermodynamic) limit* of large particle numbers, $N \gtrsim 10^{23}$. The collective behavior of many particles over long distances is a central property that characterizes a phase.

While macroscopic quantities like the critical temperature T_c or the heat of fusion are well-known for many systems and a broad range of conditions, many details on the atomic level still remain open. For this reason, the study of mesoscopic systems with a much lower particle number $10 \lesssim N \lesssim 10^4$ is very appealing. Since they contain large, optically visible microparticles, dusty plasmas allow for a tracing of all individual particles and hence, for a detailed investigation of structure, spatial correlations, and collective motion. Such an investigation on the single-particle level is unthinkable for macroscopic systems. A detailed discussion on particle diagnostics in dusty plasmas has been given in Refs. [6, 7].

In a plasma, the typical interaction energy of ions and electrons is small compared to their typical kinetic energy which implies that they are weakly coupled. Therefore, one would not expect to find strong correlations in the plasma state. However, when micro-meter sized spheres made from plastic or hollow glass are placed into the plasma, under usual experimental conditions, these particles accumulate a high negative charge $Q = \mathcal{O}(10,000)e_0$ [2, 3, 8]. These micro-meter sized spheres are usually referred to as *dust grains* or *dust particles*, and a plasma containing such particles is called a *dusty* or *complex plasma*. Due to their exceptional high charge, the dust grains can be *strongly coupled* even at room temperature. While their interaction would be described by a pure Coulomb interaction potential in vacuum, the interaction is modified by the ambient plasma. The mobile charges in a plasma, i.e., electrons and ions, have the ability to screen the electric field of a charge Q. The basic effect can be understood in a simple picture: When a mesoscopic negative charge Q is placed into the plasma, the electron density of its vicinity is reduced while the density of positive ions is enhanced. However, the quantitative treatment of the screening requires extensive calculations especially in a streaming plasma. A detailed discussion including wake-effects and non-reciprocal dust interactions is given in the next Chapters "*Introduction to Streaming Complex Plasmas A & B*" [9–11]. In weakly streaming plasmas or in the direction perpendicular to the ion flow, the *static screening* approach is adequate. In this case, the dust–dust interaction is described by a *Yukawa* or *Debye-Hückel* potential,

$$V\left(\mathbf{r}_i, \mathbf{r}_j\right) = V\left(\left|\mathbf{r}_j - \mathbf{r}_i\right|\right) = \frac{Q_i Q_j}{4\pi\epsilon_0} \frac{1}{\left|\mathbf{r}_j - \mathbf{r}_i\right|} e^{-\frac{\left|\mathbf{r}_j - \mathbf{r}_i\right|}{\lambda_\mathrm{D}}} . \tag{1.1}$$

Both electrons and ions contribute to the total *Debye screening length*

$$\lambda_\mathrm{D} = \left(\frac{n_\mathrm{e}^0 q_\mathrm{e}^2}{\epsilon_0 k_\mathrm{B} T_\mathrm{e}} + \frac{n_\mathrm{i}^0 Z_\mathrm{i}^2 q_\mathrm{e}^2}{\epsilon_0 k_\mathrm{B} T_\mathrm{i}}\right)^{-\frac{1}{2}} , \tag{1.2}$$

where n_e^0 (n_i^0) is the unperturbed density of electrons (ions) and Z_i is the charge number of the ions. The temperatures of electrons and ions T_e and T_i may differ due to the heating mechanism of the plasma. Here, we assume one single species of ions. However, (1.2) is readily extended to several species of ions by replacing the second summand by a sum over all species. In the following, we use the *screening parameter* $\kappa = \lambda_\mathrm{D}^{-1}$ as the inverse Debye length. The *Coulomb* potential can be interpreted as a special case of the Yukawa potential with zero screening, $\kappa = 0$.

Despite the Debye screening, the force between the equally charged dust grains is purely repulsive. Hence, an external potential is necessary to confine finite dust clusters. 2D dust clusters are usually formed by confining the dust grains in the plasma sheath above the lower electrode. The electric field in the sheath provides the strong vertical confinement that allows to form a mono-layer. The horizontal confinement is realized by milling a parabolic cavity into the lower electrode. The plasma sheath follows the shape of the electrode and the dust cluster is confined in this "bowl". Further effort is required to implement spherical confinement in 3D. Here, the dust is located in the plasma bulk rather than in the sheath. The horizontal confinement can be realized by placing a cubic glass box on the electrode. This glass box causes an electric field that allows for a harmonic horizontal entrapment. A harmonic potential in vertical direction can be tailored using a thermophoretic force. Therefore, the lower electrode is heated, and a vertical temperature gradient is set up [12]. In this way, a spherical *Yukawa ball* can be formed, where the confinement potential $V(\mathbf{r}_i) = \frac{m\omega^2}{2}\mathbf{r}_i^2$ is isotropic [13, 14].

In the following, we consider dust particles of equal mass m and equal charge Q. The Hamiltonian of the N particle dust system reads

$$H = \sum_{i=1}^{N} \frac{\mathbf{p}_i^2}{2m} + \sum_{i=1}^{N} \frac{m\omega^2}{2}\mathbf{r}_i^2 + \sum_{i=1}^{N-1} \sum_{j=i+1}^{N} \frac{Q^2}{4\pi\epsilon_0} \frac{1}{r_{ij}} e^{-\kappa r_{ij}}, \tag{1.3}$$

where we introduced the short notation $r_{ij} = |\mathbf{r}_j - \mathbf{r}_i|$ for the pair distance. The first term describes the classical kinetic energy of the dust grains. The external confinement energy in the second term is relevant only for finite systems. In the last term, we find a double sum over all $N(N-1)/2$ particle pairs. The number of summands scales quadratically with the particle number making this part the most expensive one in computer simulations.

At the beginning of this chapter, we give a brief introduction to the physics of dusty plasmas. We discuss the experimental realization and the impact of strong correlations. The first section deals with the question of how to identify a phase transition. Several parameters are subject to the investigation of melting processes: the particle trajectories, the radial pair correlation function $g(r_{ij})$, the specific heat capacity c_V and the bond order parameter.

1.1.1 Melting as a Loss of Order

The phase transition from the crystalline state to the liquid-like state is connected with a loss of long-range spatial order. While the dust grains arrange themselves in a highly ordered crystalline structure below a critical temperature, this order is destroyed at higher temperatures.

We illustrate this phase transition with the example of the plane 2D Yukawa dust crystal. One (clipped) snapshot of each particle configuration for three different temperatures $T_a < T_b < T_c$ is shown in Fig. 1.1. The configurations were sampled in periodic boundary conditions Monte Carlo simulations with $N = 1000$ particles. The underlying lattice is visualized utilizing the *Voronoi* analysis,[1] where the particles form the so called seeds of the Voronoi diagram. The dual graph is called *Delaunay* triangulation with the particle positions as vertices. The edges of the latter can be

[1] The Voronoi diagram is constructed from the particles' positions by the Computational Geometry Algorithms Library (CGAL) available at `www.cgal.org` as c++ library.

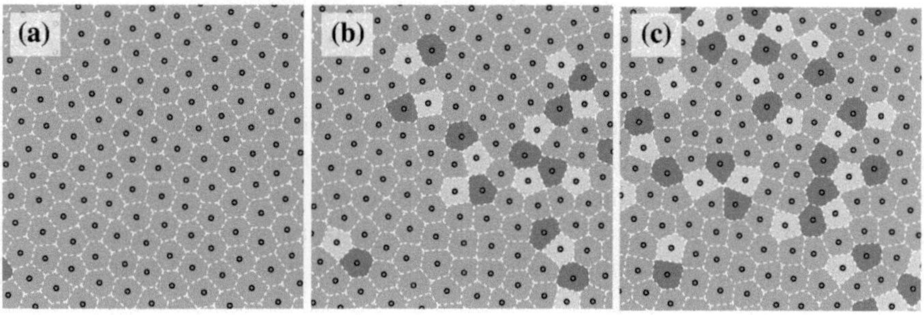

Fig. 1.1 Example of the spatial arrangement of the dust grains in a plane 2D plasma crystal for three different temperatures. The temperature increases from (**a**) to (**c**), i.e. the coupling strength decreases. The *dashed white lines* show the borders of the Voronoi cells, see text. Each cell's color corresponds to the coordination number n of the particle, i.e. the number of neighbors: *gray* for $n = 6$, *white* for $n = 5$ and *dark* for $n = 7$

interpreted as "bonds" between adjacent particles. The Voronoi cell that belongs to a seed $\mathbf{r}_i$ is defined as all points whose distance to $\mathbf{r}_i$ is not greater than the distance to any other seed $\mathbf{r}_j$. It can, in that sense, be interpreted as the "volume that belongs to one particle". The Voronoi diagram allows to define nearest neighbors clearly as those particles whose cells have a common edge. For each particle, its *coordination number n* is defined as the number of nearest neighbors.

An almost perfect hexagonal lattice is found for the dust crystal at the lowest temperature, Fig. 1.1a. Each particle has $n = 6$ nearest neighbors in this arrangement. The Voronoi cells consists of hexagons only, although these hexagons are partially deformed. When the temperature is increased, more and more lattice defects appear as particles with more or fewer than six nearest neighbors. These defects form groups with an equal number of $n = 5$ and $n = 7$ particles. When the temperature is further increased, filaments of defects thread throughout the whole "crystal".

Without going into detail, this example gives us a basic idea of the structural properties of different phases in a dusty plasma. In the following section, we discuss how the coupling strength between the dust grains is characterized and address the Coulomb coupling parameter Γ_C as well as different modifications for Yukawa systems. Then, we describe several melting parameters that allow to quantify phase transitions. The focus lies on their applicability to dusty plasmas.

1.1.2 Coupling Strength and Coupling Parameter

When the thermodynamic phase of a dusty plasma is characterized, it is appropriate to introduce a parameter that describes the coupling strength between the plasma constituents, in particular the coupling strength of the dust particles. This classical coupling parameter is commonly denoted by Γ and gives the ratio between the typical interaction energy of neighboring particles and their typical thermal energy.

It was originally defined for pure Coulomb interaction as

$$\Gamma_C = \frac{Q^2}{4\pi\epsilon_0 \cdot a \cdot k_B T}, \tag{1.4}$$

where a is a typical pair distance [15]. Commonly, the Wigner-Seitz-radius a_{WS}, i.e., the radius of a virtual sphere (circle in 2D) that is occupied by one particle, is used for a. In other works [16], the mean inter-particle spacing $a = \sqrt[3]{n^{-1}} = \sqrt[3]{V/N}$ ($a = \sqrt{n^{-1}}$ in 2D) is used as pair distance. For a constant dust particle density n and a constant charge Q, the coupling parameter Γ_C and the dust kinetic temperature T are inversely proportional.

The coupling parameter allows for a characterization of the physical stationary state of a one-component plasma (OCP). In a multi-component plasma, the coupling strength generally differs for different constituents, since the interaction energy depends on the charge of a certain constituent and on its density. We expect a *solid-like* behavior at strong coupling $\Gamma_C > \Gamma_M$, where $\Gamma_M = 175$ (137) in a macroscopic 3D (2D) Coulomb system and a *liquid-like* behavior in moderately coupled and in weakly coupled systems with $\Gamma_C < \Gamma_M$.

Due to the Coulomb scaling law, the same coupling parameter describes systems at very different conditions. A high coupling ($\Gamma_C \gg 1$) can be realized in different ways. The first way is to realize a low temperature as in experiments with ultra-cold ions [17]. The second way is to prepare an ionic system at extremely high density. Finally, in complex plasmas, the high coupling parameter of the dust component results from the high charge of the dust grains. Note that the coupling strength scales quadratically with the charge Q.

When electric screening due to ions and electrons becomes important, the Coulomb potential has to be replaced by (1.1). Clearly, we cannot expect a system with $\kappa > 0$ to behave as a Coulomb system at the same density and temperature. The simplest approach to introduce a coupling parameter for Yukawa systems is to multiply Γ_C by the exponential factor,

$$\Gamma_Y = \Gamma_C \cdot e^{-\kappa}. \tag{1.5}$$

However, different systems with equal Γ_Y do not necessarily show similar behavior in general [18–20] but can behave liquid-like or solid-like depending on the screening parameter κ. This observation is not surprising and can be understood in the Lindemann picture of melting (see Sect. 1.5.1), where in the solid phase each particle is oscillating in the potential minimum formed by the surrounding particles. In this picture, melting occurs when the amplitude of the oscillation reaches a significant fraction of the inter-particle distance. For this reason, the curvature of the energy landscape around the equilibrium position is important, rather than the absolute potential energy. Vaulina and Khrapak proposed a modified coupling parameter

$$\Gamma_V = \Gamma_C \cdot e^{-\kappa} \cdot \left(1 + \kappa + \frac{\kappa^2}{2}\right). \tag{1.6}$$

During the derivation, they considered a one-dimensional model where one particle is trapped between two other particles. They expanded the potential energy to the second order for both Coulomb and Yukawa interactions and compared the local frequency of oscillation around the equilibrium position. When they applied their parameter to the melting curve of a 3D Yukawa bulk system [21], they found good agreement with a constant coupling $\Gamma_{\mathrm{V}}^{\mathrm{crit}} \approx 106.6$ along the melting curve for κ between 0 and 5 [19]. At this point, we have to pay some attention to the typical distance a that is used to define the coupling parameter Γ and the screening parameters κ. The application of a_{WS} instead of

$$a = \sqrt[3]{n^{-1}} = \sqrt[3]{4\pi/3} \cdot a_{\mathrm{WS}} \quad \text{(3D)} \tag{1.7}$$

$$a = \sqrt{n^{-1}} = \sqrt{\pi} \cdot a_{\mathrm{WS}} \quad \text{(2D)} \tag{1.8}$$

rescales Γ_{C} and κ. The critical value $\Gamma_{\mathrm{V}}^{\mathrm{crit}}(a) = 106.6$ corresponds to $\Gamma_{\mathrm{V}}^{\mathrm{crit}}(a_{\mathrm{WS}}) = 171.8$ when the Coulomb coupling parameter is defined via a_{WS} instead of a.

In 2D systems, Γ_{V} is not constant along the melting line. Therefore, different empiric scaling functions $f(\Gamma_{\mathrm{C}}, \kappa)$ have been applied to define a universal coupling parameter

$$\Gamma^* = f(\Gamma_{\mathrm{C}}, \kappa) \cdot \Gamma_{\mathrm{C}} , \tag{1.9}$$

which is constant along a line in the $(\Gamma_{\mathrm{C}}, \kappa)$ space with constant height of the first peak of the pair distribution function, see Sect. 1.1.4. The utilization of this peak height in order to identify similar states at different screening allows to compare different systems not only at the melting line. Hartmann et al. [20, 22, 23] found the scaling law

$$f_{\mathrm{H}}(\kappa) = 1 - 0.388\,\kappa^2 + 0.138\,\kappa^3 - 0.0138\,\kappa^4 , \tag{1.10}$$

that is independent of Γ_{C}. This scaling holds for moderate screening parameters $\kappa \in [0 : 3]$ and breaks for weakly coupled systems.

Ott et al. [18] found a scaling law between the effective coupling parameter Γ_{O}^* and the Coulomb coupling parameter Γ_{C} that holds for an extended range of screening κ and also for lower coupling strength. Their inverse scaling function f^* defined by

$$\Gamma_{\mathrm{C}}\left(\Gamma_{\mathrm{O}}^*, \kappa\right) = f^*\left(\Gamma_{\mathrm{O}}^*, \kappa\right) \Gamma_{\mathrm{O}}^* \tag{1.11}$$

depends on Γ. By symbolic regression, they found the relation

$$f^*\left(\Gamma_{\mathrm{O}}^*, \kappa\right) = \exp\left(\frac{\kappa^2}{A + B\left(\Gamma_{\mathrm{O}}^*\right)\kappa}\right) , \text{ with} \tag{1.12}$$

$$B\left(\Gamma_{\mathrm{O}}^*\right) = B_1 + B_2/\Gamma_{\mathrm{O}}^* , \tag{1.13}$$

and with fit constants $A = 2.37221$, $B_1 = 0.55515$ and $B_2 = 5.56585$. For details concerning the regression and the inversion of the scaling law in (1.11), we refer the reader to the original article [18].

Although the functional form of this scaling law is not deduced from first principles, it can describe the effective coupling strength in an extended 2D Yukawa system. Furthermore, it allows for a non-invasive measurement of both κ and Γ_{O}^* in an experiment that measures the pair distribution function (PDF) and the velocity auto-correlation function [18].

1.1.3 The Specific Heat Capacity

The *specific heat capacity* (or simply *specific heat*) at constant volume c_V is widely used as an indicator for a phase transition in macroscopic systems. In this context, it is defined as the total heat capacity C_V divided by the number of particles N. Since the dust component is treated on an atomic level, it is reasonable to define the heat capacity per particle rather than per mass. The total heat capacity C_V of an N-particle system is defined as the derivative of the internal energy U with respect to the temperature

$$C_V = \left(\frac{\partial U}{\partial T}\right)_V = -\frac{1}{k_B T^2}\left(\frac{\partial U}{\partial \beta}\right)_V, \tag{1.14}$$

where $\beta = 1/(k_B T)$ is the inverse temperature. The heat capacity is exactly the amount of heat ΔQ which has to be transfered to the system to increase its temperature by ΔT at constant volume. This relation already suggests a way to determine C_V experimentally: The studied system is isolated against particle exchange and its volume is fixed. It is further isolated against heat exchange except for a heat source, e.g., a heater wire, that allows a well defined heating power $\dot{Q}$. While the system is slowly heated, its temperature is recorded and the heat capacity is determined by the ratio of $\dot{Q}$ and $\dot{T}$. Since the temperature increases during the measurement, one obtains the temperature curve $C_V(T)$.

When treating a dusty plasma on atomic level, such as in a particle based computer simulation, one has access to the internal energy of every configuration (microstate) $(\mathbf{P}_i, \mathbf{R}_i)$ during the simulation. The multi vectors $\mathbf{P}_i$ and $\mathbf{R}_i$ contain the momenta and the positions of all N particles. When these states are Boltzmann distributed as in thermal equilibrium, the relations,[2]

$$U = \langle H \rangle = \frac{1}{Z}\sum_{\mathbf{P}_i,\mathbf{R}_i} H\left(\mathbf{P}_i, \mathbf{R}_i\right) \cdot e^{-\frac{H(\mathbf{P}_i,\mathbf{R}_i)}{k_B T}} \quad \text{and} \tag{1.15}$$

[2] Z is the partition function arising from the normalization of the canonical distribution $\frac{1}{Z}\sum_{\mathbf{R}_i} e^{-H(\mathbf{R}_i)} = 1$, see also (1.33, 1.34).

$$\frac{\partial U}{\partial \beta} = \left(\frac{\partial}{\partial \beta}\frac{1}{Z}\right) \sum_{\mathbf{P}_i,\mathbf{R}_i} H\left(\mathbf{P}_i,\mathbf{R}_i\right) \cdot e^{-\frac{H(\mathbf{P}_i,\mathbf{R}_i)}{k_B T}}$$
$$+\frac{1}{Z} \sum_{\mathbf{P}_i,\mathbf{R}_i} H^2\left(\mathbf{P}_i,\mathbf{R}_i\right) \cdot e^{-\frac{H(\mathbf{R}_i,\mathbf{R}_i)}{k_B T}} \quad (1.16)$$
$$= \langle H\rangle^2 - \langle H^2\rangle \quad (1.17)$$

allow us to calculate C_V from the first and second moment of the internal energy during the simulation [24],

$$C_V = \frac{1}{k_B T^2}\left(\langle H^2\rangle - \langle H\rangle^2\right) . \quad (1.18)$$

Since both internal energy U and heat capacity C_V for a system far away from the critical point are extensive quantities, they both scale with the particle number $\propto N$. However, both $\langle H^2\rangle$ and $\langle H\rangle^2$ scale with N^2 and hence their relative difference scales as N^{-1}. Therefore, this way of calculating C_V is ill-conditioned for macroscopic systems with $N = \mathcal{O}(10^{23})$ particles. Since we will often consider particle numbers below $N = 100$, in confined dust balls, and up to $N = \mathcal{O}(10^5)$ particles, in extended dust clouds, we can use (1.18), whenever we have access to the internal energy of each configuration with sufficient precision. An example of the specific heat capacity from a simulation of a 2D Yukawa system with periodic boundary condition is shown in Fig. 1.4 in Sect. 1.2.2 and compared with the heat capacity of a Yukawa ball.

1.1.3.1 Calculation of C_V in a Computer Simulation

In order to calculate C_V, we have to compute the energy $U_i = H\left(\mathbf{P}_i,\mathbf{R}_i\right)$ for each sampled microstate $(\mathbf{P}_i,\mathbf{R}_i)$. In the simplest implementation, one sums up U_i and U_i^2 separately for N_S sampled configurations and obtains the mean-values,

$$U \approx \overline{U} = \frac{1}{N_S}\sum_{i=1}^{N_S} U_i \quad \text{and} \quad \overline{U^2} = \frac{1}{N_S}\sum_{i=1}^{N_S} U_i^2 , \quad (1.19)$$

when the sampled configurations are distributed according to the probability distribution of microstates (1.33), also see the Chapter *"Introduction to Configuration Path Integral Monte Carlo"*. The heat capacity is estimated from those means as

$$C_V = \frac{1}{k_B T}\frac{N_S}{N_S - 1}\left(\overline{U^2} - \overline{U}^2\right) \approx \frac{1}{k_B T}\left(\overline{U^2} - \overline{U}^2\right) , \quad (1.20)$$

where the factor $N_S/(N_S - 1)$ takes into account that $\overline{U}$ itself is not independent of the single measurements U_i. The reason for this is that the expectation value of the estimated variance which is calculated as

$$\overline{\sigma_y^2} = \frac{1}{N_\mathrm{S}} \sum_{i=1}^{N_\mathrm{S}} (y_i - \overline{y})^2 \qquad \text{with} \quad \overline{y} = \frac{1}{N_\mathrm{S}} \sum_{i=1}^{N_\mathrm{S}} y_i \tag{1.21}$$

from N_S measurements of the random variable y is a factor $(N_\mathrm{S}-1)/N_\mathrm{S}$ smaller than the variance σ_y^2 of the random variable. This bias is compensated by by the factor $N_\mathrm{S}/(N_\mathrm{S}-1)$ in (1.20). For a large number of samples, this factor can be neglected.

However, we can improve this straight-forward implementation with respect to rounding errors. Taking into account that the fixed-point part of a floating point number can store p bits, we can estimate the resolution of a number f as $\epsilon_f = f \cdot 2^{1-p}$. The absolute rounding error depends on the magnitude of the number itself. Let us suppose we know the mean energy U. We can then write the single measurements as $U_i = \overline{U} + \Delta U_i$. As explained above, the variation of the single measurements ΔU_i becomes small compared to the mean value, when the particle number is large or when the temperature is low. The averaging procedure (1.19), "wastes" a lot of precision to sum up the constant part $\overline{U}$. Hence, it is recommended to calculate the average and the variance of ΔU_i instead of U_i.

Even though U is a priori unknown at the beginning of the simulation, we can estimate it before the main simulation run. For that purpose, the system is initialized and equilibrated first. Then, a series of a few thousand *pre-calculation* measurements is performed to estimate the mean internal energy U. Afterward, this value is used to calculate $\overline{\Delta U}$ and $\overline{\Delta U^2}$ during the data production.

If it is not required to calculate C_V on-the-fly, we can store all single measurements U_i. We are then able to calculate $\overline{U}$ in a first run and, with that, we can calculate $\overline{\Delta U}$ in a second run through the data set.

1.1.4 The Radial Pair Distribution Function

The *radial pair distribution function* (PDF) $g(r)$, or simply *pair distribution function*, is an important structural property to characterize the order in macroscopic and mesoscopic system, in particular in 2D and in 3D dusty plasmas. This quantity describes the probability of finding a second particle at a distance r with respect to an arbitrary first particle. It is calculated from the full two-particle density[3] $n_2(\mathbf{r}_1, \mathbf{r}_2)$ by integration,

$$g(r) = \frac{1}{N n_0} \frac{1}{4\pi r^2} \left\langle \int_V \mathrm{d}^3 R \int_{|\mathbf{r}|=r} \mathrm{d}^3 r \, n_2\left(\mathbf{R}, \mathbf{R}+\mathbf{r}\right) \right\rangle \qquad \text{(3D)} \tag{1.22}$$

$$g(r) = \frac{1}{N n_0} \frac{1}{\pi r} \left\langle \int_V \mathrm{d}^2 R \int_{|\mathbf{r}|=r} \mathrm{d}^2 r \, n_2\left(\mathbf{R}, \mathbf{R}+\mathbf{r}\right) \right\rangle \qquad \text{(2D),} \tag{1.23}$$

[3] The full two-particle density $n_2(\mathbf{r}_1, \mathbf{r}_2)$ is proportional to the probability to find one particle at $\mathbf{r}_1$ and a second particle at $\mathbf{r}_2$ simultaneously. For a system of N point particles, it is given by $n_2(\mathbf{r}_1, \mathbf{r}_2) = \left\langle \sum_{i=1}^N \sum_{j \neq i}^N \delta(\mathbf{r}_1 - \mathbf{r}_i)\delta(\mathbf{r}_2 - \mathbf{r}_j) \right\rangle$, where $\mathbf{r}_i$ and $\mathbf{r}_j$ are particle positions.

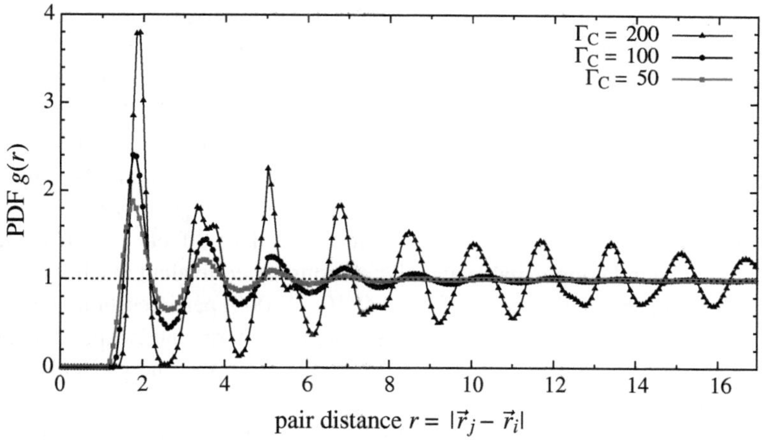

Fig. 1.2 Radial pair distribution function in an extended 2D Yukawa ($\kappa = 1$) system. The data were obtained from an MC simulation with $N = 1000$ particles and periodic boundary conditions. Units of length: a_{WS}

where $n_0 = N/V$ is the (areal) particle number density. The brackets $\langle \ldots \rangle$ denote the ensemble average over many configurations. In an uncorrelated, homogeneous system, the two-particle density $n_2(\mathbf{r}_1, \mathbf{r}_2) = n_0^2$ is constant, and we can readily show that $g(r) = 1$ for all distances. Also, $g(r)$ converges to unity for large distances in liquid-like phases as shown in Fig. 1.2.

Since the particle density is discrete in a complex plasma, the PDF is calculated from a histogram. For each sampled configuration, all pair-distances are sorted into bins with a finite width dr. The histogram counts are normalized by the volume of a spherical shell (a ring in 2D) of width dr. Figure 1.2 shows examples of the PDF in a Yukawa system for different coupling strengths Γ_C. The long-range order at $\Gamma_C = 200$ manifests itself in clear peaks even at large distances. When Γ_C is decreased, i.e. the system is heated up, the PDF becomes flat after a few Wigner-Seitz radii. Ott et al. found that the height of the first peak in the PDF shows non-continuous temperature behavior. When the 2D dust system is cooled, a jump occurs at the crystallization temperature, i.e. at the melting point Γ_M [18].

1.2 Phase Transitions in Finite Systems

By now, we understand how phase transitions can be identified in a macroscopic system. In this section, we investigate phase transitions in finite systems. These systems are far below the thermodynamic limit and exhibit fundamental differences compared to extended systems and, therefore, require new melting parameters.

Despite the absence of true phase transitions in finite systems, analogous structural transitions are observed upon the variation of the coupling parameter $\Gamma = \langle U \rangle / \langle T \rangle$– the ratio of mean interaction and mean kinetic energy of the particles. Finite size effects show up in a broadened transition region, in an N-dependence of the transition temperature (or Γ) and pronounced geometry effects [3, 25, 26]. Finally, melting is often a multi-stage process where radial order, the angular ordering inside a shell and between different shells are lost not simultaneously, e.g. Ref. [27]. An introduction to structural and dynamical properties of dusty plasmas is found in Part III of Ref. [3].

When we describe extended systems, the number of particles and the volume are considered to be large with $N \to \infty$, $V \to \infty$ and $n = \frac{N}{V} = const$. This is the *thermodynamic limit* that allows us to neglect boundary effects. The particles are considered to be indistinguishable and all positions are equivalent in a homogeneous extended system. In small systems of $N = \mathcal{O}(10-10000)$ particles, boundary effects play a non-negligible role. For example, the potential energy landscape for a particle at the center of a Yukawa ball differs substantially from the energy landscape of a particle at the boundary.

Some structural processes occur in finite clusters but not in extended systems. Coulomb and Yukawa balls usually consist of several nested spherical shells around the trap center. Three processes contribute to the phase transition of such a cluster:

(1) The *radial melting* concerns the radial motion of the dust grains [25] and is hence connected to the radial width of the shells. This width is typically small in the solid-like regime. Between two shells, the radial density $n(r)$ drops to zero, meaning that no particle transitions between two shells are possible. In contrast, the shells begin to overlap in the liquid-like regime indicating frequent particle transitions between the extended shells. At even lower coupling, only a weak modulation of the otherwise constant radial density remains.

(2) The *intra-shell* melting in 3D clusters describes the loss of order within a spherical shell. Below a critical temperature, the particles arrange themselves in a regular lattice similar to the hexagonal lattice known from extended 2D systems. However, even at infinite coupling, a certain number of sites with a coordination number below six is required due to the topology of a sphere. When the temperature is increased, this lattice is destroyed and the arrangement within the shell becomes disordered.

(3) Finally, *inter-shell* melting in both 3D and 2D clusters is attributed to the relative positions of particles in different, typically adjacent, shells. At low temperatures and at proper occupation numbers of the two shells, their orientation can be locked, while above a critical temperature, correlations between particle positions on one and another shell vanish. This effect crucially depends on the exact number of particles in both shells. In Sect. 1.2.4, we present the Center-Two-Particle (C2P) correlation function that allows for an investigation of both intra-shell and inter-shell correlations.

Figure 1.3 shows the intra-shell Voronoi diagrams for the outer shell of a Yukawa ball as an illustration of the loss of intra-shell order. During the calculation of these diagrams, particles from the radial range that corresponds to the investigated shell are selected and projected onto a unit sphere. This sphere is divided into several overlapping spherical caps before the coordinates of each cap are projected into a

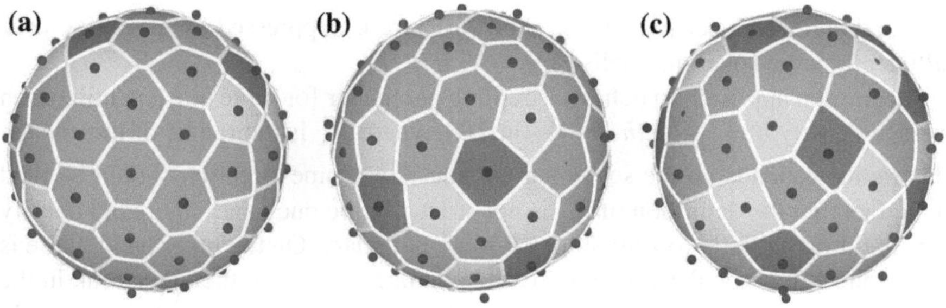

Fig. 1.3 Example of the spatial arrangement of the dust grains on the outer shell of a spherical Yukawa ball, with $N = 150$ particles in total, at three different temperatures. The temperature increases from (**a**) to (**c**), i.e. the coupling strength decreases. The *white lines* show the borders of each particle's Voronoi cell on a sphere with the radius that corresponds to the mean radial coordinate of all particles on the shell. The cell color corresponds to the coordination number n, i.e. the number of neighbors: *gray* for $n = 6$, *white* for $n = 5$ and *dark* for $n = 7$. At higher temperatures the particles (*dark dots*) begin leaving the shell (*radial melting*). (Inner particles are hidden by the colored Voronoi cells.)

plane using the stereographic projection. In these projections, the Voronoi diagrams are constructed individually and fit together in the back projection. The important property of stereographic projection is that it conserves circles. If position D is outside the circle through A, B and C, so its projection D' is outside the circle through the projections A', B' and C'. Since this property conserves the *Delaunay* triangulation, which is dual to the Voronoi diagram, the stereographic projection is well suited to construct the Voronoi diagram for a sphere in flat geometry. However, some care has to be taken when fitting together the diagrams from different sectors.

In the following, we discuss the difficulty of defining a universal coupling parameter Γ in finite systems. Thereafter, different melting parameters known from macroscopic systems as well as specifically designed melting parameters for spherically confined systems are discussed.

1.2.1 Coupling Parameter of Finite Systems

For simulations of finite dust clusters, we use the temperature in units of the characteristic energy of the harmonic oscillator, see Sect. 1.3.1, as an input parameter. In these units, the inverse temperature becomes equivalent to the Coulomb coupling parameter Γ_C for large 3D Coulomb clusters. As discussed in Sect. 1.1.2, the Coulomb coupling parameter Γ_C, as well as advanced coupling parameters for Yukawa systems, require knowledge about the typical particle spacing a. In macroscopic systems, where the number density n is fixed, $a = c\sqrt[3]{n^{-1}}$ (in 3D) or $a = c\sqrt{n^{-1}}$ (in 2D) is well defined except for a constant factor. In finite dust clusters, the number density generally has a radial dependence even if the shell structure is suppressed by averaging over sufficiently large spherical shells [28].

A common approach to define a coupling parameter for finite clusters utilizes an average density $\bar{n} = \frac{3N}{4\pi R_C^3}$ ($\bar{n} = \frac{N}{\pi R_C^2}$ in 2D), where R_C is the radius of the cluster. This radius depends on the screening parameter κ. Some attention should be paid to the fact that the definition of R_C is not strict. On the one hand, the radial density does not drop abruptly to zero at the cluster's boundary. On the other hand, there is a difference, in particular for small particle numbers, whether the outer peak in the radial density is used as R_C or whether one half of the inter-shell spacing is added.

Since R_C is a priori unknown at the beginning of an MD or an MC simulation, a coupling parameter that takes into account the actual extension of the cluster cannot simply be used as an input parameter. An increased screening results in smaller clusters, the mean particle spacing a decreases and hence the Coulomb coupling parameter Γ_C increases. The relation between advanced coupling parameters, see Sect. 1.1.2, and the screening parameter becomes more sophisticated.

Whenever one states a coupling parameter for a finite dust cluster, one should specify which length a is used as a characteristic particle spacing. In the following, we use the Coulomb coupling parameter Γ_C with $a = l_0 = \left(\frac{Q^2}{4\pi\epsilon_0 \cdot m\omega^2}\right)^{1/3}$ as a typical pair distance to characterize the coupling strength in *small Yukawa clusters*. One advantage of this definition is the independence from the screening.

Increasing the screening has two opposing effects: On the one hand, the oscillation frequency ω_d at the potential energy minimum between two particles decreases at constant distances. On the other hand, increasing κ shrinks the cluster resulting in a smaller inter-particle distance and thus, in an increased ω_d. Hence, both effects partially compensate each other. We should note that this parameter is not necessarily universal in the sense that clusters with different N and κ behave similarly along a line of constant Γ_C. However, this parameter allows for a qualitative classification.

1.2.2 The Specific Heat Capacity of Finite Systems

The specific heat capacity c_V is a well suited parameter to characterize melting in extended systems. The phase transition of a large system appears as a sharp peak at T_{crit} in the $c_V(T)$ curve. Though a sharp c_V-peak is not found for finite dust clusters, we find a peak in computer simulations for these systems as well which is broadened.

In Fig. 1.4, the specific heat capacity c_V of a Yukawa ball with $N = 150$ particles is compared to c_V of an extended system. We do not intend to give a quantitative but rather a qualitative comparison of the two systems concerning the behavior of $c_V(T)$ at T_{crit}. The absolute Γ_C position of the c_V-peak of the Yukawa ball depends on the characteristic length a in the definition of Γ_C. We used the length l_0 as defined in Sect. 1.2.1. However, other conventional definitions of a characteristic length like the Wigner-Seitz radius within the outer shell, differ by a factor of the order of unity and, hence, cause no qualitative difference. While the c_V-peak of the Yukawa ball

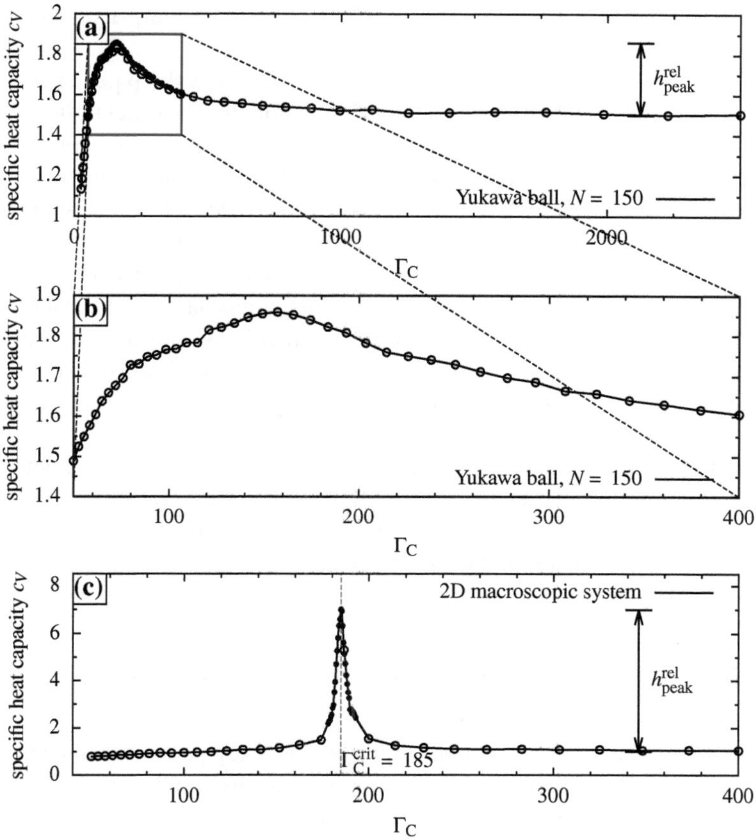

Fig. 1.4 **a** The specific heat capacity of a spherical 3D Yukawa ball ($\kappa = 1$) over a wide range of the coupling strength. **b** Shows an enlarged view of the transition range in part (**a**). **c** The heat capacity of an extended 2D system, which has a sharp peak at the phase transition, is shown for comparison. The data were obtained in MC simulations (periodic boundary for (**c**))

appears to some degree sharp when looking at a wide range of coupling strengths (Fig. 1.4a), the broad width of the peak, $\Delta\Gamma_C = \mathcal{O}(100)$, is apparent at a closer look at the transition region (Fig. 1.4b). The second striking difference between extended systems and finite clusters is the relative height h^{rel}_{peak} of the c_V-peak above the high coupling value $c_V = 1.0$ and $c_V = 1.5$ in 2D and 3D systems, respectively. This height is more than one order of magnitude larger for the extended system compared to the finite Yukawa ball. We should note that in both cases the heat capacity contribution from the kinetic energy is omitted. This contribution is constant for classical particles in the absence of velocity-dependent forces, like, e.g., the Lorentz force in a magnetic field. From a comparison of the peak positions of c_V in Fig. 1.4, one notices that the $N = 150$ cluster has a different melting point than a macroscopic system. Typically, the melting temperature decreases when N is reduced. This behavior is well known from simulations [29] and also from atomic clusters, e.g., [30].

Although the c_V-curve of a finite cluster does not show a sharp peak at the phase transition, it is still suitable to assign a transition region. In experiments, however, the heat capacity c_V is virtually inaccessible. Calorimetric probes, see Chapter *"Calorimetric Probes for Energy Flux Measurements in Process Plasmas"*, measure the heat capacity of the plasma itself. Even though the dust particles interact with the plasma and, therefore, contribute to the total heat capacity, their contribution is negligible due to their small number compared to the large number of electrons, ions and neutral gas atoms. Moreover, the bulk material of the dust grains has a heat capacity ${c_V}^{\mathrm{bulk}}$ itself. The relevant heat capacity for the structural melting is the one that results from the Yukawa interaction energy of the dust grains. In order to determine this c_V in an experiment, a method would have to be able to resolve this interaction energy with sufficiently high resolution.

In the following sections, we discuss several different melting parameters with focus on structural properties of a dust cluster. While the exact interaction energy is virtually inaccessible, the particle positions can be recorded with high spatial and temporal resolution in experiments [6, 31].

1.2.3 Radial Pair Distribution Function in Finite Systems

The *radial pair distribution function* (PDF) $g(r)$ is a key parameter to characterize the order in extended systems, see Sect. 1.1.4. Let us discuss the application of this quantity to finite systems. The basic idea of the definition of $g(r)$ is the fact that the full two-particle density $n_2(\mathbf{r}_1, \mathbf{r}_2)$ does not depend on $\mathbf{r}_1$ and $\mathbf{r}_2$ individually. Instead, due to the translational invariance of a homogeneous system, n_2 depends on the difference vector $\mathbf{r} = \mathbf{r}_2 - \mathbf{r}_1$ only. Furthermore, the ensemble average of $n_2(\mathbf{r})$ is invariant under rotation if the Hamilton function is invariant under rotation. This leads to the definition of PDF that depends on the radial pair distance only, Eqs. (1.22) and (1.23).

In a finite system, there is no translational invariance and the pair density depends on the absolute position of the first particle besides the distance. Figure 1.5 demonstrates that the number of particles in a ring of equal size differs for particles in the center compared to those at the cluster's boundary.

Although we can still bin all particle pairs into a histogram, it is much more complicated to compare $g(r)$ for different cluster sizes. The first problem is that, in contrast to the particle number N, the density n_0 is not fixed as an input parameter. Even worse, the density is not spatially constant in general, but depends on the distance from the trap center. Even in a cluster with approximately constant density inside a cut-off radius R_C, the PDF does not reproduce the behavior $g(r) \to 1$ for large distances known from extended systems in the liquid of gas phase (low coupling).

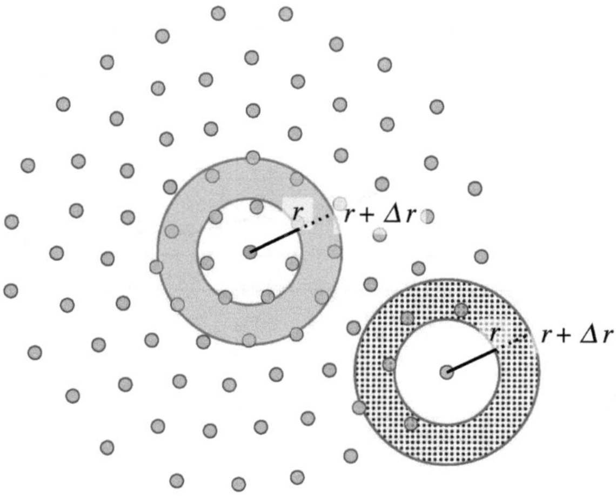

Fig. 1.5 Example for the configuration of a finite Yukawa cluster consisting of $N = 85$ particles in a harmonic 2D trap. For a particle at the cluster boundary, all particles with a distance between r and $r + \Delta r$ are highlighted by the *dotted ring*. Due to the cluster's finite size, the *right* half of this ring is empty. For a particle in the cluster's *center*, we find other particles in all directions (*gray ring*). We further notice that the nearest neighbor distance at the *center* is smaller than at the cluster's boundary

One possible modification was proposed by Wrighton, Kählert et al. and applied to finite size Coulomb clusters [32]. This modification takes into account the radial modulation of the one-particle density $n(r)$. Introducing a center-of-mass and a relative coordinate, $\mathbf{R} = \frac{1}{2}(\mathbf{r}_1 + \mathbf{r}_2)$ and $\mathbf{r} = \mathbf{r}_2 - \mathbf{r}_1$, respectively, an *uncorrelated pair-density* $\rho_2^{\text{uncorr}}(r_{ij})$ is defined as

$$\rho_2^{\text{uncorr}}(r_{ij}) = \int \mathrm{d}^3 R \int_{|\mathbf{r}|=r_{ij}} \mathrm{d}^3 r \; n\left(\left|\mathbf{R} + \frac{\mathbf{r}}{2}\right|\right) \cdot n\left(\left|\mathbf{R} - \frac{\mathbf{r}}{2}\right|\right) . \tag{1.24}$$

This is the pair density which one would measure in a system that has no correlation beyond the shell structure, i.e. a cluster with homogeneous angular density within the shells. The correlated pair density $\rho_2^{\text{corr}}(r_{ij})$ is still calculated by binning all pair distances into a histogram. Finally, the "pair distribution function" for a spherical cluster is defined as the correlated pair density relative to the "uncorrelated" pair density,

$$g_{\text{K}}(r_{ij}) = \frac{\rho_2^{\text{corr}}(r_{ij})}{\rho_2^{\text{uncorr}}(r_{ij})} . \tag{1.25}$$

As shown in Fig. 1.6, this modified PDF shows the low-coupling behavior $g_{\text{K}}(r_{ij}) \to 1$ known from extended systems. Due to the normalization by ρ_2^{uncorr}, calculated from the radial density, the absolute height of $g_{\text{K}}(r_{ij})$ is well defined. Here, no assumptions about the average density have to be made, which is important when making statements about the peak heights. When the peak heights were dependent on the average density n_0, general statements would suffer from the ambiguous

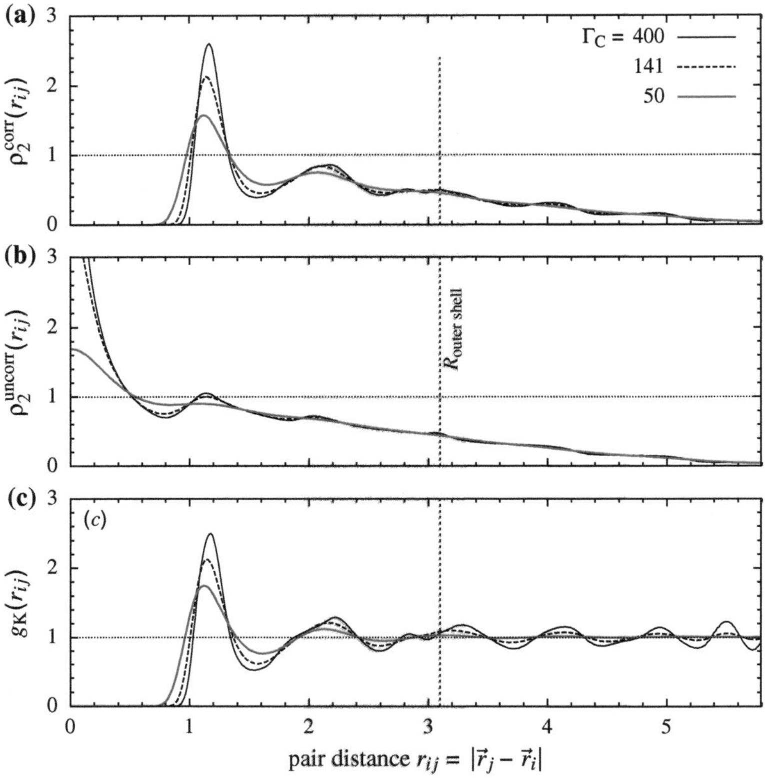

Fig. 1.6 **a** Radial pair density of a 3D Yukawa ($\kappa = 1$) ball with $N = 150$ particles $\rho_2^{corr}(r_{ij})$, scaled by the normalization factor from (1.22). **b** Uncorrelated pair density calculated from the radial density, scaled by the same normalization factor. **c** Modified PDF for spherical clusters $g_K(r_{ij})$ (1.25). The data were sampled in an MC simulation

definition of n_0 for finite clusters. However, all pair distances r_{ij} are treated equally, although nearest neighbor distances are dependent on the position $\mathbf{R}$ of the pair within the cluster, see Fig. 1.5.

We note that this quantity is not suited to distinguish between *intra-shell correlations*, where both particles of a pair are located on the same shell, and *inter-shell correlations* between particles from different shells. Such a distinction is achieved by the *Center-Two-Particle* correlation function discussed in Sect. 1.2.4.

1.2.4 The Center-Two-Particle Correlation Function

Here, we present a pair correlation function which is adapted to spherically confined systems. As shown before, the radial pair distribution function $g\left(r_{ij}\right)$ follows from the full two-particle density $n_2\left(\mathbf{r}_1, \mathbf{r}_2\right)$ assuming a homogeneous, isotropic system. Obviously, a confined system is not homogeneous. Hence, when transforming the two-particle density to center-of-mass $\mathbf{R}$ and relative coordinate $\mathbf{r}$, the dependence on $\mathbf{R}$ remains besides the dependence on the distance.

The goal is to find the set of coordinates $\{q_1, \ldots, q_k\}$ that are required to describe the pair density. Therefore, the first step is to find the symmetries in the Hamilton function (1.3). The Hamiltonian of a 2D Yukawa cluster is invariant under a common rotation of all coordinates $\mathbf{r}_i' = R_\alpha \mathbf{r}_i$. The rotation is described by the rotation matrix

$$R_\alpha = \begin{pmatrix} \cos\alpha & -\sin\alpha \\ \sin\alpha & \cos\alpha \end{pmatrix}, \tag{1.26}$$

depending on one angle α. All radial coordinates, as well as all pair distances and kinetic energy are conserved under such a rotation, and thus, H is invariant. The Hamiltonian of a 3D Yukawa cluster is again invariant under any common rotation. However, in 3D, this transformation is described by the three *Euler angles* or by the Tait-Bryan angles *yaw*, *pitch* and *roll*. The explicit transformation is omitted at this point as it is of minor interest for the following considerations. The important point is that also the *long-term average* of $n_2(\mathbf{r}_1, \mathbf{r}_2)$ and higher order n-particle densities are invariant under any transformation T that conserves H. Here, we should point out that n_2, for a single realization of particle coordinates, is not invariant under rotation, but the long-term average has no preferred direction in space.

Thus, the average two-particle density of both a 2D and a 3D Yukawa cluster depends on three coordinates:

- $r_{\mathrm{I}} = |\mathbf{r}_1|$: the radial coordinate of the first position vector
- $r_{\mathrm{II}} = |\mathbf{r}_2|$: the radial coordinate of the second position vector
- $\vartheta = \sphericalangle(\mathbf{r}_1, \mathbf{r}_2)$: the angle between both vectors with respect to the trap center

The *correlated pair density* $\rho_2(r_{\mathrm{I}}, r_{\mathrm{II}}, \vartheta)$ is calculated by binning all particle pairs of every sampled configuration into a three-dimensional histogram. The *uncorrelated pair density* is defined in an analogous way to Sect. 1.2.3, (1.24), for the modified radial pair distribution function, by

$$\rho_2^{\mathrm{uncorr}}(r_{\mathrm{I}}, r_{\mathrm{II}}, \vartheta) = \int_{|\mathbf{r}_1|=r_{\mathrm{I}}} \mathrm{d}^3 r_1 \int_{|\mathbf{r}_2|=r_{\mathrm{II}}, \vartheta(\mathbf{r}_1,\mathbf{r}_2)=\vartheta} \mathrm{d}^3 r_2\, n(|\mathbf{r}_1|) \cdot n(|\mathbf{r}_2|) \tag{1.27}$$

$$= \begin{cases} 4\pi r_{\mathrm{I}} n(r_{\mathrm{I}})\, r_{\mathrm{II}} n(r_{\mathrm{II}}) & \text{(2D)} \\ 8\pi^2 r_{\mathrm{I}}^2 n(r_{\mathrm{I}})\, r_{\mathrm{II}}^2 n(r_{\mathrm{II}}) \sin\vartheta & \text{(3D)} \end{cases} \tag{1.28}$$

taking into account the radial one-particle density $n(r)$. The term "uncorrelated" in this context means that no correlations beyond the radial density profile are captured. However, a system of fully uncorrelated and therefore necessarily non-interacting particles would not form a shell structure. The *Center-Two-Particle* (C2P) correlation function is defined as the ratio of the sampled *correlated* and *uncorrelated* pair densities in the new coordinates

$$g_2(r_{\mathrm{I}}, r_{\mathrm{II}}, \vartheta) = \frac{\rho_2^{\mathrm{corr}}(r_{\mathrm{I}}, r_{\mathrm{II}}, \vartheta)}{\rho_2^{\mathrm{uncorr}}(r_{\mathrm{I}}, r_{\mathrm{II}}, \vartheta)} . \tag{1.29}$$

This quantity was used by Rhemus et al. to investigate electron correlations in 2-electron atoms [33, 34]. In order to calculate reduced, i.e., *integrated* quantities, e.g., $g_2(r_{\mathrm{II}}, \vartheta)$, both ρ_2^{corr} and ρ_2^{uncorr} are integrated over a certain range of the coordinates that are removed, e.g. r_{I} in this case. In this section, two special cases of the integrated C2P correlation function are presented. Besides $g_2(r_{\mathrm{II}}, \vartheta)$, the angular pair correlation function $g_2(\vartheta)$ is subject to the further discussion.

The singly integrated C2P correlation function $g_2(r_{\mathrm{II}}, \vartheta)$ shows *intra-shell* and *inter-shell* correlations. If a radial range that corresponds to the outer shell is chosen for the first integration of r_{I}, this means that at least one particle of a pair is located on the outer shell. The integrated correlated density $\rho_2(r_{\mathrm{II}}, \vartheta)$ is proportional to the probability of finding another particle with the distance r_{II} from the trap center and under the angle ϑ with respect to the trap center. This density basically reproduces the radial density $n(r) \cdot 4\pi r^2$ [$n(r) \cdot 2\pi r$ in 2D] that increases from inside to outside. Moreover, the highest density ρ_2 in 3D clusters is found at angles around $\vartheta \approx 90°$, independent of the coupling strength. These two volume effects are compensated by the division by ρ_2^{uncorr}. The resulting correlation function $g_2(r_{\mathrm{II}}, \vartheta)$ describes the probability of finding a second particle in a volume at r_{II} under an angle ϑ from the first particle compared to the probability of finding a second particle in a spherical shell with radius r_{II} that has the same volume.

The first particle is chosen from the outer shell of a Yukawa cluster by the r_{I} integration range in Fig. 1.7a, while it is chosen from the middle shell in part (b) of the same figure. Intra-shell correlations appear as a sequence of minima (red) and maxima (blue) in the horizontal direction, where r_{II} corresponds to the same shell as the r_{I} integration range. This sequence starts with a correlation hole around $\vartheta = 0°$. Although the intra-shell correlations are the dominant observation, correlations with particles from other shells appear as minima and maxima in the horizontal sequence with radius around 1.5. The inter-shell correlations between particles from middle and inner shell appear as a horizontal line at $r_{\mathrm{II}} \approx 1.2$ in Fig. 1.7b. The amplitude of the deviation from $g_2^{\mathrm{uncorr}} = 1$ decays much faster compared to the intra-shell correlations. At radii between two shells, the error of $g_2(r_{\mathrm{II}}, \vartheta)$ becomes large, since the number of samples with at least one particle between two shells is small, especially at strong coupling. Thus, both ρ_2 and ρ_2^{uncorr} are small numbers between two shells and the statistical error of their ratio is large. In the example, noise is visible in the region between outer and middle shell, and between middle and inner shell.

Pure intra-shell correlations on shell R are investigated by $g_2(\vartheta)$ when the integration range is the same for both radial coordinates r_{I} and r_{II} and further corresponds to the shell R. The intra-shell angular pair correlation function on the outer shell of a Yukawa ball is shown in Fig. 1.7c. At high coupling ($\Gamma_{\mathrm{C}} = 400$) the slow decay of the peak height indicates a *long-range order* within the shell. In contrast, the peak height decays fast with ϑ in the case of medium coupling strength ($\Gamma_{\mathrm{C}} = 50$). The angular pair correlation function shows striking similarities with the usual pair correlation function for extended systems, compare Fig. 1.2. The correlation hole

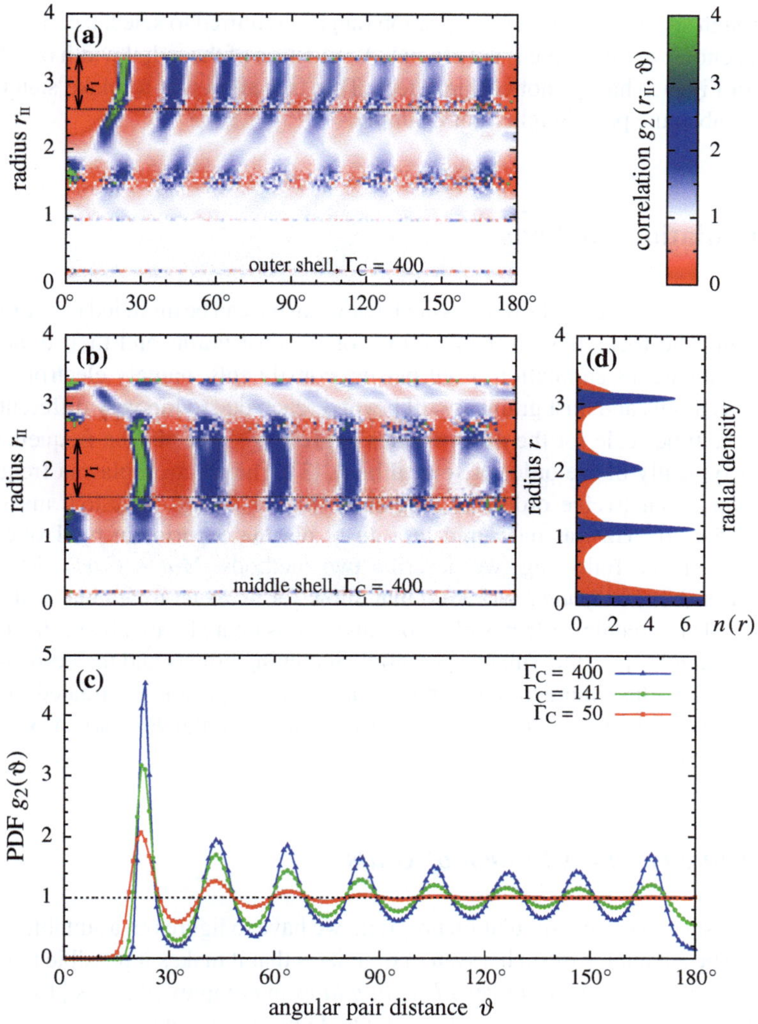

Fig. 1.7 C2P correlation function in a 3D Yukawa cluster ($\kappa = 1$) with $N = 150$ particles. **a** and **b** Show the integrated C2P $g_2(r_{II}, \vartheta)$ where the first radial coordinate r_I is integrated over a range (*vertical arrow*) that corresponds to the outer shell (*middle shell* for (**b**)). **c** Shows the angular pair correlation function $g_2(\vartheta)$ within the outer shell for different Coulomb coupling parameters Γ_C, see Sect. 1.2.1. **d** Shows the radial density for comparison for two values of Γ_C

at $r_{ij} = 0$ and $\vartheta = 0°$, respectively, appears in both quantities even at moderate coupling strength. At moderate coupling strength, both $g(r_{ij})$ and $g_2(\vartheta)$ decay to 1 after a few maxima already. Therefore, $g_2(\vartheta)$ is a suitable extension of $g(r_{ij})$ for finite spherical clusters. However, the splitting of the second neighbor's peak, that is characteristic for a hexagonal lattice in 2D, is not visible for this cluster. One reason for this is the relatively small number of $\langle N_3 \rangle = 86$ particles on the outer shell.

Although specifying a radial integration range is required to select a certain shell, the exact choice of this range is not crucial. A reduction of the radial width by a factor of two in Fig. 1.7 has no noticeable effect. This fact allows us to make statements about the absolute peak heights, for example.

1.3 Simulation Methods

In this section, we describe how a classical dusty plasma can be modeled in a computer simulation. The focus lies on the positions of the dust grains and their dynamics. A full simultaneous simulation of all plasma constituents, namely electrons, ions, neutral gas atoms and dust grains is still beyond reach due to the very different time scales. The time scale for the dust particles is given by their plasma frequency ω_{pd} which is typically of the order of several hertz [3]. The electron plasma frequency is however about twelve orders of magnitude higher due to their low mass, e.g. $\omega_{\text{pe}} \approx 3.5 \times 10^9$ Hz was measured in a dusty plasma experiment by Trottenberg et al. [35]. In the following, we describe two methods, *Monte Carlo* (MC) and *Langevin molecular dynamics* (LMD) that allow for an exact treatment of the dust dynamics. The screening effect of electrons and ions is treated statically via the Debye screening length λ_{D} in the Yukawa potential. Screening effects and the derivation of a modified dust-dust interaction potential in a streaming plasma are treated in detail in the next two Chapters "*Introduction to Streaming Complex Plasmas A & B*" and in Refs. [9–11].

1.3.1 Dimensionless System of Units

Before we start to write a simulation program, we have to figure out a suitable system of units. The advantage of such a system of units is that it makes it possible to reduce the parameter space due to *universal scaling laws* in complex plasmas [19]. Often, physically completely different systems coincide in suitable units.

As a simple example, let us consider one classical particle with mass m and charge Q in a static electric field $\mathbf{E}$. While both m and Q enter the equation of motion, the dynamics only depends the specific charge $q = Q/m$. Hence, we can reduce the two-dimensional parameter space from $\{Q, m\}$ to the one-dimensional one $\{q\}$.

In order to describe a system of classical trapped dust particle, we introduce

$$l_0 = \left(\frac{Q^2}{4\pi\epsilon_0 \cdot m\omega^2}\right)^{\frac{1}{3}}, \quad E_0 = \left(\frac{Q^2 \cdot m\omega^2}{(4\pi\epsilon_0)^2}\right)^{\frac{1}{3}}, \quad t_0 = \frac{1}{\omega} \tag{1.30}$$

as units of length, energy and time. Since the radius R_{C} of a large Coulomb ball is $R_{\text{C}} \approx \sqrt[3]{N-1} \cdot l_0$ as shown by Henning et al. [28], the unit length l_0 represents the *Wiger-Seitz radius* $a_{\text{WS}} = \sqrt[3]{3/(4\pi n)}$ in this case. The particle density is denoted by

Table 1.1 Important quantities in confined dust clusters and their units (not complete)

Quantity	Unit
Length	$l_0 = \left(\frac{Q^2}{4\pi\epsilon_0 \cdot m\omega^2}\right)^{1/3}$
Energy	$E_0 = \left(\frac{Q^2 \cdot m\omega^2}{(4\pi\epsilon_0)^2}\right)^{1/3}$
Temperature	$T_0 = k_B/E_0$
Time	$t_0 = 1/\omega$
Force	$F_0 = m\omega^2 l_0$
Momentum	$p_0 = m\omega l_0$
Electric field	$D_0 = m\omega^2 l_0/Q$
Screening parameter	$\kappa_0 = 1/l_0$

$n = N/V$. As the reduced temperature unit we use the unit of energy divided by the Boltzmann constant $T_0 = E_0/k_B$. Since the energy unit

$$E_0 = \frac{Q^2}{4\pi\epsilon_0 r_0}, \tag{1.31}$$

can be interpreted as the Coulomb interaction energy of two particles with separation r_0, the inverse temperature in these units represents the Coulomb coupling parameter Γ_C for large Coulomb balls. The units for other quantities, like force, momentum, etc. are reduced by l_0, E_0, t_0 and m, Q as units for mass and charge respectively, see Table 1.1.

In the chosen units, the dimensionless Hamilton function, see (1.32), of a dust cluster takes a very convenient form,

$$H = \sum_{i=1}^{N} \frac{\mathbf{p}_i^2}{2} + \sum_{i=1}^{N} \frac{\mathbf{r}_i^2}{2} + \sum_{i=1}^{N-1} \sum_{j=i+1}^{N} \frac{1}{r_{ij}} e^{-\kappa r_{ij}}, \tag{1.32}$$

where $r_{ij} = |\mathbf{r}_j - \mathbf{r}_i|$ is the pair distance. This Hamiltonian describes an unperturbed system. When we model laser heating experiments, see Sect. 1.4.3, we have to include the acceleration by different laser spots as time-dependent forces. Furthermore, the collisions of electrons, ions and neutral gas atoms with the dust grains have to be included. Their effect can be added in a Langevin model, see Sect. 1.3.4.

Note that this choice of dimensionless units is not unique. Also, the length $l_0' = \sqrt{2} \cdot l_0$ is often used in the dusty plasma community, e.g., Refs. [7, 25].

1.3.2 Metropolis Monte Carlo (MC) Sampling

In this chapter, we give a short introduction to the Metropolis MC technique used to model dusty plasmas. MC methods have been actively applied to dusty plasmas, e.g., Refs. [14, 36–39]. Extensions of MC to quantum system are subject of the

Chapter *"Introduction to Configuration Path Integral Monte Carlo"*. Growth and transport processes are simulated with the *Kinetic Monte Carlo* (KMC) extensions which is subject of the Chapter *"Kinetic Monte Carlo Simulations of Cluster Growth and Diffusion in Metal-Polymer Nanocomposites"*. An introduction to classical and quantum Monte Carlo is also given in Ref. [40].

While molecular dynamics simulations aim to calculate the *phase space trajectory* of a many body system exactly, MC simulations aim to calculate the correct *configuration space distribution.* The fact that the full trajectories are of minor interest when it comes to thermodynamic properties motivates this approach. The main advantage of the MC technique compared to the MD technique is the possibility to sample the accessible phase space more effectively. When the physical particle trajectories are numerically integrated, the system is easily trapped in a (meta-)stable state. The energy barriers between different metastable states can be overcome more easily in an MC simulation where the sequence of configurations is not restricted to the physical time evolution. Global updates, as for example configuration swaps between systems at different temperatures, are allowed in these simulations.

Quantities known from the thermodynamic limit such as the internal energy U, the heat capacity c_V or the radial density $n(r)$ in the *canonical ensemble* are all calculated from the probability distribution of the microstates

$$P(\mathbf{P},\mathbf{R}) = \frac{1}{Z} e^{-\frac{H(\mathbf{P},\mathbf{R})}{k_\mathrm{B}T}}, \tag{1.33}$$

where $\mathbf{P} = (\mathbf{p}_1, \mathbf{p}_2, \dots, \mathbf{p}_N)$ and $\mathbf{R} = (\mathbf{r}_1, \mathbf{r}_2, \dots, \mathbf{r}_N)$ are multi-vectors that contain the momenta and the positions of all N particles. While the Boltzmann factor can easily be calculated for each microstate $(\mathbf{P},\mathbf{R})$, the partition function

$$Z = \frac{1}{N!h^{3N}} \int_\Omega \mathrm{d}^{3N}P \, d^{3N}R \, e^{-H(\mathbf{P},\mathbf{R})/k_\mathrm{B}T} \tag{1.34}$$

involves a $6N$-dimensional ($4N$-dimensional for 2D systems) integration over the phase space Ω. Here, we replaced the summation over discrete microstates by an integration. This requires the division by the Planck constant h as the phase space volume of a single microstate.

Since the Hamiltonian of a trapped dust cluster (1.32), contains no velocity dependent forces, it splits into $H(\mathbf{P},\mathbf{R}) = H_P(\mathbf{P}) + H_R(\mathbf{R})$. The probability distribution and the partition function can be then factorized into a *spatial* and a *momentum* part,

$$P(\mathbf{P},\mathbf{R}) = \frac{1}{Z} e^{-\frac{H_P(\mathbf{P})+H_R(\mathbf{R})}{k_\mathrm{B}T}} = \frac{1}{Z_P} e^{-\frac{H_P(\mathbf{P})}{k_\mathrm{B}T}} \cdot \frac{1}{Z_R} e^{-\frac{H_R(\mathbf{R})}{k_\mathrm{B}T}} = P_P(\mathbf{P}) \cdot P_R(\mathbf{R}) \, . \tag{1.35}$$

The momentum part of the partition function is readily found as

$$Z_P = \frac{1}{\sqrt{N!h^{3N}}} \int_{\Omega_P} \mathrm{d}^{3N}P \, e^{-\frac{\mathbf{P}^2}{2mk_\mathrm{B}T}} = \frac{1}{\sqrt{N!}} \left(\frac{mk_\mathrm{B}T}{\hbar} \right)^{\frac{3N}{2}} . \tag{1.36}$$

Due to the particle interactions, the spatial part Z_R cannot be calculated analytically for most systems.

The Metropolis MC algorithm describes a procedure that randomly generates a series of microstates $\{\mathbf{R}_1, \mathbf{R}_2, \mathbf{R}_2, \ldots\}$ which are distributed according to the configurational part of (1.33), i.e. $P_R(\mathbf{R})$, the so called *Markov chain* of states. For the states to be properly distributed, it is sufficient, even though not essential, to ensure that the transition operator W fulfills the *detailed balance condition*, see also the Chapter *"Introduction to Configuration Path Integral Monte Carlo"*.

The transition operator W takes an input state $\mathbf{R}_i$ and generates an output state $\mathbf{R}_j$ from it. W consists of the proposal of a certain MC move and the acceptance probability P_{acc} of this move. For each move, for example, one particle is randomly chosen and the displacement $\Delta x = \xi_x \cdot \Delta r_{\max}$ (Δy, Δz analogous) is proposed by generating uniformly distributed random numbers $\xi \in [-1 : 1]$. The maximum absolute displacement in each direction is given by $r_{\max}$, see below.

Since the move from $\mathbf{R}_i$ to $\mathbf{R}_j$ is proposed with the same probability as the reverse move, the detailed balance condition reads

$$P_R(\mathbf{R}_i) \cdot P_{\mathrm{acc}}\left(\mathbf{R}_i \mapsto \mathbf{R}_j\right) = P_R\left(\mathbf{R}_j\right) \cdot P_{\mathrm{acc}}\left(\mathbf{R}_j \mapsto \mathbf{R}_i\right) . \tag{1.37}$$

When we insert the Boltzmann probability of the two microstates $\mathbf{R}_i$ and $\mathbf{R}_j$ from (1.35), we find that the partition function cancels and we obtain

$$\frac{P_{\mathrm{acc}}\left(\mathbf{R}_i \mapsto \mathbf{R}_j\right)}{P_{\mathrm{acc}}\left(\mathbf{R}_j \mapsto \mathbf{R}_i\right)} = e^{-\frac{H_R(\mathbf{R}_j)-H_R(\mathbf{R}_i)}{k_B T}} = e^{-\frac{\Delta E_{ij}}{k_B T}} \tag{1.38}$$

as condition for the acceptance probability. As proposed by Metropolis et al. in 1953 [41], one appropriate choice is

$$P_{\mathrm{acc}}\left(\mathbf{R}_i \mapsto \mathbf{R}_j\right) = \begin{cases} 1 & : \Delta E_{ij} \leq 0 \\ e^{-\frac{\Delta E_{ij}}{k_B T}} & : \Delta E_{ij} > 0 \end{cases} . \tag{1.39}$$

It means that the proposed displacement of one particle is always accepted if it decreases the total energy. On the other hand, displacements that increase the total energy are also accepted with a finite probability. It is essential to accept some of these moves in order to obtain a series of Boltzmann distributed states. If our MC algorithm rejected all of these moves, the series of generated configurations would "move" towards a local energy minimum and would be stuck in this configuration.

Moreover, the efficiency of the MC sampling depends on the maximum displacement $r_{\max}$. If it is too large, most of the proposed displacements are rejected and the system remains in the same configuration for many MC steps. On the other hand, if $r_{\max}$ is too small, most of the proposed displacements are accepted but it takes a large number of displacements to significantly change the configuration. Therefore, the number of MC steps that are required to sample a representative set of configurations from the accessible phase space is again large. A good choice can be made by adjust-

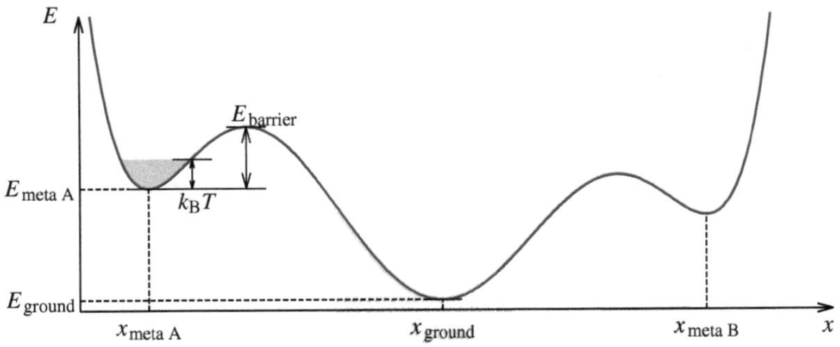

Fig. 1.8 Sketch of a one-dimensional energy landscape with the ground state at x_{ground} and two metastable states

ment of $r_{\max}$ during the equilibration phase. After several steps of MC integration with an initial value $r_{\max}$ of typically a few percent of the particle spacing, the ratio f_{acc} of accepted and proposed displacements is calculated. If f_{acc} is significantly, let us say 5 %, above (below) one half, $r_{\max}$ is increased (decreased) by a few percent. This procedure is consecutively repeated for a few hundred to thousand MC steps. The arising $r_{\max}$ is larger for high temperature compared to low temperatures.

1.3.2.1 Parallel Tempering Monte Carlo

The Metropolis MC algorithm can become inefficient at low temperatures when the configuration space is divided into different regions by *energy barriers*. Figure 1.8 shows a very simplified (one-dimensional) example of the energy landscape. In general the configuration space of a 3D system has $3N$ coordinates. However, the sketch illustrates that an energy barrier has to be overcome to cross between the ground state and the two metastable states. Overcoming this barrier requires in general several consecutive MC steps increasing the total energy. The accumulated acceptance probability for any path between two minima scales exponentially with the ratio of barrier height and temperature, $P_{\text{acc}}^{\text{cross}} \propto \exp\{-E_{\text{barrier}}/k_B T\}$.

At low temperatures, the kinetic energy is small compared to the height of the energy barrier. Consequently, many configurations within a local energy minimum are sampled before the barrier is overcome by chance. The whole accessible configuration space is no longer sampled efficiently in this case.

One method that enhances the sampling efficiency is called *Parallel Tempering* or *Exchange MC Method* [42]. The system, which is a cluster of N dust particles in our case, is simulated at different temperatures ($T_1 < T_2 < \ldots < T_M$) simultaneously. For each cluster k_{MC} MC steps are performed individually. Then swapping moves are performed that mix configurations at different temperatures, see Fig. 1.9.

In each swapping move, a system $(\mathbf{R}_m, T_m)$, $m \in [1, M-1]$ is chosen randomly. It is compared to the system with the next higher temperature $(\mathbf{R}_{m+1}, T_{m+1})$. The

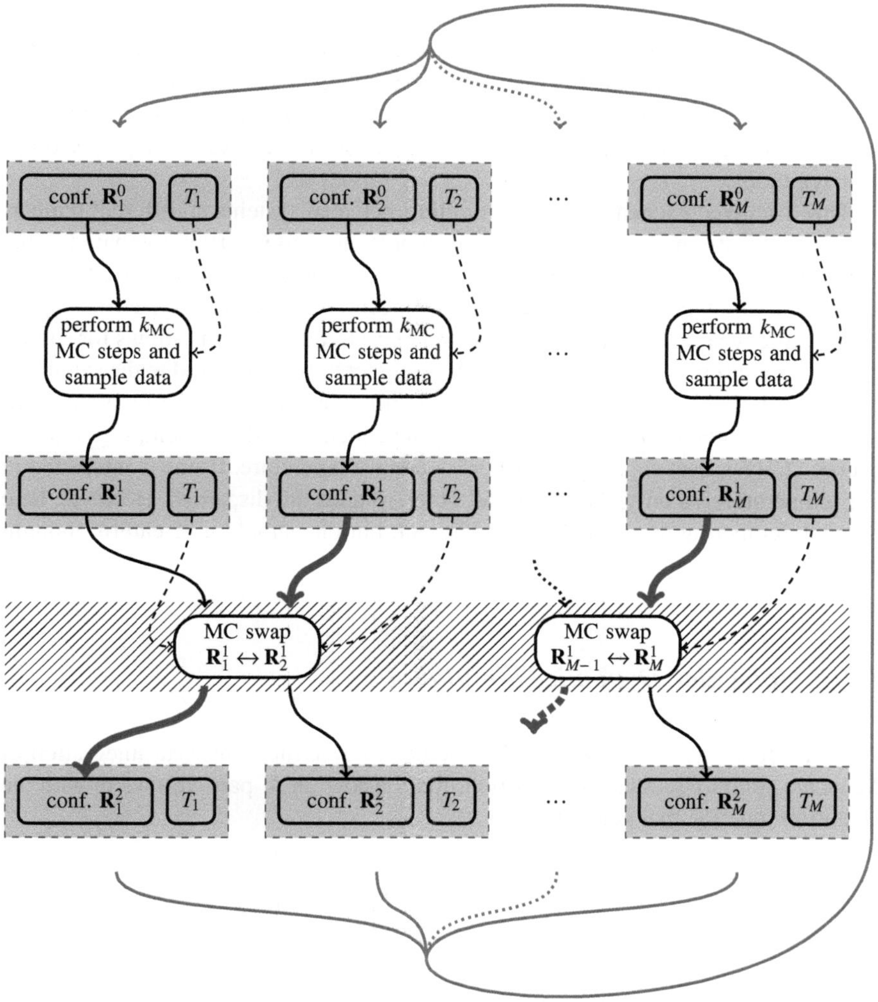

Fig. 1.9 Scheme of an MC simulation with *Parallel Tempering*. M configurations at different temperatures are processed in parallel. After k_{MC} MC steps with local updates only, the configurations are mixed in the swapping moves (*shaded box*). Then, individual MC steps are repeated for all configurations

acceptance probability for swapping the configurations $\mathbf{R}$ of the two systems is

$$P_{\mathrm{acc}}^{\mathrm{swap}} = \begin{cases} 1 & : E_m \geq E_{m+1} \\ \exp\left\{(E_m - E_{m+1}) \cdot \left(\frac{1}{k_{\mathrm{B}} T_m} - \frac{1}{k_{\mathrm{B}} T_{m+1}}\right)\right\} & : E_m < E_{m+1} \end{cases} . \quad (1.40)$$

This means the configuration swap is always excepted if the hotter system has a lower internal energy. The probability to accept a move that turns the configuration with higher energy to the colder system, decays exponentially with the product of the difference in energy and inverse temperature. A sufficiently high rate of accepted swapping moves requires the energy distributions at consecutive temperatures $p(E, T_m)$ and $p(E, T_{m+1})$ to have significant overlap. Good results are achieved with equal temperature steps on a logarithmic scale with about ten to twenty steps per order of magnitude.

This discussed exchange of configurations between systems at different temperatures represents an alternative way to overcome an energy barrier compared to the sequence of single particle moves that is necessary in pure Metropolis MC. The transition between two metastable configurations can take place in a system of higher temperature, where these transitions happen more frequently. This allows for a more effective sampling of the phase space at low temperatures compared to the same MC algorithm with local updates only. Parallel Tempering is in particular recommended when one is interested in the temperature dependence of the system, e.g. internal energy $\langle E\rangle(T)$, heat capacity $c_V(T)$ etc., versus temperature. If one is interested in one temperature T_0 only, the data of all other systems are discarded. However, here Parallel Tempering can still be more efficient, but one should give careful thought about the performance.

1.3.3 Molecular Dynamics Simulations

The molecular dynamics (MD) technique is based on the numerical integration of *Newton's equations*. As a starting point, we consider an N-particle system with the Hamilton function that generalizes (1.3)

$$H = \sum_{i=1}^{N} \frac{\mathbf{p}_i^2}{2m_i} + \sum_{i=1}^{N} V(\mathbf{r}_i) + \sum_{i=1}^{N-1} \sum_{j=i+1}^{N} W(\mathbf{r}_i, \mathbf{r}_j)\,, \tag{1.41}$$

where $V(\mathbf{r}_i)$ is an arbitrary, external potential and $W(\mathbf{r}_i, \mathbf{r}_j)$ is a particle-particle interaction potential. In this section, we constrain the discussion to conservative forces without explicit velocity dependence (like a friction term or the magnetic force). Instead, frictional forces are included in Langevin MD simulations discussed in the next section. From the Hamilton function (1.41), we obtain Newton's equations–as a set of coupled first-order differential equations,

$$\frac{d}{dt}\mathbf{r}_i = \frac{\mathbf{p}_i}{m_i} \quad \text{and} \tag{1.42}$$

$$\frac{d}{dt}\mathbf{p}_i = \mathbf{F}_i \quad \text{with} \quad \mathbf{F}_i = -\nabla_{\mathbf{r}_i}\left(\sum_{i=1}^{N} V(\mathbf{r}_i) + \sum_{i=1}^{N-1} \sum_{j=i+1}^{N} W(\mathbf{r}_i, \mathbf{r}_j)\right)\,, \tag{1.43}$$

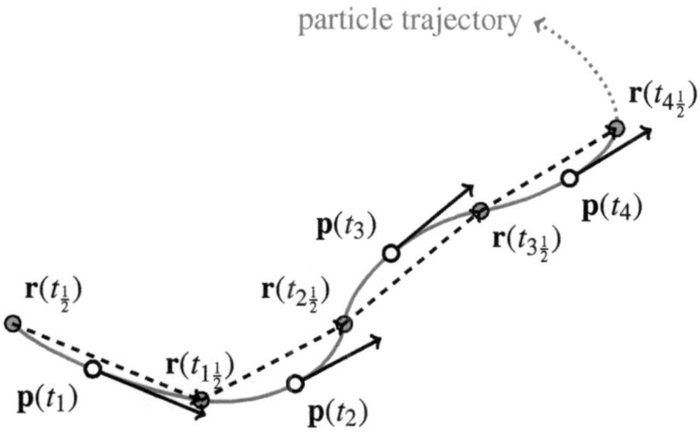

Fig. 1.10 Illustration of an exact particle trajectory (*light gray*) and the approximate integration path of the Leapfrog algorithm (*dashed*). The momentum (*arrows*) is evaluated at the full time steps, while the spatial position (*gray points*) is evaluated at the half time steps

with the particle index i from 1 to N. A simple though efficient integration scheme is the "Leapfrog" method. It is derived by Taylor expansion of the equation of motion for $\mathbf{r}_i(t)$ truncated after the second order term,

$$\begin{aligned}\mathbf{r}_i(t+\Delta t) &\approx \mathbf{r}_i(t) + \frac{1}{m_i}\mathbf{p}_i(t)\,\Delta t + \frac{\mathbf{F}_i(t)}{2m_i}\Delta t^2 \\ &= \mathbf{r}_i(t) + \frac{1}{m_i}\mathbf{p}_i\left(t+\frac{1}{2}\Delta t\right)\Delta t\,, \end{aligned} \tag{1.44}$$

$$\text{with}\quad \mathbf{p}_i(t) \approx \mathbf{p}_i\left(t-\frac{1}{2}\Delta t\right) + \mathbf{F}_i(t)\,\frac{\Delta t}{2} \tag{1.45}$$

$$\text{and}\quad \mathbf{p}_i\left(t+\frac{1}{2}\Delta t\right) \approx \mathbf{p}_i(t) + \mathbf{F}_i(t)\,\frac{\Delta t}{2}\,. \tag{1.46}$$

Combining the last two steps (1.45) and (1.46), in one step, we obtain

$$\mathbf{p}_i\left(t+\frac{1}{2}\Delta t\right) \approx \mathbf{p}_i\left(t-\frac{1}{2}\Delta t\right) + \mathbf{F}_i(t)\,\Delta t\,. \tag{1.47}$$

Hence, we can consider this integration scheme as a two-step method with shifted time grids for position $\mathbf{r}_i$ and momentum $\mathbf{p}_i$. For consistency reasons, we shift the time variable by half a time step $t \to t + \frac{1}{2}\Delta t$. Then, the momenta are evaluated at full time steps $\Delta t, 2\Delta t, 3\Delta t, \ldots$ and the positions are evaluated at half time steps. $\frac{1}{2}\Delta t, 1\frac{1}{2}\Delta t, 2\frac{1}{2}\Delta t, \ldots$. The integrated trajectory of one particle is sketched in Fig. 1.10.

In order to calculate observables like the total energy, we need the momentum information and the spatial information at the same time points. Thus, we use an integration scheme where the time propagation of the spatial coordinate is split up

```
// initialize the system randomly
For i=1 ... N
    r[i] = rand_r();              // random positions
    p[i] = rand_p();              // random momenta
t=0.0;                            // initialize the clock

// main MD simulation loop
While t < time_to_end

    // update positions by half a time step
    For i=1 ... N
        r[i] += p[i]/m*0.5*dt;

    // calculate all accellerations
    For i=1 ... N
        // external confinement
        a[i] = force_V(r[i]);

        // calculate interaction forces
        //with all other particles
        For j=1 ... N; j!=i
            a[i] += force_W(r[i],r[j]);

    // update momenta by a full time step
    For i=1 ... N
        p[i] += a[i]*dt;

    // update positions by half a time step
    For i=1 ... N
        r[i] += p[i]/m*0.5*dt;

    t += dt;                      // update the clock
    record_data(writefile);       // write output
```

Fig. 1.11 Pseudocode illustrating the integration scheme of the Leapfrog algorithm. The external forces `force_V(r[i])` and the interaction forces `force_W(r[i],r[j])` are calculated according to (1.43). Note that each variable `r[i]`, `p[i]` contains a vector with the dimension according to the system dimension

into two steps with half the time step. The pseudocode of this MD algorithm is presented in Fig. 1.11. Note that in this implementation, each pair-interaction force is calculated twice. For reciprocal interaction potentials, this can easily be optimized saving half the computation time for the calculation of the interactions. Since this part will consume most of the computation time due to the quadratic scaling with the particle number, optimization is important. Moreover, the interaction forces can be calculated in parallel.

We described the Leapfrog scheme with two half time steps for the spatial propagation since this scheme can directly be extended to a quasisymplectic integrator for the Langevin equation of motion in the next section. Symplectic integration schemes attempt to preserve two-forms $dq_i \times dp_i$ of the generalized coordinate q_i and the corresponding conjugate momentum p_i. For an introduction to sympletic integrators, we refer the reader to Refs. [43–45]. The Leapfrog scheme is equivalent to the so-called *velocity Verlet* or *Swope algorithm* [46, 47].

1.3.4 Langevin Molecular Dynamics Simulations

As a further important simulation technique, we describe the Langevin molecular dynamics (LMD) simulation. It can be seen as a step between fully deterministic time propagation of Newtons equations in MD and the stochastic approach in MC. The dynamics of the dust component are treated on the atomic level. However, the exact treatment of the electrons, ions and the neutral gas atoms in the plasma would require a temporal resolution several orders of magnitudes beyond the time scale of the dust dynamics. This large difference in time scales makes it impossible to treat both heavy dust, light ions and neutral atoms and even lighter electrons on an atomic level in the same simulation. The main effect of the light charged particles is to charge the dust particles and to screen the Coulomb interaction. Therefore, the interaction of the dust grains is modeled by a Yukawa potential or by improved potentials in streaming plasmas which are treated in detail in the Chapter "*Introduction to Streaming Complex Plasmas B: Theoretical Description of Wake Effects*".

The neutral gas interacts with the dust via collisions causing frequent exchanges of momentum. Paul Langevin [48] presented a model that includes the gross effect of the collisions via a friction term proportional to $\mathbf{p}$ and a stochastic force (ξ) in Newtons equations,

$$\frac{d}{dt}\mathbf{p}_i = \mathbf{F}_i - \gamma \mathbf{v}_i + \xi_i(t)\,. \tag{1.48}$$

The stochastic force ξ is often approximated by a Gaussian probability distribution and its amplitude is proportional to the temperature T (Gaussian white noise). In this case, the correlation function has to obey the property (Fluctuation-Dissipation theorem),

$$\left\langle \xi_{i,\alpha}(t)\xi_{j,\beta}(t')\right\rangle = 2m\gamma k_{\mathrm{B}}T\,\delta_{i,j}\delta_{\alpha,\beta}\delta(t-t'), \tag{1.49}$$

where the Latin indices $i, j = 1\ldots N$ denote the particle number and the Greek indices $\alpha, \beta = x, y(, z)$ denote the spatial direction. The Gaussian white noise further has zero average

$$\langle \xi_{i,\alpha}(t) \rangle = 0\,, \tag{1.50}$$

for a non-streaming background gas since no net momentum is transfered. As shown by Mannella in 2004 [43], friction and stochastic force can be included in a quasisymplectic integrator by modification of the momentum update in the Leapfrog scheme, Fig. 1.11,

$$p_i\,(t+\Delta t) = c_2\left[c_1\mathbf{p}_i\,(t) + \Delta t\mathbf{F}_i\left(t+\frac{\Delta t}{2}\right) + d_1\right], \tag{1.51}$$

where is a three-dimensional (two-dimensional for 2D systems) vector of independent Gaussian random variables with zero average and unit standard deviation. The constants are

$$c_1 = 1 - \frac{\gamma\Delta t}{2}\,, \qquad c_2 = \frac{1}{1-\frac{\gamma\Delta t}{2}}\,, \qquad d_1 = \sqrt{2k_\mathrm{B}T\gamma m\Delta t}\,. \tag{1.52}$$

This algorithm is referred to as *Symplectic Low Order* (SLO). Due to the choice of the constants c_1, c_2 and d_2, energy loss and gain by the friction term and stochastic force balance each other at the system temperature T. After a sufficient number of integration steps M with $M\Delta t \gg \gamma^{-1}$, the memory of the initial value of the total energy is lost, while in pure MD the initial energy is conserved. In contrast, in LMD, the energy is not constant but is an observable that fluctuates around the mean. Therefore, thermodynamic properties such as the heat capacity, see Sect. 1.1.3, can be calculated directly in these simulations.

Three most important physical ensembles are known from statistical thermodynamics. The *micro-canonical ensemble* $\Omega\,(N, V, E)$ describes a system with fixed particle number N confined in a constant volume V and with constant total energy E. We can simulate a micro-canonical ensemble with pure MD. When a system can exchange energy with a heat bath, the total energy is no longer fixed but the system adopts the temperature T of the heat bath at thermal equilibrium. Such a system is described by the *canonical ensemble* $Z\,(N, V, T)$. This ensemble is approximated by LMD, where the friction and the stochastic force simulate the heat bath. Finally, the *grand-canonical ensemble* $\Xi\,(T, V, \mu)$ describes a system which can exchange particles besides energy with the surrounding bath. The particle number is not fixed for such a system but the system is characterized by the chemical potential μ instead. This ensemble can be simulated by an *Adaptive-Resolution Technique* with one region where the particles are described with high resolution and a second region where particles are described in a coarse-grained model. As particles can interchange between both regions, the latter particles act as a particle bath [49]. For a detailed overview of molecular dynamics simulations in dusty plasmas, we refer to the chapter *"Molecular Dynamics Simulations of Strongly Correlated Dusty Plasmas"* in Ref. [3]. These simulation methods allows to calculate structural properties [50] and dynamical properties such as the eigenmode spectra [51] of confined dust cluster as well as static and dynamic properties of extend dust systems [52].

1.4 Experimental Realization of Heating and Phase Transitions

This section is devoted to the experimental realization of heating methods for dust clusters. First, we give a short overview of the heating methods, which have been used in experimental studies so far. Then, we introduce a sophisticated setup which uses laser manipulation as a tool for heating monolayers of particles, which, for example, allows detailed investigations on the melting behavior of dust clusters. Additionally, a way of modeling this laser heating method in simulations is presented at the end of this section.

1.4.1 Variation of the Dust Kinetic Temperature in Experiments

There is a close link between phase transitions and heating. In the first experiments, a variation of Γ was achieved by changing the plasma parameters, namely rf-power and neutral gas pressure. Thus, a solid-fluid transition could be reached especially for extended 2D particle systems [53–55]. This method requires no additional technical effort, which makes it simple to use in experiments. However, these systems were not true 2D systems. Depending on the plasma parameters, there is a chance of finding additional particles in a layer below the actual particle system. These particles will mostly drive the system into a non-equilibrium phase transition based on the Schweigert instability [56]. In general, changing the plasma conditions results in disadvantages for many investigations. Besides an influence on particle charge and screening length of the plasma, the trapping conditions of the systems are modified. Further, there is no precise control of the dust kinetic temperature, and due to the localized particle-interaction, the heating effect will be localized in some cases. Finally, this method is not applicable to 3D systems. Due to these drawbacks, a different method is desirable for systematic studies of phase transitions in dusty plasma experiments.

For many different experiments, laser manipulation turned out to be a versatile tool for the manipulation of dusty plasmas at the individual particle scale. The radiation pressure of intense laser sources is sufficient to push microparticles through the plasma environment. Hence, it can be utilized to drive waves and oscillations [57–60], Mach-cones [61], shear flows [58, 62] and, most importantly for the investigation of phase transitions, it can be used for laser heating. Its significance for the (kinetic) particle heating arises from the fact that it allows a precisely controllable and individual manipulation of single particles without affecting the surrounding plasma conditions. Wolter et al. were the first to demonstrate the capability of a single-sided laser system to melt finite particle systems [63]. Later, the heating method was extended to a single-axis, two-sided system and applied to an extended single-layer of particles by Nosenko et al. [2]. In principle, the laser heating can be applied to 3D systems as well [64]. Here, in experimental situations, it is considerably more challenging to realize an ideal heating effect due to the reduced accessibility of the particle systems inside the vacuum vessel, which arises from the space required for the particle imaging system as well as the need for three perpendicular axes of laser manipulation to gain an isotropic heating effect.

1.4.2 Experimental Realization of the Laser Heating of Dusty Plasmas

This section describes the latest experimental development for the laser heating of 2D systems, which was published recently [65]. For the purpose of the investigation of phase transitions in finite dust clusters, we designed the LaMa-Experiment, which essentially is a natural improvement of previous laser heating experiments [2, 63]. It aims at the realization of a true thermostat by using laser manipulation combined with an adequate heating method. The core piece of the setup is a plasma chamber, which allows us to generate plasma conditions typical for 2D plasma crystal experiments. We use an asymmetric capacitively coupled rf-discharge with argon at low discharge powers of about $1-10\,\mathrm{W}$ and neutral gas pressures in the range of $1-100\,\mathrm{Pa}$. This allows us to perform measurements at different magnitudes of friction. The microparticles are trapped in the plasma sheath above the lower electrode, which features a spherical depression to constitute the horizontal confinement for the dust clusters. As a distinctive feature, the chamber provides eight windows at the side and one top window. These numerous possibilities to access the dust clusters from many sides are crucial for the illumination, imaging and manipulation system.

Through one side window the layer of particles is illuminated by a red laser fan. The top window is used for video microscopy with a camera featuring a frame rate of up to 200 frames per second at full resolution of 4 megapixels. With a standard consumer macro lens plus additional extension tubes, the particles are imaged with high spatial and temporal resolution. A red interference filter is used to significantly increase the contrast of particles and background. After recording the images, typical particle tracking velocimetry algorithms are used to obtain the particle trajectories with subpixel resolution for further processing [67]. This high spatial and temporal resolution is sufficient to record thermal fluctuations of the particle positions without artifacts like, e.g., pixel locking [67]. Therefore, the diagnostic is very sensitive, which makes it necessary to mount the vacuum vessel with the optical diagnostics system on a vibration control table to achieve maximum data quality.

The optical setup of the laser manipulation system is illustrated in Fig. 1.12. It mainly consists of four two-axis galvanometer mirror systems, which are arranged around the plasma chamber. They are individually driven and controlled with an

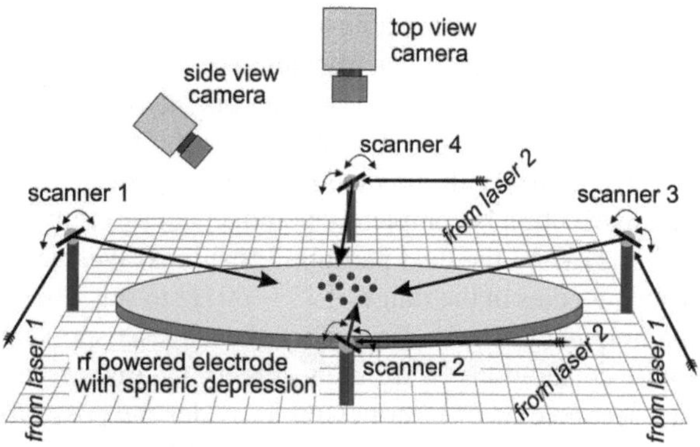

Fig. 1.12 Sketch of the optical setup for laser heating in the LaMa-Experiment

8-channel analog output board and positioned a few centimeters above the average height of the trapped particles. This allows us the possibility of steering a laser beam through the complete area of dust particles and to access individual particles in the layer for each galvanometer scanner. Due to the small incident angle the horizontal component of the radiation force onto any particle, which enters the beam, stays large, while the vertical force is small compared to the strong vertical trapping of the particles. Using four equivalent scanner systems results in two benefits for the heating process. First, using two opposing systems allows for a compensation of the net force to the particle cluster. And second, using two perpendicular axes of laser manipulation will enable the manipulation of the particles in either direction, which is important for isotropy. For this setup, four laser beams are needed in total. Therefore, we use two adjustable green lasers of about 5 W peak output power, which are both split into two beams by beam splitter cubes. Each laser's output power is monitored by laser power meters and actively controlled to keep the output power constant. To monitor the laser manipulation, a secondary camera with a green interference filter is installed in front of a side window with a slanted view from above.

Now, for the laser heating, it is important how we move the laser spots through the particle system. On the one hand, the method must meet the technical conditions, which basically means that the maximum speed of spot motion is limited to the maximum scanning speed of the galvanometer mirrors. On the other hand, for true heating in a thermostatic sense, a somehow randomized scanning pattern is desired. On account of this, we contrived the random frequency sweep method. The idea is to randomly choose a constant speed within a certain interval for each channel (two per galvanometer set) to move the spot from one to the other side of the dust cluster. After the spot has reached a predefined limit, the direction of motion is reversed and a new speed is chosen randomly again. Like this, each channel is driven by a triangular signal, which changes its frequency after each half period. The limits in which the four spots are moved have to be defined slightly larger than the cluster dimensions for two reasons: it will ensure that all particles can be hit by the laser spots at any time even if they will perform larger excursions. And, mechanically seen, the galvanometer mirrors are not able to execute any sharp change in direction due to their inertia, which will result in a truncation of the laser spots' trajectories at the turning points. Per se, this would be no issue, but any nonlinear motion results in an inhomogeneous mean intensity distribution, as the use of sine functions for a scanning pattern would do. With the random frequency sweep method, the time-averaged intensity distribution is completely homogeneous except for the outermost region, which will be beyond the dust particles' positions. We use scanning speeds corresponding to frequencies in the range of $15 - 60\,\mathrm{Hz}$ to stay above typical values of the dust plasma frequency. The heating effect of each laser spot is mainly directed forward, while any accelerations in different directions arise from thermalization effects. Hence, the second axis of laser heating provides the possibility of isotropic heating. In fact, it requires an adjustment of power of the two lasers prior to each measurement to match the resulting temperatures. The isotropy of the heating effect can be seen in the velocity distributions for a cluster at different stages of heating in Fig. 1.13a. For any laser power the slope of the distributions is the same for x- and y-direction, indicating an isotropic heating effect. The velocity distributions compare very well to ideal Maxwellian distributions (which would be a straight line in this diagram) for all laser heating powers. A fit of a Maxwellian distribution to the experimental data results in temperatures plotted in the inset of Fig. 1.13a. As predicted in [63], the temperature goes roughly with the square of the applied laser power. However, one would not expect a perfect match as for example a variation of laser output power always involves a change of the beam profile in real applications.

Figure 1.13b–d show the trajectories of all particles of the cluster corresponding to the velocity distributions shown in (a) from the unmanipulated cluster (b); for moderate laser heating (c), to strong laser manipulation (d), over a period of 66 s. In the case of no laser manipulation, the particles only perform a slight thermal motion around their equilibrium positions, which can be considered fixed. As the intensity of laser heating is increased, the thermal motion increases until particles start to change places occasionally. If the laser power is further increased, the rate of particle hopping goes up. With the laser power available in our experiment we are able to heat clusters up to the melting transition and beyond.

Still, the laser manipulation introduces a general challenge, namely a distinctive particle aging effect. In experiments, it is observed that the reaction of particles to the laser beam changes with time. This is primarily attributed to the outgassing of water [68, 69] and surface modifications by thermal effects [69]. For our experiments we expose the particles to moderate laser manipulation for about half an hour and then wait for the same time without any manipulation. This artificial aging is sufficient to keep the particle properties constant during experiments.

In conclusion, laser heating can be regarded as a true thermostat, which features a precise control of the dust kinetic temperature. Using the random frequency sweep method with a two-axis setup results in a homogeneous and isotropic heating effect. It allows us to drive dust clusters from a solid state into a fluid-like state. Thus, a detailed investigation of the phase transition is possible in experiments. One subject of interest is the dependence of the critical temperature on the particle number for example. Recently, we used the instantaneous normal mode analysis (INM) as a new

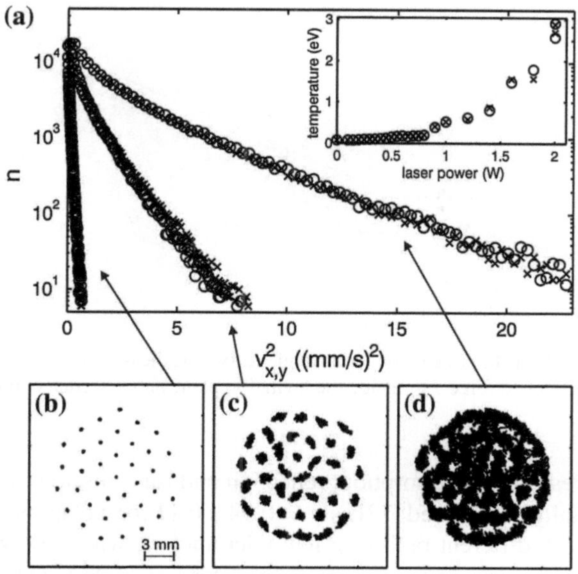

Fig. 1.13 **a** Velocity distributions of an $N = 38$ particles cluster in x-direction (*crosses*) and y-direction (*circles*) for the unmanipulated cluster, a moderate laser power, and maximum heating. The inset shows the resulting dust kinetic temperature obtained by a Maxwellian fit to the velocity distributions for different laser powers. **b–d** Trajectories of the cluster corresponding to the shown velocity distributions

approach to determine the critical temperature of a large set of data obtained in the LaMa-Experiment [70, 71].

1.4.3 Modeling Laser Heating

In this section, we show how the above discussed laser heating experiments can be modeled in a computer simulation. At this point, we do not consider the interaction between laser radiation and dust particle in detail. The momentum transfer is considered in the beam direction and proportional to the local laser intensity at the position of the particle.

The amplitude of the force on a dust particle due to the radiation pressure

$$F_{\text{laser}} = q \cdot \frac{n_1 \pi r_{\text{p}}^2 I_{\text{laser}}}{c}\,, \tag{1.53}$$

is dependent on the refractive index n_1 of the plasma surrounding the particle, the radius r_{p} of the particle, the intensity of the laser I_{laser} and a quality factor q. The constant c is the vacuum speed of light. The dimensionless quality factor q depends on the

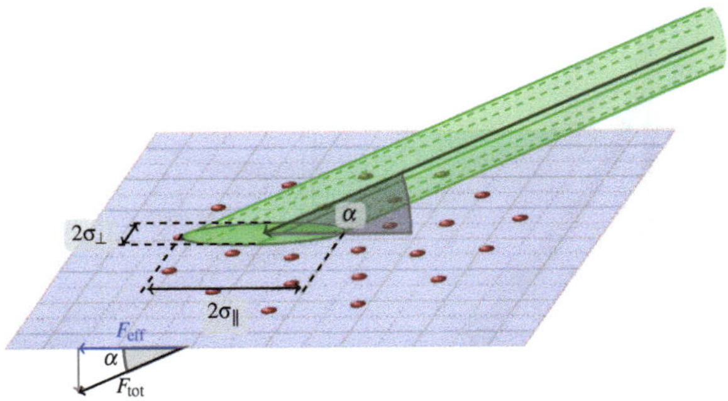

Fig. 1.14 Striking of one laser beam under an angle α. Below, the decomposition of the total laser force $F_{\rm tot}$ into an effective force $F_{\rm eff}$ inside the levitation plane and a normal component is shown

optical properties, namely absorption, reflection and transmission of the particle. It was experimentally determined [72] as $q = 0.94 \pm 0.11$ for melamine-formaldehyde microspheres with different radii of a few micrometers, which are widely used in dusty plasma experiments. The intensity of the laser $I_{\rm laser}$ is derived from a Gaussian beam profile with width σ. Since the beam strikes the levitation plane under an angle α of a few degrees, the cut face becomes anisotropic. The width $\sigma_{\parallel} = \sigma/(\sin\alpha)$ parallel to the in-plane beam direction becomes larger, while the width $\sigma_{\perp} = \sigma$ in perpendicular direction is unaffected.

The time-dependent force on a dust particle at $\mathbf{r} = (x, y)$ by a laser spot is included in the simulation as (l labels the laser spots)

$$\mathbf{F}_l\,(x, y, t) = \frac{F_0}{2\pi\sigma_x\sigma_y}\mathbf{e}_l \exp\left\{-\frac{(x - x_l(t))^2}{2\sigma_x^2} - \frac{(y - y_l(t))^2}{2\sigma_y^2}\right\}, \qquad (1.54)$$

where $\mathbf{e}_l$ is the beam direction and $\mathbf{r}_l(t) = (x_l(t), y_l(t))$ describes the trajectory of the spot in the levitation plane. The intersection of the laser spot and the levitation plane is always wide in beam direction and narrow in perpendicular direction. Which one of σ_x and σ_y is the parallel spot width depends on the beam direction. For beams in the $\pm x$-direction, it is $\sigma_x = \sigma_{\parallel}$ and $\sigma_y = \sigma_{\perp}$ and vice versa. The force amplitude

$$F_0 = q \cdot \frac{n_1 \pi r_{\rm p}^2 P_{\rm laser}}{c} \cos\alpha\,, \qquad (1.55)$$

is proportional to the power $P_{\rm laser}$ of a beam, and it is converted to dimensionless units introduced in Sect. 1.3.1. Note, that the dimension of F_0 is *force* $\times$ *area*, $[F_0] = N \cdot m^2$, since it becomes a force when divided by the spot size $2\pi\sigma_x\sigma_y$. The factor $\cos\alpha$ takes into account that only the in-plane component of the force is relevant for the acceleration of a dust grain, see Fig. 1.14.

We use the dimensionless heating power F_0 as an input parameter to our simulations. Some care about the power $P_{\rm laser}$ has to be taken, when comparing simulations with experimental results. Due to losses at mirrors, lenses, beam splitters and absorption in the neutral gas, the actual power P_{laser} at the dust cluster will be significantly smaller than the output power of the laser source [65, 73].

Figure 1.15 shows the velocity distribution and trajectories from a LMD simulation with manipulation lasers included. Although the experimental parameters are not matched exactly, this figure allows for a qualitative comparison with Fig. 1.13. The simulation confirms the quadratic scaling between the increase of the dust temperature and heating power. As it was found in the experiment, the velocity distribution retains a Maxwellian shape which is important for a thermal heating. A full quantitative comparison is difficult due to the variety of parameters and their uncertainties in the experiment.

The key benefit of the experiment oriented simulations is the precise control over parameters such as laser spot sizes, relative positions of the spots and the trap frequency. A specific scan of one single parameter can easily be performed in the simulation while this scan is often hindered due to the temporal drift of the plasma parameters in the experiments. As a result, these simulations allow us to suggest suitable spot sizes and spot velocities for an optimal heating effect [65, 73]. Moreover, the simulation may serve as preparatory work for planning promising experiments, e.g., the investigation of heat transport by inhomogeneous heating [74].

1.5 Melting Parameters with Focus on Application in Simulations and Experiments

1.5.1 Lindemann-Type Melting Parameters

When it comes to melting in few particle systems, one finds several parameters attributed to Lindemann's picture of melting [75]. In this picture, the solid phase is characterized by a crystalline structure. Each particle oscillates around its equilibrium position with an amplitude that is small compared to the lattice spacing. In the liquid phase, in contrast, each particle moves through the system and is not fixed to a lattice site. A parameter which is applicable also to 2D systems[4] is the relative inter-particle distance fluctuations (IDF) parameter [76]

$$u_{\text{rel}} = \frac{2}{N \cdot (N-1)} \sum_{i<j} \sqrt{\frac{\langle r_{ij}^2 \rangle}{\langle r_{ij} \rangle^2} - 1} \, . \tag{1.56}$$

This parameter relates the fluctuation of the pair distance $r_{ij} = |\mathbf{r}_j - \mathbf{r}_i|$ to the mean value of the pair distance. It is averaged over all particle pairs. The brackets $\langle \ldots \rangle$ denote the average over a block of sampled configurations.

[4] Lindemann himself considered the ratio $\langle r^2 \rangle / a^2$, where a is the lattice spacing. This quantity is small in a crystal and has a jump at the melting point. However, it is divergent in 2D.

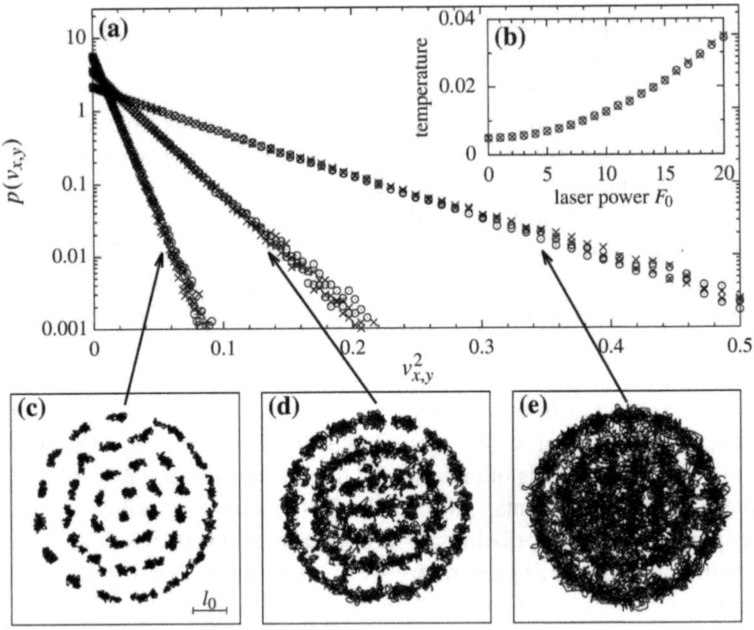

Fig. 1.15 Velocity distribution and trajectories from an LMD simulation for comparison with experimental findings in Fig. 1.13. **a** Shows the velocity distributions in x (*crosses*) and y-direction (*circles*) for the unmanipulated cluster and for two heating powers. The inset shows the resulting dust kinetic temperature obtained by a Maxwellian fit to the velocity distributions. **c**–**e** Show the trajectories of the clusters corresponding to the shown velocity distributions. Parameters: $N = 38$, $\kappa = 1.0$, $T_{\mathrm{LMD}} = 0.005$, $\Delta t = 200$ (trajectories)

The unphysical dependence of u_{rel} on the simulation's length is eliminated by calculation of u_{rel} on blocks with constant length M [76]. The whole simulation is divided into N_{B} blocks and $u_{\mathrm{rel}}(s)$ is calculated separately on each block s according to (1.56). From these different measurements of u_{rel}, we can now compute the average and the variance,

$$\bar{u}_{\mathrm{rel}} = \frac{1}{N_{\mathrm{B}}} \sum_{s=1}^{N_{\mathrm{B}}} u_{\mathrm{rel}}(s) \quad \text{and} \quad \sigma_{u_{\mathrm{rel}}}^2 = \frac{1}{N_{\mathrm{B}} - 1} \sum_{s=1}^{N_{\mathrm{B}}} \left(u_{\mathrm{rel}}(s) - \bar{u}_{\mathrm{rel}}\right)^2 . \qquad (1.57)$$

The expected observation is that the mean value $\bar{u}_{\mathrm{rel}}$ monotonically grows with the temperature. The variance $\sigma_{u_{\mathrm{rel}}}$ (VIDF), however, is small in both liquid and solid regime but large in the transition region.

While the VIDF well reflects the structural transition between solid-like and liquid-like clusters, one should always keep in mind that this parameter measures the *auto-correlation time* τ_r of the inter-particle distances. When this time is small compared to the blocklength, $\tau_r \ll M$, as in the liquid regime, many particle "jumps"

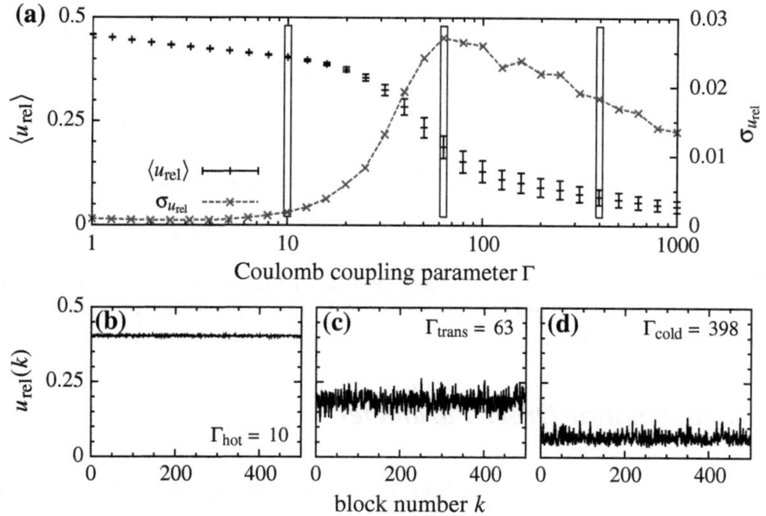

Fig. 1.16 **a** Block averaged relative inter-particle distance fluctuations (IDF) in a 2D Yukawa cluster ($\kappa = 1$) with $N = 19$ particles. The variance (VIDF) is shown as error bars and separately as *dashed curve*. It has a maximum at the transition region. **b–d** IDF on single bocks of length $M = 10^4$ samples during LMD simulations for three representative coupling strengths in the liquid, at the transition and in the solid regime

between lattice sites happen in every block. Hence, $\bar{u}_{\rm rel}$ is large but $\sigma_{u_{\rm rel}}$ is small. When the correlation time is large $\tau_r \gg M$ as in the solid regime, no "jumps" occur in most blocks. Both, mean value $\bar{u}_{\rm rel}$ and variance $\sigma_{u_{\rm rel}}$ are small then. Finally, when both time-scales are similar $\tau_r \approx M$, the number of jumps within one block fluctuates between none and a few. Since each jump increases $u_{\rm rel}$ on the block where it occurs, this means that the value of $u_{\rm rel}$ strongly fluctuates between different blocks dependent on the number of jumps. Hence, the variance $\sigma_{u_{\rm rel}}$ is large in this regime. The VIDF is well suited to characterize the melting region in small dust clusters, where many other melting parameters exhibit weaknesses. Two examples, one from a Langevin molecular dynamics simulation and one from the laser manipulation experiment are shown in, Figs. 1.16 and 1.17, respectively. While the $u_{\rm rel}$ show a monotonic increase with the temperature, the variance shows a maximum when the cluster melts.

Whenever τ_r is a reasonable indicator for the phase transition, as it is, e.g., in experiments and in MD simulations, see Sects. 1.3.3 and 1.3.4, so is $\sigma_{u_{\rm rel}}$. In MC simulations, on the other hand, an MC step can not necessarily be mapped onto a time step Δt. In particular, global updates like swapping moves in the Parallel Tempering algorithm (see Sect. 1.3.2.1) cannot be mapped to a process that takes a certain time. The application of $\sigma_{u_{\rm rel}}$ as melting parameter is valid only if the relation MC step $\mapsto \Delta t$ is known.

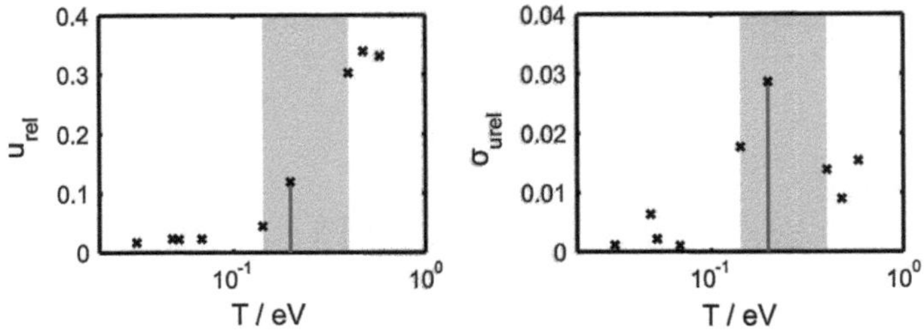

Fig. 1.17 *Left* block averaged relative inter-particle distance fluctuations (IDF) in the LaMa experiment with $N = 19$ particles at gas pressure 16.5 Pa, 5W rf-power and nine different laser heating powers. *Right* the variance (VIDF) shows a maximum at the transition region

1.5.2 Bond Angular Order Parameter

In this section, we briefly discuss the bond order parameter g_6 for flat 2D dust systems. At strong coupling, the dust grains of a Coulomb or a Yukawa bulk form a hexagonal lattice. The bond angle ϑ_{jk} is defined as the angle of the bond between two adjacent particles j and k with respect to a fixed reference direction. In a perfect lattice, all bond angles are equal when taken modulo 60°. The complex six-fold order parameter is defined as

$$\Psi_6\left(\mathbf{r}_j\right) = \frac{1}{c_j} \sum_{k \in \mathrm{nn}(j)} e^{\mathrm{i}6\vartheta_{jk}}, \tag{1.58}$$

where c_j is the coordination number of the jth particle, $\mathrm{nn}(j)$ are the nearest neighbors to this particle, and i is the imaginary unit [77, 78]. The neighbors $\mathrm{nn}(j)$ are typically calculated in a Delaunay triangulation that results in a Voronoi diagram, see Sect. 1.1.1. We can readily verify that $|\Psi_6| = 1$ for any particle with a perfect hexagonal arrangement of neighbors. The phase of Ψ_6, however, depends on the reference direction. Beyond the local symmetry, the *global bond order parameter* $g_6 = |\langle\Psi_6\left(\mathbf{r}\right)\rangle|$, as the average over all particle positions, approaches unity for a perfect lattice. On the other hand, g_6 approaches zero in a disordered lattice. In the latter case, there is no global bond-orientational order and so the single contributions of Ψ_6 cancel under the sum due to their varying phase. We should note here that Ψ_6 is averaged over all particle positions before taking the modulus. If the modulus is taken before averaging over particle positions, the *local bond order parameter* is calculated. This parameter measures the local symmetry at each lattice site but the relative orientation of bonds at different lattice sites is not captured. Gier and Murray found a sharp increase of g_6 at the freezing point of colloidal fluids that were cooled down [78]. When applied to spherical dust balls, the high symmetry of a *magic number* cluster is apparent in a high local bond order parameter [50]. Examples for the global and the local bond order parameter in a 2D Yukawa system are shown in Fig. 1.18.

The application of the global bond order parameter to finite dust clusters faces several issues. The structure of small 2D dust clusters with $N \lesssim 100$ is typically

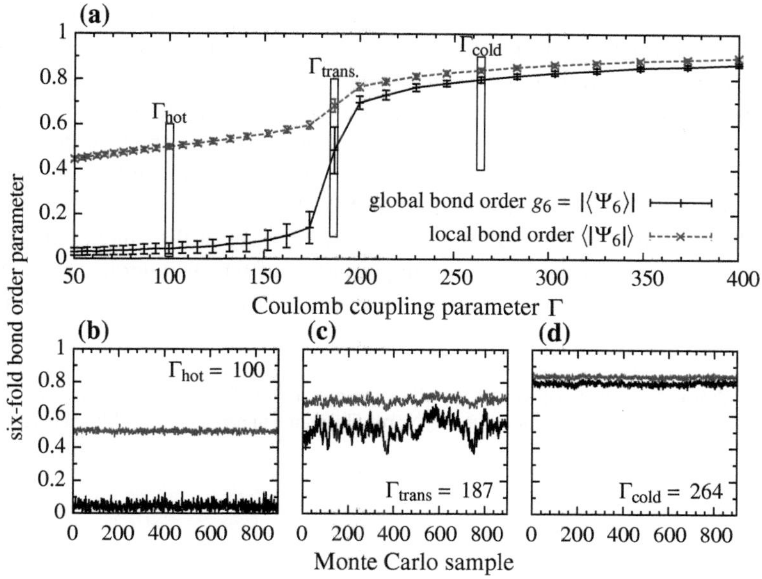

Fig. 1.18 **a** Global and local six-fold bond order parameter in a 2D extended Yukawa system ($\kappa = 1$) versus the Coulomb coupling parameter. The variance (*error bars*) is determined by averaging over several configurations from an MC simulations. **b–c** Bond order parameter of single configurations during the simulation for three representative coupling strengths in the liquid, at the transition and in the solid regime. One hundred MC steps are performed between consecutive samples

governed by concentric shells rather than by a hexagonal lattice. For this reason, a parameter that evaluates a global hexagonal structure is not optimally suited for a finite 2D cluster. The second idea is to apply g_6 on the intra-shell structure of spherical 3D Yukawa clusters and is motivated by the observation of a large fraction of quasi-hexagonal lattices on the outer shell of these clusters, see Fig. 1.3. While the evaluation of the local bond order can be carried out, the definition of a global bond order suffers from lack of a fixed reference direction on a sphere.

1.6 Summary and Conclusion

Dusty plasmas have proven to be a well-suited model system to study phase transitions and collective behavior on single particle level. Both molecular dynamics and Metropolis Monte Carlo simulations are appropriate methods to model interacting dust grains at different temperatures. The Parallel Tempering method enhances the effective sampling, especially when a dust cluster has many metastable configurations.

On the one hand, complex plasmas provide insight into the time resolved single particle trajectories in contrast to macroscopic materials. On the other hand, familiar melting parameters such as the specific heat are usually inaccessible in experiments due to the heat capacity of the surrounding plasma. For this reason, melting is better characterized by structural parameters like the radial pair distribution function. Spherically confined clusters require a parameter that takes into account the spherical symmetry. The Center-Two-Particle correlation function is appropriate to resolve both inter-shell and intra-shell correlations. For large spherical clusters, the function reveals striking similarities with the PDF in flat 2D systems.

The laser heating method allows for selective control of the kinetic temperature of the dust system without affecting the properties of the surrounding plasma. The possibility to investigate phase transitions in finite dust clusters in experiments has been demonstrated.

Acknowledgments This work was financially supported by the DFG via SFB/TR 24 project A3, A7, and A9. CPU time was granted at the North-German Supercomputing Alliance HLRN via grants SHP006.

References

1. A. Melzer, B. Buttenschön, T. Miksch, M. Passvogel, D. Block, O. Arp, A. Piel, Finite dust clusters in dusty plasmas. Plasma Phys. Controlled Fusion **52**(12), 124028 (2010)
2. V. Nosenko, J. Goree, A. Piel, Laser method of heating monolayer dusty plasmas. Phys. Plasmas **13**(3), 032106 (2006)
3. M. Bonitz, C. Henning, D. Block, Complex plasmas: a laboratory for strong correlations. Rep. Prog. Phys. **73**(6), 066501 (2010)
4. M. Himpel, C. Killer, B. Buttenschön, A. Melzer, Three-dimensional single particle tracking in dense dust clouds by stereoscopy of fluorescent particles. Phys. Plasmas **19**(12), 123704 (2012)
5. A. Celsius, *Kungliga Svenska Vetenskapsakademiens Handlingar* (Lorentz Ludewig Grefing, Stockholm, 1742). Observationer om twänne beständiga. Grader på en Thermometer
6. D. Block, A. Melzer, *Introduction to Complex Plasmas* (Springer, Heidelberg, 2010). Imaging diagnostics in dusty plasmas
7. M. Bonitz, N. Horing, P. Ludwig (eds.), *Introduction to Complex Plasmas* (Springer, Heidelberg, 2010)
8. V.R. Ikkurthi, K. Matyash, A. Melzer, R. Schneider, Computation of charge and ion drag force on multiple static spherical dust grains immersed in rf discharges. Phys. Plasmas **17**(10), 103712 (2010)
9. P. Ludwig, W.J. Miloch, H. Kählert, M. Bonitz, On the wake structure in streaming complex plasmas. New J. Phys. **14**(5), 053016 (2012)
10. P. Ludwig, H. Kählert, M. Bonitz, Ion-streaming induced order transition in three-dimensional dust clusters. Plasma Phys. Controlled Fusion **54**(4), 045011 (2012)
11. D. Block, J. Carstensen, P. Ludwig, W. Miloch, F. Greiner, A. Piel, M. Bonitz, A. Melzer, Wake formation and wake field effects in complex plasmas. Contrib. Plasma Phys. **52**(10), 804812 (2012)
12. O. Arp, D. Block, M. Klindworth, A. Piel, Confinement of Coulomb balls. Phys. Plasmas **12**(12), 122102 (2005)
13. O. Arp, D. Block, A. Piel, A. Melzer, Dust Coulomb balls: three-dimensional plasma crystals. Phys. Rev. Lett. **93**(16), 165004 (2004)
14. M. Bonitz, D. Block, O. Arp, V. Golubnychiy, H. Baumgartner, P. Ludwig, A. Piel, A. Filinov, Structural properties of screened Coulomb balls. Phys. Rev. Lett. **96**(7), 075001 (2006)

15. D.H.E. Dubin, T.M. O'Neil, Computer simulation of ion clouds in a Penning trap. Phys. Rev. Lett. **60**(6), 511 (1988)
16. V.E. Fortov, O.S. Vaulina, O.F. Petrov, M.N. Vasiliev, A.V. Gavrikov, I.A. Shakova, N.A. Vorona, Y.V. Khrustalyov, A.A. Manohin, A.V. Chernyshev, Experimental study of the heat transport processes in dusty plasma fluid. Phys. Rev. E **75**(2), 026403 (2007)
17. D.J. Wineland, J.C. Bergquist, W.M. Itano, J.J. Bollinger, C.H. Manney, Atomic-ion Coulomb clusters in an ion trap. Phys. Rev. Lett. **59**(26), 2935 (1987)
18. T. Ott, M. Stanley, M. Bonitz, Non-invasive determination of the parameters of strongly coupled 2D Yukawa liquids. Phys. Plasmas **18**(6), 063701 (2011)
19. O. Vaulina, S. Khrapak, G. Morfill, Universal scaling in complex (dusty) plasmas. Phys. Rev. E **66**(1), 016404 (2002)
20. P. Hartmann, G.J. Kalman, Z. Donkó, K. Kutasi, Equilibrium properties and phase diagram of two-dimensional Yukawa systems. Phys. Rev. E **72**(2), 026409 (2005)
21. S. Hamaguchi, R.T. Farouki, D.H.E. Dubin, Triple point of Yukawa systems. Phys. Rev. E **56**(4), 4671 (1997)
22. G.J. Kalman, P. Hartmann, Z. Donkó, M. Rosenberg, Two-dimensional Yukawa liquids: correlation and dynamics. Phys. Rev. Lett. **92**(6), 065001 (2004)
23. P. Hartmann, G.J. Kalman, Z. Donkó, Two-dimensional Yukawa liquids: structure and collective excitations. J. Phys. A: Math. Gen. **39**(17), 4485 (2006)
24. D.P. Landau, K. Binder, *A Guide to Monte-Carlo Simulations in Statistical Physics*, 1st edn. (Springer, Heidelberg, 2009)
25. S.W.S. Apolinario, F.M. Peeters, Melting transitions in isotropically confined three-dimensional small Coulomb clusters. Phys. Rev. E **76**(3), 031107 (2007)
26. M. Bonitz, P. Ludwig, H. Baumgartner, C. Henning, A. Filinov, D. Block, O. Arp, A. Piel, S. Käding, Y. Ivanov, A. Melzer, H. Fehske, V. Filinov, Classical and quantum Coulomb crystals. Phys. Plasmas **15**(5), 055704 (2008)
27. T. Ogawa, H. Totsuji, C. Totsuji, K. Tsuruta, Two-step melting of spherical Yukawa and Coulomb clusters. J. Phys. Soc. Jpn. **75**(12), 123501 (2006)
28. C. Henning, H. Baumgartner, A. Piel, P. Ludwig, V. Golubnichiy, M. Bonitz, D. Block, Ground state of a confined Yukawa plasma. Phys. Rev. E **74**, 056403 (2006)
29. J.P. Schiffer, Melting of crystalline confined plasmas. Phys. Rev. Lett. **88**(20), 205003 (2002)
30. I. Shyjumon, M. Gopinadhan, O. Ivanova, M. Quaas, H. Wulff, C.A. Helm, R. Hippler, Structural deformation, melting point and lattice parameter studies of size selected silver clusters. Eur. Phys. J. D **37**(3), 409 (2006)
31. E. Thomas, J.D. Williams, J. Silver, Application of stereoscopic particle image velocimetry to studies of transport in a dusty (complex) plasma. Phys. Plasmas **11**(7), L37 (2004)
32. J. Wrighton, H. Kählert, T. Ott, P. Ludwig, H. Thomsen, J. Dufty, M. Bonitz, Charge correlations in a harmonic trap. Contrib. Plasma Phys. **52**(1), 4548 (2012)
33. P. Rehmus, C.C. Roothaan, R. Stephen, Visualization of electron correlation in ground states of He and H-. Chem. Phys. Lett. **58**(3), 321 (1978)
34. P. Rehmus, R. Berry, Visualization of electron correlation in a series of helium S states. Chem. Phys. **38**(2), 257 (1979)
35. T. Trottenberg, B. Brede, D. Block, A. Piel, Resonance cones in a dusty magnetized plasma. Phys. Plasmas **10**(12), 4627 (2003)
36. H. Baumgartner, H. Kählert, V. Golobnychiy, C. Henning, S. Käding, A. Melzer, M. Bonitz, Shell structure of Yukawa balls. Contrib. Plasma Phys. **47**(4–5), 281290 (2007)
37. H. Kählert, P. Ludwig, H. Baumgartner, M. Bonitz, D. Block, S. Käding, A. Melzer, A. Piel, Probability of metastable configurations in spherical three-dimensional Yukawa crystals. Phys. Rev. E **78**(3), 036408 (2008)
38. H. Baumgartner, D. Asmus, V. Golubnychiy, P. Ludwig, H. Kählert, M. Bonitz, Ground states of finite spherical Yukawa crystals. New J. Phys. **10**(9), 093019 (2008)
39. H. Baumgartner, D. Block, M. Bonitz, Structure and phase transitions of Yukawa balls. Contrib. Plasma Phys. **49**(4–5), 281302 (2009)

40. M. Bonitz, D. Semkat (eds.), *Introduction to computational methods in many body physics* (Rinton Press, Princeton, NJ, 2006)
41. N. Metropolis, A.W. Rosenbluth, M.N. Rosenbluth, A.H. Teller, E. Teller, Equation of state calculations by fast computing machines. J. Chem. Phys. **21**(6), 1087 (1953)
42. K. Hukushima, K. Nemoto, Exchange Monte Carlo method and application to spin glass simulations. J. Phys. Soc. Jpn. **65**(6), 1604 (1996)
43. R. Mannella, Quasisymplectic integrators for stochastic differential equations. Phys. Rev. E **69**(4), 041107 (2004)
44. J.M. Sanz-Serna, Symplectic integrators for Hamiltonian problems: an overview. Acta Numerica **1**, 243 (1992)
45. R. Mannella, *Supercomputation in Nonlinear and Disordered Systems* (World Scientific, Singapore, 1997)
46. L. Verlet, Computer "Experiments" on classical fluids. I. Thermodynamical properties of lennard-jones molecules. Phys. Rev. **159**(1), 98 (1967)
47. W.C. Swope, H.C. Andersen, P.H. Berens, K.R. Wilson, A computer simulation method for the calculation of equilibrium constants for the formation of physical clusters of molecules: application to small water clusters. J. Chem. Phys. **76**(1), 637 (1982)
48. P. Langevin, On the theory of Brownian motion, C. R. Acad. Sci. (Paris) **146**, 530 (1908)
49. H. Wang, C. Hartmann, C. Schütte, L.D. Site, Grand-canonical-like molecular-dynamics simulations by using an adaptive-resolution technique, Phys. Rev. X **3**(1), 011018 (2013)
50. P. Ludwig, S. Kosse, M. Bonitz, Structure of spherical three-dimensional Coulomb crystals. Phys. Rev. E **71**(4), 046403 (2005)
51. H. Kählert, M. Bonitz, Collective excitations of a spherically confined Yukawa plasma. Phys. Rev. E **83**(5), 056401 (2011)
52. T. Ott, M. Bonitz, Diffusion in a strongly coupled magnetized plasma. Phys. Rev. Lett. **107**(13), 135003 (2011)
53. J. Chu, Direct observation of Coulomb crystals and liquids in strongly coupled rf dusty plasmas. Phys. Rev. Lett. **72**(25), 4009 (1994)
54. A. Melzer, A. Homann, A. Piel, Experimental investigation of the melting transition of the plasma crystal. Phys. Rev. E **53**(3), 2757 (1996)
55. H.M. Thomas, G.E. Morfill, Melting dynamics of a plasma crystal. Nature **379**(6568), 806 (1996)
56. V. Schweigert, I. Schweigert, A. Melzer, A. Homann, A. Piel, Plasma crystal melting: a non-equilibrium phase transition. Phys. Rev. Lett. **80**(24), 5345 (1998)
57. A. Homann, A. Melzer, S. Peters, A. Piel, Determination of the dust screening length by laser-excited lattice waves. Phys. Rev. E **56**(6), 7138 (1997)
58. S. Nunomura, D. Samsonov, J. Goree, Transverse waves in a two-dimensional screened-Coulomb crystal (dusty plasma). Phys. Rev. Lett. **84**(22), 5141 (2000)
59. A. Homann, A. Melzer, A. Piel, Measuring the charge on single particles by laser-excited resonances in plasma crystals. Phys. Rev. E **59**(4), R3835 (1999)
60. A. Piel, A. Homann, M. Klindworth, A. Melzer, C. Zafiu, V. Nosenko, J. Goree, Waves and oscillations in plasma crystals. J. Phys. B: At. Mol. Opt. **36**(3), 533 (2003)
61. D. Samsonov, J. Goree, Z. Ma, A. Bhattacharjee, H. Thomas, G. Morfill, Mach cones in a Coulomb lattice and a dusty plasma. Phys. Rev. Lett. **83**(18), 3649 (1999)
62. V. Nosenko, J. Goree, Shear flows and shear viscosity in a two-dimensional Yukawa system (dusty plasma). Phys. Rev. Lett. **93**(15), 155004 (2004)
63. M. Wolter, A. Melzer, Laser heating of particles in dusty plasmas. Phys. Rev. E **71**(3), 036414 (2005)
64. A. Schella, T. Miksch, A. Melzer, J. Schablinski, D. Block, A. Piel, H. Thomsen, P. Ludwig, M. Bonitz, Melting scenarios for three-dimensional dusty plasma clusters. Phys. Rev. E **84**(5), 056402 (2011)
65. J. Schablinski, D. Block, A. Piel, A. Melzer, H. Thomsen, H. Kählert, M. Bonitz, Laser heating of finite two-dimensional dust clusters: a. Experiments. Phys. Plasmas **19**(1), 013705 (2012)

66. H. Thomsen, P. Ludwig, M. Bonitz, J. Schablinski, D. Block, A. Schella, A. Melzer, Controlling strongly correlated dust clusters with lasers. J. Phys. D: Appl. Phys (2014). http://arxiv.org/pdf/1402.7182
67. Y. Feng, J. Goree, B. Liu, Accurate particle position measurement from images. Rev. Sci. Instrum. **78**(5), 053704 (2007)
68. V. Nosenko, A.V. Ivlev, G.E. Morfill, Laser-induced rocket force on a microparticle in a complex (dusty) plasma. Phys. Plasmas **17**(12), 123705 (2010)
69. J. Carstensen, H. Jung, F. Greiner, A. Piel, Mass changes of microparticles in a plasma observed by a phase-resolved resonance method. Phys. Plasmas **18**(3), 033701 (2011)
70. A. Melzer, A. Schella, J. Schablinski, D. Block, A. Piel, Instantaneous normal mode analysis of melting of finite dust clusters. Phys. Rev. Lett. **108**(22), 1 (2012)
71. A. Melzer, A. Schella, J. Schablinski, D. Block, A. Piel, Analyzing the liquid state of two-dimensional dust clusters: the instantaneous normal mode approach. Phys. Rev. E **87**(3), 033107 (2013)
72. B. Liu, J. Goree, V. Nosenko, L. Boufendi, Radiation pressure and gas drag forces on a melamine-formaldehyde microsphere in a dusty plasma. Phys. Plasmas **10**(1), 9 (2003)
73. H. Thomsen, H. Kählert, M. Bonitz, J. Schablinski, D. Block, A. Piel, A. Melzer, Laser heating of finite two-dimensional dust clusters: b. Simulations. Phys. Plasmas **19**(2), 023701 (2012)
74. G. Kudelis, H. Thomsen, M. Bonitz, Heat transport in confined strongly coupled two-dimensional dust clusters. Phys. Plasmas **20**(7), 073701 (2013)
75. F.A. Lindemann, The calculation of molecular vibration frequencies. Physik. Z. **11**, 609 (1910)
76. J. Böning, A. Filinov, P. Ludwig, H. Baumgartner, M. Bonitz, Y.E. Lozovik, Melting of trapped few-particle systems. Phys. Rev. Lett. **100**, 113401 (2008)
77. A. Zippelius, B.I. Halperin, D.R. Nelson, Dynamics of two-dimensional melting. Phys. Rev. B **22**(5), 2514 (1980)
78. D.G. Grier, C.A. Murray, The microscopic dynamics of freezing in supercooled colloidal fluids. J. Chem. Phys. **100**(12), 9088 (1994)

Chapter 2
Introduction to Streaming Complex Plasmas A: Attraction of Like-Charged Particles

André Schella, André Melzer, Patrick Ludwig, Hauke Thomsen and Michael Bonitz

Abstract Like-charged particles usually interact via a repulsive force. However, in streaming dusty plasmas one can observe that two negatively charged dust particles may attract each other. This is explained by accumulation of positive ions below the dust particles (with respect to the streaming direction). In this chapter, we describe the dependence of this ion focus and the resulting wakes on *discharge rf-power*, *pressure* and *thermophoretic force*, as the three key parameters, that can be varied in dusty plasma experiments. Moreover, we discuss the impact of this attractive force on the collective properties of many dust particles, in particular, on the structure and on the dynamics of spherically confined clusters.

2.1 Introduction

Trapped charged particle systems are ideally suited model systems to study strongly coupled matter, see Refs. [1–4] and references therein. When the particle number is low, less than $N < 1000$ say, the system is said to be finite. These systems usually require an external confinement potential to be stable. In such systems the properties

A. Schella (✉) · A. Melzer
Institut für Physik, Ernst-Moritz-Arndt-Universität Greifswald, 17489 Greifswald, Germany
e-mail: schella@physik.uni-greifswald.de

A. Melzer
e-mail: melzer@physik.uni-greifswald.de

P. Ludwig · H. Thomsen · M. Bonitz
Institut für Theoretische Physik und Astrophysik, Christian-Albrechts-Universität zu Kiel, 24098 Kiel, Germany
e-mail: ludwig@theo-physik.uni-kiel.de

H. Thomsen
e-mail: thomsen@theo-physik.uni-kiel.de

M. Bonitz
e-mail: bonitz@theo-physik.uni-kiel.de

M. Bonitz et al. (eds.), *Complex Plasmas*, Springer Series on Atomic, Optical, and Plasma Physics 82, DOI: 10.1007/978-3-319-05437-7_2,

may strongly vary with the particle number and depend on the symmetry (for example in the case of "magic" configurations). Similar effects are known from other fields such as nuclei or clusters. Dusty plasmas provide a unique way to analyze and control these effects. Since the dust particles in a plasma usually interact via a screened potential, where the screening parameter depends on the plasma environment, dusty plasmas bridge the gap between systems with long-range Coulomb interaction (such as trapped ions) and systems with a hard-core potential (such as granular matter, nuclear matter or neutral Fermi gases).

A common characterization of the coupling strength in dusty plasmas is the dimensionless Coulomb coupling parameter. This parameter is defined as the ratio of the typical interaction energy of neighboring particles characterized by the distance b_{WS} and the typical kinetic, thermal energy,

$$\Gamma = \frac{Q_{\mathrm{d}}^2}{4\pi\varepsilon_0 b_{\mathrm{WS}}} \frac{1}{k_{\mathrm{B}}T}. \tag{2.1}$$

Here, Q_{d} is the dust charge, ε_0 the vacuum dielectric permittivity, b_{WS} the Wigner-Seitz-distance and $k_{\mathrm{B}}T$ the thermal energy. When the coupling parameter exceeds unity, the system is said to be strongly coupled. With increasing Γ, correlation effects and long-range order develop, eventually leading to liquid and solid behavior. For an infinite three-dimensional (3D) system the phase transition from a liquid into a solid state occurs when this parameter exceeds a value of about 172...175 [5–7]. Although Γ is usually defined for infinite systems, the coupling parameter can still be used to characterize the coupling strength of finite 3D dust clouds [8–10]. In this case, however, the melting point may strongly vary with the particle number N, and phase transition-like behavior typically extends over a finite transition interval [11, 12].

In situations where wakefield effects are negligible, dust particles in harmonic confinement potential are found to arrange themselves to clusters with nested spherical shells [9]. Due to Yukawa-type inter-particle forces, these crystals are called "Yukawa balls" [13–16]. A very successful model to describe Yukawa balls turned out to be the one-component plasma model (Yukawa OCP) which assumes that ion-flow effects can be neglected (which is the case if the 3D dust cloud is in the plasma bulk). The Hamiltonian for the ground state (i.e. in the limit of zero kinetic energy) of an ensemble of N identical dust grains with a time independent charge[1] Q_{d} and mass m_{d} is then given by [17, 18]

$$E = \sum_{i=1}^{N} \frac{m_{\mathrm{d}}}{2}\omega_0^2 {r_i}^2 + \sum_{i>j}^{N} \frac{Q_{\mathrm{d}}^2}{4\pi\varepsilon_0} \frac{e^{-r_{ij}/\lambda_{\mathrm{D}}}}{r_{ij}}, \tag{2.2}$$

[1] While the light plasma particles have a universal charge (electrons q_{e}, ions q_{i}), the grain charge Q_{d} is subject to dynamical plasma processes and may fluctuate (this will not be taken into account in this chapter). Therefore, in order to denote this difference, we use small and capital letters, respectively.

where $r_i = |\mathbf{r}_i|$ denotes the distance of dust particle i from the trap center and $r_{ij} = |\mathbf{r}_j - \mathbf{r}_i|$ denotes the pair distance of two particles i and j. The trap frequency enters via ω_0. The first term in (2.2) describes the external harmonic trap which is, for simplicity, assumed to be isotropic. The second term represents the Yukawa-type interaction potential Φ_Y of the dust particles. Shielding by the light plasma species is taken into account by the Debye length

$$\lambda_\mathrm{D} = \left(\frac{q_\mathrm{e}^2 \bar{n}_\mathrm{e}}{\varepsilon_0 k_\mathrm{B} T_\mathrm{e}} + \frac{q_\mathrm{i}^2 \bar{n}_\mathrm{i}}{\varepsilon_0 k_\mathrm{B} T_\mathrm{i}} \right)^{-1/2}, \tag{2.3}$$

where $q_\mathrm{e,i}$, $\bar{n}_\mathrm{e,i}$, and $T_\mathrm{e,i}$ denote charge, density, and temperature of electrons and ions, respectively.

Despite its simplicity, model (2.2) serves as a good starting point to analyze finite dust clouds as long as wakefield effects play a minor role [19–21]. It can be used to study the structural properties [13, 16, 22–27] as well as collective dynamical features and excitation behavior, see Refs. [28–36]. Dust clouds confined in anisotropic traps were studied as well [20, 37–40]. (For a more detailed discussion of the structural properties and the melting behavior of dust crystals, we refer the reader to the chapter "Phase Transitions in Dusty Plasmas").

While the Yukawa OCP model (2.2) is adequate for spherical dust crystals (e.g. the plama bulk), an ion streaming motion leads to a different type of ordering–the formation of flow-aligned strings[41–43]. This is attributed to attractive interactions [44–53] between the dust grains caused by the ion focussing below the upper grain, as described below.

In this chapter, we specially address experiments with spherically confined dust clusters with particle numbers between $N = 2$ and $N \approx 50$. Of particular interest is the transition between a spherical plasma crystal with a nested shell structure (an "artificial atom") and flow-aligned strings when the ion flow speed is increased. The goal is to gain insight into the influence of the relevant plasma parameters on the inter-particle forces, the structure of the dust clouds and their dynamics. An in-depth theoretical analysis and the computation of the ion-wake field can be found in the Chap. 3.

2.2 Ion Focus and Wakefield

In the very first dusty plasma experiments, particles were embedded in the sheath of the discharge. There, they formed extended crystalline structures which are known as plasma crystals [54–56]. In the horizontal plain, the particles formed a hexagonal lattice, see Refs. [54–56] and Fig. 2.1a. It was early realized that in the vertical direction, instead of a close-packed crystal structure, the particles often aligned in vertical strings, cf. Fig. 2.1b and Refs. [41, 42, 57, 58]. Such a structure was one of the first hints towards ion focusing effects. Similar observations were made for finite

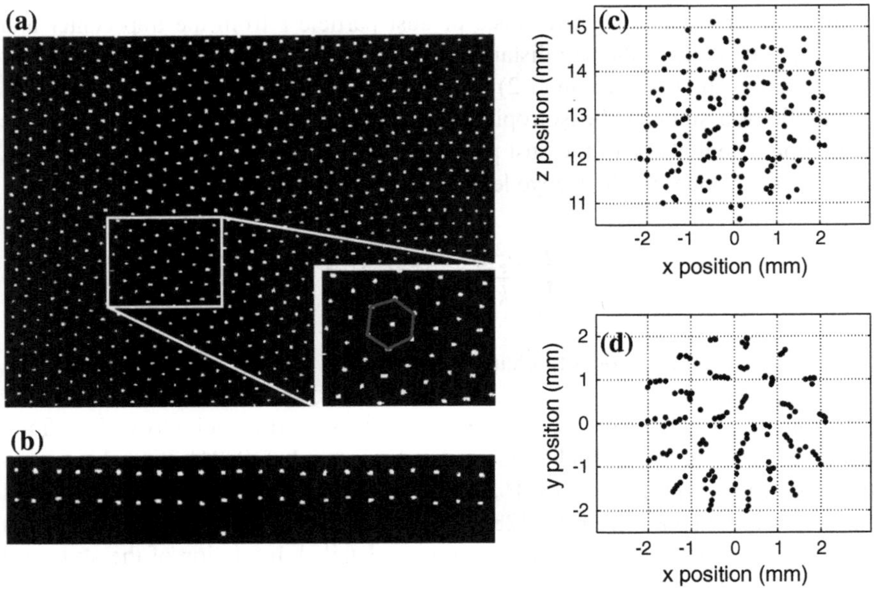

Fig. 2.1 **a** *Top view* of a plasma crystal in a rf-discharge and **b** *side view* of the same crystal. Horizontally, the particles arrange in *hexagonal* patterns, whereas the particles are aligned *vertically*. **c** and **d** *Top* and *side view* of a finite 3D dust cloud at low pressure (Reprinted with permission from [19]. Copyright [2010], AIP Publishing LLC.). Here, the particles are aligned in vertical chains due to wakefield effects

3D dust clouds at low pressure, where it was revealed that the particles tend to form aligned particle chains in vertical direction as depicted in Fig. 2.1d [19].

2.2.1 Phenomenological Description of the Ion Focus

In this section, we present a phenomenological explanation for the formation of the ion wakefields and the connected attractive forces. Wakefield and ion focus are often used synonymously because they describe the same physical process where the first refers more to a wave-like picture and the second to a more particle-like picture.

A sketch of the ion focusing effect is shown in Fig. 2.2. Since the very first observations, many different approaches faced the question towards the nature of the ion wakefield attraction. In the wave-like approach, the dust particle is considered to excite ion acoustic waves that superimpose to form regions of increased positive potential. For this purpose, the dielectric response function of the ions in the vicinity of the dust particle is computed [44, 45, 59, 60]. The solution for the total (dynamically screened) potential of the dust particle is an oscillatory wake potential, the ion

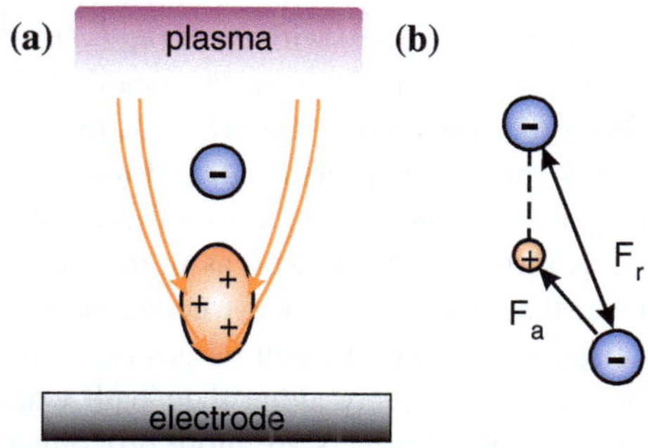

Fig. 2.2 Sketch of the wakefield mechanism; **a** The ions are accelerated by strong electric fields in the sheath region. When streaming towards the lower electrode, they are deflected by the large grain charge that has a focusing effect giving rise to a positive space charge below the grain. **b** Reduced model of the ion focus: the space charge is replaced by a positive point charge rigidly attached to the upper dust particle. While the like-charged grains repel each other with an isotropic force F_r, grains downstream are attracted to the positive space charge by F_a. Consequently, the ion focus leads to an effective non-reciprocal grain attraction and causes a vertical grain alignment.

wakefield. The model is capable of explaining the formation of aligned particles as well as instabilities, even in the presence of collisions [48, 61–63].

The particle-based models start with a microscopic description of the ions [46, 64]. Often, the ion trajectories are solved by means of Monte-Carlo [46] or Particle-In-Cell simulations [65, 66] allowing also to study nonlinear effects [67, 68]. These models also find an enhanced ion density region, the ion focus, downstream to an upper grain leading to attractive forces on the lower particles. To explain the mechanism, we restrict ourselves to the particle-based picture. A description in form of the linear dielectric plasma response is given in the Chap. 3 by Ludwig et al.

In the sheath of the discharge, positive ions are streaming towards the electrode. When passing by a dust particle, they are attracted by the potential of the highly negatively charged dust grain, see Fig. 2.2a. The fast ions are deflected and focused below this grain creating there a positive space charge region, the ion focus [46]. A second particle placed downstream experiences the presence of this ion focus of the upper grain, see Fig. 2.2b. This leads to a superposition of the repulsive dust–dust and the attractive dust–ion interaction that can cause an alignment of the particles. In contrast, the upper particle only feels a repulsive force from the lower particle. Hence, the interaction is non-reciprocal. The non-reciprocal nature of the ion focus was proven in many experiments [19, 49–52]. Moreover, the alignment of the particles can become unstable below a critical neutral gas pressure [46]. As a result, the particles are heated by this ion-focus induced instability, that can even cause phase transitions from ordered, solid-like arrangements to liquid and gas-like situations [10, 41, 69–74].

Models suggest that there are two key parameters that strongly influence the amplitude, shape and range of the wakefield potential, namely the ion streaming velocity and the neutral gas pressure. The ion streaming velocity u_i can be described in terms of the Mach number $M = u_\mathrm{i}/c_\mathrm{s}$, with $c_\mathrm{s} = (k_\mathrm{B}T_\mathrm{e}/m_\mathrm{i})^{1/2}$ being the sound (Bohm) speed [75]. From experiments deep in the sheath of the discharge, where the ion flow is supposed to be supersonic, i. e. $M \geq 1$, the particles inside a dust crystal are found to be on aligned particle chains [42]. However, recent investigations have demonstrated that particle chains even form at moderate ion flow velocities $M < 1$ in 3D dust clouds [20, 76, 77], as well as in microgravity situations [78]. The second key parameter that affects the wake structure is the ion-neutral collision frequency ν_i which is closely related to the neutral gas pressure. At low neutral gas pressure, streaming ions can move freely along their path without (or with only a few) collisions. Here, the wakefield effects are more pronounced, as confirmed by many experiments [19, 20, 42, 58, 71]. At higher damping rates, simulations suggest that the effect of the ion focus becomes weaker [63, 68].

2.2.2 *Biasing Wakes in the Experiment*

The experiments to study wakefield effects in dusty plasmas are manifold, see Refs. [19, 20, 42, 49, 50, 71, 77, 79–82], just to name a few. Often, one has the possibility to influence the dust–dust-interaction and the dust-confinement by varying several parameters that also affect the ion wakefield:

- By changing the neutral gas pressure p, one can change the friction force that acts on the dust particles and the ion-neutral collision rate of the streaming ions. A high pressure seems to reduce the ion focusing effect. Additionally, changing the pressure leads to changes in the local plasma environment of the dust grains.
- By adjusting the temperature of the lower electrode, one can induce an upward themorphoretic force that can be used to change the vertical position of the dust cluster in the plasma. This changes the local dust charge and the streaming velocity of the ions.
- By varying the rf-power, one changes the energy input into the discharge. This affects the densities of the plasma species and their temperatures which, in turn, has an intricate effect onto the dust charge, the confinement and the local Mach number.
- By external laser fields, one can feed energy to the dust system but leave the plasma unaffected. Moreover, driving dust clouds to fluid states by means of lasers hinders the establishment of particle alignment [10].
- By strong magnetic fields, the ions can be magnetized. This influences the wakefield attraction as well [82].

In contrast to the simulations, the Mach number is not directly controllable in the experiments. Therefore, in the experiments presented in the following, the first three parameters—(i) *neutral gas pressure*, (ii) *thermophoresis* and (iii) *rf-power*—are modified in order to study wake-affected 3D dust clusters. The parameters are closely related to Mach number M, ion-neutral collision frequency ν_i and relative temperature $T_\mathrm{e}/T_\mathrm{i}$, which serve as an input for the simulations [63]. Thus, the great benefit results from the possibility to compare the experimental findings with simulations presented in the Chap. 3.

2.3 Wakefield Effects in Dusty Plasma Experiments

In dusty plasma experiments, typically, micron-sized dust particles are immersed into a discharge plasma where they attain a high negative charge due to the steady flux of electrons and ions onto each particle's surface. Moreover, due to their high mass compared to ions and electrons, the charge to mass ratio becomes sufficiently low to trace the individual particles by means of high-speed video cameras on the kinetic level, see Refs. [2, 4, 83] and references therein.

2.3.1 Experimental Setup

A typical experimental setup is shown in Fig. 2.3a. Since the dust particles are all negatively charged, their mutual repulsion (even in the presence of wake effects there is repulsion between the particles that are not flow-aligned) has to be balanced by an external confinement. Trapping the dust particles inside a cubic glass box (cuvette), as shown in Fig. 2.3b, provides the great benefit to achieve harmonic confinement by a superposition of several forces acting on the individual grains [3, 9, 10, 14, 16, 19, 20, 24, 34, 39, 77, 84–87]. Radial confinement is established by the electric field, due to the dielectric glass walls. To compensate for gravity, the sheath electric field force together with a thermophoretic force from a heated lower electrode levitates the particle cloud upwards. In that manner all these forces provide a harmonic confinement where the particles form Yukawa balls, see Fig. 2.3c and Ref. [9].

Despite the harmonic confinement, the particle arrangemnet is influenced by ion wakefields. As described above, experiments upon Yukawa balls at high pressure reported a well established shell structure [9, 14, 16, 24]. At sufficiently low neutral gas pressure, wakefield effects due to the streaming ions become important [19, 20].

For all experiments presented in this chapter, melamine formaldehyde particles with a diameter of 4.86 μm (corresponding to a dust mass of $m_\mathrm{d} = 9.1 \cdot 10^{-14}$ kg) were trapped in a cubic glass box of 2.5 cm wall length. The argon plasma is operated in a capacitively coupled radio-frequency (rf) discharge at rf-powers in the range 1–4 W and a neutral gas pressure of 4–10 Pa. Due to the low pressure used here, the establishment of wakefields can be expected in the experiments.

The dust particles are observed by imaging the scattered laser light from the grains via three orthogonal high speed cameras. These share at least one common axis and thus easily allow us to reconstruct the 3D positions. Here, particle clouds with up to hundred particles can be observed. The cameras are typically operated at frames rates up to 500 frames per second. To identify the particle positions in each image, standard

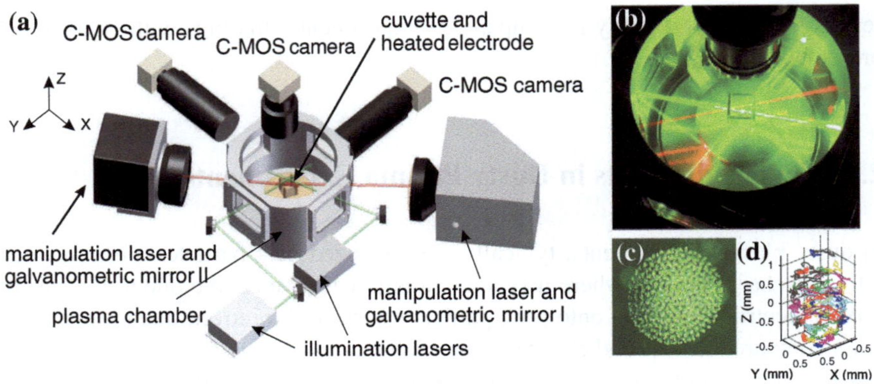

Fig. 2.3 **a** Experimental setup used to investigate and to manipulate 3D dust clusters. **b** View into the plasma chamber; **c** Photograph of a Yukawa ball (Käding 2008); **d** Reconstructed 3D trajectories of a laser heated cluster consisting of $N = 48$ particles

particle identification techniques are used [88]. Afterwards, the individual particles are digitally tracked and their 3D dynamics can be retrieved, see Fig. 2.3d and Refs. [10, 16]. Hence, the full 3D particle dynamics are available at the kinetic level.

An overview about the different approaches to image and reconstruct three-dimensional dust clouds has been given in Ref. [3]. Established methods include scanning video microscopy [9, 89, 90], stereoscopic imaging [16, 91], digital in-line-holography [92], the color-gradient method [28, 93] and, more recently, the single exposure technique [94].

All dust systems in the experiments presented here were investigated with a stereoscopic imaging system pictured in Fig. 2.3a that allows us to track all particles in 3D [10, 16].

An additional surplus of our setup is the possibility to manipulate dust clusters by focused laser beams. Laser heating has become the most common method to manipulate dust ensembles [10, 95–101]. Among others, it allows us to experimentally study dust–dust-interactions [49], phase transitions [10, 97, 102] and to excite waves in dusty plasmas [103–105].

The laser beams leave the plasma itself unaffected. The dust particles respond to the momentum transferred by the radiation pressure of the beam [95, 106]. Random laser excitation of the dust particles mimics a heating process leading to increased kinetic temperatures of the dust particles [99, 100]. Thus, controlling the laser intensity allows us to change the dust temperature and, with it, vary the coupling parameter $\Gamma \propto 1/T$ very precisely without varying other plasma properties. With the setup shown here, it is possible to cover a wide range of coupling strengths, see Fig. 2.4.

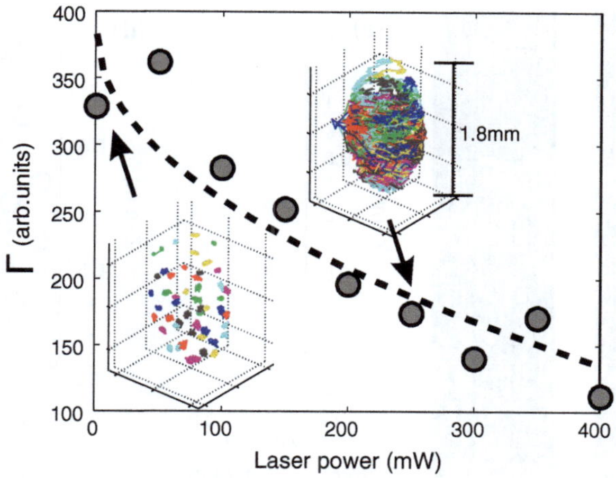

Fig. 2.4 Coupling parameter as a function of the applied laser power (*large dots*). The inset shows typical particle trajectories for two characteristic values of Γ. The *dashed line* is a guide to the eye

2.3.2 Structure of Ion-Focus Affected 3D Dust Clusters

The influence of the ion focus is omnipresent in low pressure dusty plasma experiments [19, 20, 58, 71, 77, 79]. As a first insight into the complex wakefield mechanisms, we start with a brief phenomenological introduction into the structure of wake-affected 3D dust clusters. In order to study the fundamental particle–particle interaction, it is sufficient to start with a simple system of a single elongated 1D dust string.

Figure 2.5a shows a "dust molecule" formed by two particles inside the cubic glass box. The particles were confined at 6 Pa and 1.6 W rf-power. Evidently, the like-charged particles are vertically aligned. This situation is found to be very stable (see below). Further, a low rf-power is favorable to confine single dust strings [39]. Figure 2.5b shows a camera image of a vertical dust string made of five particles. The particles are aligned along the ion streaming direction. The dust chain was confined at 0.8 W rf-power at a pressure of 10 Pa. In Fig. 2.5c, the reconstructed 3D trajectories of the particles are depicted. At these experimental conditions, the particles within the cluster are nearly at rest. In general, single aligned particle chains, as shown in Fig. 2.5a–c, can even be formed for like-charged particles that obey purely repulsive interactions if the vertical confinement is much weaker than the horizontal confinement. As a representative example of a larger dust cloud, a cluster of $N = 30$ particles with a trend to form aligned particle chains is shown in Fig. 2.5d. The highlighted particles are aligned along the ion streaming direction. This cluster structure hints towards the presence of the ion focus, because the particles in the glass box are in a nonequilibrium environment.

The competition between repulsive dust–dust interaction and attractive wakefield–dust interaction leads to a rich variety in structure and shape of the dust clouds.

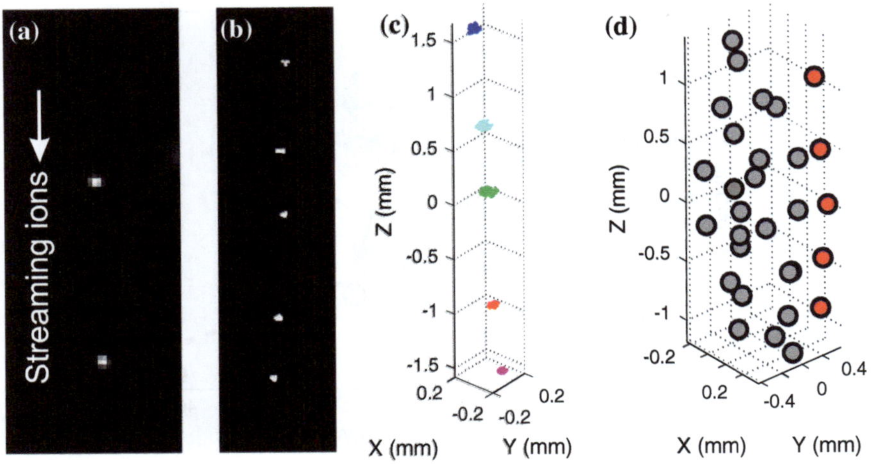

Fig. 2.5 **a** Still video image of two vertically aligned particles ("dust molecule"). **b–c** Processed video image of a single particle chain formed by $N = 5$ dust grains and reconstructed 3D trajectories. **d** 3D positions of a $N = 30$ cluster with trend to chain formation. A single particle chain is highlighted

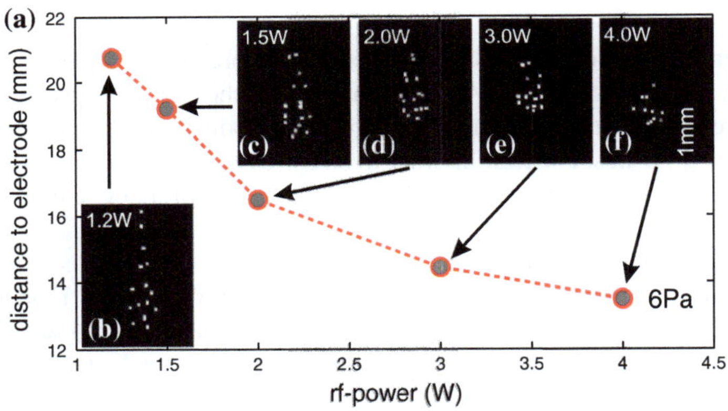

Fig. 2.6 **a** Cluster height inside the glass box as a function of the applied rf-power for a neutral gas pressure of 6 Pa. **b–f** Processed camera images of a $N = 16$ cluster for different rf-powers. The shape changes from droplet-like to spheroidal with increasing rf-power

Therefore, it is tempting to get a deeper insight about cluster structure at different experimental conditions.

Figure 2.6a shows the distance of the center of mass of the dust cluster from the lower electrode as a function of the applied rf-power. The neutral gas pressure was kept constant at 6 Pa. At fixed thermophoretic levitation force, increasing the discharge power leads to a lowering of the cluster position inside the glass box. An increase of the discharge power causes an increased density of electrons and ions.

This, in turn, leads to a smaller Debye length and hence to a smaller sheath width and to the observed sinking of the dust clusters.

In Fig. 2.6b–f snapshots of a dust cluster formed by $N = 16$ particles for different rf-powers are depicted. For a low discharge power of 1.1 W, Fig. 2.6b, the cluster is strongly elongated along the ion streaming direction. The uppermost particles form a single chain and the cluster looks like a droplet. By increasing the discharge power to 1.5 and 2.0 W, Fig. 2.6c–d, the inter-particle distance reduces, but the droplet-like shape remains. A further increase of the discharge power to 3.0 and 4.0 W leads to a transitions from the droplet-like shape to a more spheroidal shape of the dust cloud [20], see Fig. 2.6e–f.

The ions are streaming towards the lower electrode. As discussed before, below the upper particles, the streaming ions are being focused producing a positive space charge region when the mean free path length for ion-neutral collisions becomes large [75]. This positive charge leads to an attractive force on the downstream particles and favors vertical particle alignment. The strength of the wakefield attraction depends on the Mach number M of the streaming ions. By varying the height of the cluster inside the glass box, not only this local Mach number is changed. Furthermore, the dust charge crucially depends on the local plasma parameters, which change dramatically with the vertical position of the cluster [14]. Reliable calculations of the dust charge in flowing plasmas are only starting [63, 65, 66, 107], therefore, it is still a challenging task to simulate the dust charge in a flowing environment self-consistently. From the experiment point of view, understanding the complex physical mechanisms for the cluster structure under the influence of wakefields is still an open issue. Thus, the cluster structure can serve as a hint at wakefield effects in the experiment. However, further detailed investigations of the particle dynamics are required.

2.3.3 *Dynamics of Ion-Focus Affected 3D Dust Clusters*

2.3.3.1 Attraction of Two Like-Charged Particles

Two-particle dust systems are the simplest accessible system to study the interaction between like-charged particles. They have become a very attractive playground for experimentalists to study interactions and correlations [79, 108, 109] as well as wake effects [19, 49, 50, 80–82, 110].

Figure 2.7a, b shows still camera images of two dust particles inside a confining glass box are shown that form a "dust molecule" under different experimental conditions. The particles are trapped without additional thermophoretic levitation at a neutral gas pressure of 6 Pa. Thus, their equilibrium position is located where the sheath electric field force compensates gravity. In Fig. 2.7a, the particles are aligned horizontally. This is due to the high rf-power of 4.5 W which leads to a strong vertical confinement.

It is well known that reducing the discharge power weakens the vertical confinement to the dust particles in a glass box, see Refs. [20, 39] and the previous section.

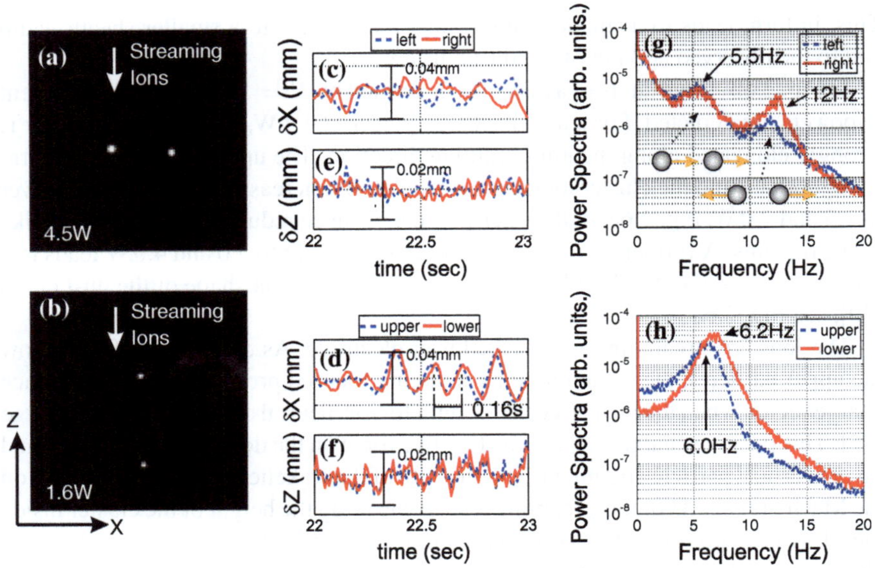

Fig. 2.7 Still video images of two dust grains confined at 6 Pa and 4.5 W (**a**) and 1.6 W (**b**). **c–f** Fluctuations of the trajectories over a time span of one second. **g–h** Power spectra of the transverse movement of particles 1 and 2

Here, by reducing the discharge power to 1.6 W, the dust particles tend to align under each other along the ion streaming direction in Fig. 2.7b. This is due to the previously discussed ion focusing effect.

Figure 2.7c–f shows the horizontal (perpendicular to the ion flow) and vertical fluctuations of the particles' movement over a time interval of one second. In the case of the horizontally aligned particles, Fig. 2.7c, d, the particles' motion is not correlated, neither in horizontal nor in vertical direction. This changes drastically for the vertically aligned particles. In Fig. 2.7e, the correlated oscillatory motion is clearly visible. Moreover, the downstream particle lags behind the upper one but with a slightly larger amplitude. This means that the lower particle tries to follow the upper particle's motion. The oscillation period can be found as $\Delta t = 0.16$ s. Consequently, the frequency of this oscillation is $f = 1/\Delta t = 1/0.16\,\mathrm{s} = 6.25\,\mathrm{Hz}$. Interestingly, the fluctuations along the z-direction in Fig. 2.7f do not show well established oscillations. Thus, restoring forces due to the ion focus seem to be mainly in the horizontal direction [67, 111].

Figures 2.7g, h show the power spectra of the transverse motion computed using trajectories along a time span of more than ten minutes. In the case of the horizontally aligned dust particles, both particles show two peaks in the spectra. The peak at 5.5 Hz can be identified with a center of mass oscillation in the confining potential, while the peak at 12 Hz is associated with the so called "breathing mode"-type oscillation, a radial expansion and compression of the dust system [109]. Both modes are sketched in the inset of Fig. 2.7g.

The corresponding power spectra for the particle system under the influence of an ion wakefield is shown in Fig. 2.7h. Here, only one peak in the power spectra is found. Interestingly, the peak for the upper particle is found at 6 Hz, whereas the peak of the downstream particle is found at a slightly higher frequency of 6.2 Hz. That already shows the sensitivity of dust particles to the local plasma environment. The two-particle system can serve as plasma probe, for instance to measure the sheath electric field or the dust charge [81, 112, 113].

A clear oscillatory movement and the phase delay between upper and lower particle are a direct indication of the ion focus. The downstream particle feels the attraction by this positive ion focus. In contrast, the upper particle feels the repulsive force of the downstream particle [46]. The observed aligned situation is preferred, because the lower particle follows the upper one. The source of the oscillation is not the confining potential, but the wakefield attraction. The particles experience friction due to the neutral gas. The oscillations are nonetheless persistent against this friction. Thus, to perform an oscillatory movement, the particles must gain their energy from the ion focus.

We can conclude that a two-particle systems can be used to experimentally study wake effects with high accuracy. The particles are sensitive to even little changes in the local plasma parameters. Thus, they may also serve as a valuable reference for wakefield simulations.

2.3.3.2 Wake-Affected Cluster Dynamics

For the sake of a broader picture, in the following, the dynamics of larger 3D dust clusters will be investigated. In order to emphasize the role of the ion wake for the cluster dynamics, experiments with a cluster consisting of $N = 11$ particles at a pressure of 4.3 Pa are presented, see Fig. 2.8a–c. (In Fig. 2.8c, the cluster consists of $N = 12$ particles due to an accidentally trapped additional particle, letting our conclusions unaffected). The discharge power was varied from 0.8 W to 2.4 W in order to increase the energy of the dust subsystem. The shape of the cluster changes from elongated and droplet-like, in (a), to an almost perfect sphere, in (c), which is due to the different confinement conditions, as was discussed in Sect. 2.3.2. The particles of the cluster perform oscillatory motions for all three discharge powers. At low power, Fig. 2.8a, the three uppermost particles perform a movement that resembles a wiggling fish tail.

To investigate the collective behavior in more detail, in the following, the oscillations of the three uppermost particles will be analyzed further. Figure 2.8d–f shows a time series of the transverse fluctuations (with $\rho = (x^2 + y^2)^{1/2}$) of the three uppermost particles over a time span of one second. Oscillatory movement is found for all the cases. For a large inter-particle spacing and a low discharge power (0.8 W), Fig. 2.8a, the fluctuations seem to be somewhat oscillatory. At intermediate discharge powers of 1.4 W, Fig. 2.8e, the oscillations become well pronounced: The amplitude grows for particles placed downstream, and one finds a constant phase delay from top to bottom. Following the discussion in Sect. 2.3.3.1, these findings can be attributed

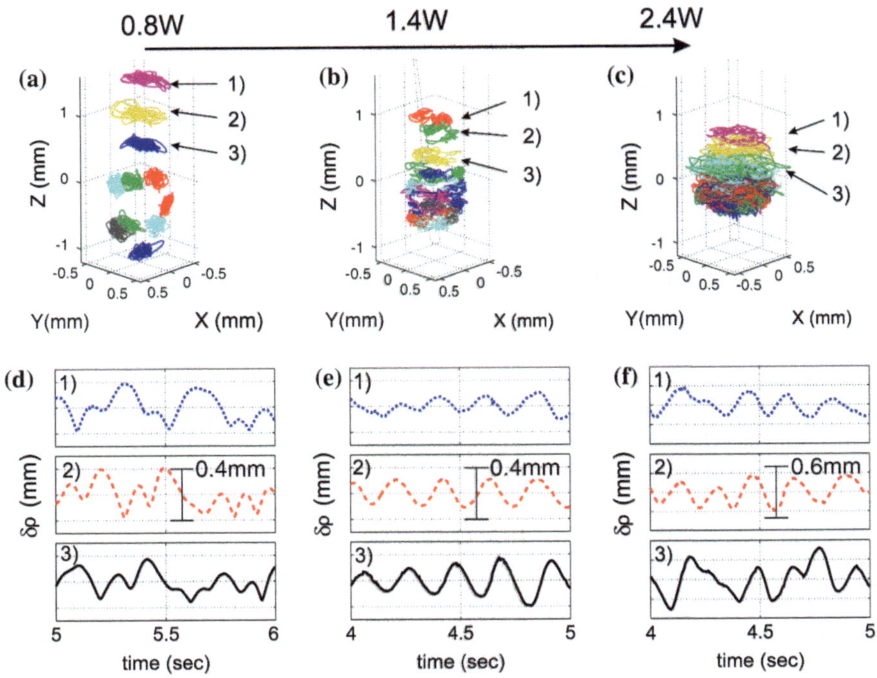

Fig. 2.8 **a–c** Trajectories of a $N = 11$ (12) cluster over a time span of 7 s for different discharge rf-powers. **d–f** Transverse fluctuations of the three uppermost particles 1–3 corresponding to (**a**)–(**c**)

to the same type of oscillation as for the two-particle system. At a high discharge power of 4.0 W, the cluster is clearly molten. The amplitude of the oscillations had grown further. Nonetheless, it seems that the motion of the particles 1–3 is not as correlated as in the previous case. A reason can be that at these high discharge powers the fluid-like particle motion hinders the establishment of coherent oscillations.

Finally, we draw the attention to the dynamics of entire dust clusters. An established technique to retrieve the dynamical properties of a finite system is the normal mode analysis (NMA) [18, 32, 34, 37, 114–116]. Here, it is assumed that the dust particles perform small oscillations around their equilibrium positions. This justifies an expansion of the system's Hamiltonian, see (2.2) up to the second order, to obtain the so-called dynamical matrix

$$\mathrm{H} = \frac{\partial^2 E}{\partial r_{\alpha,i}, \partial r_{\beta,j}}, \tag{2.4}$$

with $\alpha, \beta = x, y, z$ and i and j denoting the different particles. The solution of the corresponding eigenvalue problem of the dynamical matrix defines the $3N$ eigenfrequencies, as well as the corresponding eigenvectors $\mathbf{e}_{i,l}$ of the system that define the mode pattern of eigenmode l. The mode-resolved power spectral density (PSD), i. e.

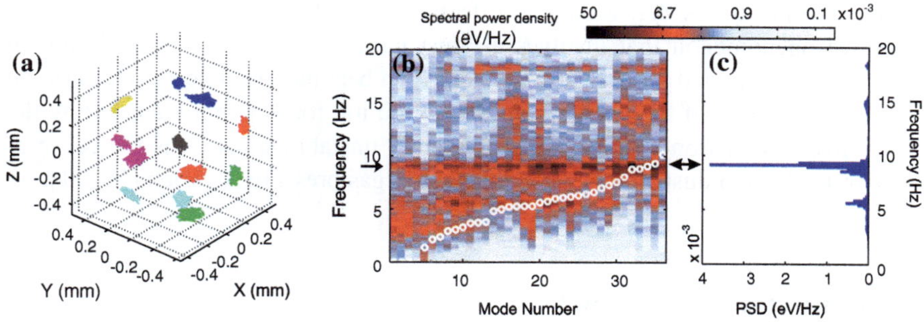

Fig. 2.9 **a** Trajectories of a $N = 12$ cluster over a time span of about 20 s; **b** derived power spectral density per mode and **c** corresponding PSD. *Darker colors* correspond to a higher energy density. The *open circles* correspond to calculated mode frequencies. An unstable oscillation occurs at 9.2 Hz within all modes (marked by the *arrows*)

the energy density per frequency interval of the lth eigenmode is calculated via [18]

$$S_l(\omega) = \frac{2}{T}\left|\int_0^T \mathrm{d}t \left(\sum_{i=1}^{N} \mathbf{v}_i(t)\mathbf{e}_{i,l}\right) e^{i\omega t}\right|^2 . \tag{2.5}$$

In other words, the velocity of the i-th particle $\mathbf{v}_i(t)$ is mapped onto the l-th eigenmode and transformed into Fourier space. The PSD $S(\omega)$, reflecting the energy density in the frequency domain, can be obtained by summation over the mode number l, i. e. $S(\omega) = \sum_l S_l(\omega)$.

A NMA from the trajectories of a cluster consisting of $N = 12$ particles was performed. The cluster was trapped at a rf-power of 3.6 W and a pressure of 5.7 Pa. From the trajectories, Fig. 2.9, it can be seen that the cluster is spherical in shape and that the cluster particles only perform small oscillations around their equilibrium positions. Thus, (2.4) is a suitable approximation. In Fig. 2.9b, the mode-resolved power spectral density is shown. The mode spectra recovers the features of a weakly damped Yukawa system [32, 116], where the normal mode frequencies rise steadily with increasing mode number. Moreover, the mode spectrum has additional features: One finds a high amount of energy stored in a narrow frequency interval at approximately 9.2 Hz in most of the modes. The dominant contribution in the PSD comes exactly from this frequency domain. This phenomenon was noticed previously for flat dust clouds under the influence of an ion focus [71]. In those dust clouds, unstable oscillations at a fixed frequency were attributed to the ion-focus instability [46]. The ion-focus instability manifests itself in a distinct frequency, the oscillation is not an eigenmode of the equilibrium system of (2.2). Hence, the oscillation can be seen in almost all of the equilibrium modes. One might be up to include wakefield effects in the energy equation (2.2) to account for these modes. However, due to the asymmetry in the forces between upper and lower particle, this cannot be done in a simple manner and is a topic of ongoing investigations.

Thus, within the NMA, wakefield effects due to the streaming ions arise as a collective phenomenon that involves all dust particles of the 3D dust cluster. In larger systems, aligned particle arrangements can become unstable below a critical neutral gas pressure if the energy gain due to the ion focus is not balanced by the energy loss via friction [46, 70]. The inherently unstable situation is the cause of phase transitions in dust clouds when the neutral gas pressure is varied [58, 71].

2.4 Summary and Outlook

Dust clusters in streaming plasmas exhibit configurations that differ qualitatively from spherical Yukawa balls that are observed in the plasma bulk where streaming effects are negligible. In particular, the *chain formation* in low-pressure dusty plasma experiments points to the presence of an ion focus in the wakefield below the individual dust grains. This positive space charge can give rise to a net attractive force between two negatively charged dust grains. These forces cause an alignment and lead to the formation of vertical particle strings.

More precisely, when two particles are vertically aligned, the lower particle directly follows the upper particle's motion, via the attractive ion-mediated force. In contrast, the oscillatory motion of two horizontally aligned particles is essentially uncorrelated. By varying the parameters of the discharge, namely rf-power and pressure, the structure of the dust cluster changes dramatically from a spheroidal to chain-like form with (unstable) particle oscillations. This complexity reflects the complex nature of the dust-plasma interaction in a flowing environment.

Due to its non-reciprocal nature, the ion wake force is capable of feeding energy into the dust system. Below a critical neutral gas pressure, the energy gain due to the streaming ions cannot be dissipated by friction and, thus, leads to the ion-focus instability. This instability manifests itself in unstable oscillations at a certain frequency in the power spectral density. To conclude, the actual particle interaction force depends on the pressure, the rf-power and on the thermophoresis which controls the vertical position of the cluster and, therefore, on the local Mach number. Thus, the experimental findings are a challenging input for theory and simulations.

Acknowledgments We would like to thank M. Mulsow for help with the data processing. We acknowledge financial support of the Deutsche Forschungsgemeinschaft via SFB-TR24, projects A3, A7 and A9.

References

1. G.E. Morfill, A.V. Ivlev, Complex plasmas: an interdisciplinary research field. Rev. Mod. Phys. **81**, 1353 (2009)
2. M. Bonitz, C. Henning, D. Block, Complex plasmas: a laboratory for strong correlations. Rep. Prog. Phys. **73**, 066501 (2010)

3. M. Bonitz, N. Horing, P. Ludwig (eds.), in *Introduction to Complex Plasmas*. Springer Series on Atomic, Optical, and Plasma Physics (Springer, Heidelberg, 2010)
4. A. Ivlev, H. Löwen, G. Morfill, C.P. Royall, in *Complex Plasmas and Colloidal Dispersions: Particle-Resolved Studies of Classical Liquids and Solids*. Series in Soft Condensed Matter, vol 5 (World Scientific Pub Co, Singapore, 2012)
5. S. Ichimaru, Strongly coupled plasmas: high density classical plasmas and degenerate electron liquids. Rev. Mod. Phys. **54**, 1017 (1982)
6. H. Ikezi, Coulomb solid of small particles in plasmas. Phys. Fluids **29**, 1764 (1986)
7. S. Hamaguchi, R. Farouki, D.H.E. Dubin, Triple point of Yukawa systems. Phys. Rev. E **56**, 4671 (1997)
8. J. Schiffer, Melting of crystalline confined plasmas. Phys. Rev. Lett. **88**, 205003 (2002)
9. O. Arp, D. Block, A. Piel, A. Melzer, Dust Coulomb balls: three-dimensional plasma crystals. Phys. Rev. Lett. **93**, 165004 (2004)
10. A. Schella, T. Miksch, A. Melzer, J. Schablinski, D. Block, A. Piel, H. Thomsen, P. Ludwig, M. Bonitz, Melting scenarios for 3D dusty plasma clusters. Phys. Rev. E **84**, 056402 (2011)
11. A. Filinov, M. Bonitz, Y. Lozovik, Wigner crystallization in mesoscopic 2D electron systems. Phys. Rev. Lett. **86**, 3851 (2001)
12. J. Böning, A. Filinov, P. Ludwig, H. Baumgartner, M. Bonitz, Y.E. Lozovik, Melting of trapped few-particle systems. Phys. Rev. Lett. **100**, 113401 (2008)
13. M. Bonitz, D. Block, O. Arp, V. Golubnychiy, H. Baumgartner, P. Ludwig, A. Piel, A. Filinov, Structural properties of screened Coulomb balls. Phys. Rev. Lett. **96**, 075001 (2006)
14. O. Arp, D. Block, M. Klindworth, A. Piel, Confinement of Coulomb balls. Phys. Plasmas **12**, 122102 (2005)
15. H. Baumgartner, H. Kählert, V. Golobnychiy, C. Henning, S. Käding, A. Melzer, M. Bonitz, Shell structure of Yukawa balls. Contrib. Plasma Phys. **47**, 281 (2007)
16. S. Käding, D. Block, A. Melzer, A. Piel, H. Kählert, P. Ludwig, M. Bonitz, Shell transitions between metastable states of Yukawa balls. Phys. Plasmas **15**, 073710 (2008)
17. V.M. Bedanov, F.M. Peeters, Ordering and phase transitions of charged particles in a classical finite two-dimensional system. Phys. Rev. B **49**, 2667 (1994)
18. A. Melzer, M. Klindworth, A. Piel, Normal modes of 2D finite Coulomb clusters in complex plasmas. Phys. Rev. Lett. **87**, 115002 (2001)
19. M. Kroll, J. Schablinski, D. Block, A. Piel, On the influence of wakefields on three-dimensional particle arrangements. Phys. Plasmas **17**, 013702 (2010)
20. C. Killer, A. Schella, T. Miksch, A. Melzer, Vertically elongated 3D Yukawa clusters in dusty plasmas. Phys. Rev. B **84**, 054104 (2011)
21. A. Schella, M. Mulsow, A. Melzer, J. Schablinski, D. Block, From transport to disorder: thermodynamic properties of finite dust clouds. Phys. Rev. E **87**, 063102 (2013)
22. H. Totsuji, T. Kishimoto, C. Totsuji, K. Tsuruta, Competition between two forms of ordering in finite Coulomb clusters. Phys. Rev. Lett. **88**, 125002 (2002)
23. P. Ludwig, S. Kosse, M. Bonitz, Structure of spherical three-dimensional Coulomb crystals. Phys. Rev. E **71**, 046403 (2005)
24. D. Block, M. Kroll, O. Arp, A. Piel, S. Käding, Y. Ivanov, A. Melzer, C. Henning, H. Baumgartner, P. Ludwig, M. Bonitz, Structural and dynamical properties of Yukawa balls. Plasma Phys. Controlled Fusion **49**, B109 (2007)
25. D. Block, S. Käding, A. Melzer, A. Piel, H. Baumgartner, M. Bonitz, Experiments on metastable states of three-dimensional trapped particle clusters. Phys. Plasmas **15**, 040701 (2008)
26. H. Baumgartner, V. Golobnychiy, D. Asmus, P. Ludwig, M. Bonitz, Ground states of finite spherical Yukawa crystals. New J. Phys. **10**, 093019 (2008)
27. P. Ludwig, H. Thomsen, K. Balzer, A. Filinov, M. Bonitz, Tuning correlations in multi-component plasmas. Plasma Phys. Controlled Fusion **52**, 124013 (2010)
28. T. Antonova, B.M. Annaratone, D.D. Goldbeck, V. Yaroshenko, H.M. Thomas, G.E. Morfill, Measurement of the interaction force among particles in 3D plasma clusters. Phys. Rev. Lett. **96**, 115001 (2006)

29. T. Antonova, B.M. Annaratone, H.M. Thomas, G.E. Morfill, Energy relaxation and vibrations in small 3D plasma clusters. New J. Phys. **10**, 043028 (2008)
30. S.W.S. Apolinario, F. Peeters, Melting transitions in isotropically confined three-dimensional small Coulomb clusters. Phys. Rev. E **76**, 031107 (2007)
31. A. Melzer, B. Buttenschön, T. Miksch, M. Passvogel, D. Block, O. Arp, A. Piel, Finite dust clusters in dusty plasmas. Plasma Phys. Controlled Fusion **52**, 124028 (2010)
32. C. Henning, H. Kählert, P. Ludwig, A. Melzer, M. Bonitz, Spectral properties of spherically confined dusty plasma crystals. J. Phys. A **42**, 214023 (2009)
33. H. Kählert, P. Ludwig, H. Baumgartner, M. Bonitz, D. Block, S. Käding, A. Melzer, A. Piel, Probability of metastable configurations in spherical three-dimensional Yukawa crystals. Phys. Rev. E **78**, 036408 (2008)
34. Y. Ivanov, A. Melzer, Modes of three-dimensional dust crystals in dusty plasmas. Phys. Rev. E **79**, 036402 (2009)
35. H. Kählert, M. Bonitz, How spherical plasma crystals form. Phys. Rev. Lett. **104**, 015001 (2010)
36. C. Henning, K. Fujioka, P. Ludwig, A. Piel, A. Melzer, M. Bonitz, Existence and vanishing of the breathing mode in strongly correlated finite systems. Phys. Rev. Lett. **101**, 045002 (2008)
37. A. Melzer, Zigzag transition of finite dust clusters. Phys. Rev. E **73**, 056404 (2006)
38. S.W.S. Apolinario, F. Peeters, Melting of anisotropically confined Coulomb balls. Phys. Rev. B **78**, 024202 (2008)
39. J. Kong, T.W. Hyde, L. Matthews, K. Qiao, Z. Zhang, A. Douglass, One-dimensional vertical dust strings in a glass box. Phys. Rev. E **84**, 016411 (2011)
40. A. Radzvilavičius, O. Rancova, E. Anisimovas, Dimensional transitions in small Yukawa clusters. Phys. Rev. E **86**, 016404 (2012)
41. A. Melzer, V.A. Schweigert, I.V. Schweigert, A. Homann, S. Peters, A. Piel, Structure and stability of the plasma crystal. Phys. Rev. E **54**, 46 (1996)
42. K. Takahashi, T. Oishi, K. Shimomai, Y. Hayashi, S. Nishino, Analyses of attractive forces between particles in Coulomb crystal of dusty plasmas by optical manipulations. Phys. Rev. E **58**, 7805 (1998)
43. J. Hammerberg, D. Lemons, M. Murillo, D. Winske, Molecular dynamics simulations of plasma crystal formation including wake effects. IEEE Trans. Plasma Sci. **29**(2), 247 (2001)
44. S.V. Vladimirov, M. Nambu, Attraction of charged particulates in plasmas with finite flows. Phys. Rev. E **52**, R2172 (1995)
45. F. Melandsø, J. Goree, Polarized supersonic plasma flow simulation for charged bodies such as dust particles and spacecraft. Phys. Rev. E **52**, 5312 (1995)
46. V.A. Schweigert, I.V. Schweigert, A. Melzer, A. Homann, A. Piel, Alignment and instability of 'dust' crystals in plasmas. Phys. Rev. E **54**, 4155 (1996)
47. O. Ishihara, S.V. Vladimirov, Wake potential of a dust grain in a plasma with ion flow. Phys. Plasmas **4**(1), 69 (1997)
48. D. Winske, W. Daughton, D.S. Lemons, M.S. Murillo, Ion kinetic effects on the wake potential behind a dust grain in a flowing plasma. Phys. Plasmas **7**(6), 2320 (2000)
49. A. Melzer, V. Schweigert, A. Piel, Transition from attractive to repulsive forces in dust molecules in a plasma sheath. Phys. Rev. Lett. **83**, 3194 (1999)
50. G.A. Hebner, M.E. Riley, Measurement of attractive interactions produced by the ion wakefield in dusty plasmas using a constrained collision geometry. Phys. Rev. E **68**, 046401 (2003)
51. A.A. Samarian, S.V. Vladimirov, Dust particle alignments in a plasma sheath. Contrib. Plasma Phys. **49**(4–5), 260 (2009)
52. A. Melzer, Laser-experiments on particle interactions in strongly coupled dusty plasma crystals. Phys. Scr. **T89**, 33 (2001)
53. M. Kroll, J. Schablinski, D. Block, A. Piel, On the influence of wakefields on three-dimensional particle arrangements. Phys. Plasmas **17**(1), 013702 (2010)
54. J.H. Chu, I. Lin, Direct observation of Coulomb crystals and liquids in strongly coupled rf dusty plasmas. Phys. Rev. Lett. **72**, 4009 (1994)

55. Y. Hayashi, K. Tachibana, Observation of Coulomb crystal formation from carbon particles grown in a methane plasma. Jpn. J. Appl. Phys. **33**, L804 (1994)
56. H. Thomas, G.E. Morfill, V. Demmel, J. Goree, B. Feuerbacher, D. Möhlmann, Plasma crystal—Coulomb crystallization in a dusty plasma. Phys. Rev. Lett. **73**, 652 (1994)
57. T. Trottenberg, A. Melzer, A. Piel, Measurement of the electric charge on particulates forming Coulomb crystals in the sheath of an rf plasma. Plasma Sources Sci. Technol. **4**, 450 (1995)
58. A. Melzer, A. Homann, A. Piel, Experimental investigation of the melting transition of the plasma crystal. Phys. Rev. E **53**, 2757 (1996)
59. M. Nambu, S.V. Vladimirov, P.K. Shukla, Attractive forces between charged particulates in plasmas. Phys. Lett. A **203**, 40 (1995)
60. F. Melandsø, Heating and phase transitions of dust plasma crystals in a flowing plasma. Phys. Rev. E **55**, 7495 (1997)
61. M. Lampe, G. Joyce, G. Ganguli, Interactions between dust grains in a dusty plasma. Phys. Plasmas **7**, 3851 (2000)
62. M. Lampe, G. Joyce, G. Ganguli, Structure and dynamics of dust in streaming plasma: dust molecules, strings, and crystals. IEEE Trans. Plasma Sci. **33**, 57 (2005)
63. P. Ludwig, W.J. Miloch, H. Kählert, M. Bonitz, On the wake structure in streaming complex plasmas. New J. Phys. **14**(5), 053016 (2012)
64. A. Piel, Alignment of dust particles by ion drag forces in subsonic flows. Phys. Plasmas **18**, 073704 (2011)
65. W.J. Miloch, M. Kroll, D. Block, Charging and dynamics of a dust grain in the wake of another grain in flowing plasmas. Phys. Plasmas **17**, 103703 (2010)
66. V.R. Ikkurthi, K. Matyash, A. Melzer, R. Schneider, Computation of charge and ion drag force on multiple static spherical dust grains immersed in rf discharges. Phys. Plasmas **17**, 103712 (2010)
67. I.H. Hutchinson, Forces on a small grain in the nonlinear plasmawake of another. Phys. Rev. Lett. **107**, 095001 (2011)
68. I.H. Hutchinson, C.B. Haakonsen, Collisional effects on nonlinear ion drag force for small grains. Phys. Plasmas **20**, 083701 (2013)
69. H. Thomas, G.E. Morfill, Melting dynamics of a plasma crystal. Nature **379**, 806 (1996)
70. V.A. Schweigert, I.V. Schweigert, A. Melzer, A. Homann, A. Piel, Plasma crystal melting: a nonequilibrium phase transition. Phys. Rev. Lett. **80**, 5345 (1998)
71. Y. Ivanov, A. Melzer, Melting dynamics of finite clusters in dusty plasmas. Phys. Plasmas **12**, 072110 (2005)
72. S.K. Zhdanov, A.V. Ivlev, G.E. Morfill, Mode-coupling instability of two-dimensional plasma crystals. Phys. Plasmas **16**, 083706 (2009)
73. L. Couëdel, V. Nosenko, A.V. Ivlev, S.K. Zhdanov, H.M. Thomas, G.E. Morfill, Direct observation of mode-coupling instability in two-dimensional plasma crystals. Phys. Rev. Lett. **104**, 195001 (2010)
74. T.B. Röcker, A.V. Ivlev, R. Kompaneets, G.E. Morfill, Mode coupling in two-dimensional plasma crystals: role of the wake model. Phys. Plasmas **19**, 033708 (2012)
75. A. Piel, *Plasma Physics: An Introduction to Laboratory, Space, and Fusion Plasmas* (Springer, Heidelberg, 2010)
76. P. Ludwig, H. Kählert, M. Bonitz, Ion-streaming induced order transition in three-dimensional dust clusters. Plasma Phys. Controlled Fusion **54**, 045011 (2012)
77. L. Wörner, C. Räth, V. Nosenko, S.K. Zhdanov, H.M. Thomas, G.E. Morfill, J. Schablinski, D. Block, String structures in driven 3D complex-plasma clusters. Europhys. Lett. **100**(3), 35001 (2012)
78. O. Arp, J. Goree, A. Piel, Particle chains in a dilute dusty plasma with subsonic ion flow. Phys. Rev. E **85**, 046409 (2012)
79. V. Steinberg, R. Sütterlin, A.V. Ivlev, G. Morfill, Vertical pairing of identical particles suspended in the plasma sheath. Phys. Rev. Lett. **86**, 4540 (2001)
80. A.A. Samarian, S.V. Vladimirov, B. James, Wake-induced symmetry-breaking of dust particle arrangements in a complex plasma. JETP Lett. **82**, 858 (2005)

81. J. Carstensen, F. Greiner, D. Block, J. Schablinski, W.J. Miloch, A. Piel, Charging and coupling of a vertically aligned particle pair in the plasma sheath. Phys. Plasmas **19**, 033702 (2012)
82. J. Carstensen, F. Greiner, A. Piel, Ion-wake-mediated particle interaction in a magnetized-plasma flow. Phys. Rev. Lett. **109**, 135001 (2012)
83. G.E. Morfill, A.V. Ivlev, H.M. Thomas, Complex (dusty) plasmas kinetic studies of strong coupling phenomena. Phys. Plasmas **19**, 055402 (2012)
84. S. Käding, A. Melzer, Three-dimensional stereoscopy of Yukawa (Coulomb) balls in dusty plasmas. Phys. Plasmas **13**, 090701 (2006)
85. T.M. Flanagan, J. Goree, Observation of the spatial growth of self-excited dust-density waves. Phys. Plasmas **17**, 123702 (2010)
86. L. Wörner, V. Nosenko, A.V. Ivlev, S.K. Zhdanov, H.M. Thomas, G.E. Morfill, M. Kroll, J. Schablinski, D. Block, Effect of rotating electric field on 3D complex (dusty) plasma. Phys. Plasmas **18**, 063706 (2011)
87. W.D.S. Ruhunusiri, J. Goree, Synchronization mechanism and Arnold tongues for dust density waves. Phys. Rev. E **85**, 046401 (2012)
88. Y. Ivanov, A. Melzer, Particle positioning techniques for dusty plasma experiments. Rev. Sci. Instrum. **78**, 033506 (2007)
89. J. Pieper, J. Goree, R. Quinn, Three-dimensional structure in a crystallized dusty plasma. Phys. Rev. E **54**, 5636 (1996)
90. M. Zuzic, A.V. Ivlev, J. Goree, G.E. Morfill, H.M. Thomas, H. Rothermel, U. Konopka, R. Sütterlin, D.D. Goldbeck, Three-dimensional strongly coupled plasma crystal under gravity conditions. Phys. Rev. Lett. **85**, 4064 (2000)
91. J.E. Thomas, J.D. Willimas, J. Silver, Application of stereoscopic particle image velocimetry to studies of transport in a dusty (complex) plasma. Phys. Plasmas **11**, L37 (2004)
92. M. Kroll, D. Block, A. Piel, Digital in-line holography of dusty plasmas. Phys. Plasmas **15**, 063703 (2008)
93. B.M. Annaratone, T. Antonova, D.D. Goldbeck, H.M. Thomas, G.E. Morfill, Complex-plasma manipulation by radiofrequency biasing. Plasma Phys. Controlled Fusion **46**, B495 (2004)
94. P. Hartmann, I. Donkó, Z. Donkó, Single exposure three-dimensional imaging of dusty plasma clusters. Rev. Sci. Instrum. **84**, 023501 (2013)
95. M. Wolter, A. Melzer, Laser heating of particles in dusty plasmas. Phys. Rev. E **71**, 036414 (2005)
96. V. Nosenko, J. Goree, A. Piel, Laser method of heating monolayer dusty plasmas. Phys. Plasmas **13**, 032106 (2006)
97. Y. Feng, J. Goree, B. Liu, Evolution of shear-induced melting in a dusty plasma. Phys. Rev. Lett. **104**, 165003 (2010)
98. P. Hartmann, M.C. Sándor, A. Kovács, Z. Donkó, Static and dynamic shear viscosity of a single-layer complex plasma. Phys. Rev. E **84**, 016404 (2011)
99. J. Schablinski, D. Block, A. Piel, A. Melzer, H. Thomsen, H. Kählert, M. Bonitz, Laser heating of finite two-dimensional dust clusters: A. Experiments. Phys. Plasmas **19**, 013705 (2012)
100. H. Thomsen, H. Kählert, M. Bonitz, J. Schablinski, D. Block, A. Piel, A. Melzer, Laser heating of finite two-dimensional dust clusters: B. Simulations. Phys. Plasmas **19**, 023701 (2012)
101. V. Nosenko, A.V. Ivlev, G.E. Morfill, Anisotropic shear melting and recrystallization of a two-dimensional complex plasma. Phys. Rev. E **87**, 043115 (2013)
102. V. Nosenko, S.K. Zhdanov, A.V. Ivlev, C.A. Knapek, G.E. Morfill, 2D melting of plasma crystals: equilibrium and nonequilibrium regimes. Phys. Rev. Lett. **103**, 015001 (2009)
103. A. Piel, A. Homann, A. Melzer, Laser-excited waves in a plasma crystal. Plasma Phys. Controlled Fusion **41**, A453 (1999)
104. A. Melzer, S. Nunomura, D. Samsonov, J. Goree, Laser-excited Mach cones in a dusty plasma crystal. Phys. Rev. E **62**, 4162 (2000)
105. V. Nosenko, J. Goree, Z.W. Ma, A. Piel, Observation of shear-wave Mach cones in a 2D dusty-plasma crystal. Phys. Rev. Lett. **88**, 135001 (2002)
106. B. Liu, J. Goree, V. Nosenko, L. Boufendi, Radiation pressure and gas drag forces on a melamine-formaldehyde microsphere in a dusty plasma. Phys. Plasmas **10**, 9 (2003)

107. W.J. Miloch, D. Block, Dust grain charging in a wake of other grains. Phys. Plasmas **19**, 123703 (2012)
108. U. Konopka, G. Morfill, L. Ratke, Measurement of the interaction potential of microspheres in the sheath of a rf discharge. Phys. Rev. Lett. **84**, 891 (2000)
109. A.K. Mukhopadhyay, J. Goree, Two-particle distribution and correlation function for a 1D dusty plasma experiment. Phys. Rev. Lett. **109**, 165003 (2012)
110. G.A. Hebner, M.E. Riley, Structure of the ion wakefield in dusty plasmas. Phys. Rev. E **69**, 026405 (2004)
111. I.H. Hutchinson, Intergrain forces in low-Mach-number plasma wakes. Phys. Rev. E **85**, 066409 (2012)
112. J. Beckers, T. Ockenga, M. Wolter, W.W. Stoffels, J. van Dijk, H. Kersten, G.M.W. Kroesen, Microparticles in a collisional Rf plasma sheath under hypergravity conditions as probes for the electric field strength and the particle charge. Phys. Rev. Lett. **106**, 115002 (2011)
113. G. Schubert, R. Basner, H. Kersten, H. Fehske, Determination of sheath parameters by test particles upon local electrode bias and plasma switching. Eur. Phys. J. D **63**, 431–440 (2011)
114. V.A. Schweigert, F. Peeters, Spectral properties of classical two-dimensional clusters. Phys. Rev. B **51**, 7700 (1995)
115. A. Melzer, Mode spectra of thermally excited two-dimensional dust Coulomb clusters. Phys. Rev. E **67**, 016411 (2003)
116. A. Schella, M. Mulsow, A. Melzer, H. Kählert, D. Block, P. Ludwig, M. Bonitz, Crystal and fluid modes in three-dimensional finite dust clouds, New J. Phys. **15**, 113021 (2013)

Chapter 3
Introduction to Streaming Complex Plasmas B: Theoretical Description of Wake Effects

Patrick Ludwig, Christopher Arran and Michael Bonitz

This chapter is dedicated to Glenn R. Joyce
(June 24, 1939—December 4, 2011)

Abstract A key problem in the description of non-ideal, multi-component plasmas is the drastic difference in the characteristic length and time scales of the different particle species. This challenging multiscale problem inherent to studying streaming complex plasmas can efficiently be tackled by a statistical ansatz for the light plasma constituents in combination with first-principle Langevin dynamics simulations of the heavy and strongly correlated dust component. Of crucial importance in this scheme is the quality of the dynamically screened Coulomb potential. For this purpose, we introduce Kielstream, a new high-performance computer code for the computation of three-dimensional plasma wakefields and the resulting electric fields. The optimization techniques used and the handling of competing numerical errors are discussed in detail. Results are presented for the wakefield around a single dust grain as well as multiscale simulations of a correlated ensemble of grains revealing fundamental structural changes when wake effects take charge.

P. Ludwig (✉)
Institut für Theoretische Physik und Astrophysik, Christian-Albrechts-Universität zu Kiel, 24098, Kiel, Germany
e-mail: ludwig@theo-physik.uni-kiel.de

C. Arran
Emmanuel College, Cambridge, CB2 3AP, UK
e-mail: cda25@cam.ac.uk

M. Bonitz
Institut für Theoretische Physik und Astrophysik, Christian-Albrechts-Universität zu Kiel, 24098, Kiel, Germany
e-mail: bonitz@theo-physik.uni-kiel.de

M. Bonitz et al. (eds.), *Complex Plasmas*, Springer Series on Atomic, Optical, and Plasma Physics 82, DOI: 10.1007/978-3-319-05437-7_3,

Table 3.1 Typical parameters of a multi-component complex plasma (argon ions, singly charged)

Plasma constituents	Charge [e_0]	Density [cm^{-3}]	Mass [kg]	Time scale [s]
Electrons	-1	3×10^8	9.1×10^{-31}	1×10^{-9}
Ions	1	3×10^8	6.6×10^{-26}	3×10^{-7}
Neutrals	0	6×10^{14}	6.6×10^{-26}	4×10^{-6}
Dust particle (Ø 10 μm)	-10000	3×10^3	7.9×10^{-13}	3×10^{-2}

The ionization fraction of the partially ionized plasma is on the order of 10^{-6}. The huge mass asymmetry of the particular plasma components leads to large differences in the relevant time scales which makes a scale-overarching selfconsistent simulation impossible. The relevant time scales of the charged components are given by the inverse plasma frequency $\omega_{p\alpha}^{-1} = (\varepsilon_0 m_\alpha / n_\alpha q_\alpha^2)^{1/2}$. The time scale of neutral atoms is characterized by the collision time, i.e., the ratio of mean free path and thermal velocity. In order to study collective dynamical processes of the dust grains the simulation time needs to cover several seconds

3.1 Introduction

Non-equilibrium processes can generate intriguing, counterintuitive many-particle phenomena such as the attraction between like charged particles as we have seen in the previous chapter. Indeed, our whole non-static universe consists to a great extent of matter in a non-equilibrium plasma state. The theoretical description of non-ideal plasmas in non-equilibrium is, however, very challenging. For the sake of simplification non-equilibrium and streaming effects are often neglected. In this chapter we present a *multiscale concept* that adequately incorporates the interplay of streaming ions, electrons, neutral atoms, and strongly coupled dust grains.

Streaming effects are often connected to a fascinating wave phenomenon: the creation of *wakefields*. This collective dynamical effect is probably best known from the V-shaped wake pattern behind objects (ships, dugs, etc.) moving on a calm water surface. As it was demonstrated in the previous chapter, similar wake effects also do exist in complex plasmas—a heterogeneous composition of electrons, ions, neutral atoms, and highly charged micrometer-sized "dust-grains", see Table 3.1. Remarkably enough, plasma wakes can even be made visible. It is reported that when an additonal object, a 'cm'-sized massive projectile is placed in a dusty plasma chamber, the dust grains arrange themselves within the ion wakefield of the projectile and make it visible to the naked eye (see sketch in Fig. 3.1).

The most remarkable wake effect is, however, found at a smaller scale when sub-nm-sized ions are streaming against the much heavier (and thus much more inert) dust grains. Here, the wakefield gives rise to an effective *attractive interaction* between like-charged micro dust particles that in the static case strongly repel each other.

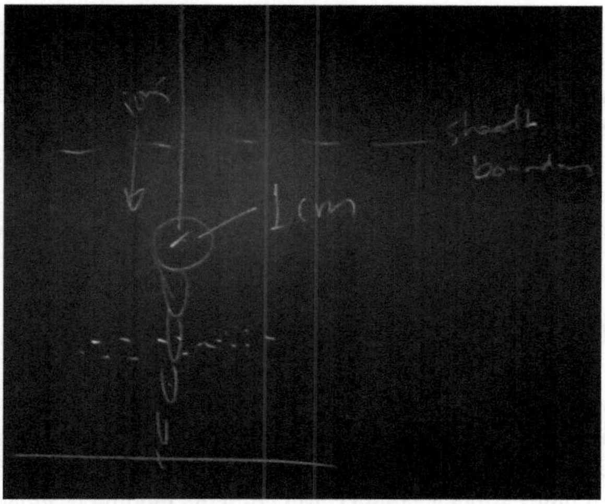

Fig. 3.1 Chalkboard sketch of a macroscopically large plasma wake which is created by fast ions streaming around a 1 cm sized projectile (which is attached to a string). The wake becomes visible to the naked eye when micrometer sized dust grains are added, which populate the 3D wake structure. Unfortunately, no picture of this experiment is known to exist. Drawing by Murillo [4]

3.2 Simulation Approaches to Streaming Multi-Component Plasmas

Highly charged polymer micro-spheres carrying about ten thousand electron charges are giving rise to the strongest correlations known in nature—even at room temperature conditions [1]. Therefore, complex (dusty) plasmas are readily used as a reference system allowing for a precise study of collective dynamical and structural properties of the strongly correlated dust component under the influence of the partially ionized plasma background. The results are of direct relevance for other type of plasmas, e.g., for dense plasmas (also called "warm dense matter") where the impact of streaming effects on the mutual interplay of quantum degenerate electrons, classical ions and neutral atoms has recently become of larger interest [2, 3].

An accurate theoretical description of a partially ionized complex plasma raises a big challenge. In order to achieve a basic theoretical understanding of streaming effects and the fundamental interactions leading to collective behavior, we have to deal with the correlated dynamics of a *non-ideal, partially ionized multi-component plasma* in a *non-equilibrium streaming situation.*

Table 3.1 reveals that the main issue in the description of a real plasma lies in the huge mass asymmetry of the different particle species which leads to drastic differences in the relevant space and time scales which must be adequately taken into account. A dust-to-ion mass ratio of about 10^{13} results in a time scale gap of five orders of magnitude that makes selfconsistent simulations numerically utmost expensive. Even worse, in order to capture collective dynamical effects long time series of up to several tens of seconds must be captured.

In dusty plasma physics there are two complementary approaches which help us to overcome these numeric limitations: (i) large-scale PARTICLE- IN- CELL (PIC) simulations, and (ii) hybrid MOLECULAR DYNAMICS (MD) simulations based on the DYNAMICAL SCREENING APPROACH. Both approaches have turned out to be very powerful in the kinetic simulation of plasma streaming and wake effects.

3.2.1 Particle-in-Cell Simulations of Plasma Wakes

Kinetic particle simulations such as the PIC method have proven to be the most efficient way of solving the Vlasov equation for arbitrary particle distributions in classical plasmas, e.g. Refs. [5–8]. To take into account short range collisions (beyond the Vlasov mean field), often a Monte Carlo collision module is added (PIC-MCC). Large-scale PIC simulations allow for an accurate description of ion-streaming effects including nonlinear phenomena and have been mostly applied to study a single dust grain or two in a complex plasma and occasionally larger systems. Furthermore PIC simulations allow for a selfconsistent description of the grain charging process.

While PIC is close to a first-principle description, in case of a complex plasma it faces severe limitations and difficulties. The mentioned huge mass asymmetry of dust grains and ions causes a many order of magnitude difference of characteristic time scales, see Table 3.1. Therefore, it is practically impossible to resolve the ionic dynamics and the dust dynamics simultaneously, i.e., with the PIC method the dust can only be treated statically. In particular for subsonic plasma conditions, $M < 1$, PIC simulations suffer serious convergence problems due to substantial numerical noise even for (insufficient) low grid-resolutions. In order to reduce this issue, often relatively large grain sizes (giving rise to large grain charges) are considered which, however, typically exaggerate the contribution of nonlinearities.

Therefore, for studying many-body effects in complex plasmas, in particular, to analyse the dust particle dynamics and compare to experimental single-particle trajectories, this method is presently not suitable. The alternative is the multi-scale DYNAMICAL SCREENING APPROACH which allows us to essentially extend the simulation time to the scale of the dust dynamics. PIC simulations (where they are available) will be used to carefully test and benchmark the linear response based simulation results.

3.2.2 Multiscale Approach to the Dynamics of Complex Plasmas

The challenging *multi-scale problem* addressed above can be effectively reduced by the DYNAMICAL SCREENING APPROACH which enables us to directly include the focussing effect of a dust grain on streaming ions by means of an effective *plasma-mediated dust potential* derived from plasma kinetic theory, (see Fig. 3.2) [2, 9–13].[1] The statistical description of the lighter plasma constituents on the basis of a *dynamic dielectric response function* allows for an accurate description of essential equilibrium and non-equilibrium plasma properties including screening, wakefield oscillations, ion and electron thermal effects as well as ion-neutral collisions and Landau damping [11, 14].

In the considered multiscale approach, the strong Coulomb correlations of the highly negatively charged dust component are treated exactly on first-principles by classical LANGEVIN MOLECULAR DYNAMICS (LMD) simulations. This allows us to analyse the dynamics of individual dust grains on a kinetic level with very high precision. The damping effect of neutrals on the grain dynamics, i.e., dust-neutral collisional effects, is taken into account by a *Langevin-Fokker-Planck approximation*.

The multiscale simulation approach consists basically of two parts:

Part I Computation of the electrostatic potential around a single grain (Fig. 3.2) from the dynamic dielectric function in a linear response approximation. The plasma-mediated potential represents an effective dust–dust Coulomb interaction which is dynamically screened by streaming ions.

Part II Use of these *wake potentials* in N-particle LMD simulations that yield the "exact" trajectory of each particle in the classical phase space [23].

The N-particle dust dynamics simulations will allow us to study the real dynamics of finite or extended dust ensembles, their self-organised structure formation as well as wake-field driven phase transitions with high numerical resolution and without having to numerically resolve the fast plasma time scales.

The results depend on three dimensionless simulation input parameters: (i) the Mach number $M = |\mathbf{u}_i|/c_s$ defined as the the ratio of ion drift velocity to the sound velocity $c_s = (k_B T_e/m_i)^{1/2}$, (ii) the electron-to-ion temperature ratio $T_r = T_e/T_i$, and (iii) the ion collision frequency ν_i which is directly proportional to the neutral gas pressure [6].

3.2.2.1 Effective Dust–Dust Interaction

The main idea of the DYNAMICAL SCREENING APPROACH is to retain the first-principle character of MD simulations for the dust particles and, at the same time, to turn the huge mass asymmetry between the dust and light particles into an advantage. In fact, even in the case of strong coupling[2] of the dust component ($\Gamma_{dd} \gg 1$, as in the case of crystallization), usually the dust-ion interaction is weak, $\Gamma_{di} \leq 1$. This means that the ion kinetic energy exceeds the dust-ion interaction energy, and

---

[1] In turn, the response of the perturbed electron-ion plasma to the presence of the dust grains determines entirely the (dynamically screened Coulomb) interaction between the grains and hence the collective dust dynamics.

[2] The coupling strength is characterized by the parameter Γ that is the ratio of mean interaction energy and dust kinetic energy. For details see Chap. 1 by H. Thomsen et al.

the ion dynamics are only weakly affected by the dust.[3] Then, to first order in Γ_{di}, the ion dynamics can be treated on the mean field level and by neglecting nonlinear effects. This provides the opportunity of using the semi-analytical results for the ions obtained from the solution of the linearized Vlasov equation. Due to their high thermal energy, it can be assumed that the electrons are completely unaffected by the presence of the dust particles, $\Gamma_{de} \ll 1$.

Considering a weak (linear) response of the plasma to the presence of the dust grain, in real space the electrostatic potential of a dust grain screened by ions streaming with the uniform and time-independent velocity $\mathbf{u}_i$ becomes [9]

$$\Phi(\mathbf{r}) = \frac{1}{(2\pi)^3} \int \mathrm{d}^3\mathbf{k}\, \frac{\phi_C(k)}{\varepsilon(\mathbf{k}, \mathbf{k} \cdot \mathbf{u}_i)} e^{i\mathbf{k}\cdot\mathbf{r}}, \tag{3.1}$$

with $\phi_C(k) = Q_d/\varepsilon_0 k^2$ being the Fourier transform of the bare Coulomb potential and $\varepsilon(\mathbf{k}, \omega)$ being the dynamic dielectric function (Fig. 3.2). In the limit $\mathbf{u}_i \to 0$ the potential (3.1) reduces to the isotropic Yukawa (Debye) potential Φ_Y.[4] The dynamically screened potential (3.1) is the main ingredient in the present multiscale approach. The idea to use this potential for the dynamics of dust grains in a complex plasma was first realized by Murillo and co-workers [15, 16] as well as by Joyce and Lampe [11, 17].

They demonstrated that, in fact, wake effects and net attractive dust interactions can be reproduced by this approximation. Despite the success of this model, comparisons with PIC simulations revealed substantial quantitative discrepancies [7]. We have, therefore, developed an independent substantially improved realization of this multi-scale approach that involves a high precision numerical computation of the screened dust potential. The code shows excellent agreement with PIC results (as discussed below).

In the following we briefly summarize the derivation of the dust potential and then discuss how to use it in N-particle MD simulations. We start from Maxwell's equations for a polarizable medium

$$\mathrm{div}\mathbf{D}(\mathbf{r}, t) = \sum_a q_a n_a(\mathbf{r}, t), \tag{3.2}$$

where $n_a(\mathbf{r}, t)$ is the number density of particles with charge q_a. In Fourier space, using the electrodynamic definition of the dielectric tensor $(i, j = x, y, z)$, $D_i(\mathbf{k}, \omega) = \sum_j \varepsilon_{ij}(\mathbf{k}, \omega) E_j(\mathbf{k}, \omega)$, we obtain for the electrostatic potential,

---

[3] This is not necessarily the case for all ions, in particular not in the vicinity of dust grains and for slow ions. Comparisons with independent PIC simulations reveal where nonlinear corrections are required.

[4] The Yukawa potential $\Phi_Y = Ze_0 \cdot e^{-r/\lambda_D}/4\pi\varepsilon_0 r$ is very successful in describing the screened Coulomb interaction of charged particles (e.g., ions or dust grains) in a "static" plasma, where the first term Ze_0 describes the effect of "bound" states and the second the exponential screening of the bar Coulomb potential by weakly interacting "free" states in the plasma. See also Chap. 1.

$$\phi(\mathbf{k},\omega) = \frac{q_a n_a(\mathbf{k},\omega)}{\sum_{ij} k_i k_j \varepsilon_{ij}(\mathbf{k},\omega)}. \tag{3.3}$$

For the case of unmagnetized electrons and ions streaming with velocity $\mathbf{u}_i$ we take into account the longitudinal plasma response, $\varepsilon_{ij} \to k_i k_j \varepsilon / k^2$, and obtain the screened potential of a single point-like grain "l" of charge Q_l, initially located at position $\mathbf{r}_{0l}$ and moving with velocity $\mathbf{v}_l$ [9]

$$\Phi_l(\mathbf{r}-\mathbf{r}_{0l};t) = Q_l \int \frac{\mathrm{d}^3\mathbf{k}}{2\pi^2} \frac{1}{k^2} \frac{e^{i\mathbf{k}(\mathbf{r}_{0l}-\mathbf{v}_l t)}}{\varepsilon\left(\mathbf{k}, \mathbf{k}\cdot[\mathbf{v}_l(t)-\mathbf{u}_i]\right)}, \tag{3.4}$$

recovering expression (3.1). The longitudinal dynamic dielectric function follows from solving kinetic equations for the electrons and ions. Approximating ion collisions by a relaxation time approximation (Bhatnagar-Gross-Krooks collision term, e.g., [18]) with the collision frequency ν_i and neglecting dynamics effects for the electrons[5] one obtains

$$\varepsilon(\mathbf{k},\omega;\mathbf{R}) = 1 + \frac{1}{k^2\lambda_{De}^2} + \frac{\omega_i^2}{k^2} \frac{\displaystyle\int \mathrm{d}^3\mathbf{v}\, \frac{\mathbf{k}\cdot\nabla_\mathbf{v} f_{i0}(\mathbf{v};\mathbf{R})}{\omega - \mathbf{k}\cdot\mathbf{v} + i\nu_i}}{\displaystyle 1 - i\nu_i \int \mathrm{d}^3\mathbf{v}\, \frac{f_{i0}(\mathbf{v};\mathbf{R})}{\omega - \mathbf{k}\cdot\mathbf{v} + i\nu_i}}, \tag{3.5}$$

where $\lambda_{De} = \sqrt{\varepsilon_0 k_B T_e / q_e^2 \bar{n}_e}$ denotes the electron Debye length, $f_{i0}(\mathbf{v};\mathbf{R})$ the unperturbed ion velocity distribution function and $\bar{n}_{e(i)}$ and $T_{e(i)}$ refer to the mean electron (ion) density and temperature, respectively. The result (3.5) includes Landau damping and collisional damping where ν_i is typically dominated by ion-neutral collisions. Note that, in the above expressions, one may allow for a space dependence of the ion parameters $\nu_i = \nu_i(\mathbf{R})$.

In most situations of interest the ions will be in (local) thermodynamic equilibrium having a local Maxwellian distribution function f_{i0}. Assuming spatially homogeneously distributed electrons, ions and neutrals,[6] the dielectric function (3.5) is readily computed

$$\varepsilon(\mathbf{k},\omega) = 1 + \frac{1}{k^2\lambda_{\mathrm{De}}^2} + \frac{1}{k^2\lambda_{\mathrm{Di}}^2} \left[\frac{1 + \zeta_i Z(\zeta_i)}{1 + \frac{i\nu_i}{\sqrt{2}k v_{Ti}} Z(\zeta_i)} \right], \tag{3.6}$$

where we employed the standard substitution

$$\zeta_i = [\mathbf{k}\cdot\mathbf{u}_i + i\nu_i]/[\sqrt{2}k v_{Ti}], \tag{3.7}$$

---

[5] Screening by the electrons is purely static since their thermal speed exceeds their field-induced drift.

[6] The back-action of the presence of the dust grains on the ion distribution is neglected, which is justified for the present low dust densities n_d (Table 3.1). The effect of the dust grains on the electron and neutrals is considerably smaller than on the ions.

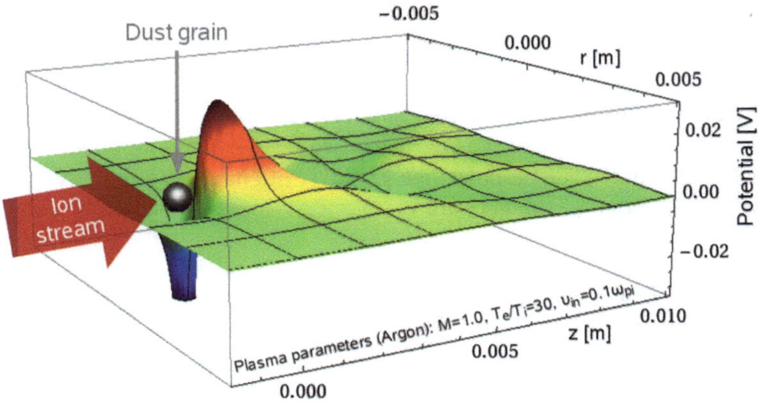

Fig. 3.2 Wake potential $\Phi(\mathbf{r})$ around a negatively charged dust grain due to streaming ions. Close to the grain there is a repulsive Yukawa-type potential (*blue*). Positively charged ions are focused in the streaming direction and create a positive space charge region in the wakefield (*black-red*) which attracts other grains downstream and gives rise to an non-reciprocal grain interaction

with the *thermal velocity*

$$v_{T\alpha} = (k_B T_\alpha / m_\alpha)^{1/2}, \tag{3.8}$$

and introduced the *plasma dispersion function* [19, 20]

$$Z(z) = i\sqrt{\pi}e^{-z^2}\mathrm{erfc}(-iz), \tag{3.9}$$

where $\mathrm{erfc}(z)$ is the complementary error function.

Using expression (3.6), we can compute the dynamically screened dust potentials (3.4) very accurately via a fast Fourier transform on large wave number grids, see Sects. 3.3 and 3.4.1. In a second step of the multiscale scheme, we will use the dust (wake) potentials for N-particle LMD simulations.

3.2.2.2 N-particle Langevin Dynamics Simulations

Our first-principle computational approach to treat strongly correlated dust particles at a finite neutral gas temperature is based on Langevin molecular dynamics simulations. In contrast to the Yukawa one-component plasma model considered in Chap. 1 by Thomsen et al., we now use the dynamically screened dust potentials, (3.4), and we add neutral gas friction (collision frequency ν_{dn}) and a stochastic noise term. The equation of motion for the kth dust grain in an external harmonic trap has the form

$$m_d\ddot{\mathbf{r}}_k = -\nabla_k V_k^{\mathrm{eff}}(\mathbf{r}_k, t) - \omega_0^2 m_d \mathbf{r}_k - \nu_{dn} m_d \dot{\mathbf{r}}_k + \mathbf{f}_k(t), \tag{3.10}$$

where f_k is a Gaussian white noise, the intensity of which is determined by the neutral gas temperature (fluctuation-dissipation relation, e.g., Ref. [21]). A screening-independent parabolic confinement is explicitly included in (3.10). It mimics the combined effect of gravity, thermophoresis, and electric fields, which in the experiments with Yukawa balls is used to achieve an approximately isotropic harmonic confinement for the dust grains [22].[7] Depending on the experimental condition, in particular rf-power, this potential is replaced by an anisotropic oscillator potential. The key ingredient in (3.10) is, however, the total dynamic potential acting on the kth grain

$$V_k^{\mathrm{eff}}(\mathbf{r}_k, t) = Q_d \sum_{l \neq k}^{N} \Phi(\mathbf{r}_k - \mathbf{r}_l, \mathbf{t}), \tag{3.11}$$

where, in accord with the present linear response approximation, V_k^{eff} is the sum of the dynamically screened potentials (3.4) created by all other $(N - 1)$ grains. In order to make N-particle simulations feasible, an expensive re-computation of the complicated screened potential Φ has to be avoided. Instead, this potential is typically pre-computed and stored in a large table.

3.3 Introduction to the Kielstream Program

Calculation of the linearized potential (3.4) crucially relies on a three dimensional discrete Fourier transform (3D DFT) of a grid of the plasma response function which is defined in k-space. Including streaming effects in the model, however, requires the plasma dispersion function (closely related to the Faddeeva function) and makes the calculation non trivial. In 2009 an original code was written [6] using `Mathematica` [24], as there was no implementation of the Faddeeva function in C/C++ that was reliable across a broad range, see also Ref. [20]. The `Mathematica` platform has 'sophisticated built-in automatic numeric precision and accuracy control', that allows for very high accuracy and virtually arbitrary precision computation. However, due to the complexities of calculating both the plasma polarisation function and a 3D DFT, this code is slow, with runtimes of up to several hours for a medium sized $(256 \times 256 \times 512)$ grid. In order to study broader ranges of physical parameters, including high electron-to-ion-temperature systems, higher grid resolutions are necessary, and a faster code is required. Thanks to publication of a new library for calculating the Faddeeva function, `libcerf` [25], in 2013, which we validated against `Mathematica`, it is now possible to reliably calculate the plasma response function in C/C++.

[7] For applications to macroscopic systems, the confinement potential is dropped, and periodic boundary conditions have to be applied, see Ott et al. [23].

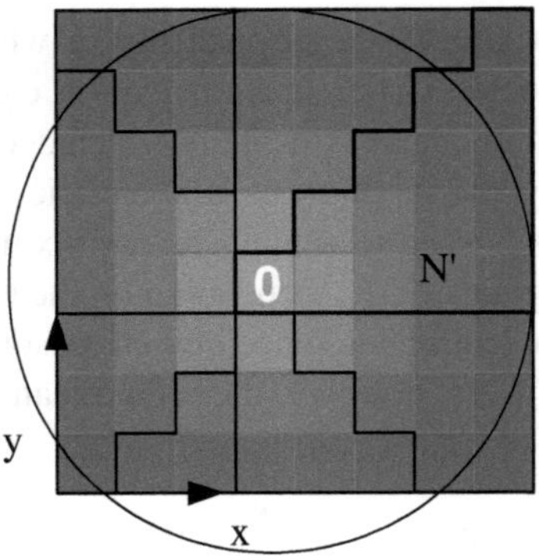

Fig. 3.3 Radial symmetry of the response function implies there are eight symmetrical triangular sectors, hence we must calculate only one triangular sector, area N', of the $x - y$ cross section, asymptotically reducing calculation time eightfold. Note, the even number of grid elements forces the origin to lie off centre, introducing boundary cases which become less noticeable with increasing N_r. This introduces a shift by only one element but has significant repercussions on the Fourier transform, which is very sensitive to symmetry

Here we therefore introduce `Kielstream`,[8] a new linear response program written in C++ to calculate the electric potential around a charged dust particle in a streaming plasma. The code uses the new `libcerf` library and the C++ `FFTW` library [26] to calculate the plasma response function and its Fourier transform respectively. Using a combination of symmetry and software optimisations and parallelisation, the code runs for the same grid resolution up to 10000 times faster than the previous `Mathematica` code. Its results agree well with theory and with the `Mathematica` code to closer than 1 %.

3.3.1 Code Structure

The high performance of `Kielstream` is achieved partly by exploiting the cylindrical symmetry of the problem in different ways:

Separating the transform: Firstly, the 3D DFT assumes linear separability across the three k-space dimensions. This allows us to split the transform into two stages, (i) a 2D transformation on each $x - y$ cross section and (ii) 1D transformations in the z-direction (Fig. 3.4). Radial symmetry implies only a 2D subspace, one $x - z$ halfplane, is required for the output. This allows us to discard all unnecessary data following the 2D transformation. We retain only data in a spoke along the positive

[8] Derived from the German word Kielwasser, meaning a wake, the name is particularly appropriate as this program was written at Christian-Albrechts-Universität, Kiel.

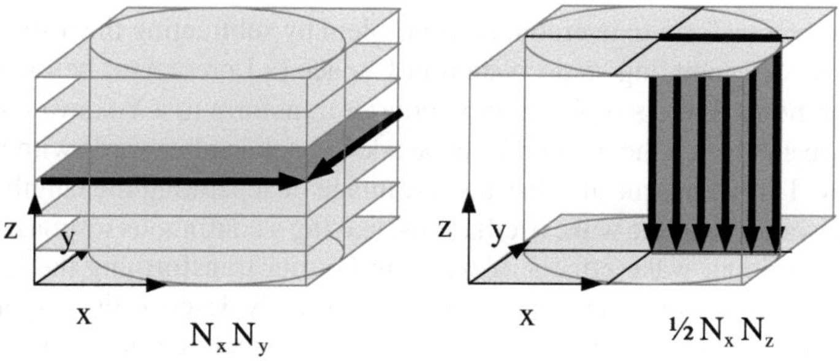

Fig. 3.4 Method of splitting the 3D DFT first into N_z 2D transforms and then subsequently $N_r = N_x/2$ 1D transforms. The area of the data arrays required and hence transform time and space requirements are dramatically reduced

x axis, which greatly reduces the number of 1D transforms required. Furthermore, this splitting means we must only hold one of the N_z 2D cross sections in memory at a time, reducing the size of the problem from $\mathcal{O}(N_x N_y N_z)$, for the whole grid, to $\mathcal{O}(N_x N_y + N_x N_z)$, for one cross section and one half plane (Fig. 3.4). This optimization greatly enhances the performance and enables both larger grid sizes to be used without extensive memory allocation, and the straightforward introduction of parallelisation on the N_z independent 2D slices, with each array created, populated with the plasma response function and Fourier transformed independently of the others. Even a single call of the `OpenMP` library [27] improves times (almost) proportionally to the number of processors. Overall, splitting the 3D Fourier transform into two stages and performing the z-transforms only along the positive x axis reduces the number of 1D DFTs by $2N_x$, leaving the N_z 2D DFTs as the main contributors to runtime. Asymptotic complexity of the transform is reduced from $\mathcal{O}(N_z N_r^2 \log(N_z N_r^2))$ to $\mathcal{O}(N_z N_r^2 \log(N_r^2) + N_r N_z \log(N_z))$.

Exploiting cross sectional symmetry: Secondly, the grid on which the program operates is cuboidal,[9] but both the plasma polarisation and its Fourier transform depend only on a $k \cdot v$ scalar product and therefore must be cylindrically symmetric. This means for every z value ($x - y$ cross section), only one triangular section, one eighth of each of the cross sections, is unique (Fig. 3.3). The plasma polarisation function for the seven other symmetrical sections can be found by simply changing the x and y indices (e.g. x↔y or x↔-x, see pseudocode below). Populating the 3D grid with the plasma polarisation function is typically a very time consuming process but is drastically reduced by taking advantage of the eightfold symmetry. This leaves the 2D Fourier transforms as the rate determining step.

Removing the static part: The singularity at the origin in the screened Coulomb potential can be a problem when performing the numerical transform, for instance causing sampling errors. Following G. Joyce [17], we can exploit the linearity of

[9] Radial Fourier transforms in cylindrical coordinates lead to strong oscillations and poor convergence.

the Fourier transform to overcome this problem by subtracting the response function for a corresponding static system in k-space (a Lorentzian) before the DFT. This Lorentzian has a simple analytic Fourier transform to a Yukawa potential in real space and hence the potential can be added back in afterwards with very high accuracy. This technique also has the advantage of separating the length scales of the Yukawa part and the wake oscillations, leaving a continuous response function purely describing wake effects. If we were Fourier transforming the Lorentzian we would need a larger range in k-space to accurately describe the response function, whereas by subtracting it we can focus on the small region around the origin in k-space where high resolution is needed.

Further improvements: Another opportunity offered by the symmetry of both the input and the output is calculating the x-transformation only upon the positive y halfplane and copying the results into the negative y halfplane (vice versa for the y-transformation). This technique was however not employed as the overheads involved outweighed the benefits for grids of $N_x \times N_y \times N_z = 1024 \times 1024 \times 2048$ and smaller.

By default `Kielstream` performs all operations in dimensionless units scaled by the electron Debye length, with lengths in units of λ_{De} and wavevectors in units of its reciprocal. This means both k-space and r-space ranges are of the order $10 - 100$. Assuming a net neutral plasma, with equal ion and electron densities ($Z_e = Z_i$), dimensionless units allow the program to be controlled with merely three *dimensionless parameters*: the Mach number M of the ion flow relative to the plasma sound speed, the ion collision frequency ν_i in terms of the ion plasma frequency, and the relative electron temperature $T_r = T_e/T_i$. These parameters are all of the order $0.01 - 100$, maximising floating point accuracy.

The code algorithm is organised as shown in Fig. 3.5. The main procedure can be summarised as follows:

(i) Within a (parallelized) loop calculate the 2D DFTs for each (consecutive) $x - y$ section along the z-direction:

- Create a 2D temporary array, reducing memory requirements compared to the entire 3D grid.
- Populate 2D array with polarisation function using `libcerf`, evaluating only for one triangular section of unique values.
- Calculate 2D DFT using the same `FFTW` algorithm, saving only data along the positive x axis.

(ii) After the loop perform all 1/2 N_x 1D DFTs in the z-direction.

3.3.2 Performance and Runtimes

The improvement in speed between `Mathematica` and `Kielstream` is stunning. C++ is a faster lower level language without built in accuracy checks, and the use

```
//SUBROUTINE-Populates array with response function
find_integrand(temporary_array, at z index iz)
{
  for (x index, ix=0 to Nx/2)
    for (y index, iy=0 to ix) //iterate over triangular section
    {
      f = response_function(kr[ix,iy], kz[iz]);
          //use libcerf to calculate response function
          //then copy into all eight segments
      for (eight symmetrical segments, octant number n)
      {
        switch octant number n
        {
          case n=1: mx=ix; my=iy
          case n=2: mx=iy; my=ix;
          case n=3: mx=Nx-1-ix; my=iy;
          ...
          case n=8: mx=Ny-1-iy; my=Nx-1-ix;
        }
        temporary_array[mx, my, iz] = f;
      }
    }
}

//MAIN PROGRAM
Kielstream(input_file)
{
     //define system and grid parameters
  initialise_parameters(from input_file);
  FFTW_2D_plan(size (Nx,Ny), on empty array, in_place);
  new data_array(Nx/2,Nz);

     //loop over slices-this is parallelised}
  parallel for (every x-y cross section, z index iz)
  {
    new temporary_array(Nx,Ny);
    find_integrand(temporary_array, iz);   //populate array
    execute_2D_plan(on temporary_array);   //do 2D transform
    for (x >= 0, x index ix=0 to Nx/2)
      data_array[ix,iz] = temporary_array[ix,iy=0];
        //only save data on positive x axis
  }

  FFTW_1D_plans(Nx/2 plans, size Nz, on data_array, in_place)
  execute_1D_plans(on data_array) //do the 1D transforms

  reorder(data)                   //put the origin in the centre
  makegrids()                     //prepare the x and z grids
  print()
}
```

Fig. 3.5 Example pseudocode describing the algorithms used by `Kielstream`. Note the optimization of `find_integrand` and the separation of the 3D DFT into 2D and 1D components. Using a temporary data array inside the loop of 2D DFTs makes the process thread safe

of the techniques mentioned above allow further order of magnitude improvements. The final speed is up to 1000 times faster with no parallelisation on the *Intel Core i5*

Table 3.2 Effect of varying grid sizes on runtimes for `Kielstream`, using an *Intel Core i5* (4×3.1 GHz)

Runtimes for `Kielstream` in seconds

		Radial resolution $N_r = N_x = N_y$				
		256	*512*	*1024*	*2048*	*4096*
Axial	*256*	0.3	1	7	34	120
Resolution	*512*	**0.7**	3	14	61	230
N_z	*1024*	1	6	27	120	450
	2048	2	12	**55**	230	890
	4096	5	22	97	430	1800

Runtime is approximately quadratic in radial resolution and linear in axial resolution, as expected discounting log factors. Highlighted (bold) are the gridsizes and runtimes for the previously used standard grid size 256 × 256 × 512 [6], and the new standard grid size 1024 × 1024 × 2048. The largest grid size used was 4096 × 4096 × 16384 for weakly damped systems

processor. With parallelisation, using the `OpenMP` routines, this speed is increased by the number of processors. This means the time taken to do the 256 × 256 × 512 grid used as standard in the `Mathematica` code is reduced from over an hour to under a second with the parallelized `C++` code (Table 3.2). A grid of 64 times the size of `Mathematica`'s, 1024 × 1024 × 2048, was used, with a runtime of just under a minute. This significantly larger grid size greatly extends the range of physical parameters for which `Kielstream` gives a high accuracy description of the system. This allows us to overcome the numerical limits formerly required for convergence (e.g. in the collisionless case $T_r \leq 25$), which were given in the previous work [6].

The larger number of grid points possible with `Kielstream` allows for a finer resolution that prevents undersampling in k-space and, by Nyquist's theorem, aliasing of downstream features to upstream of the dust particle. Causally, it is impossible for wakefeatures to appear upstream and aliasing is easily spotted giving rise to pseudo periodicity effects as shown in Fig. 3.6.[10]

The original `Mathematica` code allows for a systematic test of `Kielstream`'s reliability (specifically for smaller grid sizes).[11] Replicating the `Mathematica` grid and physical parameters with `Kielstream`, the resulting electric potentials generally show good agreement, to better than 1 %, between the real parts of the `Mathematica` and the `Kielstream` output (Fig. 3.8). It is, however, clear from Fig. 3.7 that at high Mach numbers and low damping (i.e., low Landau damping $T_i/T_e \rightarrow 0$ and low frictional damping $\nu_i \rightarrow 0$) the codes diverge, despite the agreement of `libcerf` with `Mathematica` (up to 16 significant figures) and the same grids and physical parameters.

[10] Showing the imaginary part; Fig. 3.7 shows the corresponding real part.

[11] Note that the `Mathematica` code is in SI units rather than units of the electron Debye length, which `Kielstream` uses. By default, Mathematica uses a dust charge of $10^4 e_0$, hence `Kielstream` results must be scaled for comparison.

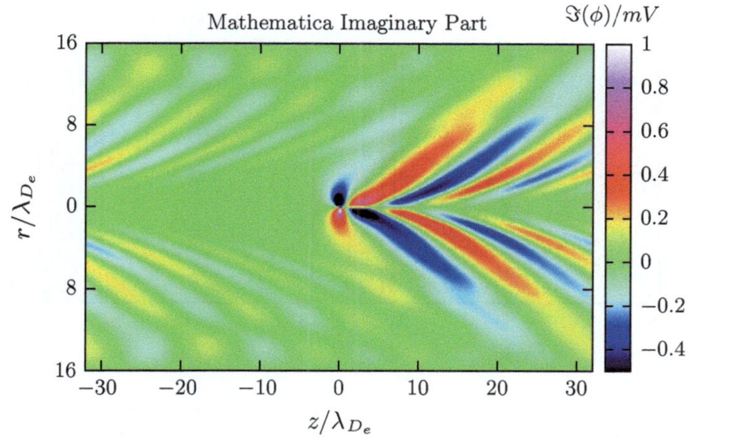

Fig. 3.6 Demonstration of aliasing in the imaginary part of the `Mathematica` results, for $M = 1.5$, $T_r = 100$, $\nu_i = 0.1$ at a grid size of $256 \times 256 \times 512$. Physically impossible pseudo-periodicity effects appear upstream of the dust particle, due to undersampling in k-space. `Kielstream`'s capacity for higher resolution allows this effect to be strongly reduced. Note also that the imaginary part has odd radial symmetry, which is caused by a real part in reciprocal-space which has odd symmetry in k_r, implying a single element shift in the grid indexing. See Fig. 3.7 for a plot of the corresponding real part of the potential. Take notice of the different colour scale

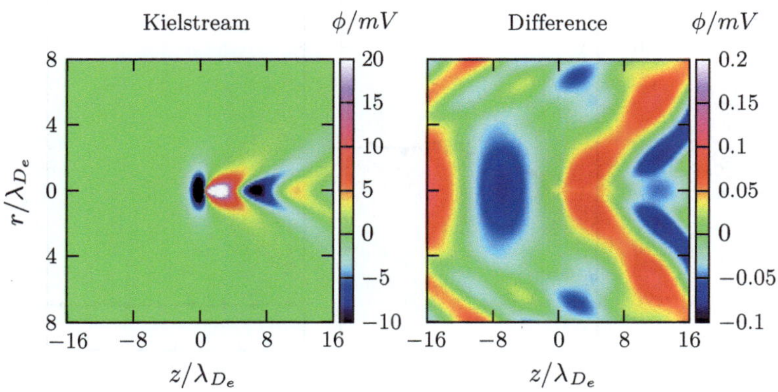

Fig. 3.7 Comparison between the resulting (real) potential from `Kielstream` and `Mathematica` for the same parameters as in Fig. 3.6, using the same (low) grid resolution and ranges. The *left* panel shows the results from `Kielstream` and the *right* the difference compared with `Mathematica`. Differences are greater than 1 % for a high relative temperature. At these small grid sizes, both `Kielstream` and `Mathematica` have pseudo-periodicity effects on the 0.1 mV scale

The difference in the underlying implementations can also be seen in a non-zero imaginary part of the result for the `Mathematica` code (Fig. 3.6), which suggests that the `Mathematica` implementation has an index shift error, as an imaginary electric potential is unphysical. This is due to the DFT routine expecting the origin at

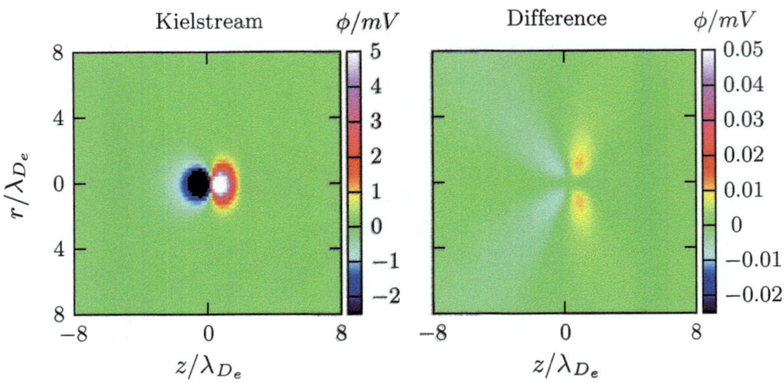

Fig. 3.8 Comparison between the resulting (real) potential from `Kielstream` and `Mathematica` for $M = 0.5$, $T_r = 10$, $\nu_i = 0.1$, using the same (low) grid resolution and ranges as in Figs. 3.6 and 3.7. The *left* panel shows the results from `Kielstream` and the *right* the difference compared with `Mathematica`. This shows agreement between `Kielstream` and `Mathematica` to better than 1 % as Landau damping is higher. The difference must derive from intrinsic differences in the programs, which use different data indexing and DFT routines

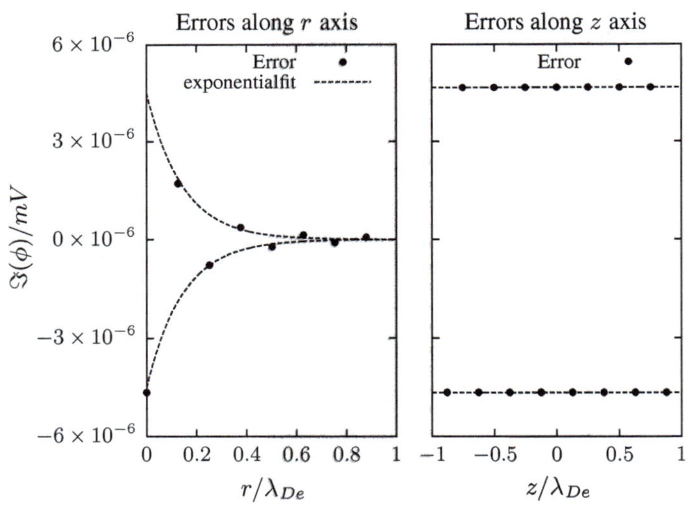

Fig. 3.9 Imaginary part of the electric potential along $z = 0$ (*left*) and $r = 0$ (*right*) for a grid size of $512 \times 512 \times 1024$, demonstrating a sharp decay along r and constancy along z, with sign alternating along both axes. This marks the error caused by a discontinuity at the z boundary of the odd imaginary part of the function in reciprocal space. The amplitude is negligible compared to the real part

the first array index. The resulting imaginary part is also purely odd in r, whereas the physical system is cylindrically symmetric in both reciprocal-space and real-space,

Table 3.3 Effect of varying axial resolution N_z and axial range on the amplitude of errors originating from edge effects

Maximum Amplitude of Imaginary Part [mV]

		Axial range / λ_{De}^{-1}		
		50	*100*	*200*
Axial	*256*	2e-5	8e-7	6e-8
resolution	*512*	1e-5	4e-7	3e-8
N_z	*1024*	5e-6	2e-7	1e-8

As the feature is caused by a single discontinuity in the k_z-direction, increasing axial grid size N_z linearly reduces the error (but not the signal), while increasing the axial range moves the discontinuity to regions where its amplitude is dramatically smaller. N_r has no effect on the amplitude. See also Fig. 3.9

indicating an error.[12] `Kielstream` therefore appears more reliable, as its imaginary part is almost everywhere zero, as discussed below (cf. Fig. 3.9).

3.3.3 Estimation of Numerical Errors

A rough estimate of the errors induced by the finite grid size is given by the imaginary part of the result of the Fourier transform, which should be zero everywhere for a physical electric potential. As opposed to the `Mathematica` code, the imaginary part in `Kielstream` does not follow the real wake profile (as in Fig. 3.6) but is instead sharply peaked around the z axis (up to three or four elements from the origin) and is of constant amplitude in z (Fig. 3.9). The oscillating sign with a period of two points along both r and z indicates by the Fourier shift theorem that the error is an artifact arising from a feature at the boundary, while the constant amplitude in z implies that it originates from a discontinuity in k-space. This is due to the inherent numerical limits of a finite integration interval, where the function must be cut off before it reaches zero. When the function is made periodic, this results in a discontinuity in the imaginary part when going from positive to negative k_z values. This effect can be reduced by increasing the physical range in the k_z-direction over which the plasma polarisation function is evaluated, or else by increasing the axial resolution, as shown in Table 3.3.

For a $512 \times 512 \times 1024$ grid with a k_z range of $50\,\lambda_{De}^{-1}$ (to be precise, $[-25 : 25]$), the maximum imaginary part is of order 10^{-6} mV, and hence compared to data of order 10 mV is negligible. Furthermore, it can be effectively reduced to below 10^{-8} mV by using a sufficiently large k_z range. Note, as the real part of the function has even symmetry, it has no such discontinuity in reciprocal space and the imaginary part is merely an indication of the amplitude of the error in real space, and not of its precise form.

[12] It is remarkable that a single element difference in the index causes such a profound change in the symmetry.

3.4 Simulation Results

Wake effects are well known from objects moving in water (birds, ships). Behind the object a periodic modulation of the water surface is observed, connected with a characteristic opening angle. In Cartesian coordinates, the height of the surface gravity wake is approximately given by

$$h(r, z) = \int_{-\pi/2}^{\pi/2} \cos[(\tilde{z} - \tilde{r}\tan(\vartheta))/\cos(\vartheta)]\,\mathrm{d}\vartheta, \tag{3.12}$$

with $\tilde{z} = g \cdot z/u_z^2$ and $\tilde{r} = g \cdot r/u_z^2$, where u_z denotes the ship's speed and $g = 9.81\,\mathrm{m/s^2}$ the gravitational acceleration [28]. This pattern (Kelvin's wave pattern) is shown in the bottom part of Fig. 3.10. In the following, we will see that dust particles in a streaming plasma show a very similar behavior.

3.4.1 Wake Structure Around a Single Grain

The presence of the dust grain perturbs the plasma around it. Using `Kielstream`, we now can evaluate the impact of the directed ion flow on the response of the plasma, (3.4), with very high numerical accuracy, even resolving the long tail of the wakefield in a collisionless plasma, see Fig. 3.10. At finite ion flow velocities, $M > 0$, the streaming ions are deflected by attraction to the grain giving rise to an oscillatory wakefield with regions of an enhanced (respectively reduced) ion density. Therefore, the dynamically screened potential around the grain is highly anisotropic and depends on the spatial orientation relative to the plasma flow. In particular, the strong trailing peak in the ion wake pattern can be sufficient to overcome the mutual repulsion of charged grains giving rise to a net attractive grain-grain interaction and the formation of particle strings.

Figure 3.11 gives an overview of the effect of increasing Mach number on the electric grain potential for representative plasma parameters. The relative temperature T_r is 30 in the upper half of the figures and 100 in the lower half. It is immediately clear that the isotropic Yukawa potential fails dramatically at finite flow velocities. Even in the subsonic regime, $M < 1$, streaming distorts the potential away from an isotropic Yukawa shape and towards a conical wavepacket. The severity of the distortion depends on the electron/ion temperature ratio T_r, and is stronger at the higher temperature ratio (corresponding to weak Landau damping), which supports a much longer wavepacket. Equally spaced peaks reach higher amplitude, as increasing ion flow, with greater ion kinetic energy, leads to a more pronounced wake structure with several maxima and minima. As the stream becomes supersonic, the wavefront $\phi = 0$ shows the information barrier along cones defined by the Mach angle, which in contrast to Kelvin's wave pattern, Fig. 3.10, does depend on M.

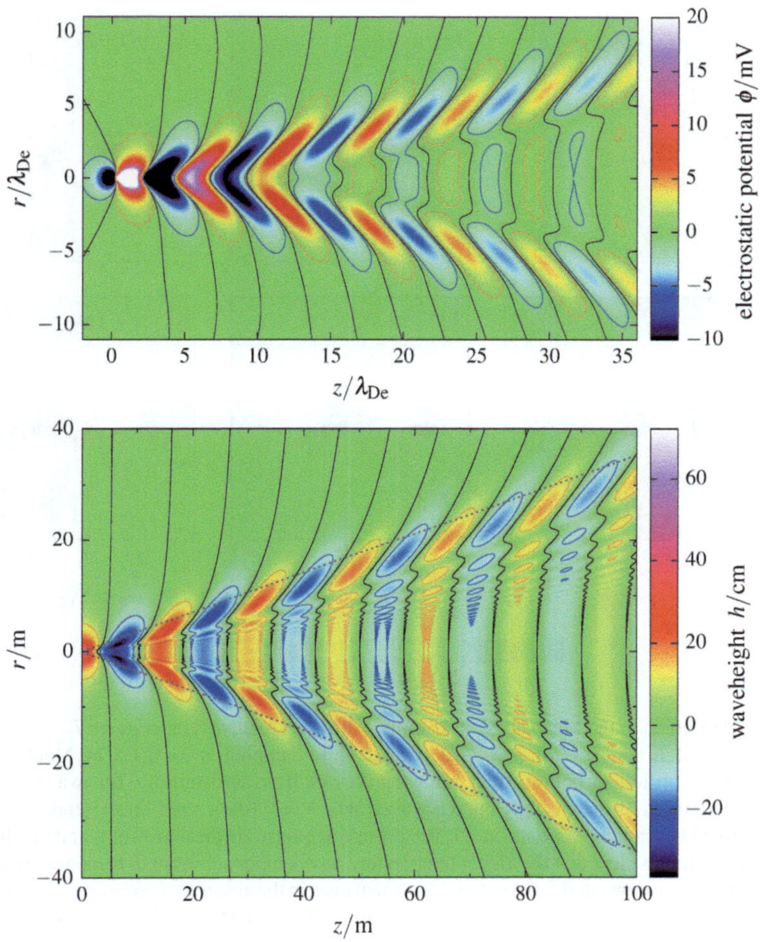

Fig. 3.10 Top: longitudinal 3D plasma wake $\phi(r, z)$ computed with `Kielstream` on a 4096×16384 element sized grid for $M = 0.7$, $T_r = 100$, and $\nu_i = 0.0$. The point-sized dust grain is located at the origin. The stationary flow of ions along the z-direction leads to strong deviations from the Yukawa potential by giving rise to an anisotropic, oscillating wake structure behind the grain. The solid blue, black, and orange curves are equipotential lines at $\phi = \{-1, 0, 1\}$ mV, respectively. The angle of the V-shaped wake depends on the Mach number M, cf. Fig. 3.11
Bottom: *Kelvin's wave pattern* of a ship located at the origin and moving with constant speed $|u_z| = 5$ m/s on deep water (in the reference frame where the ship is at rest). Equipotential lines are shown for the wave heights $h = \{-10, 0, 10\}$ cm, respectively. The V-shaped pattern has a characteristic angle $\varphi = \arcsin(1/3) \approx 19.5°$ (*dashed gray lines*) which is generally considered to be independent of the ship's speed [29]

It is interesting to note that in the static case, $M = 0$, the negative potential of the dust grain is predominantly screened by a positively charged ion cloud, which isotropically forms around the grain. The Debye screening length of the ions, $\lambda_{Di} =$

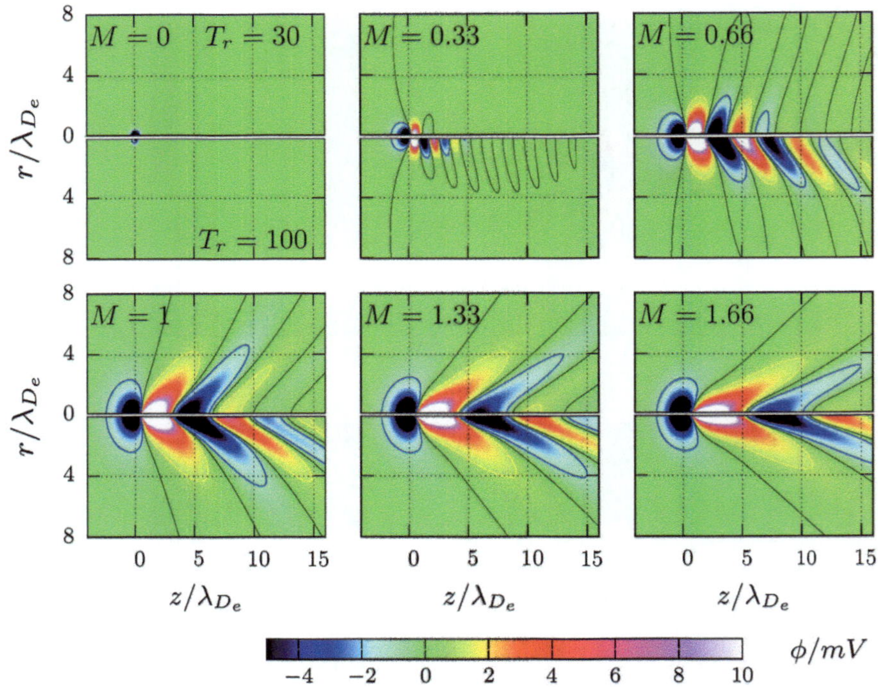

Fig. 3.11 The effect of the Mach number M for two different relative temperatures $T_r = 30$ (*upper half figures*) and $T_r = 100$ (*lower half figures*) at finite damping $\nu_i = 0.1$. The Mach number increases from top left to bottom right, showing change from an isotropic Yukawa potential at $M = 0$ to formation of increasingly curved wavefronts $\phi = 0$ (dark solid lines), tending towards the distinctive Mach angle. Contours at $-1, 0$ and 1 mV demonstrate the wavefronts and the shape of the wake pattern. It is clear that the higher temperature ratio supports a stronger response, especially in producing a more elongated wake pattern. Note that even in the absence of collisions with neutrals, the amplitude of the wake oscillations damps out due to Landau damping which naturally limits the range of the wakefield

$(\varepsilon_0 k_B T_i/\bar{n}_i q_i^2)^{1/2}$, is significantly shorter (and contribute more to screening) than the electron Debye length $\lambda_{De} = (\varepsilon_0 k_B T_e/\bar{n}_e q_e^2)^{1/2}$. Note, particularly in the top left panel of Fig. 3.11, that the total Debye length $\lambda_D^{-2} = \lambda_{De}^{-2} + \lambda_{Di}^{-2} = \lambda_{De}^{-2}(1 + T_r)$, and thus the range of the Yukawa potential, varies for the different relative temperatures. With increasing M, the trajectories of fast moving ions are less affected by the grain and the plasma ions contribute less to the screening. The effective Debye length $\lambda_D = \lambda_D(M)$ decreases. In the supersonic limit, $M \to \infty$, the spherical symmetry of the Debye screening is restored and entirely determined by the electrons [6].[13]

[13] As mentioned before, the field-induced electron flow can be neglected, since the electron thermal velocity exceeds the field-induced drift by far, $v_{Te} \gg |\mathbf{u}_e|$.

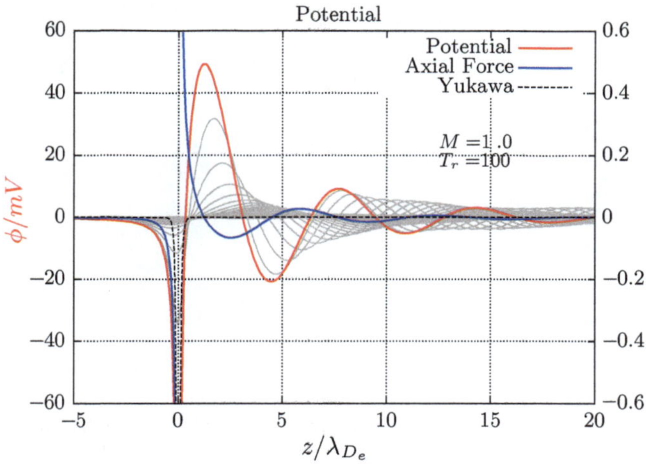

Fig. 3.12 Plot of electric potential (*red*) and resulting axial force acting upon a second particle (*blue*) against position for a nearly collisionless system, with $M = 1.0$, $T_r = 100$ and $\nu_i = 0.01$. Potential along the z-axis (*red*) changes from the electron only screened Yukawa potential in the upstream direction to the ion dominated wake oscillations downstream. Near the dust particle at the origin the deep potential minimum resembles the isotropic electron and ion screened Yukawa potential of the corresponding static case ($M = 0$, *dashed black*). The electric force in the z-direction (*blue*) attracts a second dust particle to equally spaced points in the wake of the first (around $z = 1.2, 7.7$ and $14.2\lambda_{De}$). At this low damping the potential off-axis (*gray*, spaced every $0.3\lambda_{De}$ in radius) does not decay rapidly and instead is shifted downstream compared to the on-axis oscillations

Figure 3.12 demonstrates the electric potential for a typical charge[14] of $-10\,000e_0$ and the resulting axial force upon a second identical particle for weak damping ($M = 1.0$ with low Landau damping $T_r = 100$ and friction $\nu_i = 0.01$). The force is simply the z-gradient of the potential multiplied by a negative charge of $10^4 e_0$ and is calculated in a subroutine of `Kielstream`. Note the smoothness in the calculated force function, which is crucial for accurate MD simulations. The wake oscillations are strongly pronounced, both on-axis and off-axis (spaced every $0.3\lambda_{De}$ radially), with the force attracting the second dust particle to regularly spaced points in the wake (the zeros of the force function).

3.4.1.1 Comparison of Linear Response and PIC Results

At this point, it is instructive to compare these results obtained in the framework of linearized-response kinetic-theory for a point charge with fully nonlinear PIC simulations of the stationary plasma flow past a finite-sized grain charge, see Figs. 3.13 and 3.14. We find that the linearized potential agrees very well with collisionless (non-linear) *DiP3D-PIC simulations* by W. Miloch [5, 6]. Both complementary ap-

[14] A dust grain with a diameter of $10\,\mu$m typically acquires a negative charge on the order of 10^4 elementary charges. See Table 3.1 for further parameters.

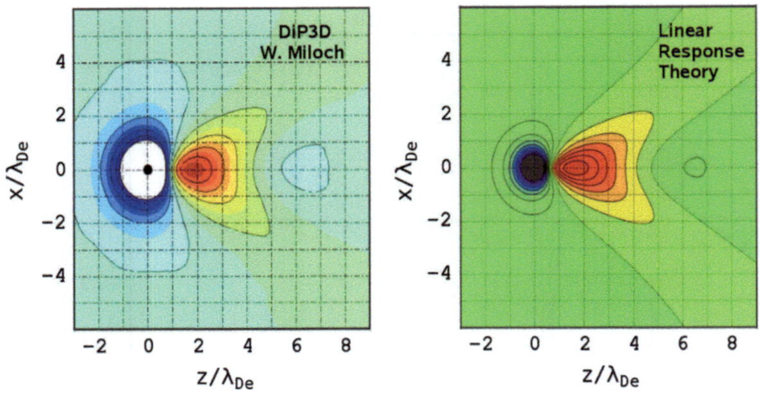

Fig. 3.13 Contour plots of the screened grain potential obtained from full 3D PIC (*left*) and the corresponding linear response result (*right*) for a collisionless plasma with $T_r = 10$. The grain is marked by the black dot and has a radius $a = 0.185\lambda_{De}$ (PIC only). The supersonic ion flow ($M = 1.5$) is from *left* to *right*. Attractive potential values are shown in *yellow* and *red*. The agreement is excellent, see also Fig. 3.14. The PIC simulations are due to Miloch [6]. Reprinted with permission from [35]

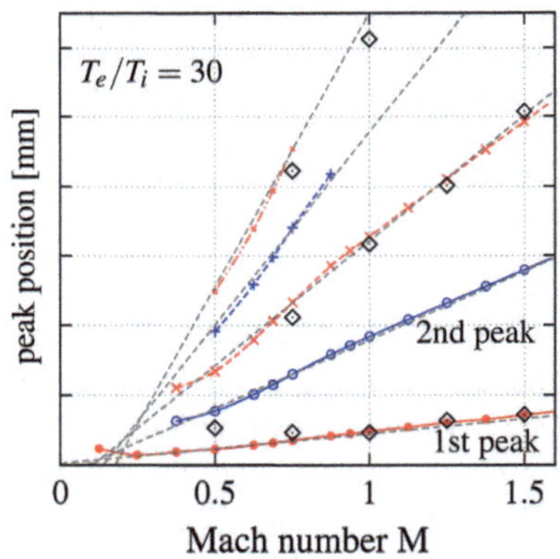

Fig. 3.14 Peak positions of the wake potential as function of Mach number M. *Red* (*blue*) lines correspond to a *positive* (*negative*) space charge. The *dashed* lines show the linear behaviour over broad ranges. *Diamonds*: 3D PIC results of Miloch. From Ref. [6]. Reprinted with permission from [6]

proaches yield the same number of potential extrema where the positions of the maxima and minima, the peak heights as well as the topological structure of the wakefield including the characteristic angle of the wave front are correctly reproduced. This qualitative and quantitative agreement is even more impressive when one takes into account that, for computational reasons, the PIC simulations used a grain charge of several times larger than the experimental values, which naturally increases nonlinear effects.

Extensive systematic comparisons show that linear response is able to yield reliable results for the effective potential of dust grains in non-equilibrium complex plasmas with streaming ions. Due to its high numerical accuracy, the present DYNAMICAL SCREENING APPROACH is well suited for a detailed theoretical analysis of correlated dust systems and for quantitative comparisons with experiments. Moreover, the comparison of both simulation methods is also a valuable test for PIC, since the DYNAMICAL SCREENING APPROACH does not suffer the noise and convergence problems of PIC codes for realistic grain charges. On the other hand, PIC helps to estimate an upper bound for the importance of nonlinear effects setting a natural limit of applicability of our present linear response approach. In particular for low Mach-numbers $M < 0.5$, deviations are expected to increase, since then the ions are more strongly affected by the dust.

3.4.2 Many-Particle Langevin Dynamics Simulations

Having obtained high-quality linear response results for the dynamically screened dust potential we are now in the position to investigate many-body properties of these systems as discussed in the beginning, Sect. 3.2.2. In particular, we can evaluate the structural and thermodynamic properties, phase transitions and even time-dependent properties in a system of many strongly correlated dust particles in the presence of streaming ions, plasma wakes and attractive forces.

Using the multiscale concept based on (3.10) we have performed Langevin molecular dynamic simulations for $N = 46$ dust grains and plasma parameters that match the typical experimental conditions: grain charge $Q_d = -6000\,e_0$ (radius $R_d = 2.43\,\mu\text{m}$), electron temperature $T_e = 2.5\,\text{eV}$, ion temperature $T_i = 0.03\,\text{eV}$, and argon gas pressure $p = 15\,\text{Pa}$ which corresponds to an ion collision frequency (in units of the ion plasma frequency) of $\nu_i/\omega_{pi} = 0.2$. A dust-neutral collision frequency of $\nu_{dn} = 19.1\,\text{Hz}$ drives the dust temperature toward the neutral gas temperature $T_n = T_i$. Assuming singly charged argon ions, the electron density equals the ion density $\bar{n}_e = \bar{n}_i = 10^8\,\text{cm}^{-3}$, and we obtain for the total Debye screening length $\lambda_{Di} = 128.0\,\mu\text{m}$, which is mainly due to the ions ($\lambda_{Di} = 128.8\,\mu\text{m}$, $\lambda_{De} = 1175\,\mu\text{m}$). The external confinement is isotropic and harmonic with the trap frequency being $\omega_0 = 7.0$ [30]. We want to point out that all input parameters are closely related to the experimental parameters. There are no free "fitting" parameters in the simulations resulting in a predictive power that is comparable to that of PIC simulations.

Varying the Mach number M, our multi-scale simulation approach reproduces the experimentally observed cluster configurations in the two limiting cases of the ion flow velocity see Fig. 3.15a and d:

1. Without streaming, $M = 0$, we reproduce strongly correlated *Yukawa balls* [1, 23] consisting of several nested spherical shells. The number of shells and even the exact shell population numbers observed in the experiments are perfectly reproduced by the Yukawa interaction [31].
2. At a supersonic streaming velocity, $M = 1$, a completely different type of dust structure is observed: under the influence of an attractive wakefield, the grains align behind each other forming particle strings.

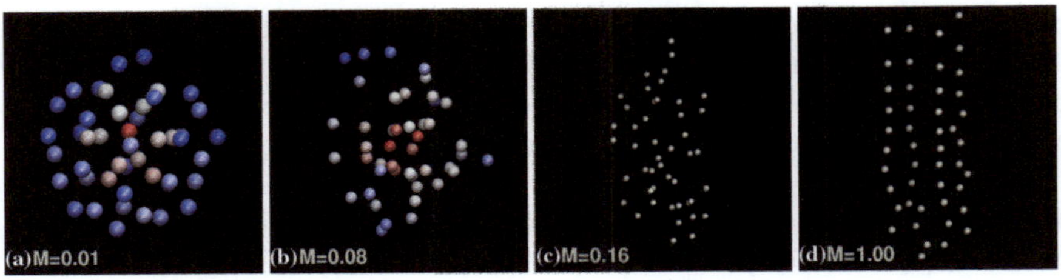

Fig. 3.15 Simulation snapshots of $N = 46$ dust grains in a spherically confined complex plasma in presence of ions streaming from top to bottom. Depending on the ion Mach number M completely different self-organized dust structures are observed: **a** Yukawa ball with a (1,12,33) shell configuration (cluster size $a = 2.2$ mm), **b** ion flow destabilises the plasma crystal; the grains are spatially delocalized ($a = 3.0$ mm), **c** formation of 'dust molecules'; vertically aligned pairs move through a weakly coupled amorphous state ($a = 6.3$ mm), **d** wake attraction leads to formation of flow-aligned particle chains ($a = 7.0$ mm). The color code for $M = 0.01, 0.08$ denotes the particle's radial distance from the trap center. Reprinted with permission from [30]

In the latter case, the streaming ions couple kinetic energy into the non-Hamiltonian dust system[15] which leads to self-excited vibrations which cause a destabilization of longer strings. As observed in the experiments, lower grains are directly following the motion of the upper ones due to the non-reciprocal wakefield attraction. The amplitude of the particle oscillations grows in the streaming direction. This can be seen in Fig. 3.15d where the upper part of the particle strings are highly ordered while in the lower part grain dislocations appear due to the downward increasing, ion streaming-induced dust temperature [14, 16, 30].

Moreover, our simulations reveal a discontinuous transition between these two limits where the grains are found to from regular structures. In fact, a relatively moderate ion flow, $M \approx 0.1$, destabilizes the highly ordered Yukawa balls and initiates an abrupt order-disorder transition. For the considered plasma conditions, the critical value of M for this abrupt, non-equilibrium melting process is found to be independent of cluster size. Increasing M more and more grains line up behind each other forming flow-aligned particle complexes and finally, at larger M values, elongated chains as shown in Fig. 3.15.

In both mentioned limiting cases (i) and (ii), the grains are strongly coupled, while in an intermediate range of Mach numbers they are in a weakly coupled amorphous state. The reason for this surprising finding is that in the static case, $M = 0$, the strong mutual grain repulsion is compensated by an external confining potential. Just this stabilizing interplay allows for the strong interparticle correlations and even crystallization. At low M values, however, the grain potential in downstream direction becomes non-monotonic and relatively flat. Consequently, the repelling force between the grains vanishes and the stabilizing force balance breaks down giving rise to an amorphous many-particle state. Only with further increase of the ion flow, a new order mechanism of local nature takes charge. This mechanism, however,

[15] Since the grain interaction is mediated by the flowing ions which couple kinetic energy into the dust subsystem, neither the total energy nor momentum of the dust grains are conserved quantities.

does not stabilize the system as a whole, but rather, individual grains agglomerate to larger complexes due to the wakefield attraction. That is, a global stabilization mechanism is replaced by a local one.

Interestingly, we find that already a relatively moderate ion flow is sufficient for wake attraction and the onset of string formation. This behavior was also observed in recent experiments [32]. Current PIC simulations suggest, however, that wakefield effects disappear for M less than approximately 0.3 [8]. This discrepancy for low Mach numbers shows that details of the associated processes are still poorly understood and also underlines the demand for conclusive experimental studies.

3.5 Synopsis

Only a few years ago, it was a widely accepted that plasma wakefields and an effective particle attraction exist only if the speed of the ion flow exceeds the ion-acoustic velocity, e.g., Ref. [12], and until today there are only very few theoretical calculations for the subsonic regime. However, this regime is of broad interest for many current experiments where grains are confined in the pre-sheath region or the plasma bulk [33–35].

In order to correctly describe the non-equilibrium nature of a realistic plasma, including ion flows, we presented a multi-scale DYNAMICAL SCREENING APPROACH for partially ionized complex plasmas. This linear-response/molecular dynamics hybrid approach provides the unique opportunity to study the impact of plasma wakes in stationary flowing plasmas on the dynamical, correlated interplay of many dust grains with unrivalled precision [14, 16, 30].

The scheme is based on a linearized kinetic-theory treatment of the ion flow past the grain charge. The quality of the plasma wakefield depends fully on the precise numerical evaluation of rather complex integrals on large three-dimensional grids which even today is not routine. In order to improve the computational efficiency and to allow for higher grid resolutions, `Kielstream`, a parallelized high-performance linear response code has been developed utilizing the FFTW library [26] for the discrete Fourier transform and a very recent implementation of the plasma dispersion function by Zaghloul et al. (algorithm 916, [20, 25]).

Our multi-scale simulations reveal a remarkable structural transition from spherical Yukawa balls ($M = 0$) to flow-aligned strings when the ion Mach number M is increased. Recent experiments on confined systems [35] allow to systematically analyse the competition of spherical order which is induced by the external confinement, on the one hand, and ion-focus induced attraction of like-charged particles and chain formation under subsonic flow conditions, on the other hand. Since subsonic flows occur in a broad range of situations this is one of the central questions that has importance far beyond the dusty plasma community, in particular for warm dense matter [2, 3].

Acknowledgments We thank Glenn Joyce and Martin Lampe for stimulating discussions at the early stage of this project as well as Hanno Kählert for a fruitful collaboration on the dielectric theory. The authors wish to thank Hauke Thomsen for the computation of the Kelvin wake pattern and preparation of Fig. 3.10. We also acknowledge helpful discussions with Wojciech Miloch and Ian Hutchinson and comparisons with their PIC results. Financial support by the Deutsche Forschungsgemeinschaft via SFB-TR24 project A9 and the German Academic Exchange Service (DAAD) via the RISE program is acknowledged.

References

1. M. Bonitz, C. Henning, D. Block, Complex plasmas–a laboratory for strong correlations. Rep. Prog. Phys. **73**, 066501 (2010)
2. P. Ludwig, M. Bonitz, H. Kählert, J.W. Dufty, Dynamics of strongly correlated ions in a partially ionized quantum plasma. J. Phys. Conf. Ser. **220**, 012003 (2010)
3. F.R. Graziani et al., Large-scale molecular dynamics simulations of dense plasmas: the cimarron project. High Energy Density Phys. **8**, 105 (2012)
4. Private communication with Michael Murillo (Los Alamos National Laboratory) at the workshop Computational Methods in High Energy Density Plasmas, University of California (UCLA), Los Angeles (2012)
5. W.J. Miloch, Wake effects and Mach cones behind objects. Plasma Phys. Control. Fusion **52**, 124004 (2010)
6. P. Ludwig, W.J. Miloch, H. Kählert, M. Bonitz, On the wake structure in streaming complex plasmas. New J. Phys. **14**, 053016 (2012)
7. I.H. Hutchinson, Nonlinear collisionless plasma wakes of small particles. Phys. Plasmas **18**, 032111 (2011)
8. I.H. Hutchinson, Intergrain forces in low-Mach-number plasma wakes. Phys. Rev. E **85**, 066409 (2012)
9. A.F. Alexandrov, L.S. Bogdankevich, A.A. Rukhadze, *Principles of Plasma Electrodynamics*, (Springer, Ney York, 1984)
10. D.S. Lemons, M.S. Murillo, W. Daughton, D. Winske, Two-dimensional wake potentials in sub- and supersonic dusty plasmas. Phys. Plasmas **7**, 2306 (2000)
11. M. Lampe, G. Joyce, G. Ganguli, V. Gavrishchaka, Interactions between dust grains in a dusty plasma. Phys. Plasmas **7**, 3851 (2000)
12. S.V. Vladimirov, K. Ostrikov, A.A. Samarian, Physics and applications of complex plasmas, (Imperial College Press, London, 2005), pp. 57,59
13. R.L. Dewar, D. Leykam, Dressed test particles, oscillation centres and pseudo-orbits. Plasma Phys. Control. Fusion **54**, 014002 (2012)
14. M. Lampe, G. Joyce, G. Ganguli, Structure and dynamics of dust in streaming plasma: dust molecules, strings, and crystals. IEEE Trans. Plasma Sci. **33**, 57 (2005)
15. J.E. Hammerberg, B.L. Holian, M.S. Murillo, D. Winske, in *Frontiers in Dusty Plasmas*, eds. by Y. Nakamura, T. Yokota, P.K. Shukla. Directional ordering and dynamics in dusty plasmas (Elsevier, Sao Paulo, 2000), p. 37
16. J.E. Hammerberg, D.S. Lemons, M.S. Murillo, D. Winske, Molecular dynamics simulations of plasma crystal formation including wake effects. IEEE Trans. Plasma Sci. **29**, 247 (2001)
17. G. Joyce, M. Lampe, G. Ganguli, Particle simulation of dust structures in plasmas. IEEE Trans. Plasma Sci. **29**, 238 (2001)
18. F. Jenko, G. Joyce, H.M. Thomas, Dressed particle simulation of dusty plasmas. Phys. Plasmas **12**, 022309 (2005)
19. M.R. Zaghloul, Comment on "a fast method of modeling spectral line". J. Quant. Spectrosc. Radiat. Transf. **109**, 2895 (2008)
20. M.R. Zaghloul and A.N. Ali, Algorithm 916: Computing the Faddeyeva and Voigt functions. ACM Trans. Math. Softw. **38**(2),15 (2011)
21. H. Kählert, P. Ludwig, H. Baumgartner, M. Bonitz, D. Block, S. Käding, A. Melzer, A. Piel, Probability of metastable configurations in spherical three-dimensional Yukawa crystals. Phys. Rev. E **78**, 036408 (2008)

22. O. Arp, D. Block, M. Klindworth, A. Piel, Confinement of Coulomb balls. Phys. Plasmas **12**, 122102 (2005)
23. T. Ott, P. Ludwig, H. Kählert, M. Bonitz, in *Introduction to Complex Plasmas*, eds. by M. Bonitz, N. Horing, P. Ludwig. Molecular dynamics simulation of strongly correlated dusty plasmas, Springer Series: Atomic, Optical and Plasma Physics, vol. 59 (Springer, Heidelberg, 2010), p. 231
24. Wolfram Research Inc, `Mathematica`, Version 8.0, Champaign (2010)
25. S. G. Johnson, A. Cervellino, J. Wuttke, `libcerf`, Numeric library for complex error functions, Version 1.1; http://apps.jcns.fz-juelich.de/libcerf (2013)
26. M. Frigo, S.G. Johnson, The design and implementation of FFTW3, Proc. IEEE **93**(2), 216 (2005); http://www.fftw.org
27. OpenMP Architecture Review Board (2008), `OpenMP` Application Program Interface Version 3.0, http://www.openmp.org/mp-documents/spec30.pdf. 2008
28. National Institute of Standards and Technology (2013), Kelvin's ship-wave pattern, in *Digital Library of Mathematical Functions*, §36.13 (2013), http://dlmf.nist.gov/36.13
29. M. Rabaud, F. Miosy, Ship wakes: Kelvin or Mach angle?, Phys. Rev. Lett. **110**, 214503 (2013)
30. P. Ludwig, H. Kählert, M. Bonitz, Ion-streaming induced order transition in 3D dust clusters. Plasma Phys. Control. Fusion **54**, 045011 (2012)
31. M. Bonitz, D. Block, O. Arp, V. Golubnychiy, H. Baumgartner, P. Ludwig, A. Piel, A. Filinov, Structural properties of screened Coulomb balls. Phys. Rev. Lett. **96**, 075001 (2006)
32. O. Arp, J. Goree, A. Piel, Particle chains in a dilute dusty plasma with subsonic ion flow, Phys. Rev. E **85**, 046409 (2012)
33. M. Kroll, J. Schablinski, D. Block, A. Piel, On the influence of wakefields on three-dimensional particle arrangements. Phys. Plasmas **17**, 013702 (2010)
34. W.J. Miloch, M. Kroll, D. Block, Charging and dynamics of a dust grain in the wake of another grain in flowing plasmas, Phys. Plasmas **17**, 103703 (2010); ibid. *Erratum: "Charging and dynamics of a dust grain in the wake of another grain in flowing plasmas"*. Phys. Plasmas **18**, 109901 (2011)
35. D. Block, J. Carstensen, P. Ludwig, W.J. Miloch, F. Greiner, A. Piel, M. Bonitz, A. Melzer, Wake formation and wake field effects in complex plasmas. Contrib. Plasma Phys. **52**, 804 (2012)

Part II
Quantum Plasmas

Chapter 4
Quantum Hydrodynamics

Shabbir A. Khan and Michael Bonitz

Abstract Quantum plasma physics is a rapidly evolving research field with a very inter-disciplinary scope of potential applications, ranging from nano-scale science in condensed matter to the vast scales of astrophysical objects. The theoretical description of quantum plasmas relies on various approaches, microscopic or macroscopic, some of which have obvious relation to classical plasma models. The appropriate model should, in principle, incorporate the quantum mechanical effects such as diffraction, spin statistics and correlations, operative on the relevant scales. However, first-principle approaches such as quantum Monte Carlo and density functional theory or quantum-statistical methods such as quantum kinetic theory or non-equilibrium Green's functions require substantial theoretical and computational efforts. Therefore, for selected problems, alternative simpler methods have been put forward. In particular, the collective behavior of many-body systems is usually described within a self-consistent scheme of particles and fields on the mean-field level. In classical plasmas, further simplifications are achieved by a transition to hydrodynamic equations. Similar fluid-type descriptions for quantum plasmas have been proposed and widely used in the recent decade. This chapter is devoted to an overview of the main concepts of quantum hydrodynamics (QHD), thereby critically analyzing its validity range and its main limitations. Furthermore, the results of the linearized QHD in unmagnetized and magnetized plasmas and a few nonlinear solutions are examined with illustrations. The basic concepts and formulation of particle–particle interactions are also reviewed at the end, indicating their possible consequences in quantum many-body problems.

S. A. Khan
National Centre for Physics, Quaid-i-Azam University Campus, Islamabad 45320, Pakistan
e-mail: sakhan@ncp.edu.pk

M. Bonitz (✉)
Institut für Theoretische Physik und Astrophysik, Christian-Albrechts Universität zu Kiel, 24098 Kiel, Germany
e-mail: bonitz@physik.uni-kiel.de

M. Bonitz et al. (eds.), *Complex Plasmas*, Springer Series on Atomic, Optical, and Plasma Physics 82, DOI: 10.1007/978-3-319-05437-7_4,

4.1 Introduction

Conventional plasmas found naturally in the visible universe (e.g. the Sun's environment, interplanetary and intergalactic media, etc.) or created in the laboratory (e.g. discharge experiments, etc.) are ionized gases in which the charged particles (electrons and different ions) move under the influence of long-range electromagnetic forces. Although, the individual particles obey the laws of quantum mechanics, the wave nature of the particles has practically no effect on the collective motion in the case of classical plasmas, due to the large interparticle distances, and the plasma can adequately be described by classical dynamical laws in the framework of Newtonian mechanics and Maxwell–Boltzmann (MB) statistics.

It has been known since long ago that the conduction electrons in metals behave very similar to gaseous plasmas and can be well treated as an electron gas. Similarly, electrons in semiconductors excited across the band gap behave very similarly to a plasma of electrons (in the conduction band) and holes (missing electrons in the valence band). This electron-hole plasma is very similar to classical two-component plasmas. However, there is a basic difference: the relevant statistics change from MB to Fermi–Dirac (FD), applicable to identical quantum particles with half-integer spin whose distribution is restricted by the Pauli exclusion principle. The quantum electron gas in metals is globally neutralized by the lattice ions whose properties are governed by various control parameters (see below, for more details, see [1]). Recent developments in ultrafast spectroscopic techniques have made it possible to monitor the collective behavior of the quantum electron gas confined in nanomaterials (nanotubes, metal clusters, nanoparticles, etc.) at the femtosecond scale. The collective electron oscillations which are principally governed by plasma effects lead to fascinating paradigm of plasmonics—a research field currently under way at a breathtaking pace [2–5]. In semiconductors, even though the electron density is much lower than in metals, the ongoing miniaturization in nanotechnology applications has made the spatial variations of the doping profiles comparable to the de Broglie wavelength of the electrons. This indicates the central role of typical quantum effects, such as tunneling, on the behavior of future electronic components.

Other realizations of quantum plasmas are obtained in high-density matter. By using various dynamic and static compression techniques (diamond anvils, gas guns, and so on), or high energy sources (intense lasers, ion beams), dense plasma conditions with densities of the order of $10^{23} \ldots 10^{25}\,\mathrm{cm}^{-3}$ have been achieved in the laboratory [1]. At the initial stage of compression, the temperature is moderate, and degenerate electrons are expected. Significance of such experiments can be seen in the warm dense matter (WDM) physics—an area getting increasing attention due to its importance for inertial confinement fusion (ICF). The advent of superintense lasers (in the petawatt range and beyond) provides tools for light matter-interactions expecting the creation of overdense plasmas in the laboratory reaching an electron number density up to $\sim 10^{26}\,\mathrm{cm}^{-3}$ [6, 7]. Such developments open the door to new avenues, making it possible to understand the physics underlying various phenomena, experimentally probing the quantum plasma regimes and, ultimately exploiting such states for applications.

Ionized quantum matter is found naturally in dense astrophysical objects such as stellar cores, white and brown dwarfs, neutron stars, and interior of giant planets (e.g. Jovian planets) in the solar system [8]. The electrons in these systems constitute a degenerate plasma that often is under extreme conditions of density. Thereby, the electrons may be non-relativistic or relativistic, depending upon the ratio of the Fermi energy to the rest energy of an electron. At extremely high densities, exceeding the nuclear density, $n(\sim 10^{39}\ \mathrm{cm}^{-3})$, nuclei break apart (Mott-like transition) giving rise to a dense system of protons and neutrons. At still higher density even protons and neutrons break up, turning into the exotic quark-gluon plasma (QGP), a very special kind of quantum plasma where the particles interact via a (color) Coulomb potential. Such plasmas are believed to having existed immediately after the Big Bang [9], and seen in Relativistic Heavy Ion Collider (RHIC) and Large Hadron Collider (LHC) experiments. For an overview on the density–temperature range, see Fig. 4.1, for a more detailed introduction to quantum plasmas, see Ref. [1].

Basic parameters. In many-particle quantum systems, the mean particle distance $\bar{r_i} = [3n_i/4\pi]^{-1/3}$ of species i is comparable to or smaller than the de Broglie wavelength associated with the particle, $\Lambda_{B_i} = h/\sqrt{2\pi m_i k_B T_i}$, where m_i is particle mass and T_i the temperature, leading to an overlap of the wave functions of spatially extended mutually penetrating quantum particles, and the quantum degeneracy parameter exceeds unity, $\chi_i = n_i \Lambda_{B_i}^3 \geq 1$. For classical systems, one can define the Coulomb coupling parameter as the ratio of the average interaction energy $\langle U_{ii} \rangle = (e_i^2/4\pi\varepsilon)(1/\bar{r_i})$ and the average kinetic (thermal) energy $\langle K \rangle = E_{Ti}$ $(\sim k_B T_i)$, i.e. $\Gamma_i = |\langle U_{ii} \rangle| / E_{Ti}$, where ε is the static background dielectric constant. But, for sufficiently cold and dense plasmas which are quantum degenerate (assuming fermions), i.e. $\chi_i > 1$, the role of kinetic energy is taken over by the Fermi energy; $E_{Fi} = \hbar^2 \left(3\pi^2 n_i\right)^{2/3} / 2m_i$. In a quantum plasma, the strength of particle correlations is measured by the Brueckner parameter; $r_{si} = \bar{r_i}/a_{Bi}$, where $a_{Bi} = (\varepsilon/e_i^2)(\hbar^2/m_i)$ is the effective Bohr radius. It is easily verified that $r_{si} \propto \Gamma_{qi} = |\langle U_{ii} \rangle| / E_{Fi} = (\hbar\omega_{pi}/E_{Fi})^2$, where Γ_q is the quantum generalization of the Coulomb coupling parameter [10], and ω_{pi} is the plasma frequency given by $\omega_{pi}^2 = n_i e_i^2/\varepsilon_0 m_i$ for particles in vacuum (in three dimensions). The ideal behavior is recovered for $\chi_i \ll 1$, $\Gamma_i \ll 1$, in classical and $\chi_i \gg 1$, $r_{si} \ll 1$, in degenerate quantum plasmas. Both these limits correspond to a structureless gas-like system and are simple to analyze theoretically. The quantum coupling parameter shows the peculiar property of dense quantum systems: they become increasingly ideal with increasing density.[1] In contrast, in classical systems, the strength of correlations increases upon compression since the interaction energy increases as $n^{1/3}$ but thermal energy remains constant. Various plasma regimes are illustrated in the density–temperature phase diagram in the following section.

[1] Note that this is different from ultrarelativistic quantum plasmas where kinetic and interaction energy have the same scaling with density.

Additional parameters. For completeness we list further parameters of relevance in quantum plasmas. The Fermi energy is related to a characteristic velocity, wavenumber and length scale: the Fermi velocity, wavenumber and wave length, $v_F^2 = 2E_F/m$, $k_F^2 = 2mE_F/\hbar^2$ and $\lambda_F = 2\pi/k_F$. Furthermore, there exists a characteristic screening length—the Thomas–Fermi length λ_{TF} that replaces the Debye screening length of a classical plasma, $[\lambda_D^2 = k_B T/(2\pi e^2 n)]$, $\lambda_{TF} = v_F/(\sqrt{3}\omega_p)$.

Theoretical concepts for quantum plasmas. In quantum plasmas, strong inter-particle interactions at de-Broglie length scale impede the use of conventional classical theoretical models. That's why the early descriptions of the most immediate quantum plasma–the quantum electron gas in metals employed different approaches based on semiclassical or quantum mechanical methods including some fundamental works of the pioneers of the field [11–16]. To find the properties of quantum plasmas obeying FD or Bose–Einstein (BE) statistics, the N-particle Schrödinger equation is the key equation which describes the evolution of a general pure quantum state arising from some initial state whereas the dynamics of the system is governed by the Hamiltonian. In addition, the solution has to be anti-symmetrized for the case of fermions (symmetrized for bosons).

For quantum plasmas, usually the description in terms of mixed states is more appropriate. Then, instead of the wave function, the system is described by the density operator $\hat{\rho}$, and the von Neumann equation is the central equation that governs the dynamics of $\hat{\rho}$. For many-particle problems, the computational tools based on the (time-dependent) Hartree–Fock (HF, TDHF) method derivable from various techniques [1, 17] provide a useful path which allows for a solution of the many-particle Schrödinger equation in an approximate way, accurately describing and simulating the quantum and spin effects at weak coupling.

If coupling is strong quantum kinetic methods become very demanding. For equilibrium properties a powerful tool is Quantum Monte Carlo, for instance Path Integral Monte Carlo (PIMC) [1, 18, 19]. This method is a very successful first-principle approach avoiding model assumptions, and is well suited for bosonic particles. At the same time, for fermions, it is limited to small systems, due to the so-called fermion sign problem. Equilibrium properties of correlated quantum systems can also be described by quantum molecular dynamics (QMD) techniques which include, for instance, the Wigner function QMD [20], or classical MD with quantum and spin effects included via effective quantum potentials [21]. For equilibrium solutions, theories like the random phase approximation (RPA) and quantum mechanical modeling by density functional theory (DFT) [22, 23] are also very successful. The DFT has a vast range of applicability from atoms, molecules, solids to classical and quantum fluids, and is generalized to deal with many different situations. We further mention ideas to map a quantum system to an effective classical one due to Dharma wardana et al. [24, 25] and Dufty et al. [26–28].

The standard description of *non-equilibrium* quantum plasmas is based on kinetic theory which involves density matrices or phase space distribution functions of coordinates and momenta. The time evolution of the distribution function is given by a quantum kinetic equation (QKE) which differs from the corresponding classical kinetic equation in the appearance of the explicit difference of arguments in the potentials, creating a nonlocal coupling due to finite spatial extension of quantum particles [29]. The self-consistent kinetic modeling is one of the main tools in quantum plasma dynamics in which the notation of phase space is provided by the Wigner representation in terms of the density matrix. The QKE is a numerically expensive, integro-differential equation which provides the basis for various semiclassical approximations and computational schemes. Furthermore, for non-equilibrium processes, the widely applicable method of non-equilibrium Green's functions (NEGF) has allowed to achieve significant progress in the past few years [29–33]. It can successfully describe the ultrafast dynamics of many-particle systems and allows for a self-consistent treatment of the collective linear and nonlinear response of correlated Coulomb electron systems and non-perturbative inclusion of external fields and systematic many-body approximations via Feynman diagrams. It also offers an alternative formulation and extension of the TDHF method in terms of a generalized one-particle density matrix $G\left(x, t; \bar{x}, \bar{t}\right)$—the Green's function, which depends upon two space-time variables (in general, also including the spin projection), whose evolution is governed by the Kadanoff–Baym (KB) equations [34]. The KB method has been used to investigate the dielectric properties of plasmon oscillation spectrum with collision effects included in a systematic and consistent way in a correlated electron gas [35]. The developments in analytical and computational tools have led to a number of excellent textbooks including [1, 18, 29, 36] and review articles, for instance [37–43]. Finally, we mention that the progress is significant, however the solution and detailed analysis of QKE or the full description of many-particle wave functions have been major challenges from a theoretical perspective for the last several decades.

Owing to the analytical complexity of the quantum kinetic approach, drastically simplified macroscopic models (e.g. semiclassical molecular dynamics or quantum hydrodynamics) have been frequently adopted in recent years which can reproduce some of the salient features of quantum plasmas, although not providing the same detailed information which can be extracted from quantum kinetic theory. However, one has a choice with the alternative of studying a physical problem microscopically—with inherent technical difficulties—or macroscopically with a less cluttered and simpler approach which usually has a more restricted applicability range.

Out of the semiclassical approaches for theoretical description of quantum systems, Bohmian quantum mechanics and quantum hydrodynamics (QHD) have been widely used. The former considers real particles in the classical sense of having their configuration space trajectories determined by the Newtonian mechanics with specific positions and momenta (the so-called hidden variables). The latter is a more general method applicable to both pure and mixed states of quantum statistical systems. The QHD equations are usually obtained by taking moments of the appropriate kinetic equation (e.g. the Wigner function equation) in analogy with the moments of the classical kinetic equation. This leads to the conservation laws for particle number, momentum and energy in terms of macroscopic variables by choosing some suitable closure scheme in an approximate way (for details, see below).

Since the early introduction by Madelung [44], various versions of QHD have been developed and applied to many-particle bosonic and fermionic systems, some of them have been mentioned above. For instance, the QHD equations have been developed to study the dynamics of the quantum electron gas in metals and thin metal films [45–49]. For electrons in metals, the typical electron density $n_0 \simeq 10^{23}\,\mathrm{cm}^{-3}$ yields the quantum coupling parameter r_s of the order of unity which apparently shows that the collisionless models are not applicable to the metallic electrons. However, the $e-e$ collision rate (inverse of the electron lifetime τ_{ee}) is controlled by the process of Pauli blocking [45]. At room temperature, $\tau_{ee} \simeq 10^{-10}\,\mathrm{s}$ which is much larger than the typical collisionless time scale τ_p, the inverse of the electron plasma frequency, i.e. $\tau_p = \omega_p^{-1} \simeq 10^{-16}\,\mathrm{s}$. In addition, the typical relaxation time scale $\tau_r \simeq 10^{-14}\,\mathrm{s}$ is also larger than τ_p. Therefore, for time scales smaller than τ_{ee}, the electron collisions can be neglected and the collisionless models are appropriate. This standard justification of QHD, however, has to be considered with great caution as it assumes that the electron gas is in thermodynamic equilibrium. For example, laser excited metals with nonequilibrium carrier distributions may have much larger e-e scattering rates, despite the Pauli blocking mechanism.

Similarly, the hydrodynamics formulation is applicable to semiconductors which provides useful explanation of resonant tunneling processes and many ultrafast phenomena at ultrasmall scales [50, 51]. The model has also been extended to plasmonics, for instance, surface-plasmon dispersion [52], plasmonic device applications [53], and so on. For low-temperature bosons (e.g. Bose–Einstein condensates (BECs) in trapped Bose gases), the elementary excitations and related phenomena can be seen by employing the Gross–Pitaevskii theory in the spirit of QHD [54, 55]. The model has also been applied to high gain free electron lasers [56], and dense astrophysical plasmas [57], with the possibility of the inclusion of effects like relativity and magnetic fields. However, here as well one has to carefully examine the applicability limits of QHD. Many of the recent predictions of QHD have to be treated with great care as long as no experimental verification is possible or tests against more accurate kinetic approaches have not been made. This has to be reiterated since many of the QHD papers are neglecting these applicability limits and do not provide the necessary tests of their results, see below.

In this chapter, we review the main concepts and limitations of QHD and its validity in various applications starting from the simple case of the weakly coupled, non-relativistic plasma in the electrostatic limit. Since the topic of QHD is not new some obvious derivations are not included and the reader is referred to appropriate references. We start the introduction to the method with a brief note on the initial proposals (Sect. 1.2) and a discussion of the main assumptions and applicability conditions discussing electron and ion plasma waves within linearized QHD. In addition, a brief overview on some nonlinear solutions of QHD as well as results for a magnetized quantum plasma are included with a focus on the relevant low-frequency modes. The basic concepts of correlations and their implications in quantum plasmas are then introduced in Sect. 1.3. Finally, we discuss some recent problems that are related to an incorrect application of QHD to hydrogen bound states and spin effects in dense quantum plasmas. The intention of this chapter is to discuss the concepts in a pedagogical manner giving the interested readers recommendations for suitable additional references and text books for a more detailed study.

4.2 Basics of Quantum Hydrodynamics

The hydrodynamic formulation of systems which demonstrate behavior implicit in quantum mechanical subsystems is almost as old as the Schrödinger equation. It started in the early days of quantum mechanics when Madelung proposed that the Schrödinger equation for spinless one-electron problems can be transformed into the form of hydrodynamics equations. By taking the complex wave function of the form $\psi = \alpha e^{i\beta}$ with time-dependent, real valued α and β, he derived the continuity equation and Euler-like equation from the Schrödinger equation. Later on, after a long pause, Bohm and others played a major role in the further developments in this direction. This so-called Madelung hydrodynamics is usually considered as a precursor of the Bohmian mechanics—a quantum theory based on causal interpretation in terms of hidden variables [58–60] in which the reinterpretation of the solution of the Schrödinger equation and associated phenomena on the lines of classical dynamics was proposed. This interpretation is also known as de Broglie–Bohm theory due to the idea of the pilot-wave by L. de Broglie carried forward by Bohm to its logical conclusion.

4.2.1 The Time-Dependent Schrödinger Equation

We begin by writing down the N-particle Schrödinger equation

$$i\hbar \frac{\partial \Psi_N}{\partial t} = \hat{H}_N \Psi_N, \tag{4.1}$$

$$\hat{H}_N = \sum_{i=1}^{N} \left\{ -\frac{\hbar^2}{2m} \Delta_i + V(\mathbf{r}_i) \right\} \tag{4.2}$$

which is supplemented by the initial condition for Ψ_0 for the wave function at $t = 0$. In (4.2), Δ_i denotes the Laplace operator (second spatial derivative with respect to the coordinate of particle “i”). The first term in the sum represents the kinetic energy of the particles and $V(\mathbf{r}_i)$ the potential energy, just like in classical mechanics. Note that in (4.2) we neglect the interaction between the particles. Similarly, we have disregarded the spin variables which will be discussed later.

Now, according to the ideas of Madelung and Bohm, the solution of (4.1) for time-dependent (in general complex) N-particle wave function is constructed with the ansatz

$$\Psi_N(\{\mathbf{r}_i\}, t) = A(\{\mathbf{r}_i\}, t) \exp\left[\frac{i}{\hbar} S(\{\mathbf{r}_i\}, t)\right], \quad \text{where} \quad \{\mathbf{r}_i\} = \{r_1, r_2, \ldots, r_N\}, \tag{4.3}$$

with $A(\{\mathbf{r}_i\}, t)$ being the real amplitude function and $S(\{\mathbf{r}_i\}, t)$, the real phase. The statistical distribution of the trajectories determined from $|\Psi_N|^2$ gives the probability density that the particles are located at the coordinates $\{\mathbf{r}_i\}$ and thus the measurable physical features of the quantum system just like in classical statistics.

Inserting the ansatz (4.3) for the wave function into the time-dependent Schrödinger equation (4.1) and separating the real and imaginary parts, one obtains the following coupled equations for the two functions A and S

$$\partial_t S + \sum_{i=1}^{N} \left[\frac{1}{2m} \left(\nabla_{\mathbf{r}_i} S\right)^2 + V(\mathbf{r}_i) + Q(\{\mathbf{r}_i\}, t)\right] = 0, \tag{4.4}$$

and

$$\partial_t A + \frac{1}{2m} \sum_{i=1}^{N} \left[2 \left(\nabla_{\mathbf{r}_i} S\right) \cdot \nabla_{\mathbf{r}_i} A + A \nabla^2_{\mathbf{r}_i} S\right] = 0, \tag{4.5}$$

where, in addition to the conventional potential energy related to $V(\mathbf{r}_i)$ there arises a new term, $Q(\{\mathbf{r}_i\}, t)$, that can be understood as effective *quantum* potential (or Bohm potential) which is absent in the corresponding classical system.[2] Bohm noticed that (4.5) describes a conservation law of the probability density whereas (4.4) has the form of the classical Hamilton–Jacobi equation with the generating function $S(\{\mathbf{r}_i\}, t)$, with an additional term given by

$$Q(\{\mathbf{r}_i\}, t) = \frac{\hbar^2}{2m} \frac{\nabla^2_{\mathbf{r}_i} A}{A}. \tag{4.6}$$

The associated "effective Hamilton function" now contains a total potential that is the sum of the external potential and the quantum potential (summed over all particles) and depends on the dynamical variables $\{\mathbf{r}_i(t), \mathbf{p}_i(t)\}$ such that the quasi-trajectories may be found from

$$m\dot{\mathbf{r}}_i = \nabla_{\mathbf{r}_i} S(\{\mathbf{r}_i\}, t), \tag{4.7}$$

to yield $r_i = r_i(t)$ with initial position r_0, where $\mathbf{p}_i = \nabla_{\mathbf{r}_i} S(\{\mathbf{r}_i\}, t)$. The quasi-trajectories evolve under the influence of classical and quantum potentials just like in classical mechanics. The key difference is that the initial state of the particles is, in general, not given in a deterministic manner, but different initial coordinates occur with a finite probability that is given by the initial continuous wave function Ψ_0. A computational implementation of this scheme then requires a suitable statistical procedure: one has to consider an ensemble of trajectories which start from

[2] Note again that the interactions between the particles are neglected, hence all energy contributions are of single-particle type. We will discuss the role of interactions later, in Sect. 4.3.

different initial conditions that have probabilities (statistical weights) according to $|\Psi_0(\{\mathbf{r}_i\})|^2$. The resulting time-dependent wave function is then obtained according to the ensemble average of the individual trajectories with the same weights. The above introduced QHD-description by a well defined wave function can be extended to more general systems within a mixed state representation given by density matrix, for more details, see [29, 61, 62].

Another important remark has to be made. In fact, for a many-particle system (even in a pure state) the Schrödinger equation (4.1) does not provide the correct quantum result. Since quantum particles are either bosons or fermions which differ by their symmetry (in particular, the spin) the N-particle wave function has to be symmetric (anti-symmetric) for bosons (fermions), i.e. we need to apply a proper (anti-)symmetrization procedure, $\Psi_N \to \Psi_N^{A/S}$, see. e.g. [29, 37, 40]. This has the well-known effect that even noninteracting quantum particles become correlated with each other (or "entangled"). This is fully included in the quantum kinetic methods and simulations that were discussed in Sect. 4.2 but is rarely discussed when applying the QHD approach. We will return to this problem below in Sect. 4.4.

4.2.2 Quantum Mixed State Description: Wigner Function

In 1932, Wigner [63] suggested the phase space formulation of quantum mechanics, a representation by means of joint distributions of probabilities (more precisely, the quasi-probabilities) for coordinates and momenta in phase space which has led to another route to QHD. Wigner's original interest was to find quantum corrections to classical statistical theory where the Boltzmann factors contain energies expressible as functions of both coordinates and momenta. The Wigner function doesn't necessarily stay nonnegative in its evolution process for some regions of phase space due to restrictions on the simultaneous measurements of coordinate and momentum by the Heisenberg uncertainty principle. Unlike the classical case, it can therefore not be interpreted as a true probability density. However, it is real, normalizable to unity and gives averages just like the classical statistical distribution function. The Wigner formalism has attracted considerable attention in various disciplines of physics, and has also been the subject of a detailed theoretical analysis, in turn motivating the efforts to formulate various versions of quantum hydrodynamics due to the analogy with classical fluid systems. For more details on quasi-probability distribution functions and Wigner function method, see [10, 29, 64–67].

The Wigner function is a function of phase space variables $(\mathbf{r}, \mathbf{v})$ and time (in the following we will use velocities instead of momenta). For simplicity, we consider the one-dimensional problem for quantum statistical mixture of states $\{\psi_i(x,t), p_i\}$, $i = 1, 2, \ldots, K$, where each wave function $\psi_i(x,t)$ is assumed with a real non-negative probability p_i $(0 \le p_i \le 1)$ satisfying the normalization condition $\sum_{i=1}^{K} p_i = 1$. The results are straightforwardly generalized to higher dimensions.

When correlations are ignored, the many-particle wave function can be written as product of one-particle functions. Although we will follow this idea below because it is at the heart of the QHD approach, one has to clearly realize that this neglects the spin properties of quantum particles. For fermions (or bosons) the N-particle wave function—even if interactions are neglected—*is not the product of single-particle wave functions* but has the form of a Slater determinant (permanent). Therefore, all QHD results so far assume that the associated exchange corrections (terms additional to the simple product form) are not important. Note that there is no guarantee for that and, depending on the problem studied, the results may be quantitatively or even qualitatively wrong, in particular if spin effects are studied, see Sect. 4.4.

In this way, the quantum mixture (ensemble) of single-particle wave functions is now represented by the density matrix (this is the coordinate representation of a more general quantity—the density operator $\hat{\rho}$ [33])

$$\rho\left(x', x'', t\right) = \sum_{i=1}^{K} p_i \, \psi_i\left(x', t\right) \psi_i^*\left(x'', t\right), \tag{4.8}$$

where '$*$' denotes the complex conjugate, and the sum extends over all states contributing to the mixture. The next step is to introduce center of mass and relative coordinates, $x = (x' + x'')/2$ and $s = x' - x''$, respectively which allows to rewrite the original coordinates appearing in (4.8) as $x' = x + s/2$ and $x'' = x - s/2$. As in classical kinetic theory, these coordinates have different meanings: x is related to the position of a particle whereas the distance s is related to the internal structure and is the Fourier adjoint of the momentum.[3] Then, the Wigner function $f_W\left(x, v, t\right)$ can be written as the Fourier transform of (4.8) leading to

$$f_W\left(x, v, t\right) = \frac{m}{2\pi\hbar} \sum_{i=1}^{K} p_i \int_{-\infty}^{\infty} ds \psi_i^*\left(x + \frac{s}{2}, t\right) \psi_i\left(x - \frac{s}{2}, t\right) e^{imvs/\hbar}, \tag{4.9}$$

where m is the particle mass and $v = p/m$ is the velocity.

As a side remark we mention that for a general N-particle system with exchange and correlations described by a statistical mixture we also can compute the single-particle Wigner function. However, then the starting point is the ensemble of the full (anti-)symmetrized N-particle ensemble wave function $\Psi_{Ni}^{S/A}(x_1, x_2, \ldots, x_N, t)$. Then the single-particle Wigner function follows from integration over the variable of particles $2, 3, \ldots N$,

$$f_W^{S/A}\left(x_1, v_1, t\right) = \int dx_2 dv_2, \ldots, dx_N dv_N f_{WN}^{S/A}\left(x_1, v_1, \ldots, x_N, v_N, t\right), \tag{4.10}$$

[3] This is seen by considering a spatially homogeneous system. Then all points x are equivalent, and the dependence on x drops out, whereas the dependence on s remains.

where the integrand contains the (anti-)symmetrized N-particle Wigner function which is the Fourier transform of the (anti-)symmetrized N-particle density matrix as in the one-particle case above [33]. Instead of the known equation of motion for the N-particle density matrix—the von Neumann equation—one can also derive a chain of coupled equations for the one-particle, two-particle etc. functions exactly like in the classical case. This hierarchy is nothing but the quantum Bogoliubov–Born–Green–Kirkwood–Yvon (BBGKY) hierarchy of equations [33]. As in the classical case, this hierarchy can only be solved in special cases. In general, one has to resort to closure approximations. In most cases one expresses the two-particle function as a functional of single-particle functions motivated by physical information about the system. The problem is drastically simplified when particle correlations are neglected such that the two-particle Wigner function is approximated as $f_{W2}^{S/A}(x_1, v_1, x_2, v_2, t) = \Lambda^{S/A}\{f_W(x_1, v_1, t)\, f_W(x_2, v_2, t)\}$, where $\Lambda^{S/A}$ is the (anti-)symmetrization operator. If further, exchange (and spin) effects are being neglected we can drop the superscript "S/A", and the two-particle function is just the product of two one-particle functions. This is nothing but the quantum Vlasov (or Hartree) approximation which is commonly used to derive the QHD equations. This approximation means that the quantum plasma is considered as an ensemble of particles interacting through a mean field potential.

The equation of evolution for the one-particle Wigner function (4.9) for a scalar potential V included in (4.4) is given by

$$\begin{aligned} &\frac{\partial f_W}{\partial t} + v\frac{\partial f_W}{\partial x} \\ &-\frac{im}{2\pi\hbar^2}\int\!\!\int ds d\bar{v} e^{im(v-\bar{v})s/\hbar}\left[V^{\text{eff}}\left(x+\frac{s}{2}\right) - V^{\text{eff}}\left(x-\frac{s}{2}\right)\right] f_W(x, \bar{v}, t) = 0, \end{aligned} \tag{4.11}$$

and is obtained from the Wigner transform (4.9) of the von Neumann equation (or quantum Liouville equation) for the single-particle density matrix [33]. Here, $V^{\text{eff}} = V + V^{ind}$ is the total selfconsistent potential that contains the mean field potential V^{ind} (in the case of Coulomb interaction it is given by the solution of Poisson's equation) exactly like in the classical Vlasov equation.[4]

The underlying idea is that the quantum transport can be seeded into generalized kinetic equation in the spirit of the Boltzmann equation, appropriately extended with terms that represent quantum corrections. However, the resulting Wigner kinetic equation gives rise to a nonlocal dependence of distribution function on momentum (for details on quantum kinetic equation, see [29]). Due to the finite spatial extension of the quantum particles, the value of the potential energy at one space point also depends on the values of V at all other points—a pure quantum effect. The classical limit is recovered in the limit of vanishing difference of arguments of the two potentials.[5]

[4] The derivation of (4.11) follows from straightforward algebra and can be found in many text books, e.g. [29], and will, therefore, not be reproduced here.

[5] Then the integral term becomes $\frac{1}{m}\frac{dV^{\text{eff}}(x)}{dx}\frac{\partial f_W}{\partial v}$, as in the classical Vlasov equation.

4.2.3 Moments of the Wigner Function: Hydrodynamics

The QHD model [46] can be obtained by taking the moments of (4.11). Since, for all hydrodynamic approaches, the j-th order moment requires the knowledge of the j+1-th moment, an infinite chain of equations is found which demands a suitable truncation scheme. Generally, the lower-order moments are related to physically relevant quantities such as the particle density, average velocity, and pressure etc. For mixed states, the pressure tensor (in higher dimensions) requires the second moment of the Wigner function equation to couple to the third moment. The closure assumption allows to establish a relationship between the electron pressure and density demanding an appropriate equation of state. This is, in general, a subtle issue, however, if the system is in thermodynamic equilibrium (or sufficiently close) the known equilibrium results for the equation of state can be used.

Defining the macroscopic variables, i.e. density, mean velocity and pressure in the usual way (σ_u^2 is the variance of the velocity):

$$n\,(x,t) = \int f_W(x,v,t)dv, \quad u\,(x,t) = \frac{1}{n}\int f_W(x,v,t)\,v\,dv, \tag{4.12}$$

$$\frac{1}{m}p\,(x,t) = \sigma_u^2 = \int f_W(x,v,t)\,v^2 dv - n(x,t)u^2(x,t) \tag{4.13}$$

and representing each single-particle orbital

$$\psi_i(x,t) = A_i(x,t)\exp(iS_i(x,t)/\hbar), \tag{4.14}$$

with real amplitude $A_i(x,t)$ and real phase $S_i(x,t)$, the first two equations become

$$\frac{\partial n}{\partial t} + \frac{\partial\,(nu)}{\partial x} = 0, \tag{4.15}$$

$$m\left(\frac{\partial}{\partial t} + u\frac{\partial}{\partial x}\right)u = -\left(\frac{\partial V^{\mathrm{eff}}}{\partial x} + \frac{1}{n}\frac{\partial p}{\partial x}\right), \tag{4.16}$$

where $V^{\mathrm{eff}} = V + V^{ind}$ is the single-particle potential[6] and p is the total scalar pressure which consists of two terms, $p = p^c + p^q$, that will be discussed below.

In the following, we consider the simple case of electrons in an electrostatic field where the ions are treated as a homogeneous background. Then $V^{\mathrm{eff}} = e\varphi$ with $\varphi = \varphi^{ext} + \varphi^{ind}$, which leads from (4.16) to

$$\left(\frac{\partial}{\partial t} + u\frac{\partial}{\partial x}\right)u = \frac{e}{m}\frac{\partial\varphi}{\partial x} - \frac{1}{mn}\frac{\partial p^c}{\partial x} - \frac{1}{mn}\frac{\partial p^q}{\partial x}, \tag{4.17}$$

[6] V is in fact non-local, which can be seen from (4.11), but this is neglected in the hydrodynamic formulation.

where the induced electrostatic potential φ^{ind} obeys Poisson's equation

$$\frac{\partial^2 \varphi^{ind}(x,t)}{\partial x^2} = \frac{e}{\varepsilon_0}\left(\int dv\, f_W(x,v,t) - n_0\right), \tag{4.18}$$

with ε_0 and n_0 being the vacuum dielectric constant and uniform background ion density, respectively.

In deriving the above QHD equations for the mean density n and mean velocity u, (4.12), it is now crucial to have a prescription how to connect them with the number density $n_i\,(x,t)$ and velocity $u_i\,(x,t)$ for each individual orbital. The latter are defined from the wave function (4.14) of each individual orbital according to $n_i\,(x,t) = |\psi_i(x,t)|^2 = A_i^2(x,t)$ and $u_i\,(x,t) = \partial_x S_i(x,t)/m$, i.e. just like in the case that the system is in a pure state ψ_i (as in Madelung's theory). This connection follows readily from the definition of the density matrix (4.8). Thus the average with the Wigner function can be expressed as an ensemble average

$$\langle \ldots \rangle = \sum_{i=1}^{K} p_i \ldots \tag{4.19}$$

where the contribution of each orbital enters with the weight p_i. Thus, for the mean density we obtain

$$n(x,t) = \int dv\, f_W(x,v,t) = \sum_{i=1}^{K} p_i \int dv\, f_{Wi}(x,v,t) = \langle n_i(x,t) \rangle, \tag{4.20}$$

and for the mean velocity follows analogously

$$\begin{aligned} n(x,t)u(x,t) &= \int dv\, v\, f_W(x,v,t) = \sum_{i=1}^{K} p_i \int dv\, v\, f_{Wi}(x,v,t) \\ &= \sum_{i=1}^{K} p_i\, n_i(x,t)u_i(x,t) = \langle n_i(x,t)u_i(x,t) \rangle. \end{aligned} \tag{4.21}$$

Finally, we obtain for the pressure from (4.13) two contributions: the first is the same as in classical hydrodynamics,

$$\frac{p^c(x,t)}{m} = \sigma_u^2 = \sum_{i=1}^{K} p_i \left(\int dv v^2 f_{Wi}(x,v,t) - n_i(x,t)u_i^2(x,t)\right), \tag{4.22}$$

and is given by the dispersion of the velocities. The orbital densities and velocities will be eliminated from this expression below by postulating a suitable equation of state, see (4.26).

In contrast to classical hydrodynamics, here appears a second contribution to the pressure that arises from the coordinate dependence of the orbital amplitudes in (4.14),

$$p^q(x,t) = \frac{\hbar^2}{2m}\sum_{i=1}^{K} p_i \left[\left(\frac{\partial A_i}{\partial x}\right)^2 - A_i \frac{\partial^2 A_i}{\partial x^2}\right], \tag{4.23}$$

which has been called *quantum pressure* [46].

Thus the mean values n, u, p^c and p^q can be computed if the wave functions of all orbitals and, hence, the density matrix (Wigner function) are known, which is in general a very difficult task. Instead one can try to get a closed set of hydrodynamic equations, by invoking an equation of state (hydrodynamic closure relation) that relates p^c and p^q to the macroscopic density $n\,(x,t)$, thereby eliminating the individual n_i. In Ref. [46] a very simple solution was proposed: The authors assumed a particular statistical mixture of states in which *all single-electron wave functions (orbitals)* ψ_i *have identical amplitudes* that are allowed to be space-dependent, i.e. $A_i\,(x) = A\,(x)\,,\ i = 1,\ldots,K$. At the same time the different ψ_i are allowed to have different phases, S_i, that are related to the mean orbital velocity u_i through the relation $mu_i = \partial S_i/\partial x$ whereas the u_i are related to the global mean velocity u via relation (4.21). This condition with the help of (4.9) and (4.13) gives the density $n = A^2$. Also, this can be understood as an assumption of uncorrelated electrons where the spatial distribution of each electron defined by the amplitude A_i doesn't depend upon the spatial distribution of the other electrons in the system [10]. This is a key assumption of QHD for a many-fermion system, and we will discuss and test it more in detail in Sect. 4.2.4.

With this assumption, the relation for p^q can be rewritten as

$$p^q(n) = \frac{\hbar^2}{2m}\left[\left(\frac{\partial}{\partial x}\sqrt{n}\right)^2 - \sqrt{n}\frac{\partial^2}{\partial x^2}\sqrt{n}\right], \tag{4.24}$$

where the last term in (4.17) turns out to be

$$\frac{1}{mn}\frac{\partial p^q}{\partial x} = -\frac{\hbar^2}{2m^2}\frac{\partial}{\partial x}\left(\frac{1}{\sqrt{n}}\frac{\partial^2\sqrt{n}}{\partial x^2}\right). \tag{4.25}$$

When compared to classical fluid equations for electrostatic plasmas, the main difference is the Bohm potential term (4.25) which takes the role of an additional pressure. It is not a true pressure in thermodynamic sense since it involves no velocity averages. In contrast, it is caused by the quantum kinetic energy (which is proportional to minus the Laplacian of the wave function) having the effect of particle spreading (quantum diffraction, tunneling) which is formally equivalent to a positive pressure.

To relate p^c with the macroscopic density, a useful choice is the equation of state for strongly degenerate (D-dimensional) fermion system in thermodynamic equilibrium (we restrict ourselves to zero-temperature)[7]

$$p^c(x) = p_F^D(n_0) \cdot \left(\frac{n(x)}{n_0}\right)^{5/3}, \tag{4.26}$$

$$p_F^D(n_0) = \frac{2}{D+2} n_0 E_F(n_0) = n_0 \frac{m v_F^2}{D+2}, \tag{4.27}$$

where $v_F = \hbar k_F/m$ is the Fermi velocity defined via the electron Fermi wave number, $k_F = \left(3\pi^2 n\right)^{1/3}$. Here $E_F(n_0)$ is the Fermi energy of electrons in a homogeneous system of density n_0, and $p_F^D(n_0)$ is the associated Fermi pressure of an ideal Fermi gas at zero temperature. Note that the relation (4.27) between pressure p_F^D and energy density $n_0 E_F$ is exact for a non-relativistic ideal Fermi gas at $T = 0$. The expression (4.26), on the other hand, extends this result to an inhomogeneous system via the local approximation.[8]

Using the assumption of identical orbital amplitudes, allows to reformulate the quantum hydrodynamic equations (4.15) and (4.17) as a nonlinear Schrödinger (NLS) equation [46]

$$i\hbar \frac{\partial \Psi}{\partial t} = -\frac{\hbar^2}{2m}\frac{\partial^2 \Psi}{\partial x^2} - e\varphi\Psi + \tilde{V}\Psi, \tag{4.28}$$

where an effective wave function $\Psi(x,t) = \sqrt{n(x,t)}\exp(iS(x,t)/\hbar)$ is defined with $mu(x,t) = \partial S/\partial x$, $n(x,t) = |\Psi|^2$ and $\tilde{V} = \frac{m v_F^2}{3 n_0^2}|\Psi|^4$. The nonlinear Schrödinger–Poisson system captures the nonlinear interaction between the electron density fluctuations and the electrostatic potential. The NLS equation is easily amenable to numerical analysis and its generalization can describe the behavior of bosonic systems as well.

A related approach to derive the hydrodynamics equations is based on the Dawson (classical) multistream model [68] which is extended to the quantum case [69] by considering a statistical mixture of K pure states representing K "streams" of particles each characterized by the same velocity. Following the Hartree representation (well known in condensed matter physics), the states each with wave functions $\psi_i(x,t), i = 1, \ldots, K$ obey K independent Schrödinger equations that are coupled via the electrostatic potential[9]

$$i\hbar \frac{\partial \psi_i}{\partial t} = -\frac{\hbar^2}{2m}\frac{\partial^2 \psi_i}{\partial x^2} - e\varphi\psi_i, \qquad i = 1, \ldots, K. \tag{4.29}$$

---

[7] The frequently used notion 'classical' for $p^c(n)$ is somewhat misleading because it contains $\hbar$ through v_F. However it is analogous in the sense of measurement of the velocity dispersion.

[8] An alternative choice is a cubic dependence on $n(x)/n_0$ which is motivated in Ref. [10] by assuming an adiabatic equation of state.

[9] We mention again, that this neglects exchange. Also the picture of "streams" of particles with the same velocity is—strictly speaking—not compatible with the Pauli principle.

Introducing the Madelung representation of wave function (4.14) in (4.29) and separating the real and imaginary parts, it reduces to the hydrodynamic equations, the continuity equation (4.15) and an Euler-like equation

$$\left(\frac{\partial}{\partial t}+u_i\frac{\partial}{\partial x}\right)u_i=-\frac{e}{m}\frac{\partial\varphi}{\partial x}+\frac{\hbar^2}{2m^2}\frac{\partial}{\partial x}\left(\frac{\partial^2\sqrt{n_i}/\partial x^2}{\sqrt{n_i}}\right), \tag{4.30}$$

where $n_i = A_i^2$. Setting $\hbar = 0$, the classical Dawson relation [68] is retrieved. Although the equation (4.30) takes into account the quantum diffraction effects, its limitations are the same as those of (4.17) and are described in the following section.

4.2.4 Examples and Test of the Assumption $A_i(x,t) = A(x,t)$

The assumption that all orbitals have the same space and time-dependent amplitude, $A_i(x,t) = A(x,t)$, is a key assumption of QHD for a many-fermion system at zero temperature.[10] It is, therefore, important to verify it. In fact, the single-particle wave functions ψ_i are often easily found, so it is possible to determine the amplitudes A_i and phases S_i, explicitly. We will do this for three typical examples where, for simplicity we consider the one-dimensional case.

I. Homogeneous free electron gas. We consider N particles in a box of length $2L$ with $-L \le x \le L$. To model a macroscopic system, periodic boundary conditions are implied, i.e. $\psi_i(-L) = \psi_i(L)$, for all i. If necessary, in the end one can take the limit $L \to \infty$ and $N \to \infty$, while maintaining a constant density $n_0 = N/2L$ =const. The solutions of the one-particle Schrödinger equation that satisfy these boundary conditions are well known ($E_l = \hbar^2 k_l^2/2m$):

$$\psi_l(x,t) = \frac{1}{\sqrt{2L}}e^{-i(E_l t - k_l x)/\hbar}, \quad k_l = l k_0, \quad k_0 = \pi/L, \quad l = \pm1, \pm2, \ldots$$

$$A_l(x,t) = A(x,t) = \frac{1}{\sqrt{2L}} = \text{const}, \quad S_l(x,t) = -E_l t + \hbar k_l x. \tag{4.31}$$

Evidently, the amplitudes of all orbitals are equal.

II. Non-interacting electrons in a deep potential well. Consider now the situation that the electrons are confined to a box of length $2L$ with $-L \le x \le L$, where the potential walls are assumed infinitely high. Then the electrons cannot penetrate into the regions $x < -L$ and $x > L$ which leads to the boundary condition on the wave functions $\psi_l(-L,t) = \psi_l(L,t) = 0$, for all l and t. Again, the solutions to this problem are well known from basic quantum mechanics. As usually, the time-dependent

---

[10] It is also an assumption for a single particle at finite T.

solution is $\psi_l(x,t) = \psi_l(x)e^{-iE_l t}$, where the stationary solution is $(k_0 = \pi/2L)$

$$\begin{aligned}
\psi_l(x) &= \frac{1}{\sqrt{L}} \cos k_l x, \qquad l = \pm 1, \pm 3, \cdots, \\
\psi_l(x) &= \frac{1}{\sqrt{L}} \sin k_l x, \qquad l = \pm 2, \pm 4, \cdots, \\
A_l(x,t) &= \psi_l(x), \qquad S_l(x,t) = -E_l t.
\end{aligned} \tag{4.32}$$

In this case, the amplitudes A_l of the orbitals are time-independent, but all completely different, in striking contrast of the main assumption of QHD.

III. Non-interacting electrons in a harmonic oscillator potential. The previous case was characterized by a discontinuous change of the external potential in space. Now we consider the case of a smooth potential that is a quadratic function of the coordinate, $V(x) = m\omega^2 x^2/2$. Then, the wave function can extend into the whole space where, due to particle number conservation (normalization condition), it should vanish sufficiently fast for $|x| \to \infty$. The solution is again well known, and the stationary wave functions are given by the Hermite polynomials H_l,

$$\begin{aligned}
\psi_l(x) &= \frac{1}{\sqrt{2^l l! \sqrt{\pi} x_0}} H_l(u) e^{-\frac{u^2}{2}}, \qquad u = \frac{x}{x_0}, \quad x_0 = \sqrt{\frac{\hbar}{m\omega}}, \quad l = 0, 1, 2, \ldots \\
A_l(x,t) &= \psi_l(x), \qquad S_l(x,t) = -E_l t.
\end{aligned} \tag{4.33}$$

As in the second case, the amplitudes of the orbitals are time-independent, but all different.

It is easy to understand the origin of this behavior. In case I the solutions are freely propagating waves described by complex wave functions, and the amplitudes are equal. In contrast, cases II and III correspond to bound states, where the electron motion is spatially restricted. Correspondingly, the stationary wave functions are real[11] and the phases S_l are just determined by the time-dependent exponential whereas the amplitudes are all different. This also affects the quasi-classical velocities given by $mu_l = dS_l/dx$. In case I we obtain $u_l = \hbar k_l/m$, whereas in cases II and III $u_l = 0$ since S_l is independent of x for all l. This is due to the fact, that bound wave functions correspond to standing waves with zero mean momentum.

The most important conclusion of these examples is that the *key assumption of QHD is not fulfilled for spatially confined electrons*. While it is fulfilled for an infinite (noninteracting and spinless) system, this case does never occur in a real plasma. Furthermore, when studying waves in quantum plasmas we are interested in the behavior of electrons in the presence of an external potential well which gives rise to spatial confinement effects. Therefore, the case of confined particles is of particular importance.

---

[11] This is not universal. For example for Coulomb bound states, the wave function is complex (the angular part).

It turns out that while the condition of equal orbital amplitudes [10, 66], $A_l(x,t) = \sqrt{n(x,t)}$, for all l, is certainly sufficient for the validity of the QHD equations (together with an appropriate equation of state), it is—most likely—not necessary. In other words, the QHD equations may also be satisfied if the condition $A_l(x,t) = \sqrt{n(x,t)}$ is not satisfied—which is practically never fulfilled. What is necessary for the QHD equations is that all ensemble averages can be replaced by mean quantities, in particular

$$\frac{2m}{\hbar^2} p^q(x,t) = \left\langle \left(\frac{\partial A_i}{\partial x}\right)^2 - A_i \frac{\partial^2 A_i}{\partial x^2} \right\rangle \Rightarrow \left(\frac{\partial}{\partial x}\sqrt{n}\right)^2 - \sqrt{n}\frac{\partial^2}{\partial x^2}\sqrt{n}, \tag{4.34}$$

where $\langle \ldots \rangle$ denotes an average over the orbitals with their weights p_i. These replacements can be understood as an averaging procedure. Furthermore, note that the terms on the left are rapidly varying in space (at least for the examples II. and III.) because the amplitudes A_l are oscillating increasingly rapidly with growing l. This is not the case for the terms on the right, i.e. for the mean density n. Therefore, these replacements imply a suitable spatial average which—in principle—is consistent with the concept of a hydrodynamic approach. For the example II, an average over the spatial period $\lambda_l = 2\pi/k_l$ of the square of the amplitude A_l^2 yields $1/2$, for all l. However, problems remain: it is not clear how to systematically choose a *single length scale* averaging over which would apply to *all* (or at least most) orbitals. In some cases, the Thomas–Fermi length λ_{TF}, (or a multiple of it) may be the proper scale as was found e.g. in Ref. [70]. Then for all orbitals with $\lambda_l \leq \lambda_{TF}$ an averaging occurs with $A_l^2 \Rightarrow 1/2$. Furthermore, the validity condition of the assumption (4.34) remains open.

4.2.5 Main Assumptions and Applicability Conditions of QHD

The set of equations (4.15), (4.17) and (4.18) constitutes a reduced model whose validity rests on several assumptions, thus imposing important limitations on the model, as described below. In the following we assume that only the electrons are quantum degenerate, so all these conditions apply to the electron component. Generalizations to several quantum components are straightforward.

(i) The plasma is ideal (weakly coupled) which means all types of interactions (or collision effects) are much weaker than the quantum kinetic energy i.e.

$$r_s \ll 1, \quad \text{or, equivalently,} \quad \Gamma_q \ll 1. \tag{4.35}$$

(ii) The interaction of the particles is treated in mean field approximation and described by the induced electrostatic potential. No electromagnetic and quantization effects are considered.

(iii) The wave phase velocities (as well as the particle velocities) are non-relativistic i.e. $\omega/k < c$.

(iv) The resolvable length scales in QHD are large ($L >$ several $\bar{r}$) which means the equations are applicable to the long wavelength limit only i.e. $\lambda \gg \lambda_{TF}$, or alternatively $k \ll \omega_p/v_F$. Length scales shorter than the Thomas–Fermi screening length $\lambda_{TF}\left(= v_F/\sqrt{3}\omega_p\right)$, obviously, cannot be resolved. This gives rise to a small parameter,[12]

$$\kappa \equiv k\lambda_{TF} \ll 1. \tag{4.36}$$

Kinetic phenomena such as Landau damping cannot be described by QHD since they depend on the details of the equilibrium Wigner function and have to be treated with kinetic theory.

(v) The equation of state of an ideal Fermi gas at $T = 0$ is used within the local approximation. The extension to space-dependent density profiles is done by introducing a factor $(n(x)/n_0)^{5/3}$, (4.26). This again imposes restrictions on the resolvable length scales. In particular, rapidly varying density profile (e.g. in the case of steep external potentials) cannot be treated properly because the non-locality would give rise to large gradient effects. The zero-temperature assumption requires that $T \ll T_F$, otherwise there will be corrections from the Sommerfeld expansion of the Fermi function, e.g. to the equation of state.

(vi) No energy transport equation is considered. This could be done by taking the second order moment of the Wigner function equation.

(vii) The model is approximate via the Bohm potential. The closed set of equations follows if the amplitudes $A_i(x, t)$ of all single-particle orbitals are equal (but not necessarily constant). These orbitals can have different phases $S_i(x, t)$ which are related to the mean orbital velocity through $mu_i = \partial S_i/\partial x$, and u_i is related to the global mean velocity u defined in (4.12). This in turn implies the same macroscopic density $n(x) = A^2(x)$. A less restrictive condition is given by (4.34), see the examples and discussion in Sect. 4.2.4.

(viii) No spin effects are taken into account. However, the inclusion of a magnetic field in QHD is straightforward by starting from a quantum kinetic equation with an electromagnetic field included (by any gauge), as is briefly discussed on the following pages.

As is often the case, a physical model may be valid even beyond its formal conditions of applicability. This is also sometimes the case with QHD which may give reasonable results even beyond the conditions listed above. However, there is no

---

[12] The long wavelength assumption is also evident from analogy with the classical case. The assumption of the classical pressure $p^c = mn\left(\langle u_i^2\rangle - \langle u_i\rangle^2\right)$ leads to the equation of state, $p^c = p^c(n)$ for dense degenerate electrons at $k_B T \ll E_F$. This in turn demands the condition $k\lambda_{TF} \ll 1$ to describe the wavelengths within the QHD just like the classical fluid condition $k\lambda_D \ll 1$ with $\lambda_D\left(= k_B T/\sqrt{4\pi n e^2}\right)$ being the Debye length.

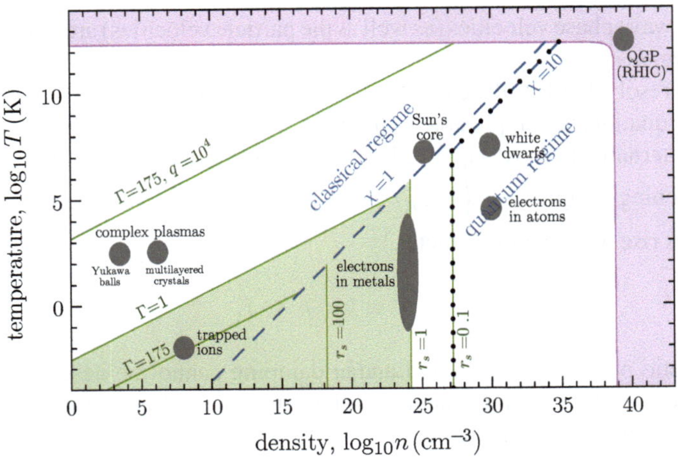

Fig. 4.1 Density–temperature phase diagram of a one-component plasma, for example electrons in a neutralizing positive background. The *green triangle* corresponds to strongly correlated electrons, below (*left* from) the line $\Gamma = 175$ ($r_s = 100$) electrons form a Wigner crystal. The *green line* $\Gamma = 175$, $q = 10^4$ corresponds to crystallization in dusty plasmas containing particles with charge $q = 10{,}000e_0$. Quantum effects are relevant to the *right* of the *dashed line* which is given by $\chi = 1$. Some occurrences of quantum plasmas are noted in the figure. QGP denotes the quark-gluon plasma that (is thought to have) existed shortly after the Big Bang and which also was produced at the relativistic heavy ion collider in Brookhaven (RIC) and at the LHC at CERN. The restriction of QHD to weak coupling, $r_s < 0.1$, corresponds to densities larger than $10^{27}\,\mathrm{cm}^{-3}$. The restriction to the ground state ($T = 0$) requires at least $\chi > 10$. Note that this makes applications of QHD to electrons in metals or to warm dense matter very questionable. This range is indicated by the *dotted line*

guarantee for this, and a careful analysis of the relevant conditions should always be performed.

In what follows, we have shown in Fig. 4.1 the regions of applicability of the QHD through the density–temperature phase diagram.

The properties of quantum electron gas can be measured with good accuracy in hydrodynamic approximation and provide an ease to search and analyze the linear waves and instabilities which gives insight of the main role of quantum effects.

4.2.6 Linearized QHD: Linear Waves of Quantum Plasmas

We begin the applications of QHD by considering the linear response of electrons in a quantum plasma to a weak external excitation. Then the QHD equations can be linearized allowing to compute a dielectric function from which the plasmon spectrum is straightforwardly obtained. The problem to study first is electron plasma oscillations where many results exist against which the QHD result can be directly tested.

4.2.6.1 Electron Plasma Waves

In order to apply the quantum hydrodynamics equations, consider a zero-temperature fermion gas in one spatial dimension with the pressure given by (4.27) as follows

$$p^c(n) = \frac{mv_F^2}{3n_0^2}n^3. \tag{4.37}$$

The electron dynamics are governed by equations (4.15), (4.17) and (4.18) whereas the ions are considered immobile, forming a neutralizing background. Linearizing the equations around the homogenous equilibrium; $n = n_0$, $u = 0$ and $\varphi = 0$, and Fourier analyzing as usual with small fluctuating quantities, n_1, u_1, ϕ_1, expressed as $\exp\left[i\left(kx - \omega t\right)\right]$, one obtains the dispersion relation

$$\omega^2 = \omega_p^2 + k^2v_F^2 + \frac{\hbar^2k^4}{4m^2} = \omega_p^2 + k^2v_F^2\left(1 + \frac{3}{16}\Gamma_q\kappa^2\right), \tag{4.38}$$

where ω is the wave frequency and k the wave number. This relation is also derivable from the NLS-Poisson equations (4.18) and (4.28) in the linear limit. While it is nice to keep in the dispersion relation (4.38) the fourth order term in k one has to clearly remember the limitations of QHD, see above. Indeed, the last equality in (4.38) shows that the k^4 term is about three orders of magnitude smaller (considering that $\Gamma_q \ll 1$ and $\kappa \ll 1$) than the k^2 term and there is no justification to retain it within QHD.

We now want to compare (4.38) with the result obtained from the Wigner–Poisson model (4.11) and (4.18). Assuming that the potential in (4.11) depends on one coordinate only, say x, we approximate the evolution equation up to $O\left(\hbar^2\right)$ given by [29]

$$\frac{\partial f_W}{\partial t} + v\frac{\partial f_W}{\partial x} + \frac{e}{m}\frac{\partial\varphi}{\partial x}\frac{\partial f_W}{\partial v} = \frac{e\hbar^2}{24m^3}\frac{\partial^3\varphi}{\partial x^3}\frac{\partial^3 f_W}{\partial v^3} + O(\hbar^4). \tag{4.39}$$

The right-hand side of (4.39) is due to the non-locality of the potential in the equation for the Wigner function (4.11). It is now easy to see that, in the limit $\hbar \to 0$, one recovers the familiar Vlasov equation for a classical collisionless plasma. The result can be found in perturbation theory and using a Fourier decomposition of the perturbations. Considering the contribution from a monochromatic perturbation proportional to $\exp\left[i\left(kx - \omega t\right)\right]$, i.e.

$$\begin{aligned} f_W\left(x, v, t\right) &= f_0\left(v\right) + f_1\left(v\right)\exp\left[i\left(kx - \omega t\right)\right], \\ \varphi(x, t) &= \varphi_1 \exp\left[i\left(kx - \omega t\right)\right], \end{aligned} \tag{4.40}$$

where f_1 and φ_1 are first order perturbed quantities and $|f_1| \ll f_0$. It leads to the dispersion relation $\varepsilon\left(\omega, k\right) = 0$, where the dielectric function ε for the Wigner–Poisson system reads,[13]

[13] Here we assume that ω contains an infinitely small imaginary part in order to assure causality (Landau pole integration). i.e. ε is understood as a retarded quantity.

$$\varepsilon(\omega, k) = 1 - \frac{m\omega_p^2}{n_0 \hbar k^2} \int \frac{f_0(v + \hbar k/2m) - f_0(v - \hbar k/2m)}{kv - \omega} dv. \tag{4.41}$$

With a suitable change of variables,[14] the dispersion relation for high frequency electron plasma oscillations becomes

$$\varepsilon(\omega, k) = 1 - \frac{\omega_p^2}{n_0} \int \frac{f_0(v)}{(\omega - kv)^2 - (\hbar^2 k^4)/4m^2} dv = 0. \tag{4.42}$$

This is just the Lindhard dispersion relation [71] which is well known in solid state physics. In the one-dimensional case, the equilibrium Wigner function for a fully degenerate Fermi gas is given by

$$f_0(v) = \frac{n_0}{2v_F}, \qquad |v| < v_F, \tag{4.43}$$

$$= 0, \qquad |v| > v_F, \tag{4.44}$$

which leads to [29]

$$\varepsilon(\omega, k) = 1 - \frac{m\omega_p^2}{2\hbar k^3 v_F} \ln \left| \frac{\omega^2 - (kv_F - \hbar k^2/2m)^2}{\omega^2 - (kv_F + \hbar k^2/2m)^2} \right| = 0. \tag{4.45}$$

In the long wavelength limit, $kv_F \ll \omega$, $\hbar k^2/2m \ll \omega$, expansion of $\varepsilon(\omega, k)$ in (4.45) gives (4.38), the limit of the kinetic dispersion relation for small wave numbers. Like for the one-dimensional case, the equilibrium function equals zero for $v > v_F$. Only for smaller velocities $|\mathbf{v}| < v_F$, the absolute value differs from 1D: $f_0(\mathbf{v}) = \frac{n_0}{\pi v_F^2}$, for $D = 2$, and $f_0(\mathbf{v}) = \frac{n_0}{4\pi v_F^3/3}$, for $D = 3$, reflecting the different normalization conditions. So, combining the results for the different dimensions, the dispersion relation for a fully degenerate Fermi gas in the long wavelength limit takes the form

$$\omega^2 = \omega_p^2 + \left(\frac{3}{D+2}\right) k^2 v_F^2 + \frac{\hbar^2 k^4}{4m^2}, \tag{4.46}$$

which should be reproduced by the hydrodynamic equations. Quantum mechanical effects enter this result in two distinct ways: the first is statistical in the sense that the equilibrium distribution is the Fermi distribution, and the second is quantum dynamical [the last term in (4.46)], arising from the energy associated with the finite momentum transfer $\hbar k$ of an electron interacting with a plasma oscillation.[15]

---

[14] i.e. Changing the integration variable according to $v \to v \mp \hbar k/2m$) and bringing both terms to a common denominator.

[15] The quantum picture describes this as scattering of an electron with a quantum particle—the plasmon. This momentum change appears in the arguments of the distribution functions in (4.41). The classical limit is obtained from formally letting $\hbar \to 0$, then the difference of distribution

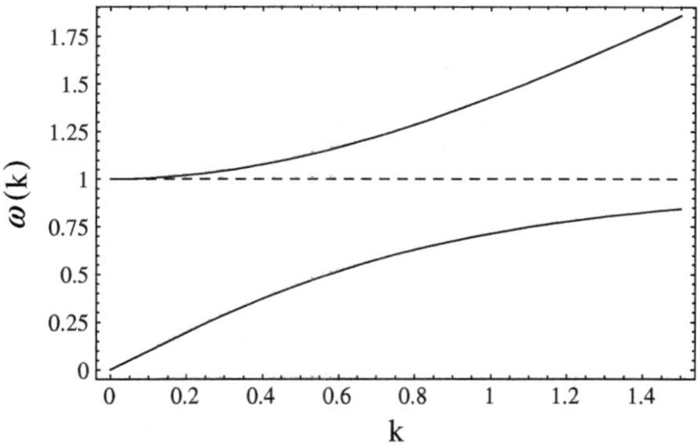

Fig. 4.2 The dispersion relations (4.38) for electrons (*upper curve*) and (4.66) for ions (*lower curve*) are shown for the case of strong degeneracy, ($E_T \ll E_F$). The wave frequency for electrons (ions) is normalized by $\omega_p(\Omega_p)$ and the wave number by $\omega_p/v_F(\Omega_p/c_q)$

We note that this type of dispersion of electron plasma oscillations (the quantum Langmuir-like wave shown in Fig. 4.2) is not new and has already been found by Klimontovich and Silin [12] by using the Wigner distribution function, and by Bohm and Pines [11] by developing canonical transformations of the Hamiltonian of the system of electrons interacting through the electrostatic force.

In analyzing the dispersive properties of quantum plasmas, the coupling and degeneracy parameters play a key role in choosing an appropriate model [72]. The quantum coupling parameter r_s is a function of density only which shows that the higher is the density of quantum particles, the weaker are the correlations in the system. Some important parameters related to typical degenerate laboratory and astrophysical plasmas are given in the following tables (Tables 4.1 and 4.2).

The significance of quantum dispersion effects of electron plasma oscillations have been observed in solid-density plasmas. The plasma compression experiments show that the plasmon frequency is a sensitive measure of the electron density and the plasmon dispersion relation includes the Fermi degeneracy effects. In these experiments the temperature is finite and the above result is not applicable. On the other hand, the plasmon dispersion of a classical plasma is well known. It starts from the plasma frequency (for $k = 0$) as well and then ω^2 increases proportional to $k^2 v_{th}^2$ [$v_{th} = (k_B T/m)^{1/2}$ is the electron thermal velocity]—the so-called Bohm–Gross dispersion. If, at finite T, quantum effects become relevant one obtains the modified Bohm–Gross relation for small k that contains quantum corrections [6]:

$$\omega^2 = \omega_p^2 + 3k^2 v_{th}^2 \left(1 + 0.088 n \Lambda_B^3\right) + \frac{\hbar^2 k^4}{4m^2}, \tag{4.47}$$

---

functions turns into a derivative with respect to momentum, and one recovers the classicial Vlasov dielectric function.

Table 4.1 The parameters of a degenerate electron gas with number densities of the order of metallic electrons with large but constant degeneracy parameter ($\chi \gg 1$)

$n(10^{23}\,\text{cm}^{-3})$	$\bar{r}(10^{-9}\,\text{cm})$	r_s	T (K)	$\omega_p(10^{16}\,\text{s}^{-1})$	$E_F(10^{-11}\,\text{erg})$	$E_T(10^{-14}\,\text{erg})$
3.0	9.2	1.10	300	2.4	2.6	4.4
3.7	8.6	1.02	350	2.7	3.0	4.8
4.6	8.0	0.95	400	3.0	3.5	5.5
5.5	7.6	0.90	450	3.3	3.9	6.2
6.5	7.2	0.84	500	3.6	4.3	6.9

$E_T = k_B T$. Since the electron gas is moderately coupled ($r_s \sim 1$), this system is, strictly speaking, not accessible to QHD

Table 4.2 Typical parameter range of a high-density degenerate electron gas found in compact astrophysical systems such as dwarf stars

r_s	n (cm^{-3})	$\bar{r}$ (10^{-10} cm)	E_F (erg)	v_F (cm/s)	$\lambda_{TF}(10^{-8}$ cm)
0.10	4.0×10^{26}	8.3	3.1×10^{-9}	2.6×10^{9}	7.2
0.08	7.8×10^{26}	6.7	4.9×10^{-9}	3.2×10^{9}	6.5
0.06	1.8×10^{27}	5.0	8.8×10^{-9}	4.3×10^{9}	5.7
0.04	6.3×10^{27}	3.3	1.9×10^{-8}	6.5×10^{9}	4.6
0.02	5.0×10^{28}	1.6	7.9×10^{-8}	1.3×10^{10}	3.2

These systems are weakly coupled and are well suited for a QHD description

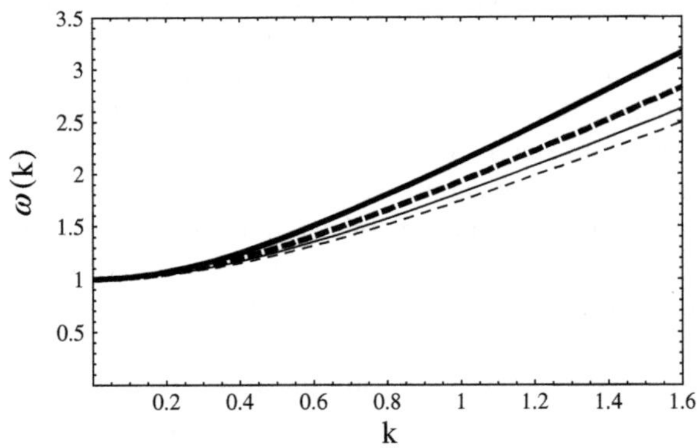

Fig. 4.3 The modified Bohm–Gross relation (4.47) is shown with variation in degeneracy parameter $\chi = n\Lambda_B^3$ for a fixed quantum coupling parameter $r_s = 0.1$. *Thick* (*thick dashed*) *line* corresponds to $T = 300$ K ($T = 500$ K) whereas *thin* (*thin dashed*) *line* is for $T = 700$ K ($T = 900$ K). The wave frequency is normalized by ω_p, and wave number by ω_p/v_F, respectively

where Λ_B is the thermal de Broglie wavelength. Since the degeneracy parameter varies with temperature, the relation shows the increase in wave dispersion with decreasing temperature (Fig. 4.3). Such noticeable effects of fermion degeneracy in dense matter at relatively high temperature provide useful information about the plasmon dispersion in future experiments, for details, see [6].

4.2.6.2 Dielectric Tensor of a Relativistic Quantum Plasma

Let us turn to the full dielectric tensor of a degenerate non-relativistic electron gas. This tensor is well known since the 1950s [12, 71] with various generalizations to the fully relativistic quantum regime. The dielectric tensor of an unpolarized isotropic electron-positron plasma can be written as [73, 74]

$$\varepsilon_{ij}(\omega, \mathbf{k}) = \delta_{ij} - \frac{4\pi e^2}{m\omega^2} \int \frac{d^3\mathbf{p}}{\gamma} \frac{\gamma^2 (\omega - \mathbf{k}\cdot\mathbf{v})^2}{\gamma^2 (\omega - \mathbf{k}\cdot\mathbf{v})^2 - Q_r^2} \times \left[\delta_{ij} + \frac{k_i v_j + k_j v_i}{(\omega - \mathbf{k}\cdot\mathbf{v})} + \frac{\left(k^2 - \omega^2/c^2\right) v_i v_j}{(\omega - \mathbf{k}\cdot\mathbf{v})^2} \right] f(\mathbf{p}), \quad (4.48)$$

where $\mathbf{p} = \gamma m \mathbf{v}$, $\gamma = \left(1 - v^2/c^2\right)^{-1/2}$, and $f(\mathbf{p}) = 2\bar{n}(\mathbf{p}) / (2\pi\hbar)^3$ with $\bar{n}(\mathbf{p})$ being the sum of the occupation numbers for electrons and positrons and

$$Q_r = \frac{\hbar}{2m}\left(\frac{\omega^2}{c^2} - k^2\right). \quad (4.49)$$

It is instructive to consider the first denominator of the integrand in (4.48) the zeroes of which contain the resonance condition for the interaction of electrons (positrons) with the electromagnetic wave:

$$\gamma^2 (\omega - \mathbf{k}\cdot\mathbf{v})^2 - Q_r^2 = [\gamma(\omega - \mathbf{k}\cdot\mathbf{v}) + Q_r][\gamma(\omega - \mathbf{k}\cdot\mathbf{v}) - Q_r]. \quad (4.50)$$

Even though the electromagnetic field is treated classically the zeroes of the two factors can be understood as arising from the emission and absorption of a field quantum by the particles. Thereby the particle energy and momentum change from E to $\acute{E}$ and $\mathbf{p}$ to $\acute{\mathbf{p}}$ by the discrete amount of $\hbar\omega$ and $\hbar\mathbf{k}$, respectively, i.e. $\acute{E} = E \mp \hbar\omega$ and $\acute{\mathbf{p}} = \mathbf{p} \mp \hbar\mathbf{k}$, where $E^2 = \mathbf{p}^2c^2 + m^2c^4$ and $\acute{E}^2 = \acute{\mathbf{p}}^2c^2 + m^2c^4$. For nonrelativistic particle velocities ($\gamma \approx 1$), the resonance condition (4.50) becomes

$$\omega - \mathbf{k}\cdot\mathbf{v} \mp \frac{\hbar}{2m}\left(\frac{\omega^2}{c^2} - k^2\right) = 0. \quad (4.51)$$

There other interesting limit is the classical limit. Then the resonance condition (Cherenkov condition) is simply $\omega - \mathbf{k}\cdot\mathbf{v} = 0$. Clearly, this limit is recovered by putting $Q_r \to 0$ which amounts to neglecting quantum effects (terms proportional to $\hbar$). The quantum correction Q_r to the classical case is frequently called "quantum recoil" [74] althouth this is slightly misleading.[16]

It is important to note that in a strictly nonrelativistic treatment, where one uses the dispersions $E = \mathbf{p}^2/2m$, the term ω^2/c^2 doesn't appear in the expressions (4.49) and

[16] The energy and momentum balance that includes the absorption and emission of photons has been written above and does not contain any additional "recoil" contribution.

(4.51), showing that a nonrelativistic treatment is valid only for $\omega^2 \ll \mathbf{k}^2 c^2$—a fact that is well-known in the theory of plasma oscillations but still often ignored, see e.g. Ref. [74]. From (4.48), the dispersion relation for longitudinal electron waves can be found as before from vanishing of the longitudinal part of the tensor, $\varepsilon_{ij}(\omega, \mathbf{k}) = 0$.

Damping of plasma waves. The use of QHD neglects certain kinetic effects such as Landau damping. This effect is easily treated taking into account that the dielectric function, (4.42), is complex since it includes a small imaginary correction to the frequency (see footnote above). So far we did only consider its real part.

In general, the poles at $v = \omega/k \pm \hbar k/2m$ have both a real and an imaginary part, so the integration has to be performed using the Landau pole integration in the complex velocity plain (analytic continuation is assumed [29]),

$$\varepsilon(\omega, k) = 1 - \frac{m\omega_p^2}{n_0 \hbar k^2}\left[\int_{C_+} \frac{f_0(v)}{k(v - \hbar k/2m) - \omega} dv - \int_{C_-} \frac{f_0(v)}{k(v + \hbar k/2m) - \omega} dv\right] = 0, \tag{4.52}$$

where the integration is performed with Landau contours $C_\pm$ passing under the poles at $v = \omega/k \pm \hbar k/2m$. Equation (4.52) is a useful starting point for the discussion of the quantum Landau damping just like the collisionless damping in classical plasmas [13, 49]. Adopting the procedure similar to the classical plasmas, and assuming small damping (or growth) rate $|\omega_i| \ll \omega$, we obtain [4, 29]

$$\omega_i = \frac{\pi\omega_p^3}{4n_0 k^2}\left[\frac{f_0(\omega/k + \hbar k/2m) - f_0(\omega/k - \hbar k/2m)}{\hbar k/2m}\right]. \tag{4.53}$$

In the limit $\hbar \to 0$, the known classical relation is recovered,

$$\omega_i = \frac{\pi\omega_p^3}{2n_0 k^2}\frac{df_0}{dv}\Big|_{v=\omega/k}, \tag{4.54}$$

which shows that (4.53) can be considered as a finite-difference generalization of the classical expression (4.54).

Generalizing (4.53) to three dimensions, the dispersion equation becomes

$$\varepsilon = 1 - \frac{m\omega_p^2}{n_0 \hbar k^2}\left[\int_{C_+} \frac{f_0(\mathbf{v})}{k(v_z - \hbar k/2m) - \omega} d\mathbf{v} - \int_{C_-} \frac{f_0(\mathbf{v})}{k(v_z + \hbar k/2m) - \omega} d\mathbf{v}\right] = 0, \tag{4.55}$$

where $f_0(\mathbf{v})$ is the equilibrium Wigner distribution function, the Landau contours are passing below the poles lying at $v_z = \omega/k \pm \hbar k/(2m)$, and the coordinate system is chosen such that the wave vector points in z-direction, $\mathbf{k} = (0, 0, k)$. Introducing $f_{0z}(v_z) = \int dv_x dv_y f_0(\mathbf{v})$, (4.55) can be integrated over the perpendicular velocity components, leading to

$$\varepsilon = 1 - \frac{m\omega_p^2}{n_0 \hbar k^2} \left[\int_{C_+} \frac{f_0(v_z)}{k(v_z - \hbar k/2m) - \omega} dv_z - \int_{C_-} \frac{f_0(v_z)}{k(v_z + \hbar k/2m) - \omega} dv_z \right] = 0. \tag{4.56}$$

This result is formally the same as (4.52), which allows one to write for the damping/growth rate in the classical limit, analogous to (4.54),

$$\omega_i = \frac{\pi \omega_p^3}{2n_0 k^2} \frac{df_{0z}}{dv_z} |_{v_z = \omega/k} . \tag{4.57}$$

Damping or growth depend on the sign of the derivative of the projected equilibrium Wigner function. In equilibrium, the distribution function is monotonically decaying with momentum and ω_i is negative, corresponding to damping of the wave. In nonequilibrium, the situation can be opposite.[17]

To apply the above result to a degenerate plasma, it is useful to start from the finite-temperature case because the zero-temperature distribution has a singular derivative. Therefore, consider the Thomas–Fermi distribution [76]

$$f_0(\mathbf{v}) = \frac{\alpha}{\exp\left[\beta\left(\frac{mv^2}{2} - \mu\right) + 1\right]}, \tag{4.58}$$

where $v^2 = v_x^2 + v_y^2 + v_z^2$, $\beta = (k_B T)^{-1}$, and the normalization constant $\alpha = 2(m/2\pi\hbar)^3$. When the temperature T approaches zero, μ approaches to the Fermi energy $\varepsilon_F = mv_F^2/2$. Then the integration over the perpendicular velocity components leads to

$$f_{0z}(v_z) = \frac{2\pi\alpha}{m\beta} \ln\left[1 + \exp\left[\beta\left(\mu - \frac{mv_z^2}{2}\right)\right]\right], \tag{4.59}$$

which has a bell shaped profile. Then, by employing (4.57) for the damping, we obtain

$$\omega_i = \frac{\pi^2 \alpha \omega_p^4}{n_0 k^3} \left[1 + \exp\left[\beta\left(\frac{m\omega_p^2}{2k^2} - \mu\right)\right]\right]^{-1}, \tag{4.60}$$

where in deriving (4.60), the replacement $\omega \simeq \omega_p$ was done. Upon analyzing (4.60), we can see that for very low temperature (large β), $\mu \simeq \varepsilon_F$. Then for $\omega_p/k > v_F$, there will be no damping because the wave phase velocity lies in a region where there are no particles. Therefore, high-frequency electron plasma oscillations of a degenerate plasma at very low temperature remain undamped (in the absence of particle collisions [77]). On the other hand, when $\omega_p/k < v_F$, the exponential term

[17] Note that the existence of instabilities depends on the system dimensionality. While a monotonic increase of f_0 leads to an instability in a one-dimensional and two-dimensional system, this is not the case in a spherically symmetric 3D system [75].

in (4.60) becomes zero for very large β and damping is significant[18] which cannot be taken into account in the QHD description. So the long wavelength assumption $\left(\omega_p/k > v_F\right)$ must hold in the QHD application to degenerate plasmas to avoid damping of waves.

4.2.6.3 Streaming Instabilities

Considering the one-stream plasma case with a single pure quantum state[19] with equilibrium solutions $n = n_0$ and $u = u_0$ at $\varphi = 0$, the Fourier decomposition of the perturbed quantities in (4.15), (4.17) and (4.18) leads to the dielectric function

$$\varepsilon\left(\omega, k\right) = 1 - \frac{\omega_p^2}{\left(\omega - ku_0\right)^2 - \hbar^2 k^4/4m^2}, \tag{4.61}$$

where the term ku_0 just represents a Doppler shift and $v_F = 0$ is assumed. Charge neutrality is provided by the motionless background ions. If the effect of quantum statistics is included in the momentum equation, the dielectric function changes to

$$\varepsilon\left(\omega, k\right) = 1 - \frac{\omega_p^2}{\left(\omega - ku_0\right)^2 - k^2 v_F^2 - \hbar^2 k^4/4m^2}. \tag{4.62}$$

Here, the frequency ω is always real, and the oscillations are stable and undamped [69]. Similarly, for a two-stream plasma, when two counter-streaming electronic populations (beams) are considered with streaming velocities $\pm u_0$, the dielectric function becomes

$$\varepsilon\left(\omega, k\right) = 1 - \sum_{a=\pm 1} \frac{\omega_p^2/2}{\left(\omega - aku_0\right)^2 - k^2 v_F^2 - \hbar^2 k^4/4m^2}. \tag{4.63}$$

The dispersion relation follows from $\varepsilon\left(\omega, k\right) = 0$ and represents a Doppler shifted spectrum [46], where $n_1 = n_2 = n_0/2$, $u_1 = -u_2 = u_0$, $\varphi = 0$, and $u_0 \neq 0$. Here, the solution for ω^2 has two branches, one of which is always positive giving stable oscillations. The other solution is negative $\left(\omega^2 < 0\right)$ which shows

$$\left[F^2 K^2 - 4\left(1 - u_F^2\right)\right]\left[F^2 K^4 - 4\left(1 - u_F^2\right) K^2 + 4\right] < 0, \tag{4.64}$$

---

[18] The region where the imaginary part of the dielectric function is non-zero and damping occurs at $T = 0$ is called "pair continuum" since in this region the plasma wave loses energy by processes where electron-hole pairs are created even when no collisions are taken into account, see e.g. [29]. At finite temperature, there always exist particles with high velocity, so the damping is always non-zero.

[19] Recall that the Pauli principle prohibits that several electrons move with exactly the same velocity. In reality even in an electron beam the particles have a finite velocity spread Δv, and the present model is to be understood as the limit of small velocity spread, $\Delta v/u_0 \ll 1$.

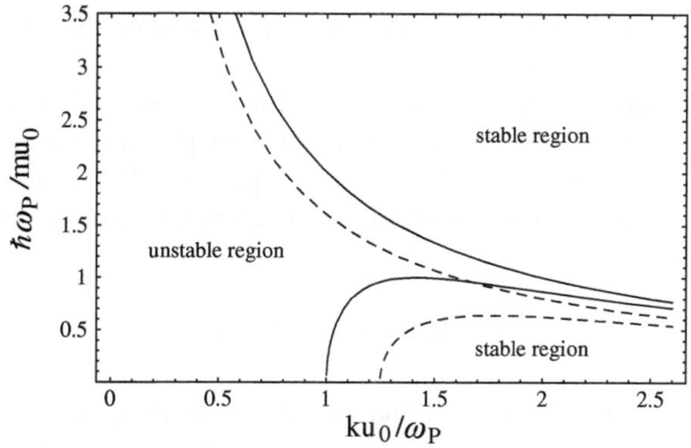

Fig. 4.4 Two-stream instability analysis in a quantum plasma by using normalized parameters. The area enclosed by the *solid curves* along the vertical axis shows an unstable region when $u_F = 0$. With increase in u_F, the region shows a shifting as seen by the *dashed lines* corresponding to $u_F = 0.6$

where the rescaled variables are $F = \hbar\omega_p/mu_0^2$, $K = ku_0/\omega_p$, and $u_F = v_F/u_0$. For $u_F < 1$, a bifurcation in (4.64) is seen for $F = 1 - u_F^2$. If $F \geq 1 - u_F^2$, the second factor of the inequality is always positive for $F^2K^2 < 4\left(1 - u_F^2\right)$ which gives rise to an instability. Similarly, if $F < 1 - u_F^2$, an instability occurs if either $0 < F^2K^2 < 2\left(1 - u_F^2\right) - 2\sqrt{\left(1 - u_F^2\right) - F^2}$, or $2\left(1 - u_F^2\right) - 2\sqrt{\left(1 - u_F^2\right)^2 - F^2} < F^2K^2 < 2\left(1 - u_F^2\right)$. The limit $\hbar \to 0$ leads to $K^2 < 1$ which is the classical instability criterion. The stability/instability region can be seen in Fig. 4.4 with a shift in the presence of nonzero Fermi speed.

The one and two stream cases show the main features of the oscillation spectrum. However, when generalized to a larger number of streams, the coherence and resonant contribution as well as collision between the groups of particles lead to additional damping or dephasing, and a kinetic treatment is required.

4.2.6.4 Longitudinal Ion Waves

When dealing with electrostatic oscillations having a frequency close to the electron plasma frequency, the response of the ion motion is very weak and does not need to be taken into account.

That's why, the ions were considered motionless in Sect. 4.2.6.3 forming a neutralizing background. However, when the wave frequency is less than the ion plasma frequency, $\Omega_p = (m/M)^{1/2}\omega_p$, the dynamics of both species have to be taken into account (M is the ionic mass). In a completely degenerate two-component electron–ion quantum plasma, the Fermi energy of the lighter species (electron) is larger than that of the ions due to smaller electron mass, $E_F \sim m^{-1}$. Similarly, the de Broglie wavelength scales as $m^{-1/2}$ and the degeneracy parameter as $\chi \sim m^{-3/2}$, thus the ion degeneracy is much smaller than the one of the electrons. So, we will continue to consider the ions classical.

In the case of classical plasmas, the longitudinal ion oscillations give rise to the low-frequency ion-acoustic wave which is modified in quantum plasmas, and a quantum ion-acoustic mode appears [78]. The wave dispersion relation of longitudinal ion waves in homogenous electron–ion plasmas is given by the zeroes of the longitudinal dielectric function,

$$1 + \chi_e^{lo}(\mathbf{k}, \omega) + \chi_i^{lo}(\mathbf{k}, \omega) = 0, \tag{4.65}$$

where χ_e^{lo} and χ_i^{lo} are the electron and ion susceptibilities, respectively. For low phase velocity, it follows $\omega \ll kv_{Fe}$, $\chi_i^{lo}(\mathbf{k}, \omega) = -\Omega_p^2/\omega^2$, resulting in

$$1 + \frac{\omega_p^2}{k^2 v_F^2 + \hbar^2 k^4/4m^2} - \frac{\Omega_p^2}{\omega^2} = 0. \tag{4.66}$$

The result for the dispersion is

$$\omega^2 = \frac{\Omega_p^2}{1 + \aleph}, \tag{4.67}$$

where $\aleph = \omega_p^2 / \left(k^2 v_F^2 - \hbar^2 k^4/4m^2\right)$. For $\aleph \gg 1$, (4.67) reduces to

$$\omega \simeq kc_q \left(1 + \frac{\hbar^2 k^4}{4m^2 \omega_p^2}\right)^{1/2}, \tag{4.68}$$

where $c_q = (E_F/M)^{1/2}$ is the speed of linear electrostatic ion waves in a quantum plasma (the so-called quantum ion-acoustic wave). In the classical limit, $\hbar \to 0$, and (4.68) gives the dispersion relation of the usual ion acoustic wave in a thermal plasma.

When using the QHD equations, the momentum equation (4.17) for ions can be written as

$$\left(\frac{\partial}{\partial t} + u_i \frac{\partial}{\partial x}\right) u_i = \frac{q_i E}{M}, \tag{4.69}$$

with u_i, and q_i being the ion velocity, and ion charge, respectively. The last two terms in (4.17) can be ignored for ions due to smallness of ionic quantum effects. Similarly, the electron inertia can be neglected in the limit $m/M \ll 1$. The space charge electric field $E = -\partial\varphi/\partial x$ couples ions with the electrons.

If there exists a drift between electrons and ions in a quantum plasma, the Buneman mode appears [79] just like the classical plasmas [80]. Taking into account collision effects in the electron and ion momentum equations with ν_{en} (ν_{in}) being the collision frequency of an electron (ion) with neutrals,[20] the dispersion relation in the reference frame of the drifting ions becomes

$$1 - \frac{\omega_p^2}{(\omega - kv_0)(\omega - kv_0 + i\nu_{en}) - \hbar^2 k^4/4m^2} - \frac{\Omega_p^2}{\omega(\omega + i\nu_{in})} = 0, \tag{4.70}$$

where the relative electron–ion equilibrium drift velocity in the presence of a static electric field E_0 is

$$v_0 = -eE_0 \left(\frac{1}{m\nu_{en}} + \frac{1}{M\nu_{in}} \right). \tag{4.71}$$

For very low frequencies, $\omega \ll \nu_{in} \ll kv_0$, $\nu_{en} \ll kv_0$, the dispersion equation predicts that the mode is unstable under the condition

$$\omega_p^2 > k^2 v_0^2 - \frac{\hbar^2 k^4}{4m^2} > 0, \tag{4.72}$$

because $\Im(\omega) > 0$, otherwise it is damped. Due to the presence of quantum effects, small wavelength oscillations are always stable. For a slow temporal dynamics, appropriate rescaling of the parameters gives rise to electron momentum equation of the form

$$v_e \frac{\partial v_e}{\partial x} = -E + \frac{F^2}{2} \frac{\partial}{\partial x} \left(\frac{\partial^2 \sqrt{n_e}/\partial x^2}{\sqrt{n_e}} \right), \tag{4.73}$$

provided $\Omega_p \ll \nu_{in}$, $\nu_{en} \ll \omega_p$, with $F = \hbar\omega_p/mv_0^2$ being the dimensionless parameter which measures the contribution of the Bohm potential. The linearization of the normalized set of equations around homogenous equilibrium leads to the dispersion relation with imaginary part of the frequency

$$\omega_i = \frac{k^2 \left(1 - F^2 k^2/4\right)}{1 - k^2 + F^2 k^4/4}. \tag{4.74}$$

When the quantum parameter $F = 0$ (classical limit), $\omega_i = k^2/\left(1 - k^2\right)$. Then, a singularity appears at $k = 1$ and a linear instability exists for $0 < k < 1$. For non-zero F, various instability conditions emerge for $F < 1$, $F = 1$, and $F > 1$. We consider one example of the growth rate in the weak coupling regime, $0 < F < 1$, as shown in Fig. 4.5, which has two asymptotic values, k_1 and k_2, given by

[20] This approximation assumes a weakly ionized plasma where neutrals dominate and, hence, collisions with neutrals play the main role. This is the case at low temperature and not too high densities below the Mott point.

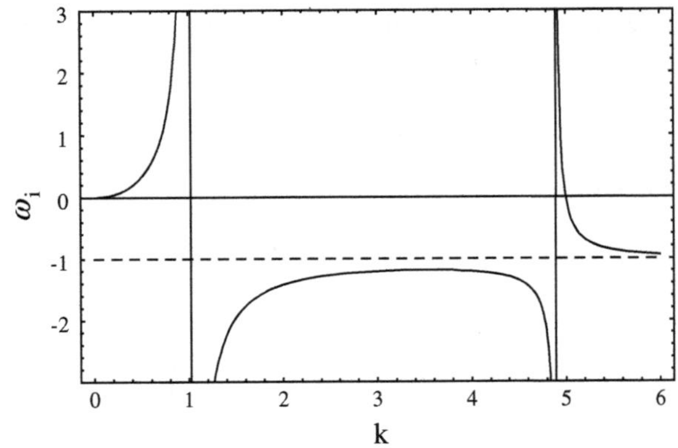

Fig. 4.5 Instability growth rate from (4.74) is plotted as a function of normalized wave number k for $F = 0.4$. Two asymptotic values of k are denoted by k_1 and k_2. Instability occurs at $0 < k < k_1$ and $k_2 < k < k_3$. The growth rate $\omega_i \to -1$ as $k \to \infty$ in both the cases of zero and nonzero F

$$k_1^2 = \frac{2}{F^2}[1 - \sqrt{1 - F^2}], \tag{4.75}$$

$$k_2^2 = \frac{2}{F^2}[1 + \sqrt{1 - F^2}]. \tag{4.76}$$

The growth rate is positive for $0 < k < k_1$, or $k_2 < k < k_3$, where

$$k_3^2 = k_1^2 + k_2^2 = \frac{4}{F^2}. \tag{4.77}$$

This case of Buneman instability in collisional quantum plasma is formally similar to the two-stream instability already discussed in Sect. 4.2.6.3 above.

4.2.7 Nonlinear Waves in Quantum Plasmas

In Sect. 4.2.6.4, we have discussed linearized QHD results following the standard procedure of linearization. When the amplitude of a wave in plasma grows sufficiently large, the nonlinearities in the QHD equations grow and cannot be neglected any more. This makes the system more complicated and its analysis more difficult. The nonlinearities in plasmas may enter through various processes like advection, trapping of particles in the wave potential, the nonlinear Lorentz force, ponderomotive force, etc. Sometimes, the nonlinearities in plasma contribute to the localization of waves giving rise to different types of interesting coherent structures, for instance solitary waves, shocks, vortices, and so on.

Due to highly nontrivial physics involved in the nonlinear regime of quantum plasmas, only a limited analysis has been done in QHD so far. As was shown above the *nonlinear Schrödinger (NLS) equation* (4.28) is equivalent to the QHD in many respects. It has many properties characterisitic of nonlinear waves, especially localized modes and solitons, beam-driven waves and instabilities. The NLS equation and its variants describe nonlinear physical systems appearing in a wide spectrum of problems in (quantum) plasmas[21] and other fields, for example, in fluids and water waves, ultrafast transmission systems, condensed matter systems, and so on. NLS contains an additional nonlinear term in the Schrödinger equation responsible for the nonlinear effects. The solution of NLS equation (4.28) also facilitates the verification of numerical solvers and aids in the stability analysis. Discrete nonlinear Schrödinger (DNLS) equations are also important in discrete lattice models in nonlinear optics, condensed matter and trapped Bose–Einstein condensates where a numerical evaluation is straightforward using e.g. the Crank–Nicolson method, e.g. [81].

The third and fourth terms on the right-hand side of (4.28) represent the nonlinearities associated with the nonlinear coupling between the electrostatic potential and the quantum statistical pressure associated with Fermi–Dirac statistics. Linearizing the NLS-Poisson system gives the frequency spectrum (4.38) where $n = |\Psi|^2$. In equilibrium, $\tilde{V} \sim n^2 = \text{constant}$, otherwise it is repulsive because $\tilde{V}$ is derived from $p^c(n)$ which is related to the dispersion of velocities in a Fermi gas. The NLS equation admits modulational wave solutions and a stability analysis can be performed by standard procedures. Depending upon the type of nonlinearity, it is also capable to provide valuable information of the quasi-stationary structures and nonlinear interaction mechanisms of waves at various length scales [42].

The fluid modeling of the nonlinear long-wave-short-wave interaction in plasmas is provided by the *Zakharov equations*, first derived by Zakharov [82], which get modified in quantum plasmas [83, 84]. The derivation of the *quantum Zakharov system* follows a two-time scales analysis of the QHD equations which becomes possible due to the presence of fast (Langmuir type) and slow (ion-acoustic) oscillations. The limitations of the model are similar to QHD and the allowed wavelengths are $\lambda \gg \lambda_{TF}$, or, equivalently, $k\lambda_{TF} \ll 1$. All QHD variables, i.e. the electron (ion) macroscopic density $n_e\,(n_i)$, velocity $u_e\,(u_i)$, and electric field E are separated into fast (subscript f) and slow (subscript s) oscillatory components

$$n_e(x,t) = n_0 + n_{es}(x,t) + n_{ef}(x,t)\,, \tag{4.78}$$

$$n_i(x,t) = n_0 + n_{is}(x,t)\,, \tag{4.79}$$

$$u_e(x,t) = u_{es}(x,t) + u_{ef}(x,t)\,, \tag{4.80}$$

$$u_i(x,t) = u_{is}(x,t)\,, \tag{4.81}$$

$$E(x,t) = E_s(x,t) + E_f(x,t)\,. \tag{4.82}$$

The slowly varying quantities are considered not significantly changing over a period of oscillation whereas the fast quantities assume zero average. In addition, the quasi-neutrality condition, $n_e \approx n_i$, $u_e \approx u_i$ is assumed, and the high frequency ion terms

[21] Always assuming that the applicability conditions of QHD are fulfilled.

are disregarded due to the smallness of m_e/m_i. This analysis is in the spirit of a classical plasma with the inclusion of the quantum (Bohm) potential for a zero-temperature electron (Fermi) gas. In 1D, this gives the quantum corrected Zakharov equations

$$i\frac{\partial E}{\partial t}+\frac{\partial^2 E}{\partial x^2}-H^2\frac{\partial^4 E}{\partial x^4}=nE, \tag{4.83}$$

$$\frac{\partial^2 n}{\partial t^2}-\frac{\partial^2 n}{\partial x^2}+H^2\frac{\partial^4 n}{\partial x^4}=\frac{\partial^2 |E|}{\partial x^2}, \tag{4.84}$$

where E and n are normalized quantities describing the slowly varying envelope field and plasma density, respectively, and quantum corrections are included via the nondimensional quantum parameter $H = \hbar\Omega_p/E_{Te}$ with Ω_p being the ion plasma frequency and E_{Te} the electron thermal energy. The system can admit periodic, chaotic or similar states, and describes nonlinear dynamics and instabilities. The extension of (4.83) and (4.84) to three dimensions makes the inclusion of electromagnetic effects possible [84].

The nonlinear effects causes the distortion of waves in plasma . Then, wave steepening can occur until some dispersive or dissipative process kicks in which broadens the profile, in turn balancing the nonlinear steepening. Haas and co-workers [78] have attempted to include the quantum effects in nonlinear ion wave excitations in the small and large amplitude regimes in the QHD framework. In the small amplitude limit, the 1D QHD equations (4.15), (4.17), (4.18) and (4.69) reduce through multiscale expansion with appropriate rescaling of parameters to some form of *Korteweg-de Vries (KdV)* equation. Assuming $m/M \ll 1$, the electron momentum equation (4.17) with the boundary condition $n_e = 1, \varphi = 0$ at infinity leads to

$$\varphi=-\frac{1}{2}+\frac{n_e^2}{2}-\frac{\Gamma_q}{2}\left[\frac{\partial_x^2\sqrt{n_e}}{\sqrt{n_e}}\right], \tag{4.85}$$

where the nondimensional quantum parameter $\Gamma_q = (\hbar\omega_p/E_F)^2$. For $\Gamma_q = 0$, the charge density is directly related to the potential by an algebraic equation. We now introduce the slowly varying stretched coordinates

$$\xi=\varepsilon^{1/2}(x-t), \qquad \tau=\varepsilon^{3/2}t, \tag{4.86}$$

where ε is a small parameter proportional to the amplitude of the perturbation which provides the basis of the scaling, merely a convention. Then, expanding the state variables into a series in powers of ε, the low orders of ε result in a KdV type equation with quantum corrections

$$\frac{\partial \Phi}{\partial \tau}+2\Phi\frac{\partial \Phi}{\partial \xi}+\frac{1}{2}\left(1-\frac{\Gamma_q}{4}\right)\frac{\partial^3 \Phi}{\partial \xi^3}=0, \tag{4.87}$$

which admits solitary ion wave and periodic solutions. The function $\Phi(\xi, \tau)$ arises from the zero order solutions and the boundary conditions.

It is important to discuss the features of the dispersive term in (4.87). The equation is obtained by employing the reductive perturbative technique and the term containing the quantum diffraction (coupling) parameter Γ_q appears from the electron equation (4.85). For $\Gamma_q = 4$, the dispersive term in (4.87) disappears. In this case, the quantum diffraction exactly matches the classical dispersion term in the KdV equation. Then, no soliton solution exists and only free streaming is possible like for a free ideal classical fluid which eventually produces a shock wave. Recall that the parameter Γ_q is related to the coupling strength of the plasma. Since the applicability of QHD (as discussed in Sect. 4.2.5 above) is limited to weak coupling, $\Gamma_q < 1$ the value $\Gamma_q = 4$ is certainly out of the scope of QHD. Nevertheless such values are often considered and we present one example below to illustrate the mathematical consequences. For $\Gamma_q \neq 4$, setting the wave-frame position variable $\eta = \xi - V_0\tau$ with constant wave phase velocity V_0 leads to akin to energy first integral for a particle of unit mass in a (pseudo) potential well (Sagdeev-like potential) whose localized solution depends upon Γ_q and V_0. A general profile of the potential $V(\Phi)$, for different values of Γ_q and $V_0 > 0$, is shown in Fig. 4.6 which exhibits the localized (soliton) structure. For $\Gamma_q < 4$ and $V_0 > 0$, some algebra leads to a solitary pulse solution of (4.87) of the form

$$\Phi(\xi, \tau) = \frac{3V_0}{2}\operatorname{sech}^2\left(\frac{\eta}{\sqrt{(4-\Gamma_q)/2V_0}}\right), \tag{4.88}$$

also shown in Fig. 4.7 as a typical case. The pulse height scales as V_0 whereas the pulse width as $V_0^{-1/2}$ which also depends upon Γ_q. The larger amplitude pulses are sharper and can propagate with higher speed.

For fully nonlinear large-amplitude localized ion waves, one can define some non-dimensional parameter playing the role of the Mach number. The wave form can be introduced via the variable,

$$\xi = (x - Mt), \tag{4.89}$$

with M being the Mach number. Then, the QHD equations reduce to the dynamical equations which can be written in the form of conservation laws leading to solitary wave solutions depending upon Γ_q and M.

4.2.8 Magnetized Quantum Plasmas

So far, we have considered only the electrostatic case. The inclusion of a magnetic field leads to a more general form of the QHD equations derivable from the

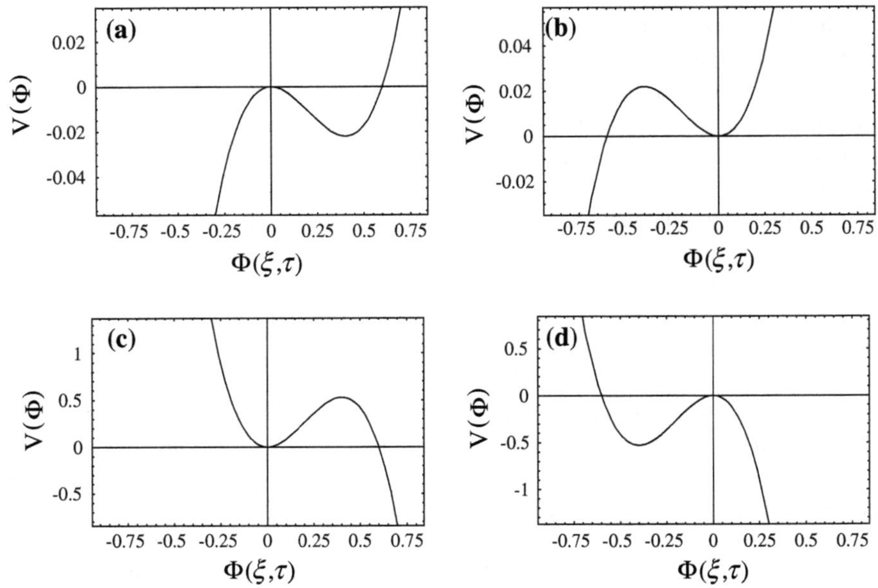

Fig. 4.6 The potential $V(\Phi)$ versus $\Phi(\xi, \tau)$ for arbitrary constant phase velocity V_0 is shown for small and large values of Γ_q. **a**: $\Gamma_q = 0.1$, $V_0 = 0.4$, **b**: $\Gamma_q = 0.1$, $V_0 = -0.4$, **c**: $\Gamma_q = 4.1$, $V_0 = 0.4$, **d**: $\Gamma_q = 4.1$, $V_0 = -0.4$. The conditions of solution; $V'(\Phi) = 0$ at $\Phi = 0$ and $\Phi = V_0$ are satisfied where prime denotes the derivative with respect to η

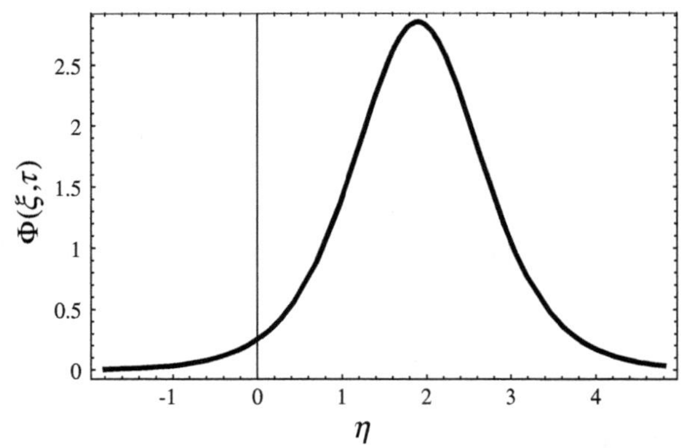

Fig. 4.7 The solitary pulse profile for $\Gamma_q = 0.1$ and arbitrary constant phase velocity $V_0 > 0$. For such small values of Γ_q, the effect on pulse width is negligibly small

electromagnetic Wigner evolution equation. Similar to the unmagnetized case, the Madelung decomposition of the ensemble wave functions allows to identify the classical and quantum parts of the pressure dyad. Considering the statistical mixture of K-states $\psi_i = \psi_i(\mathbf{r}, t)$, $i = 1, \ldots, K$, such that the probabilities $p_i \geq 0$, with $\sum_{i=1}^{K} p_i = 1$, each ψ_i, satisfies the Schrödinger equation

$$i\hbar\frac{\partial\psi_i}{\partial t}=\frac{1}{2m}\left(-i\hbar\nabla-q\mathbf{A}\right)^2\psi_i+q\varphi\psi_i, \tag{4.90}$$

where the charge carriers have mass m and charge q under the influence of self-consistent scalar and vector potentials $\varphi(\mathbf{r},t)$ and $\mathbf{A}(\mathbf{r},t)$, respectively, with choice of Coulomb gauge $\nabla\cdot\mathbf{A}=0$. Then the one-particle Wigner function in terms of coordinate $\mathbf{r}$ and momentum $\mathbf{p}=m\mathbf{v}+q\mathbf{A}$ becomes

$$f_W(\mathbf{r},\mathbf{p},t)=\frac{1}{(2\pi\hbar)^3}\sum_{i=1}^{K}p_i\int_{-\infty}^{\infty}d\mathbf{s}\psi_i^*\left(\mathbf{r}+\frac{\mathbf{s}}{2},t\right)e^{i\mathbf{p}\cdot\mathbf{s}/\hbar}\psi_i\left(\mathbf{r}-\frac{\mathbf{s}}{2},t\right), \tag{4.91}$$

which leads to the equation for f_W—the quantum Vlasov equation—obtained after a cumbersome calculation, for details, see [49]. Since, the complexity of the Wigner function equation in the electrostatic case makes it very hard to be fully examined, except for the simpler linear case, a non-zero magnetic field makes the problem even more challenging and an analysis more difficult. This motivates a hydrodynamic description.

Introducing the moment equations in the usual way, the continuity and the momentum transport equations in a magnetized plasma following from the Wigner function evolution equation can be written as

$$\frac{\partial n}{\partial t}+\frac{\partial(n\mathbf{u})}{\partial x}=0, \tag{4.92}$$

$$\frac{\partial\mathbf{u}}{\partial t}+\mathbf{u}\cdot\nabla\mathbf{u}=\frac{q}{m}\left(\mathbf{E}+\mathbf{u}\times\mathbf{B}\right)-\frac{1}{mn}\nabla p^c(n)+\frac{\hbar^2}{2m^2}\nabla\left(\frac{\nabla^2\sqrt{n}}{\sqrt{n}}\right), \tag{4.93}$$

where the closure assumption is made by defining a diagonal form of the classical pressure dyad. Since the classical part of the pressure dyad p^c can be written as the sum of average velocity dispersions, the diagonal isotropic form assumes the components

$$P_{ij}=\delta_{ij}p^c, \tag{4.94}$$

with $p^c=p^c(n)$ being a suitable equation of state. For $\hbar\to 0$, (4.93) is just like the momentum equation of classical fluids. Equations (4.92) and (4.93) together with Maxwell's equations constitute the QHD model for magnetized plasmas where the limitations of the electrostatic QHD equations are also valid in the present case. For more subtle issues like gauge invariance of Wigner equation, magnetohydrodynamic equilibria, the inclusion of spin and strong field effects (Landau quantization) etc., see [49].

4.2.8.1 Electrostatic and Electromagnetic Linear Modes

We start from the QHD equations for a two-component dense uniform magnetized plasma consisting of degenerate electrons and non-degenerate ions. Assuming the dynamics of electrons with a background of stationary ions embedded in a uniform magnetic field $B_0\hat{\mathbf{z}}$, the low-frequency (in comparison with the electron cyclotron frequency) electric and magnetic field perturbations are defined as $\mathbf{E} = -\nabla\varphi - c^{-1}(\partial A_z/\partial t)\,\hat{\mathbf{z}}$ and $\mathbf{B}_\perp = \nabla_\perp A_z \times \hat{\mathbf{z}}$, respectively. From (4.93), the linearized components of the electron and ion fluid velocity parallel and perpendicular to $\hat{\mathbf{z}}$ become

$$\frac{\partial v_{jz}}{\partial t} \simeq \delta_j \frac{e}{m_j}\left(\frac{\partial \varphi_j}{\partial z} + \frac{1}{c}\frac{\partial A_z}{\partial t}\right), \tag{4.95}$$

$$\mathbf{v}_{j\perp} \simeq \frac{c}{B_0}\left(\hat{\mathbf{z}} \times \nabla_\perp \varphi_j + \delta_j \frac{1}{\Omega_{cj}}\frac{\partial \nabla_\perp \varphi_j}{\partial t}\right), \tag{4.96}$$

where $j = e(i)$ denotes electron (ion), $\delta_e = 1$, $\delta_p = -1$, and we denoted $\phi_e = \phi - \frac{1}{n_0 e}\left(\frac{2E_F}{3} - \frac{\hbar^2\nabla^2}{4m}\right) n_e$, with $n_e \ll n_0$. Assuming low frequencies, $\omega \sim |\partial/\partial t| \ll \omega_c, ck$, but ω higher than the ion plasma frequency Ω_p, and ion cyclotron frequency, Ω_c, respectively, the dispersion relation for the shear electromagnetic mode in this case is derived by employing (4.95) and (4.96), together with the linearized electron continuity equation, Poisson's equation and Ampere's law which, upon Fourier transformation, results in [85]

$$\omega^2 = \frac{c^2 k_z^2 k_\perp^2 \left(1 + \Lambda_q^2 k^2\right)}{\left(1 + \lambda_e^2 k_\perp^2\right)\left[k^2 + k_\perp^2 \left(1 + \Lambda_q^2 k^2\right)\frac{\omega_p^2}{\omega_c^2}\right]}, \tag{4.97}$$

which shows strong dependence of the electron quantum statistical and quantum diffraction effects (Fig. 4.8), where $\lambda_e = c/\omega_p$, $\Lambda_q = v_q/\omega_p$, and $v_q = \sqrt{T_q/m}$ with $T_q = \left(\hbar^2 k^2/4m + 2E_F/3\right)$ being the quantum parameter in energy units playing the same role as effective temperature in classical plasmas. The electromagnetic mode (4.97) ceases to exist for $k_z = 0$.

Next, we consider the ion dynamics in the frequency regime $\omega \ll \Omega_c$. The ion perpendicular velocity component then consists of the electric and ion polarization drifts, $\mathbf{v_E}$ and $\mathbf{v_p}$ respectively. Then the dispersion equation with assumption of $k_z \ll k_\perp$ acquires the form

$$\left[\left(1 + K_A^2\right)\omega^2 - \frac{c_q^2 k_z^2}{\left(1 + \lambda_e^2 k_\perp^2\right)}\left(1 + K_A^2\right)\right]\omega^2$$
$$- \frac{\omega_A^2}{\left(1 + \lambda_e^2 k_\perp^2\right)}\left[\left(1 + \frac{c_q^2 k^2}{\omega_{pi}^2}\right)\omega^2 - c_q^2 k_z^2\right] = \frac{\rho_q^2 k_\perp^2 \omega_A^2}{\left(1 + \lambda_e^2 k_\perp^2\right)}\omega^2, \tag{4.98}$$

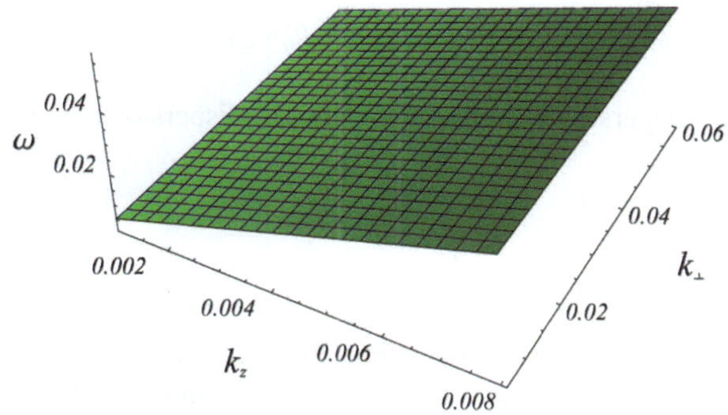

Fig. 4.8 Low-frequency electromagnetic mode in a strongly degenerate electron plasma with immobile ions. The wave frequency is normalized by ω_c and the wave numbers by v_F/ω_c

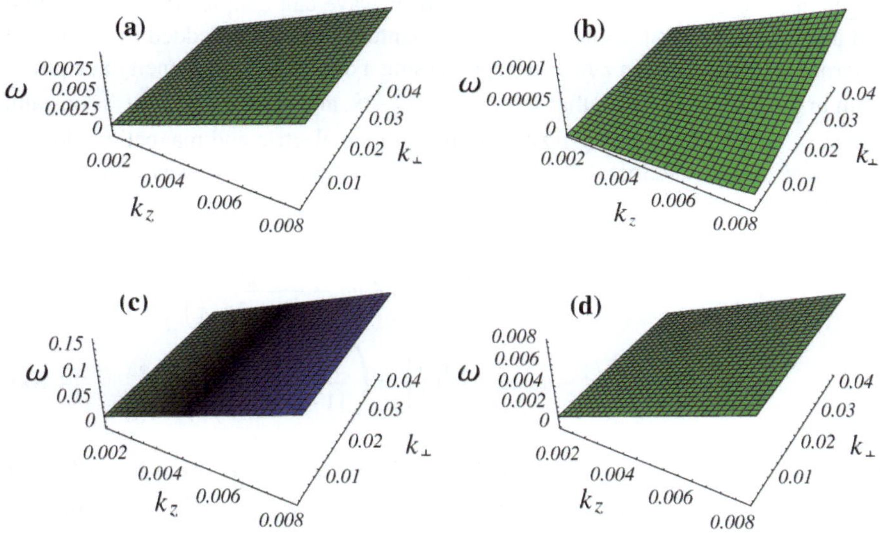

Fig. 4.9 Low-frequency linear electrostatic (**a, c**) and electromagnetic (**b, d**) modes in a magnetized quantum plasma in the absence (*upper panel*) and the presence (*lower panel*) of degeneracy pressure for typical quantum plasma of white dwarf stars taken from Table 4.2 with density $\sim 10^{26}\,\mathrm{cm}^{-3}$ and magnetic field $\sim 10^{8}$ G. The dispersion relation shows that the electrostatic (quantum ion-acoustic) mode can couple with the electromagnetic (shear Alfvén) mode in high magnetic field regions. The wave frequency is normalized by Ω_c and wave numbers by v_A/Ω_c

where $K_A = \frac{v_A}{ck_\perp/k}$, $v_A = B_0/\sqrt{4\pi n_0 M}$ is the speed of the Alfvén wave, $c_q = \sqrt{T_q/M}$ is the speed of the electrostatic ion wave, and $\rho_q = c_q/\Omega_c$. The co-existing electrostatic and electromagnetic modes (Fig. 4.9) are well separated, however, the difference of frequencies become lesser and lesser as the magnetic field is increased.

In the limiting case, for $v_A \ll ck_\perp/k$, $\lambda_e^2 k_\perp^2 \ll 1$, (4.98) reduces to

$$\omega^2 = k_z^2 v_A^2 \left(1 + \rho_q^2 k_\perp^2\right), \tag{4.99}$$

which shows the dispersive Alfvén wave where the dispersion comes from the electron quantum effects.

4.2.8.2 Drift Mode

In the presence of gradients (inhomogeneities in density, temperature, etc.), there may appear drift waves in classical as well as quantum plasmas which play an important role in transport of plasma particles and energy/momentum across the magnetic field lines. Drift waves are low frequency waves in comparison with the ion gyrofrequency Ω_c with perpendicular (with respect to the magnetic field) wave number $k_\perp$ much larger than $k_\parallel$. For relatively large $k_\parallel$, the drift wave can couple with the quantum ion acoustic wave [86]. Consider a dense quantum plasma embedded in a constant external magnetic field in z-direction possessing a density inhomogeneity at equilibrium in the x-direction such that $\nabla n_{j0} = -(\frac{dn_{j0}}{dx})\hat{\mathbf{x}}$, and $\kappa_{jn} = |\frac{1}{n_{j0}}\frac{dn_{j0}}{dx}| = \text{constant}$, with $\kappa_{jn} \ll k_\perp$. Using the QHD equations with the electric and magnetic field perturbations as given in Sect. 4.2.8.1 the dispersion relation for drift waves in quantum plasma becomes

$$\left[(\omega - \omega_q^*)\omega - \frac{c_q^2 k_z^2}{(1 + \lambda_e^2 k_\perp^2)}\right]\omega^2 - \frac{\omega_A^2}{(1 + \lambda_e^2 k_\perp^2)}\left[\omega^2 - \omega_q^*\omega - c_q^2 k_z^2\right] = \left(\frac{\rho_q^2 k_\perp^2}{(1 + \lambda_e^2 k_\perp^2)}\right)\omega_A^2\omega^2, \tag{4.100}$$

where $\omega_A = k_z v_A$ is the frequency of the Alfvén wave, $\omega_q^* = \mathbf{v}_{qD}.\mathbf{k}$ is the drift wave frequency, $\mathbf{v}_{qD} = \frac{cT_q}{eB_0}\nabla \ln n_0 \times \hat{\mathbf{z}}$ is the drift wave velocity, and $T_q = \hbar^2 k^2/4m$. The above relation has an analogy with the classical drift wave frequency which depends upon the equilibrium electron pressure defined by the ideal gas law. However, both are very different physically. The classical drift wave depends upon electron thermal energy, but in a quantum plasma, the role of thermal energy is taken over by the Fermi energy. In deriving (4.100), the drift approximation $|\partial_t| \ll \Omega_c$ is used in the limit of small Fermi pressure. In Fig. 4.10, the frequencies of the four modes are plotted against k_z. One branch of the shear Alfvén wave and one branch of the electrostatic wave are influenced by the quantum drift wave ω_q^* near $k_z \sim 4 \times 10^3\,\text{cm}^{-1}$. Since $T_i = 0$ has been assumed, therefore the second branch of the Alfvén wave remains a straight line in this figure. The second branch of the electrostatic wave has also the effects of dispersion similar to the classical case. The density matrix approach can also be used to study quantum drift wave in two component inhomogeneous plasma in a strong magnetic field under strong and weak quantum effects [87]. Here, the problem is treated semiclassically with a modified Maxwell distribution function in

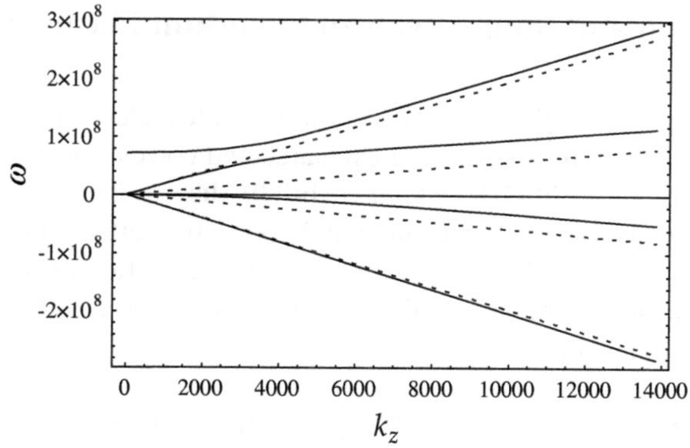

Fig. 4.10 The appearance of a drift mode in a magnetized non-uniform quantum plasma due to density inhomogeneity. The frequencies of the four low-frequency modes (in units of s^{-1}) are plotted versus the wave numbers k_z for fixed k_y (units of cm^{-1}) for a dense astrophysical plasma with typical parameters selected from Table 4.2 with density $\sim 10^{27}\,cm^{-3}$ and magnetic field $\sim 10^6$ G. One branch of the shear Alfvén wave and one branch of the electrostatic wave are influenced by the quantum drift wave ω_q^* near $k_z \sim 4 \times 10^3 cm^{-1}$. The *dashed lines* represent $\pm k_z v_A$ (*outer lines*) and $\pm k_z c_q$ (*inner lines*), respectively

order to determine the quantum effects. Such waves become unstable under some circumstances.

Concluding this section, let us briefly discuss for what systems the present results can be relevant. In the QHD approach, the effect of Fermi degeneracy and (quantum) Bohm potential are the main quantum ingredients. The applications of nonlinear waves with or without magnetic field can be found in quasi-free electron gas in metals and in the high density regimes relevant to degenerate plasmas of dense astrophysical objects (regions of white dwarfs and neutron stars). Such densities are also expected in the lab in next generation laser-plasma experiments. At such high densities, r_s will be well below unity and the QHD model may be even better applicable (Fig. 4.10).

Let us consider a typical example from Table 4.2, with density $n_0 = 4.0 \times 10^{26}\,cm^{-3}$, $\bar{r} = 8.3 \times 10^{-10}\,cm$ and $\lambda_{TF} = 7.2 \times 10^{-8}\,cm$. The wave number k should be well below $\bar{r}^{-1}$ for both electronic and ionic perturbations. For low frequency perturbations, if $k \simeq 10^6\,cm^{-1}$, the ion wave speed is $c_q \simeq 3.8 \times 10^7\,ms^{-1}$ and the frequency, from (4.65) is $\simeq 6 \times 10^{13}\,s^{-1}$. If we consider the soliton solution, (4.88), for such a plasma, the effect of the quantum parameter Γ_q is vanishingly small. In the weakly nonlinear limit, a soliton with a typical speed $\sim 0.1 c_q$ shows very small amplitude and width parameters. For a magnetized plasma with high ambient magnetic field $\left(B_0 \simeq 1 \times 10^7\,G\right)$ and low frequency electrostatic and electromagnetic perturbations, dispersion equation (4.98) with $k_z/k_\perp \simeq 0.002$ reveals that the frequency of electrostatic mode is $1.4 \times 10^{13}\,s^{-1}$. Similarly, the frequency of the shear Alfven mode is $3.1 \times 10^9\,s^{-1}$, with $v_A = 2.2 \times 10^5 ms^{-1}$ being the speed of the wave. It should be noted that the modes are well separated for low magnetic field but their frequencies get closer and closer as the magnetic field increases.

4.3 Interaction and Spin Effects in Quantum Plasmas

As was discussed above QHD assumes an (almost) ideal electron Fermi gas. There have recently been attempts to include exchange and correlation effects in order to extend the validity range of QHD which we briefly discuss below.

The first attempt to include exchange and correlation effects in QHD phenomenologically was presented by Manfredi and co-workers in Ref. [48]. Inspired by the procedure used in density functional theory (DFT) they used an additional exchange-correlation functional, V_{xc}

$$V_{xc} = 0.985(e^2/\varepsilon)n^{1/3}[1 + (0.034/a_B n^{1/3})\ln(1 + 18.37 a_B n^{1/3})], \qquad (4.101)$$

in the momentum equation that gives rise to an additional force on the electrons. The authors performed comparisons with DFT simulations for electrons in condensed matter and observed reasonable agreement.

4.3.1 *Prediction of Attractive Forces Between Protons in Quantum Plasmas*

Using the QHD with the above mentioned potential V_{xc} Shukla and Eliasson [88] considered the problem of the effective potential of a proton embedded in a dense quantum plasma. The QHD equations together with the Poisson equation for the electrostatic potential are

$$\frac{\partial n}{\partial t} + \nabla \cdot (n\mathbf{u}) = 0,$$

$$m_* \left(\frac{\partial \mathbf{u}}{\partial t} + \mathbf{u} \cdot \nabla \mathbf{u} \right) = e\nabla\phi - n^{-1}\nabla P + \nabla V_{xc} + \nabla V_B, \qquad (4.102)$$

$$\nabla^2 \phi = \frac{4\pi e}{\varepsilon}(n - n_0) - 4\pi Q \delta(\mathbf{r}), \qquad (4.103)$$

where the positive test charge Q is located at $\mathbf{r} = 0$. Quantum effects are taken into account as usual via the Bohm potential $V_B = (\hbar^2/2m_*)(1/\sqrt{n})\nabla^2\sqrt{n}$. The pressure of the ideal Fermi gas at zero temperature, $P = (n_0 m_* v_*^2/5)(n/n_0)^{5/3}$, is used and complemented by the exchange-correlation potential V_{xc}. Here, the following definitions are used: ε denotes the relative dielectric permeability of the material, $v_* = \hbar(3\pi^2)^{1/3}/m_* r_0$ is the electron Fermi speed, m_* is the effective mass of electron and $r_0 = n_0^{-1/3}$ is the Wigner–Seitz radius.

Shukla and Eliassion linearized these equations, writing $n = n_0 + n_1$ and $|n_1| \ll n_0$. Neglecting dynamic effects in the dielectric function, $\varepsilon(\mathbf{k}, \omega) = \varepsilon(\mathbf{k}, 0)$, the electrostatic potential of a proton is given by

$$\phi(\mathbf{r}) = \frac{Q}{2\pi^2} \int \frac{\exp(i\mathbf{k}\cdot\mathbf{r})}{k^2 \varepsilon(\mathbf{k})} d^3k. \tag{4.104}$$

From the linearized QHD equations they obtained for the inverse dielectric function

$$\frac{1}{\varepsilon(k)} = \frac{(k^2/k_s^2) + \alpha k^4/k_s^4}{1 + (k^2/k_s^2) + \alpha k^4/k_s^4}, \tag{4.105}$$

with the definitions

$$\alpha = \hbar^2 \omega_{pe}^2 / 4m_*^2 (v_*^2/3 + v_{ex}^2)^2, \tag{4.106}$$

$$k_s = \omega_{pe} / \sqrt{v_*^2/3 + v_{ex}^2}, \quad b_* = 1/\sqrt{4\alpha - 1}, \tag{4.107}$$

$$v_{ex} = (0.328 e^2 / m_* \varepsilon r_0)^{1/2} [1 + 0.62/(1 + 18.36 a_B n_0{}^{1/3})]^{1/2}, \tag{4.108}$$

$$k_r = (k_s/\sqrt{4\alpha})(\sqrt{4\alpha} + 1)^{1/2}, \quad k_i = (k_s/\sqrt{4\alpha})(\sqrt{4\alpha} - 1)^{1/2}. \tag{4.109}$$

The parameter α is shown in Fig. 4.11. While for $\alpha < 0.25$ the potential (4.104) is always positive, for $\alpha > 0.25$, it develops a negative (attractive) minimum. In the latter case the potential is given by [88]

$$\phi(\mathbf{r}) = \frac{Q}{r} [\cos(k_i r) + b_* \sin(k_i r)] \, e^{-k_r r}. \tag{4.110}$$

The maximum value of α is approximately 0.64. Inserting all parameters in the definition of α, existence of a negative potential in the linearized QHD is confined to a finite density interval where $0.61 \le r_s \le 26.22$, see Fig. 4.11. Here the standard quantum coupling (Brueckner) parameter has been used, $r_s = \bar{r}/a_B$, where $\bar{r}$ denotes the mean interparticle distance and a_B the Bohr radius. For weak coupling, $r_s \ll 1$, linearized QHD does not predict a negative potential.

In Ref. [88], based on the existence of a negative minimum of the proton potential (4.104), Shukla and Eliasson claimed the discovery of a novel attractive force between ions in dense quantum plasmas. They claimed that this potential would lead to novel bound states and to a proton lattice. However, as can be seen in Fig. 4.11 where we also show the depth of this potential, negative values occur only in the regime of moderate coupling, i.e. way outside the validity range of QHD which was discussed in Sect. 4.2.5 above. We note that the linearized version of QHD, obviously, is even less accurate. To verify these strong claims Ref. [70] reported ab initio density functional theory simulations. The DFT result for the effective potential of a proton in dense hydrogen, indeed, was found to exhibit an attractive minimum, in two cases: first, at low density, $r_s > 1.5$ there is a minimum corresponding to binding of two hydrogen atoms into a molecule. Second, at large distances, there are shallow oscillations of the potential which are related to Friedel oscillations (originating from the step character of the zero temperature Fermi distribution). No other cases of attractive potentials between protons were observed in the simulations. Therefore, Ref. [70] had to conclude that the predictions of Ref. [88] are wrong.

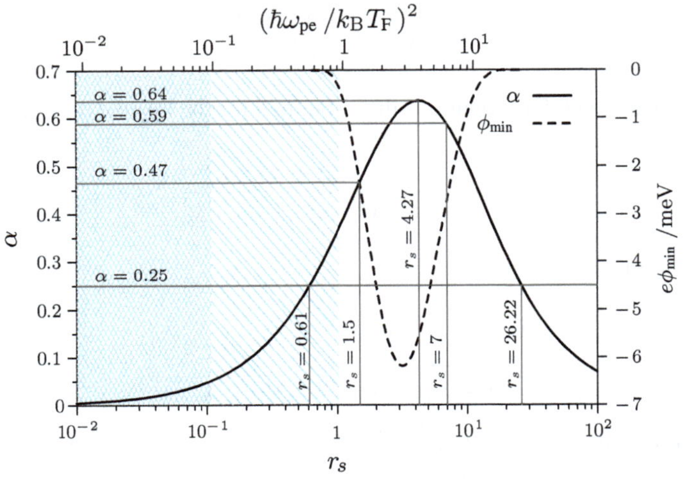

Fig. 4.11 The Shukla–Eliasson parameter α (*left axis*) versus quantum coupling parameters: the Brueckner parameter r_s (*bottom axis*) and Γ_q (*upper axis*). The potential of a proton, (4.104), derived by Shukla and Eliasson from linearized QHD becomes attractive for $\alpha > 0.25$, corresponding to $26.22 \geq r_s \geq 0.61$, at zero temperature. The two *shaded areas* denotes the range of moderate (weak) coupling given by $(\hbar\omega_{pe}/k_B T_F)^2 < 1$ (< 0.1). Also shown is the depth of the SE potential (*right axis*). From Ref. [91]

The disagreement between linearized QHD and DFT was further discussed in Refs. [89, 90]. A careful analysis of the applicability range of QHD and DFT shows that, from its construction, DFT is always more accurate. The lesson to learn from this is that the applicability limits of QHD should be taken very seriously and clearly checked in any application.

4.3.2 Spin Effects in Quantum Plasmas

In recent years attempts have been made to extend the QHD to quantum plasmas with spin effects. This is natural as spin effects are always present for quantum particles. For the case of plasmas with degenerate electrons, the effect of fermionic statistics (spin 1/2) has to be considered. The corresponding extension of QHD to spin QHD (SQHD) can be found e.g. in refs. [92–94] and references therein. These papers came to the conclusion that collective spin effects can dominate the plasma dynamics which is derived from a (possibly macroscopically large) spin magnetization current

$$\mathbf{j}_{\text{spin}} = \nabla \times 2n\mu_B \mathbf{S}, \tag{4.111}$$

where n is the total electron density, μ_B the Bohr magneton and $\mathbf{S}$ the local average "spin vector". For a high density, as is often the case in quantum plasmas and, assuming spin polarization (i.e. all spins are aligned), this current and the associated magnetization may become very large.

It has recently been pointed out [95] that this prediction is in striking contrast to standard condensed matter physics and experiments as well. In particular, the quantum theory of magnetism, e.g. [96–98] is well developed and does not predict any such gigantic magnetizations. There it is known that the magnetization arises from unpaired electron spins and is proportional to the density of spin up minus spin down electrons, $n_+ - n_-$. However, due to Pauli blocking, at low temperature this difference vanishes, it is zero for an ideal Fermi gas in the ground state. At finite temperature, a finite difference may exist which scales as [95]

$$\frac{n_+ - n_-}{n} = O\left(\frac{T}{T_F}\right), \tag{4.112}$$

i.e. in the magnetization not all electrons but only those in a small layer (of order T) around the Fermi edge participate resulting in the known moderate values for the spin magnetization of real materials. Thus, at low temperatures when the system approaches an ideal Fermi gas the spin magnetization vanishes, whereas at high temperatures it vanishes as well because all quantum effects are washed out by thermal fluctuations leading to random spin orientations.

The striking contrast between the SQHD prediction, (4.111) and (4.112) is surprising since the theoretical concepts that are used in condensed matter physics include correlations and spin effects in a much more accurate fashion than QHD.

As is pointed out in the analysis of Ref. [95] the SQHD analysis contains a major inconsistency (see our discussion above): the N-particle wave function is represented by a product of single-particle orbitals (Hartree or Vlasov approximation) whereas for fermions an anti-symmetrized ansatz has to be used. This leads to Slater determinants that guarantee the Pauli principle, in contrast to the Hartree ansatz. As a result the Fermi statistics and the Pauli principle are lost in key places of the QHD theory. In particular, kinetic effects such as the sharp Fermi surface are lost.[22]

Therefore, the predictions of exotic spin quantum effects in quantum plasmas such as spin-gradient-driven light amplification [94] have to be questioned. While there is always room for new fascinating discoveries they, obviously, have to be based on a well established theory that includes all relevant effects. An urgent next step, to resolve these conflicting predictions is, therefore, to reformulate QHD fully using anti-symmetric N-particle states, thus building in the Pauli principle from the very beginning.

---

[22] The Fermi distribution is only included via the equation of state relating pressure and density, but this introduces degeneracy effects only in an average fashion.

4.4 Conclusion and Outlook

In this chapter, we have discussed the theoretical treatment of dense quantum plasmas that are increasingly important in many laboratory and astrophysical systems. While accurate approaches to quantum plasmas have been in existence for many years—based on first-principle simulation, quantum kinetic theory and nonequilibrium Greens functions—these approaches are quite difficult, especially for magnetized plasmas. This makes it highly desirable to have at hand simpler models. Here quantum hydrodynamics has become quite popular in various scientific areas, as is evident from the vast literature that appeared in recent years. At the same time, most papers have essentially ignored the limited applicability range of QHD raising questions about the reliability of the results and of their relevance for practical applications.

In this chapter, we have discussed the main concepts of quantum hydrodynamics and its relations to quantum kinetic theory in terms of the Wigner distribution function and its equation of motion. We analyzed in detail the basic assumptions that lead to the QHD equations and their limitations. Strictly speaking, QHD applies to an ideal Fermi gas (where the quantum coupling parameters are small, i.e. $r_s \ll 1$ and $\Gamma_q \ll 1$) at zero temperature, and it entirely neglects quantum exchange effects. Furthermore, as any hydrodynamic theory, QHD is only able to resolve processes at sufficiently large length scales exceeding a threshold which is on the order of the Thomas–Fermi screening length λ_{TF}. Furthermore, QHD uses a closure of the system of hydrodynamic equations that involves an equation of state in the local approximation which again rules out strong inhomogeneities. When any of these inherent limitations is neglected, unphysical results can follow which includes the predictions of attractive forces between protons at atomic scales as well as giant magnetizations related to the electron spin.

In the linearized QHD, we have briefly reviewed some properties of electron and ion plasma oscillations for unmagnetized as well as magnetized quantum plasma. These include the linear electron plasma waves in the strong degeneracy limit at $T = 0$ as well as for finite T and their dispersion. This was compared with the results from kinetic theory which also provide information on the damping of the oscillations. We further considered the quantum streaming and Buneman instabilities, and low frequency electrostatic and electromagnetic ion modes in uniform and nonuniform quantum plasma. Finally, an overview of the nonlinear solutions of the QHD equations was given, leading to localized coherent structures and the model equations for a magnetized plasma have been presented with illustrations for the sake of generality.

Acknowledgments The authors thank Tim Schoof for valuable remarks. This work has been supported by the Deutsche Forschungsgemeinschaft via SFB-TR24.

References

1. M. Bonitz, N. Horing, P. Ludwig (eds.), *Introduction to Complex Plasmas* (Springer, Berlin, 2010) (Chapters 3 & 4)
2. S.A. Maier, *Plasmonics-Fundamentals and Applications* (Springer, New York, 2007)
3. H.A. Atwater, A. Polman, Plasmonics for improved photovoltaic devices. Nature Mat. **9**, 205 (2010). and references therein
4. M. Bonitz, R. Binder, S.W. Koch, Carrier-acoustic plasmons instability in semiconductor quantum wires. Phys. Rev. Lett. **70**, 3788 (1993)
5. M. Bonitz, R. Binder, D.C. Scott, S.W. Koch, D. Kremp, Theory of plasmons in quasi-one-dimensional degenerate plasmas. Phys. Rev. E **49**, 5535 (1994)
6. S.H. Glenzer, R. Redmer, X-ray Thomson scattering in high energy density plasmas. Rev. Mod. Phys. **81**, 1625 (2009)
7. C.P. Ridgers, C.S. Brady, R. Duclous, J.G. Kirk, K. Bennett, T.D. Arber, A.P.L. Robinson, A.R. Bell, Dense electron-positron plasmas and ultraintense γ-rays from laser-irradiated solids. Phys. Rev. Lett. **108**, 165006 (2012)
8. G. Chabrier, D. Saumon, A.Y. Potekhin, Dense plasmas in astrophysics: from giant planets to neutron stars. J. Phys. A: Math. Gen. **39**, 4411 (2006)
9. M.H. Thoma, Strongly coupled plasma in high energy physics. IEEE Trans. Plasma Sci. **32**, 738 (2004)
10. V.I. Tatarskii, The Wigner representation of quantum mechanics. Sov. Phys. Usp. **26**, 311 (1983) [Usp. Fis. Nauk. 139, 587 (1983)]
11. D. Bohm, D. Pines, A collective description of electron interactions: III. Coulomb interactions in a degenerate electron gas. Phys. Rev. **92**, 609 (1953)
12. Yu.L. Klimontovich, V.P. Silin, Zh. Eksp. Teor. Fiz. **23**, 151 (1952)
13. Yu.L. Klimontovich, V.P. Silin, in *The Spectra of Systems of Interacting Particles*, ed. by J. Drummond. Plasma Physics (McGraw-Hill, New York, 1961)
14. D. Pines, A collective description of electron interactions: IV electron interaction in metals. Phys. Rev. **92**, 626 (1953)
15. D. Pines, Quantum plasma physics: classical and quantum plasmas. J. Nucl. Energy, Part C **2**, 5 (1961)
16. P. Nozières, D. Pines, *The Theory of Quantum Liquids* (Benjamin, New York, 1966)
17. G.D. Mahan, *Many-Particle Physics*, 2nd edn. (Plenum, New York, 1990)
18. M. Bonitz, D. Semkat (eds.), *Introduction to Computational Methods for Many-Body Physics* (Rinton, Princeton, 2006)
19. V.S. Filinov et al., Monte Carlo simulations of dense quantum plasmas. J. Phys. A: Math. Gen. **39**, 4421 (2006)
20. V.S. Filinov, M. Bonitz, A. Filinov, V.O. Golubnychiy, in *Wigner Function Quantum Molecular Dynamics*, ed. by H. Fehske, R. Schneider, A. Weiβe. Computational Many-Particle Physics, Lecture Notes in Physics, vol. 739 (Springer, Berlin, 2008)
21. M. Bonitz, A. Filinov, V.O. Golubnychiy, Th Bornath, W.D. Kraeft, First principle thermodynamic and dynamic simulations for dense quantum plasmas. Contrib. Plasma Phys. **45**, 450 (2005)
22. P. Hohenberg, W. Kohn, Inhomogenous electron gas. Phys. Rev. **136**, B864 (1964)
23. W. Kohn, L.J. Sham, Self-consistent equations including exchange and correlation effects. Phys. Rev. **140**, A1133 (1965)
24. F. Perrot, M.W.C. Dharma-wardana, Spin-polarized electron liquid at arbitrary temperatures: exchange-correlation energies, electron-distribution functions, and the static response functions. Phys. Rev. B **62**, 16536 (2000)
25. M.W.C. Dharma-wardana, The classical-map hyper-netted-chain (CHNC) method and associated novel density-functional techniques for warm dense matter. Int. J. Quantum Chem. **112**, 53 (2012)
26. J.W. Dufty, S. Dutta, Classical representation of a quantum system at equilibrium: theory. Phys. Rev. E **87**, 032101 (2013)

27. S. Dutta, J.W. Dufty, Classical representation of a quantum system at equilibrium: applications. Phys. Rev. E **87**, 032102 (2013)
28. S. Dutta, J.W. Dufty, Uniform electron gas at warm, dense matter conditions. Eur. Phys. Lett. **102**, 67005 (2013)
29. M. Bonitz, *Quantum Kinetic Theory* (Teubner, Stuttgart, 1998)
30. D. Kremp, Th Bornath, M. Bonitz, M. Schlanges, Quantum kinetic theory of plasmas in strong laser fields. Phys. Rev. E **60**, 4725 (1999)
31. M. Bonitz, Th Bornath, D. Kremp, M. Schlanges, W.D. Kraeft, Quantum kinetic theory for laser plasmas. Dynamical screening in strong fields. Contrib. Plasma Phys. **39**, 329 (1999)
32. M. Bonitz, D. Kremp, D.C. Scott, R. Binder, W.D. Kraeft, H.S. Köhler, Numerical analysis of non-Markovian effects in charge-carrier scattering: one-time versus two-time kinetic equations. J. Phys.: Condens. Matter **8**, 6057 (1996)
33. K. Balzer, M. Bonitz, *Nonequilibrium Green's Function Approach to Inhomogenous Systems*, Lecture Notes in Physics, vol. 867 (Springer, Berlin, 2013)
34. L.P. Kadanoff, G. Baym, *Quantum Statistical Mechanics* (W.A. Benjamin, New York, 1962)
35. N.-H. Kwong, M. Bonitz, Real-time Kadanoff-Baym approach to plasma oscillations in a correlated electron gas. Phys. Rev. Lett. **84**, 1768 (2000)
36. D. Kremp, M. Schlanges, W.D. Kraeft, *Quantum Statistics of Nonideal Plasmas* (Springer, Berlin, 2005)
37. M. Bonitz et al., Theory and simulations of strong correlations in quantum Coulomb systems. J. Phys. A: Math. Gen. **36**, 5921 (2003)
38. M. Bonitz, J.W. Dufty, Quantum kinetic theory of metal clusters in an intense electromagnetic field I. Cond. Matt. Phys. **7**, 483 (2004)
39. M. Bonitz et al., Classical and quantum Coulomb crystals. Phys. Plasmas **15**, 055704 (2008)
40. M. Bonitz, Kinetic theory for quantum plasmas. AIP Conf. Proc. **1421**, 135 (2012)
41. S. Ichimaru, Strongly coupled plasmas: high-density classical plasmas and degenerate electron liquids. Rev. Mod. Phys. **54**, 1017 (1982)
42. P.K. Shukla, B. Eliasson, Colloquium: nonlinear collective interactions in quantum plasmas with degenerate electron fluids. Rev. Mod. Phys. **83**, 885 (2011)
43. S.V. Vladimirov, Y.O. Tyshetskiy, On description of collisionless quantum plasmas. Phys. Uspekhi **54**, 1243 (2011)
44. E. Madelung, Z. Physik **40**, 322 (1927)
45. G. Manfredi, How to model quantum plasmas. Fields Inst. Commun. **46**, 263 (2005)
46. G. Manfredi, F. Haas, Self-consistent fluid model for a quantum electron gas. Phys. Rev. B **64**, 075316 (2001)
47. G. Manfredi, P.-A. Hervieux, Y. Yin, N. Crouseilles, in *Collective Electron Dynamics in Metallic and Semiconductor Nanostructures*, ed. by C. Massobrio, H. Bulou, C. Goyhenex. Atomic-Scale Modeling of Nanosystems and Nanostructured Materials, Lecture Notes in Physics, vol. 795 (Springer, Berlin, 2010)
48. N. Crouseilles, P.-A. Hervieux, G. Manfredi, Quantum hydrodynamic models for nonlinear electron dynamics in thin metal films. Phys. Rev. B **78**, 155412 (2008)
49. F. Haas, *Quantum Plasmas-An Hydrodynamic Approach* (Springer, New York, 2011)
50. A. Jüngel, *Transport Equations in Semiconductors* (Springer, Berlin, 2009)
51. C.L. Gardner, The quantum hydrodynamic model for semiconductor devices. SIAM (Soc. Ind. Appl. Math.) J. Appl. Math. **54**, 409 (1994)
52. A. Bennett, Influence of the electron charge distribution on surface-plasmon dispersion. Phys. Rev. B **1**, 203 (1970)
53. M. Marklund, G. Brodin, L. Stenflo, C.S. Liu, New quantum limits in plasmonic devices. Europhys. Lett. **84**, 17006 (2008)
54. E.P. Gross, Structure of a quantized vortex in boson system. Nuovo Cimento **20**, 454 (1961)
55. L.P. Pitaevskii, Zh. Eksp. Teor. Fiz. **40**, 646 (1961) [Sov. Phys. JETP 13, 451 (1961)]
56. A. Serbeto, L.F. Monteiro, K.H. Tsui, J.T. Mendonca, Quantum plasma fluid model for high-gain free-electron lasers. Plasma Phys. Control. Fusion **51**, 124024 (2009)

57. A. Kendl, P.K. Shukla, Drift wave turbulence in a dense semi-classical magnetoplasma. Phys. Lett. A **375**, 3138 (2011)
58. D. Bohm, A suggested interpretation of the quantum theory in terms of "hidden" variables. I. Phys. Rev. **85**, 166 (1952)
59. D. Bohm, A suggested interpretation of the quantum theory in terms of "hidden" variables. II. Phys. Rev. **85**, 180 (1952)
60. D. Bohm, J.P. Vigier, Model of the causal interpretation of quantum theory in terms of a fluid with irregular fluctuations. Phys. Rev. **96**, 208 (1954)
61. P.R. Holland, *The Quantum Theory of Motion* (Cambridge, New York, 1993)
62. R.E. Wyatt, *Quantum Dynamics with Trajectories* (Springer, Berlin, 2005)
63. E.P. Wigner, On the quantum correction for thermodynamic equilibrium. Phys. Rev. **40**, 749 (1932)
64. J.E. Moyal, Quantum mechanics as a statistical theory. Proc. Cambridge Phil. Soc. **45**, 99 (1949)
65. P. Carruthers, F. Zachariasen, Quantum collision theory with phase-space distributions. Rev. Mod. Phys. **55**, 245 (1983)
66. M. Hillery, R.F. O'Connell, M.O. Scully, E.P. Wigner, Distribution functions in physics: fundamentals. Phys. Rep. **106**, 121 (1984)
67. H.W. Lee, Theory and application of the quantum phase-space distribution-functions. Phys. Rep. **259**, 147 (1995)
68. J. Dawson, On Landau damping. Phys. Fluid **4**, 869 (1961)
69. F. Haas, G. Manfredi, M. Feix, Multistream model for quantum plasma. Phys. Rev. E **62**, 2763 (2000)
70. M. Bonitz, E. Pehlke, T. Schoof, Attractive forces between ions in quantum plasmas: failure of linearized quantum hydrodynamics. Phys. Rev. E **87**, 033105 (2013)
71. J. Lindhard, K. Dan. Vidensk. Selsk. Mat. Fys. Medd. **28**, 8 (1954)
72. V.O. Golubnychiy, M. Bonitz, D. Kremp, M. Schlanges, Dynamical properties of plasmon dispersion of a weakly degenerate correlated one-component plasma. Phys. Rev. E **64**, 016409 (2001)
73. D.M. Melrose, in *Quantum Plasmadynamics: Unmagnetized plasmas*, Lecture Notes in Physics, vol. 735 (Springer, New York, 2008)
74. D.B. Melrose, A. Mushtaq, Quantum recoil and Bohm diffusion. Phys. Plasmas **16**, 094508 (2009)
75. M. Bonitz, Impossibility of plasma instabilities in isotropic quantum plasmas. Phys. Plasmas **1**, 832 (1994)
76. W.R. Frensley, Boundary conditions for open quantum systems driven far from equilibrium. Rev. Mod. Phys. **62**, 745 (1990)
77. A.F. Alexandrov, L.S. Bogdankevich, A.A. Rukhadze, Osnovy Elektrodinamiki Plazmy (*Principles of Plasma Electrodynamics*) (Vysshaya Shkola, Moscow, 1988) [Translated into English (Springer, Berlin, 1984)]
78. F. Haas, L.G. Garcia, J. Goedert, G. Manfredi, Quantum ion-acoustic waves. Phys. Plasmas **10**, 3858 (2003)
79. F. Haas, A. Bret, Nonlinear low frequency collisional quantum Buneman instability. Europhys. Lett. **97**, 26001 (2012)
80. O. Buneman, Dissipation of currents in ionized media. Phys. Rev. **115**, 503 (1959)
81. S. Bauch et al., in *Introduction to Complex Plasmas*. ed. by M. Bonitz, N. Horing, P. Ludwig (Springer, Berlin, 2010)
82. V.E. Zakharov, Collapse of Langmuir waves. Sov. Phys. JETP **35**, 908 (1972)
83. L.G. Garcia, F. Haas, L. de Oliveira, J. Goedert, Modified Zakharov equations for plasmas with a quantum correction. Phys. Plasmas **12**, 012302 (2005)
84. F. Haas, P.K. Shukla, Quantum and classical dynamics of Langmuir wave packets. Phys. Rev. E **79**, 066402 (2009)
85. S.A. Khan, H. Saleem, Linear coupling of Alfven waves and acoustic-type modes in dense quantum magnetoplasmas. Phys. Plasmas **16**, 052109 (2009)

86. H. Saleem, A. Ahmad, S.A. Khan, Low frequency electrostatic and electromagnetic modes of ultracold magnetized nonuniform dense plasmas. Phys. Plasmas **15**, 094501 (2008)
87. B. Shokri, A.A. Rukhadze, Quantum drift waves. Phys. Plasmas **6**, 4467 (1999)
88. P.K. Shukla, B. Eliasson, Novel attractive force between ions in quantum plasmas. Phys. Rev. Lett. **108**, 165007 (2012) [Erratum: Phys. Rev. Lett. **108**, 219902 (2012); Erratum. Phys. Rev. Lett. **109**, 019901 (2012)]
89. P.K. Shukla, B. Eliasson, M. Akbari-Moghanjoughi, Comment on "attractive forces between ions in quantum plasmas: failure of linearized quantum hydrodynamics". Phys. Rev. E **87**, 037101 (2013)
90. M. Bonitz, E. Pehlke, T. Schoof, Reply to "comment on attractive forces between ions in quantum plasmas: failure of linearized quantum hydrodynamics". Phys. Rev. E **87**, 037102 (2013)
91. M. Bonitz, E. Pehlke, T. Schoof, Phys. Scr. 88, 057001 (2013)
92. M. Marklund, G. Brodin, Dynamics of spin-$\frac{1}{2}$ quantum plasmas. Phys. Rev. Lett. **98**, 025001 (2007)
93. S.M. Mahajan, F. Asenjo, Vortical dynamics of spinning quantum plasma: helicity conservation. Phys. Rev. Lett. **107**, 195003 (2011)
94. S. Braun, F. Asenjo, S.M. Mahajan, Spin-gradient-driven light amplification in a quantum plasma. Phys. Rev. Lett. **109**, 175003 (2012)
95. G.S. Krishnaswami, R. Nityananda, A. Sen, A. Tyagaraja, A critique of recent theories of spin half quantum plasmas, arxiv:1306.1774 (2013)
96. E.M. Lifshitz, L.P. Pitaevski, *Physical Kinetics* (Pergamon, Oxford, 1981)
97. N.W. Ashcroft, N.D. Mermin, *Solid State Physics* (Harcourt Brace, New York, 1976)
98. C. Kittel, *Introduction to Solid State Physics* (John Wiley, New York, 1963)

Chapter 5
Introduction to Configuration Path Integral Monte Carlo

Tim Schoof, Simon Groth and Michael Bonitz

Abstract In low-temperature high-density plasmas quantum effects of the electrons are becoming increasingly important. This requires the development of new theoretical and computational tools. Quantum Monte Carlo methods are among the most successful approaches to first-principle simulations of many-body quantum systems. In this chapter we present a recently developed method—the configuration path integral Monte Carlo (CPIMC) method for moderately coupled, highly degenerate fermions at finite temperatures. It is based on the second quantization representation of the N-particle density operator in a basis of (anti-)symmetrized N-particle states (configurations of occupation numbers) and allows to tread arbitrary pair interactions in a continuous space. We give a detailed description of the method and discuss the application to electrons or, more generally, Coulomb-interacting fermions. As a test case we consider a few quantum particles in a one-dimensional harmonic trap. Depending on the coupling parameter (ratio of the interaction energy to kinetic energy), the method strongly reduces the sign problem as compared to direct path integral Monte Carlo (DPIMC) simulations in the regime of strong degeneracy which is of particular importance for dense matter in laser plasmas or compact stars. In order to provide a self-contained introduction, the chapter includes a short introduction to Metropolis Monte Carlo methods and the second quantization of quantum mechanics.

T. Schoof (✉) · S. Groth · M. Bonitz
Department of Physics, Kiel University, Leibnizstr. 15, 24098 Kiel, Germany
e-mail: schoof@theo-physik.uni-kiel.de

S. Groth
e-mail: groth@theo-physik.uni-kiel.de

M. Michael
e-mail: bonitz@physik.uni-kiel.de

M. Bonitz et al. (eds.), *Complex Plasmas*, Springer Series on Atomic, Optical, and Plasma Physics 82, DOI: 10.1007/978-3-319-05437-7_5,

5.1 Introduction

Interacting fermionic many-body systems are of great interest in many areas of physics. These include atoms and molecules, electron gases in solids, partly ionized and dense plasmas in strong laser fields, and astro-physical plasma or the quark-gluon plasma [1–6]. In these systems the quantum behavior of the electrons (and, at high density, also of the ions) plays an important role. Similar relevance of quantum effects is observed in low-temperature plasmas at atmospheric pressure, in particular, when the plasma comes in contact with a solid surface. For the quantum treatment of electrons in the latter case, see the chapter by Heinisch et al.

Thermodynamic properties of these system can only be calculated with huge difficulties if long range interactions as well as degeneracy of the quantum particles become important. Many models and approximations exist, including quantum hydrodynamics (QHD), cf. the chapter by Khan et al. in this book, but these are often inaccurate or not controllable. For that reason *first-principle* computer-simulation are of great importance. Among the most successful methods for the simulation of interacting many-body quantum systems are the density functional theory (DFT) [7], many-body theories like, e.g. Green functions [8–10], and quantum Monte Carlo (QMC) methods. Nevertheless, the *ab initio* simulation of fermions is still an unsolved problem.

The main idea of finite-temperature QMC methods is based on the description of the system in terms of the Feynman path integral [11]. In this formulation a quantum system in thermodynamic equilibrium can be described by classical paths in an effective "imaginary time". While, in principle, exact results can be obtained for arbitrary large particle numbers for bosons [12, 13], Monte Carlo (MC) methods for fermions suffer from the so-called *fermion sign problem* [14, 15], which leads to exponentially increasing statistical errors in dependence of the particle number. The mathematical reason for this is the alternating sign of the terms that contribute to the expectation values due to the antisymmetry of the fermionic wave function under particle permutations. The physical origin is the Pauli principle. There exist several approaches to reduce or even avoid the sign problem. The *fixed-nodes* path integral Monte Carlo approximation can be considered as one of the most successful methods for highly degenerate systems like warm dense matter [16–19]. It uses the knowledge about the nodal surface structure of a trial density matrix to completely avoid the sign problem. However, the choice of the trial density matrix introduces systematical errors that are difficult to estimate. Another approach is the multi-level-blocking algorithm [20]. Exact direct path integral Monte Carlo (DPIMC) calculations are also possible with the use of several optimizations [21, 22] but this is possible only for sufficiently high temperature (moderate degeneracy).

In this chapter we will present a different approach—the configuration path integral Monte Carlo (CPIMC) method [23]. The main idea of this approach is to evaluate the path integral not in configuration space (as in DPIMC), but in second quantization. This leads to paths in Slater determinant space in occupation number representation instead of paths in coordinate space. The idea is based on the continuous time path integral method [24, 25], which is widely used for lattice models such as the Hubbard model, see [26], and for impurity models, see [27]. These models are described by simplified Hamiltonians where the interaction is typically of short-range. However, for many systems, in particular plasmas, the long-range Coulomb interaction is of central importance. In this chapter we will show how the CPIMC idea can be used to calculate thermodynamic properties of plasmas or, more generally, spatially continuous fermionic systems with arbitrary pair interactions.

The chapter will start with a short introduction to Metropolis Monte Carlo methods and an advanced description of the estimation of statistical errors. We will then introduce the formalism of the second quantization as far it is needed to understand the presented algorithm and give a short overview over the equations of statistical quantum physics that describe the thermodynamics of the systems of interest in equilibrium. In the main part we will derive the underlying formulas of CPIMC in detail and give a complete description of the MC algorithm. At the end of the chapter a demonstrative application of the method to a one-dimensional system of Coulomb-interacting fermions in a harmonic trap will be discussed.

5.2 Monte Carlo

In the first section of this chapter, the *Metropolis algorithm*, which has been developed by Metropolis et al. in 1953 [28], shall be explained in a fairly general form. The next section, where we will briefly illustrate how to estimate the error of quantities computed with the *Metropolis algorithm*, is based on [29]. We mention that similar Monte Carlo methods are used also in the chapters by Thomsen et al. (classical Metropolis MC) and Rosenthal et al. (kinetic MC) of this book.

5.2.1 Metropolis Algorithm

Our ultimate goal is the computation of thermodynamic properties of quantum many-body systems in equilibrium. However, the following considerations are completely general and are valid for classical and quantum systems. In the classical case, we only have to remove the operator heats of the observables. Regardless of the chosen ensemble, expectation values of observables $\hat{O}$ are always of the form

$$\langle \hat{O} \rangle = \sum\!\!\!\!\!\!\int_C O(C) \frac{W(C)}{Z} \quad \text{with} \quad Z = \sum\!\!\!\!\!\!\int_C W(C). \tag{5.1}$$

The partition function Z is given by the sum over all weights $W(C)$, where the multi-variable C defines exactly one contribution to the partition function, i.e. one specific weight. Moreover, we refer to the multi-variable C as a configuration or a microstate of the system. Such a configuration is generally defined by a set of discrete and/or continuous variables, which is why the symbol $\displaystyle\sum\!\!\!\!\!\!\int_C$ was introduced in (5.1) denoting the integration or summation over all these variables. Further, $O(C)$ denotes the value of the observable in the configuration C. We can interpret $P(C) = \frac{W(C)}{Z}$ as the probability of the system to be in the configuration C if and only if the weights $W(C)$ are always real and positive. Hence, the expectation value of an observable (5.1) is given by the sum over all possible values of the observable $O(C)$ weighted with its corresponding probability.

The difference between classical and quantum systems as well as the choice of the ensemble (microcanonical, canonical or grand canonical ensemble) enter in the actual form of the weights and in the set of variables necessary to define a configuration. [In Chap. 1 by Thomsen et al. Metropolis Monte Carlo of a classical system in the canonical ensemble is described.] For a quantum system, the partition function is given by the trace over the $N-$particle density operator $\hat{\rho}$ discussed in detail in Sect. 5.4 below:

$$Z = \mathrm{Tr}\hat{\rho}. \tag{5.2}$$

Each choice of a different basis in which the trace is performed leads to a different Monte Carlo algorithm. For example, performing the trace in coordinate space results in the DPIMC method [12], whereas the occupation number representation yields the CPIMC method described in Sect. 5.5 and [23].

Now we will continue with the general consideration of the *Metropolis algorithm* for quantum and classical systems. Using the *Metropolis algorithm*, one can generate a sequence of configurations C_i which are distributed with probability $P(C_i)$. However, $P(C_i)$ involves the partition function Z (normalization of the distribution), that is typically very hard to compute. To avoid this problem, one defines a transition probability $\nu(C_i \to C_{i+1})$ for the system that characterizes the transition into the configuration C_{i+1} starting from C_i. To ensure that the configurations C_i are distributed with $P(C_i)$, the transition probability has to fulfill the *detailed balance condition* [1]:

$$P(C_i)\nu(C_i \to C_{i+1}) = P(C_{i+1})\nu(C_{i+1} \to C_i). \tag{5.3}$$

A possible solution of this equation is given by

$$\nu(C_i \to C_{i+1}) = \min\left[1, \frac{P(C_{i+1})}{P(C_i)}\right] = \min\left[1, \frac{W(C_{i+1})}{W(C_i)}\right]. \tag{5.4}$$

Now the algorithm works as follows. Suppose the system is in the configuration C_i. A Monte Carlo step then consists of proposing a certain change of the configuration converting it into C_{i+1}. After that, a random number from [0, 1) is generated, and the transition probability is calculated according to (5.4). If the random number is smaller than $\nu(C_i \to C_{i+1})$, then the change (Monte Carlo step) is accepted. Otherwise, the

---

[1] In fact, there are weaker conditions on the transition probabilities, but in practice, one uses the *detailed balance*.

system stays in the previous configuration, i.e. it is $C_{i+1} = C_i$. Starting from an arbitrary configuration C_0, after some steps, the configurations will be distributed with $P(C)$ forming a so-called *Markov chain*. We refer to the number of steps which are necessary for the initial correlations to vanish as the *equilibration time*.

In addition to the *detailed balance*, we have to make sure the transition probabilities are ergodic, i.e. it must be possible to reach every configuration with $W(C) > 0$ in a finite number of steps. In practice, one will usually need a few different Monte Carlo steps to ensure the ergodicity. These steps address different degrees of freedom of the configurations, which, of course, depend on the actual form of the partition function, (5.1). Furthermore, in the majority of cases, it will be more efficient not to propose every change of the configuration with equal probability. Therefore, we split the transition probability into a sampling probability $T(C_i \to C_{i+1})$ and an acceptance probability $A(C_i \to C_{i+1})$ for that specific change, i.e. it is

$$\nu(C_i \to C_{i+1}) = A(C_i \to C_{i+1})T(C_i \to C_{i+1}). \tag{5.5}$$

Inserting this factorization into the *detailed balance*, one readily obtains the solution for the acceptance probability

$$A(C_i \to C_{i+1}) = \min\left[1, \frac{T(C_{i+1} \to C_i)W(C_{i+1})}{T(C_i \to C_{i+1})W(C_i)}\right]. \tag{5.6}$$

This generalization is also called the *Metropolis Hastings algorithm*. The algorithm is most efficient if the acceptance probability is about 50 %. A suitable choice of the sampling probability can help optimizing the acceptance probability.

Once we have generated a *Markov chain* of length N_{MC} via the *Metropolis algorithm*, a good estimator for the expectation value of an observable is given by

$$\langle \hat{O} \rangle \approx \frac{1}{N_{MC}} \sum_{i=1}^{N_{MC}} O(C_i) =: \overline{O}, \tag{5.7}$$

which directly follows from (5.1) and the fact that all configurations C_i in the *Markov chain* are distributed with $P(C_i)$.

Unfortunately, the weight $W(C)$ is not always strictly positive, so we can not always interpret $\frac{W(C)}{Z}$ as a probability. Especially for systems of fermions, there are many sources of sign changes in the weights. However, we can still apply the *Metropolis algorithm* if we define a new partition function

$$Z' := \sum_C |W(C)|, \tag{5.8}$$

and rewrite the expectation values as follows:

$$\langle\hat{O}\rangle = \frac{\sum_C O(C)W(C)}{\sum_C W(C)} = \frac{\frac{1}{Z'}\sum_C O(C)S(C)|W(C)|}{\frac{1}{Z'}\sum_C S(C)|W(C)|} = \frac{\langle\hat{O}S\rangle'}{\langle S\rangle'}, \tag{5.9}$$

where $S(C) = \mathrm{sgn}[W(C)]$ denotes the sign of the weight. Obviously, the expectation value of $\hat{O}$ in the physical system, described by the partition function Z, is equivalent to the expectation value of $\hat{O}S$ divided by the expectation value of the sign (*average sign*), both averaged in the primed system described by Z', cf. (5.8). For that system, we can generate a *Markov chain* if we simply insert the modulus of the weights $|W(C)|$ in (5.6) when computing the acceptance probability. Combining (5.7) and (5.9), we obtain an expression for the estimator of the observable in the real system:

$$\langle\hat{O}\rangle \approx \frac{\sum_{i=1}^{N_{MC}} O(C_i)S(C_i)}{\sum_{i=01}^{N_{MC}} S(C_i)} =: \frac{\overline{OS}'}{\overline{S}'} = \overline{O}. \tag{5.10}$$

Thus, we actually simulate the primed system and calculate quantities of the physical system via (5.10).

5.2.2 *Error in the Monte Carlo Simulation*

Estimating the errors of quantities computed with the *Metropolis algorithm* is not a trivial task. First, we consider the case of strictly positive weights, where the estimator of an observable is given by (5.7). It is reasonable to interpret the values of the observable in each configuration of the *Markov chain* as a series of measurements. The estimator $\overline{O}$, which is obtained by averaging over all measurements, fluctuates statistically around the true expectation value $\langle\hat{O}\rangle$. The error of the arithmetic mean of a measurement series of length N_{MC} is defined by

$$\Delta\overline{O} = \frac{\sigma_{\overline{O}}}{\sqrt{N_{MC}}}, \tag{5.11}$$

where $\sigma_{\overline{O}}$ is the standard deviation which, for uncorrelated measurements, can be estimated by

$$\sigma_{\overline{O}} = \sqrt{\frac{1}{(N_{MC}-1)}\sum_{i=0}^{N_{MC}}(O_i - \overline{O})^2}, \quad \text{with} \quad O_i = O(C_i). \tag{5.12}$$

Certainly, the "measurement" series obtained from the *Markov chain* is correlated, since we generate each configuration from the previous one. The correlation of these configurations can be measured by the integrated auto-correlation time

$$\tau_{\text{int},O} = \frac{1}{2} + \sum_{k=1}^{N_{MC}} \frac{\overline{O_i O_{i+k}} - \overline{O}^2}{\sigma_{\bar{O}}^2}, \tag{5.13}$$

$$\text{with} \quad \overline{O_i O_{i+k}} = \frac{1}{N_{MC}} \sum_{i=1}^{N_{MC}} O_i O_{i+k}\,.$$

The auto-correlation of the measurements enhances the statistical error by a factor of $\sqrt{2\tau_{\text{int},O}}$, i.e. it is

$$\Delta\overline{O}_{\text{auto}} = \Delta\overline{O}\sqrt{2\tau_{\text{int},O}} = \frac{\sigma_{\bar{O}}}{\sqrt{N_{MC}}}\sqrt{2\tau_{\text{int},O}} = \frac{\sigma_{\bar{O}}}{\sqrt{N_{\text{eff}}}}. \tag{5.14}$$

Therefore, if we have generated a *Markov-chain* of length N_{MC}, we have effectively gained only $\frac{N_{MC}}{2\tau_{\text{int},O}}$ uncorrelated measurements. To keep the correlation small, efficient Monte Carlo steps are essential, which change many degrees of freedom of the configurations simultaneously while still having an acceptance ratio of about 50 %. In practice, acceptance ratios are much smaller than 50 %, and one therefore performs a certain number of Monte Carlo steps (a *cycle*) before again measuring the observables.

Things become more complex if the weights are not strictly positive. In that case, we have expressed quantities of the physical system by quantities of the primed system, (5.10). We simulate the primed system and calculate the r.h.s. of (5.10) to obtain $\overline{O}$. The measurements of the quantities $\overline{OS}'$ and $\overline{S}'$ are not only auto-correlated but also cross-correlated with each other. The relative statistical error can be estimated according to (cf. [29])

$$\frac{\Delta\overline{O}_{\text{auto,cross}}}{\overline{O}} = \sqrt{\left(\frac{\Delta\overline{OS}'}{\overline{OS}'}\right)^2 + \left(\frac{\Delta\overline{S}'}{\overline{S}'}\right)^2 - \frac{2}{N_{MC}}\frac{\overline{OSS}' - \overline{OS}'\,\overline{S}'}{\overline{OS}'\,\overline{S}'}2\tau_{int,OS,S}}, \tag{5.15}$$

where we have the statistical errors of $\overline{OS}'$ and $\overline{S}'$

$$\Delta\overline{OS}' = \sqrt{\frac{\overline{(OS)^2}' - \overline{(OS)}'^2}{N_{MC}}2\tau_{\text{int},OS}}\,,$$

$$\Delta\overline{S}' = \sqrt{\frac{\overline{S^2}' - \overline{S}'^2}{N_{MC}}2\tau_{\text{int},S}}, \tag{5.16}$$

each enhanced by their individual auto-correlation time and in the last term the integrated cross-correlation time

$$\tau_{\mathrm{int},OS,S} = \frac{1}{2} + \sum_{k=1}^{N_{MC}} \frac{\overline{O_i S_i S_{i+k}}' - \overline{OS}'\,\overline{S}'}{\overline{OSS}' - \overline{OS}'\,\overline{S}'}\,, \quad \text{with} \quad \overline{O_i S_i S_{i+k}}' = \frac{1}{N_{MC}} \sum_{i=1}^{N_{MC}} O_i S_i S_{i+k}\,.$$

Assuming we are working in the canonical ensemble, then, from the definition of the partition function and the *average sign*, cf. (5.9), it immediately follows that

$$\overline{S}' \approx \langle S \rangle' = \frac{Z}{Z'} = e^{-\beta N (f - f')} \leq 1\,, \tag{5.17}$$

where β is the inverse temperature, N the particle number, and $f^{(\prime)}$ the free energy per particle in the physical (primed) system. Combining (5.15) and (5.17) yields

$$\frac{\Delta \overline{O}_{\mathrm{auto,sign}}}{\overline{O}} \propto \frac{1}{\overline{S}'\sqrt{N_{MC}}} \approx \frac{1}{\sqrt{N_{MC}}} e^{\beta N (f - f')}\,. \tag{5.18}$$

The relative error of an observable grows exponentially with the product of the particle number, the inverse temperature and the difference of the free energy per particle in the physical and the primed system. Unfortunately, the error can only be reduced with the square root of the number of Monte Carlo samples (measurements). Therefore, the value of the *average sign* of a given system determines if one can compute reliable quantities via the *Metropolis algorithm* for that system. This severe limitation is called the (fermion) *sign problem*. In practice, an average sign of the order $\approx 10^{-3}$ is the limit for reliable Monte Carlo simulations. Also, one should be aware that even the sign itself has a statistical error of the form (5.14). Besides, we note that the proportionality (5.17) can be obtained by a simple Gaussian error propagation of $\overline{O}$, cf. (5.10).

The *sign problem* is strongly dependent on the actual representation of the partition function, cf. (5.1). For some special systems, the *sign problem* can be circumvented by a specific choice of the representation [30, 31], but a general solution is highly unlikely, since it has been shown to be NP-complete [15].

5.3 Second Quantization

Second quantization refers to the introduction of creation and annihilation operators for the description of quantum many-particle systems. Thereby, not only the observables are represented by operators but also the wave functions, in contrast to the standard formulation of quantum mechanics (correspondingly called "first" quantization), where only the observables are represented by operators. The proper symmetry properties of the bosonic (fermionic) many-particle states are automatically included in the (anti-)commutation relations of the creation and annihilation operators. Simultaneously, also observables can be expressed by these operators.

This unifying picture of quantum mechanics is often advantageous when it comes to the description of many-particle systems.

Since CPIMC makes extensive use of the second quantization of quantum mechanics, a brief introduction will be given in this section, where we will focus on those aspects which are necessary for the comprehension of the method. Especially the derivation of the crucial *Slater-Condon rules* shall be outlined. For a detailed introduction to the second quantization see e.g. [32].

5.3.1 (Anti-)Symmetric Many-Particle States

We consider a system of N identical, ideal (i.e. not interacting) quantum particles which is described by the Hamiltonian $\hat{H}_0$. For an ideal system, the Hamiltonian can be written as a sum of one-particle Hamiltonians:

$$\hat{H}_0 = \sum_{\alpha=1}^{N} \hat{h}_\alpha, \tag{5.19}$$

$$\hat{h}_\alpha = \frac{\hat{p}_\alpha^2}{2m} + \hat{v}_\alpha, \tag{5.20}$$

where $\hat{p}_\alpha$ denotes the momentum operator and $\hat{v}_\alpha$ the operator of the potential energy of the particle α in an external field. Thus, the subscript α indicates that those operators are acting on states from the one-particle Hilbert space $\mathscr{H}_\alpha$ of the particle α. We imply that the solutions of the eigenvalue problem

$$\hat{h}\,|i\rangle = \varepsilon_i\,|i\rangle\ , \quad \text{with} \quad i \in \mathrm{N}_0 \tag{5.21}$$

are known. The one-particle states $|i\rangle$ form a complete orthonormal system (CONS) of the one-particle Hilbert space. Further, we assume the states are arranged according to their one-particle energy, i.e. $\epsilon_i \le \epsilon_j$ for all $i \le j$. In general, the one-particle states are spin orbitals, i.e. $|i\rangle \in \mathscr{H}_{\text{coord}} \otimes \mathscr{H}_{\text{spin}}$ with the tensor product of the coordinate and spin Hilbert space. Hence, the wave function $\langle \mathbf{r}\sigma|i\rangle = (\langle \mathbf{r}\,|\,\langle\sigma|)\,|i\rangle = \phi_i(\mathbf{r},\sigma)$ depends on both the coordinate $\mathbf{r}$ and the spin projection σ of the particle.

In the case of an ideal system, the solution of the $N-$particle eigenvalue problem

$$\hat{H}_0\,|\Psi\rangle = E\,|\Psi\rangle \tag{5.22}$$

can be constructed from products of one-particle states:

$$|i_1 i_2 \ldots i_N\rangle = |i_1\rangle_1\,|i_2\rangle_2 \ldots |i_N\rangle_N\,. \tag{5.23}$$

The many-body Hilbert space of N particles $\mathcal{H}^N = \bigotimes_{\alpha=1}^{N} \mathcal{H}_\alpha$ is thus given by the tensor product of one-particle Hilbert spaces. The states $|i_1 i_2 \ldots i_N\rangle$ form a basis of $\mathcal{H}^N$, and the used notation indicates that particle α is in the state $|i_\alpha\rangle$.

Apart from some special exceptions as e.g. Axions, only totally symmetric or antisymmetric states with respect to arbitrary two-particle exchanges are physically realized, i.e. states with

$$|\ldots i_\alpha \ldots i_\beta \ldots\rangle_\pm = \pm |\ldots i_\beta \ldots i_\alpha \ldots\rangle \quad \forall \alpha, \beta. \tag{5.24}$$

Particles that are described by symmetric states (upper sign) are called bosons and those with antisymmetric states (lower sign) fermions. The reason for this symmetry of $N-$particle states lies in the indistinguishability of the quantum particles, whereby physical properties of the system cannot change under particle exchange. The spin-statistics-theorem states that fermions have half-integer and bosons integer spin.

Totally (anti-)symmetric states can be constructed from the product states (5.23) by summing up all $N!$ permutations of N particles:

$$|i_1 i_2 \ldots i_N\rangle_\pm = \frac{1}{\mathcal{N}} \sum_{\mathcal{P} \in S_N} (\pm 1)^P \hat{\mathcal{P}} \, |i_1\rangle_1 \, |i_2\rangle_2 \ldots |i_N\rangle_N \,, \tag{5.25}$$

where $\hat{\mathcal{P}}$ is the $N-$particle permutation operator that can be constructed from a composition of two-particle exchanges. The normalization factor $\mathcal{N}$ is given by

$$\mathcal{N} = \begin{cases} \sqrt{N! \prod_{i=0}^{\infty} n_i!} \,, & \text{for bosons,} \\ \sqrt{N!} \,, & \text{for fermions,} \end{cases} \tag{5.26}$$

where n_i denotes the number of one-particle states $|i\rangle$ in the product state. The hermitian operator $\hat{S}(\hat{A})$

$$\hat{S}/\hat{A} = \frac{1}{\mathcal{N}} \sum_{\mathcal{P} \in S_N} (\pm 1)^P \hat{\mathcal{P}} \,, \tag{5.27}$$

is referred to as the (anti-)symmetrization operator. For fermions, the sign of each summand is determined by the number of two-particle exchanges P in the permutation operator $\hat{\mathcal{P}}$. Further, arbitrary (anti-)symmetric states can be constructed from linear combinations of these (anti-)symmetric states, (5.25). Therefore, they form a basis in the (anti-)symmetric Hilbert space $\mathcal{H}_\pm^N \subset \mathcal{H}^N$. The operator $(\mathcal{N}/N!)\hat{S}$ or $(\mathcal{N}/N!)\hat{A}$ is a projection operator on the respective subspace, it is $\hat{S}^2 = (N!/\mathcal{N})\hat{S}$ and $\hat{A}^2 = (N!/\mathcal{N})\hat{A}$, respectively and also $\hat{S}\hat{A} = 0$.

In coordinate-spin representation, the anti-symmetric product states of fermions can be written as determinants (called *Slater determinants*):

$$\Psi(x_1, \ldots, x_N)_- = \frac{1}{N!} \begin{vmatrix} \phi_{i_1}(x_1) & \phi_{i_2}(x_1) & \cdots & \phi_{i_N}(x_1) \\ \phi_{i_1}(x_2) & \phi_{i_2}(x_2) & \cdots & \phi_{i_N}(x_2) \\ \vdots & \vdots & & \vdots \\ \phi_{i_1}(x_N) & \phi_{i_2}(x_N) & \cdots & \phi_{i_N}(x_N) \end{vmatrix} . \tag{5.28}$$

For bosons, one obtains instead permanents. From this representation of antisymmetric states it becomes obvious that there is none with a one-particle state occurring twice in the product state, for a determinant with two equal columns vanishes. This is also known as the famous *Pauli principle*. For bosons, there is no such restriction.

5.3.2 Occupation Number Representation

Due to the (anti-)symmetrization of the many-particle states (5.25), such states are entirely characterized by the occurrence frequency of each one-particle state in the product state. The number of particles n_i in the one-particle state $|i\rangle$ is called the occupation number (of the i-th state/orbital). Since the complete set of occupation numbers, denoted with $\{n\}$, defines an (anti-)symmetric many-particle state, we can also write

$$|i_1 \ldots i_N\rangle_\pm \equiv |n_0 n_1 n_2 \ldots\rangle =: |\{n\}\rangle , \tag{5.29}$$

with

$$n_i \in \begin{cases} \mathbb{N}_0, & \text{for bosons} \\ \{0, 1\}, & \text{for fermions} \end{cases} .$$

The order of the occupation numbers equals the order of the one-particle states in the product state, which can be arbitrary but must remain fixed for further calculations. The *Pauli principle* for fermions is automatically included in the restriction on the occupation numbers to be either zero or one.

If we only consider many-particle states with fixed particle number, $N = \sum_{i=0}^{\infty} n_i$, then the states $|\{n\}\rangle$ form a CONS of the (anti-)symmetric $N-$particle Hilbert space $\mathscr{H}_\pm^N$ with the orthogonality relation

$$\langle\{n\}|\{\bar{n}\}\rangle = \prod_{i=0}^{\infty} \delta_{n_i,\bar{n}_i} =: \delta_{\{n\},\{\bar{n}\}} , \tag{5.30}$$

and the completeness relation

$$\sum_{\{n\}} |\{n\}\rangle \langle\{n\}| \, \delta_{\sum_i n_i, N} = \hat{1}_N . \tag{5.31}$$

To shorten the notation, we introduced the following abbreviation:

$$\sum_{\{n\}} := \begin{cases} \sum_{n_0=0}^{\infty} \sum_{n_1=0}^{\infty} \cdots, & \text{for bosons} \\ \sum_{n_0=0}^{1} \sum_{n_1=0}^{1} \cdots, & \text{for fermions} \end{cases} . \tag{5.32}$$

In practical simulations, one has to work with a finite number of one-particle orbitals N_B. This equals the approximation

$$\sum_{\{n\}} \approx \begin{cases} \sum_{n_0=0}^{\infty} \sum_{n_1=0}^{\infty} \cdots \sum_{n_{N_B}=0}^{\infty}, & \text{for bosons} \\ \sum_{n_0=0}^{1} \sum_{n_1=0}^{1} \cdots \sum_{n_{N_B}=0}^{1}, & \text{for fermions} \end{cases} . \tag{5.33}$$

Given a certain one-particle basis, each antisymmetric state $|\{n\}\rangle$ can be uniquely identified with a *Slater determinant* (5.28). The relations (5.30) and (5.31) are consistent with the respective relations for determinants. For that reason, it is common to also refer to the states $|\{n\}\rangle$ as *Slater determinants.*

Now we drop the restriction of a fixed number of particles. The inner product in (5.30) is still well defined for states with different particle numbers; only the completeness relation is slightly modified

$$\sum_{\{n\}} |\{n\}\rangle \langle\{n\}| = \hat{1}. \tag{5.34}$$

The states $|\{n\}\rangle$ thereby form a CONS in the so-called Fock space $\mathscr{F}_{\pm} = \mathscr{H}^0 \oplus \mathscr{H}^1 \oplus \mathscr{H}^2_{\pm} \ldots$, which contains (anti-)symmetric states of varying particle number. Consequently, any state of the Fock space can be written as a linear combination of the *Slater determinants* $|\{n\}\rangle$:

$$|\Psi\rangle = \sum_{\{n\}} c_{\{n\}} |\{n\}\rangle . \tag{5.35}$$

For example, if we consider a general Hamiltonian of interacting particles

$$\hat{H} = \hat{H}_0 + \hat{W}, \tag{5.36}$$

where we added the interaction operator $\hat{W}$ to the ideal Hamiltonian $\hat{H}_0$, then the solution $|\Psi\rangle$ of the eigenvalue problem

$$\hat{H} |\Psi\rangle = E |\Psi\rangle , \tag{5.37}$$

can in general not be written in terms of a single *Slater determinant* $|\{n\}\rangle$ but will be of the form (5.35). Besides, those states do not necessarily have a defined particle number. Furthermore, the *Slater determinant* without any particles $|00\ldots\rangle = |\{0\}\rangle$

is the *vacuum state* that is normalized to one according to (5.30) and also belongs to the Fock space (The *vacuum state* should not be confused with the zero vector).

Finally, we could have chosen any arbitrary, complete one-particle set $\{|\nu\rangle\}$ in the definition of the states $|\{n\}\rangle$, cf. (5.29). Since from the states $|\{n\}\rangle$ it is not clear which one-particle basis has been used for the quantization, one usually has to explicitly specify the chosen basis.

5.3.3 Creation and Annihilation Operators

The introduction of creation and annihilation operators, which induce transitions between states of different particle number, is of crucial importance for the formalism of the second quantization. Originally, these operators have been constructed for the description of the harmonic oscillator (so-called ladder operators). In the following bosons and fermions will be treated separately.

5.3.3.1 Bosons

In the bosonic case, the action of the creation and annihilation operators on symmetric states in the occupation number representation is defined by

$$\hat{a}_i^\dagger \,|n_0 n_1 \ldots n_i \ldots\rangle = \sqrt{n_i + 1}\,|n_0 n_1 \ldots n_i + 1 \ldots\rangle \ , \tag{5.38}$$

$$\hat{a}_i \,|n_0 n_1 \ldots n_i \ldots\rangle = \sqrt{n_i}\,|n_0 n_1 \ldots n_i - 1 \ldots\rangle \, , \tag{5.39}$$

where the prefactors ensure the correct normalization. Apparently, the annihilation operator vanishes for $n_i = 0$. Hence, the creation and annihilation operators are not hermitian but pairwise adjoint. The bosonic creation and annihilation operators fulfill the commutation relations

$$\begin{aligned} [\hat{a}_i^\dagger, \hat{a}_j^\dagger] &= [\hat{a}_i, \hat{a}_j] = 0, \\ [\hat{a}_i, \hat{a}_j^\dagger] &= \delta_{i,j}, \end{aligned} \tag{5.40}$$

with the commutator $[\hat{A}, \hat{B}] = \hat{A}\hat{B} - \hat{B}\hat{A}$. Using the creation operator, we can construct arbitrary states of the form (5.29):

$$|\{n\}\rangle = \frac{1}{\sqrt{\prod_i n_i!}} \Big(\prod_{i=0}^{\infty} (\hat{a}_i^\dagger)^{n_i} \Big) \, |\{0\}\rangle \, . \tag{5.41}$$

Of particular interest is the hermitian occupation number operator of the ith orbital

$$\hat{n}_i = \hat{a}_i^\dagger \hat{a}_i . \tag{5.42}$$

Its action is given by

$$\hat{n}_i \, |\{n\}\rangle = n_i \, |\{n\}\rangle, \tag{5.43}$$

and the eigenvalues are simply the occupation numbers n_i of the ith orbital. Similarly, we have the total particle number operator

$$\hat{N} = \sum_{i=0}^{\infty} \hat{n}_i, \tag{5.44}$$

with the total particle number N of the state as its eigenvalue

$$\hat{N} \, |\{n\}\rangle = N \, |\{n\}\rangle \,. \tag{5.45}$$

Consequently, the states $|\{n\}\rangle$ are common eigenstates of the occupation number operator, the particle number operator and the ideal Hamilton operator (if we choose the ideal one-particle basis (5.21) for the quantization).

5.3.3.2 Fermions

For fermions, the states have to be antisymmetric. A definition of the fermionic creation and annihilation operators which satisfies this condition is given by

$$\begin{aligned} \hat{a}_i^\dagger \, |\{n\}\rangle &= (1 - n_i)(-1)^{\alpha_{\{n\},i}} \, |\ldots, n_i + 1, \ldots\rangle \,, \\ \hat{a}_i \, |\{n\}\rangle &= n_i (-1)^{\alpha_{\{n\},i}} \, |\ldots, n_i - 1, \ldots\rangle \,, \end{aligned} \tag{5.46}$$

where the sign is determined by

$$\alpha_{\{n\},i} = \sum_{l=0}^{i-1} n_l. \tag{5.47}$$

The sign is positive if the number of particles in the one-particle states before the ith state is even (with respect to the chosen order) and negative if it is odd. Further, the prefactors ensure that the creation (annihilation) operator vanishes if the orbital i, where the particle shall be created (annihilated), is already occupied (unoccupied), and so the definition also includes the *Pauli principle*. In contrast to the bosonic commutation relations, for fermions, the creation and annihilation operators fulfill the three anti-commutation relations:

$$\begin{aligned} \{\hat{a}_i^\dagger, \hat{a}_j^\dagger\} &= \{\hat{a}_i, \hat{a}_j\} = 0 \,, \\ \{\hat{a}_i, \hat{a}_j^\dagger\} &= \delta_{i,j} \end{aligned} \tag{5.48}$$

with the anti-commutator being defined as $\{\hat{A}, \hat{B}\} = \hat{A}\hat{B} + \hat{B}\hat{A}$. When it comes to the algebraic treatment of physical systems, the only difference between the second quantization for bosons and for fermions lies in the difference of the commutation relation (5.40) and anticommutation relation (5.48). For example, from the first anti-commutation relation directly follows that $(\hat{a}_i^\dagger)^2 = 0$, i.e. doubly occupied orbitals do not exist in the case of fermions. As a consequence of the non-commutativity of the fermionic creation and annihilation operators, we have to make sure the ordering of the creation operators equals the chosen order of the orbitals when constructing an arbitrary state from the vacuum states according to

$$|\{n\}\rangle = \left(\prod_{i=0}^{\infty} (\hat{a}_i^\dagger)^{n_i}\right) |\{0\}\rangle\,. \tag{5.49}$$

The formulas for the occupation number operator and the particle number operator (5.42)–(5.44) remain unchanged for fermions.

For further calculations, we will need the matrix elements of the creation and annihilation operators

$$\begin{aligned}
\langle\{n\}|\hat{a}_k^\dagger|\{\bar{n}\}\rangle &= (-1)^{\alpha_{\{n\},k}} \delta^k_{\{n\},\{\bar{n}\}} \delta_{n_k,1} \delta_{n_k,\bar{n}_k+1}\,, \\
\langle\{n\}|\hat{a}_k|\{\bar{n}\}\rangle &= (-1)^{\alpha_{\{n\},k}} \delta^k_{\{n\},\{\bar{n}\}} \delta_{n_k,0} \delta_{n_k,\bar{n}_k-1}\,, \\
\delta^k_{\{n\},\{\bar{n}\}} &:= \prod_{\substack{i=0 \\ i\neq k}}^{\infty} \delta_{n_i,\bar{n}_i}\,.
\end{aligned} \tag{5.50}$$

These follow directly from the definitions (5.46). Evidently, the creation and annihilation operators are defined with respect to a certain one-particle basis $\{|i\rangle\}$, the basis that we chose for the quantization. The transformation of the creation and annihilation operators to a different basis $\{|\nu\rangle\}$ can be performed as follows:

$$\hat{a}_\nu^\dagger = \sum_{i=0}^{\infty} \langle i|\nu\rangle\, \hat{a}_i^\dagger, \tag{5.51}$$

$$\hat{a}_\nu = \sum_{i=0}^{\infty} \langle \nu|i\rangle\, \hat{a}_i. \tag{5.52}$$

For example, choosing spin-coordinate space for the quantization, i.e. $|\nu\rangle = |x\rangle$, $x = \{\mathbf{r}, \sigma\}$, leads to the so-called *field operators*

$$\hat{\Psi}^\dagger(x) := \hat{a}_x^\dagger = \sum_{i=0}^{\infty} \langle i|x\rangle\, \hat{a}_i^\dagger = \sum_{i=0}^{\infty} \phi_i^*(x) \hat{a}_i^\dagger\,, \tag{5.53}$$

$$\hat{\Psi}(x) := \hat{a}_x = \sum_{i=0}^{\infty} \langle x|i\rangle\, \hat{a}_i = \sum_{i=0}^{\infty} \phi_i(x)\hat{a}_i. \tag{5.54}$$

The *field operator* $\hat{\Psi}^\dagger(x)$ creates and $\hat{\Psi}(x)$ annihilates a particle at the space point $\mathbf{r}$ with spin projection σ.

5.3.4 Operators in Second Quantization

Arbitrary first quantized operators can be expressed in the second quantization via the creation and annihilation operators. In the following, we will consider this representation for the one- and two-particle operators. For fermions, in particular, we will give the expressions for the corresponding matrix elements of the interaction potential, which leads to the *Slater-Condon-rules*.

5.3.4.1 One-Particle Operators

Though somewhat misleadingly, it is common to refer to the $N-$particle operator

$$\hat{B}_1 = \sum_\alpha \hat{b}_\alpha, \tag{5.55}$$

that is a sum of true one-particle operators $\hat{b}_\alpha$ as a "one-particle operator", too. In second quantization, that operator takes the following form:

$$\hat{B}_1 = \sum_{i,j=1}^{\infty} b_{ij}\hat{a}_i^\dagger \hat{a}_j, \tag{5.56}$$

with the one-particle integrals

$$b_{ij} = \langle i|\hat{b}|j\rangle = \int \mathrm{d}x \phi_i^*(x) b(x) \phi_j(x), \tag{5.57}$$

where the integration over $x = \{\mathbf{r}, \sigma\}$ includes an integration over the space coordinate and a summation over the spin. As usual, we have assumed local operators in coordinate space. From the matrix elements of the creation and annihilation operators (5.50), we readily obtain the matrix elements of the one-particle operators (5.56):

$$\begin{aligned}\langle\{n\}|\hat{a}_l^\dagger \hat{a}_k|\{\bar{n}\}\rangle &= \sum_{\{n'\}} \langle\{n\}|\hat{a}_l^\dagger|\{n'\}\rangle\, \langle\{n'\}|\hat{a}_k|\{\bar{n}\}\rangle \\ &= \sum_{\{n'\}} (-1)^{\alpha_{\{n'\},l}+\alpha_{\{\bar{n}\},k}} \delta^l_{\{n\},\{n'\}} \delta^k_{\{n'\},\{\bar{n}\}} \delta_{n'_l,0}\delta_{n_l,1}\delta_{\bar{n}_k,1}\delta_{n'_k,0}.\end{aligned} \tag{5.58}$$

Via case differentiation we can simplify this to

$$\langle\{n\}|\hat{a}_l^\dagger\hat{a}_k|\{\bar{n}\}\rangle = \begin{cases}(-1)^{\alpha_{\{n\},l}+\alpha_{\{\bar{n}\},k}}\delta^{kl}_{\{n\},\{\bar{n}\}}\delta_{n_l,1}\delta_{\bar{n}_k,1}\delta_{\bar{n}_l,0}\delta_{n_k,0}, & k\neq l\\ n_l\delta_{\{n\},\{\bar{n}\}}, & k=l\end{cases}. \tag{5.59}$$

Inserting this into (5.56) and rearranging the sums with respect to terms with $k = l$ and $k \neq l$ yields

$$\begin{aligned}\langle\{n\}|\hat{B}_1|\{\bar{n}\}\rangle &= \delta_{\{n\},\{\bar{n}\}}\sum_{k=0}^{\infty} b_{kk}n_k\\ &+\sum_{k=1}^{\infty}\sum_{\substack{l=1\\ l\neq k}}^{\infty} b_{lk}(-1)^{\alpha_{\{n\},l}+\alpha_{\{\bar{n}\},k}}\delta^{kl}_{\{n\},\{\bar{n}\}}\delta_{n_l,1}\delta_{\bar{n}_k,1}\delta_{\bar{n}_l,0}\delta_{n_k,0}.\end{aligned} \tag{5.60}$$

The first term only gives a contribution if both *Slater determinants* are equal; thus, these are the diagonal elements of the matrix. The second term does not vanish only if the right and left state differ in exactly two occupation numbers while conserving the total particle number.

Let $|\{n\}^p_q\rangle$ be the $N-$particle state that one obtains from the state $|\{n\}\rangle$ if one particle is removed from the qth orbital and added to the pth orbital, i.e. for $q < p$ it is

$$|\{n\}^p_q\rangle = |\ldots, n_q - 1, \ldots, n_p + 1, \ldots\rangle\,. \tag{5.61}$$

Using this notation, we can rewrite the matrix elements of a one-particle operator of the form (5.56) in a compact way:

$$\langle\{n\}|\hat{B}_1|\{\bar{n}\}\rangle = \begin{cases}\sum\limits_{k=1}^{\infty} b_{kk}n_k, & \{n\}=\{\bar{n}\}\\ b_{pq}(-1)^{\sum\limits_{l=\min(p,q)+1}^{\max(p,q)-1} n_l}, & \{n\}=\{\bar{n}\}^p_q\\ 0, & \text{else}\end{cases}. \tag{5.62}$$

Thereby, we have expressed the matrix elements of a second quantized one-particle operator (5.56) by the one-particle integrals (5.57).

5.3.4.2 Two-Particle Operators

In analogy to the one-particle operators, general two-particle operators of the form $\hat{B}_2 = \frac{1}{2}\sum_{\alpha\neq\beta=1}^{N}\hat{b}_{\alpha,\beta}$ take the following form in second quantization:

$$\hat{B}_2 = \frac{1}{2} \sum_{i,j,k,l=1}^{\infty} b_{ijkl} \hat{a}_i^\dagger \hat{a}_j^\dagger \hat{a}_l \hat{a}_k. \tag{5.63}$$

Note the exchange of the orbital indices of the annihilation operators and the two-particle integrals, which is important for fermions due to the non-commutativity of the creation and annihilation operators. The two-particle integrals are given by

$$b_{ijkl} = \langle ij|\hat{b}|kl\rangle = \int \mathrm{d}x \int \mathrm{dy}\, \phi_\mathrm{i}^*(\mathrm{x})\phi_\mathrm{j}^*(\mathrm{y})\mathrm{b}(\mathrm{x},\mathrm{y})\phi_\mathrm{k}(\mathrm{x})\phi_\mathrm{l}(\mathrm{y}). \tag{5.64}$$

The following considerations are restricted to fermions only. For the most interesting case of the pair interaction operator $\hat{W}$, one can take advantage of the fact that the interaction is symmetric under particle exchange, i.e. $w(x, y) = w(y, x)$, and real, i.e. $w^*(x, y) = w(x, y)$. Hence, for the two-particle integrals, we have

$$w_{ijkl} = w_{jilk} \,, \tag{5.65}$$

$$w_{ijkl}^* = w_{klij}. \tag{5.66}$$

Using these symmetries, we can bring the interaction operator into a more suitable form:

$$\hat{W} = \sum_{i=1}^{\infty} \sum_{j=i+1}^{\infty} \sum_{k=1}^{\infty} \sum_{l=k+1}^{\infty} w_{\overline{ij}kl}\, \hat{a}_i^\dagger \hat{a}_j^\dagger \hat{a}_l \hat{a}_k, \tag{5.67}$$

where we have introduced the antisymmetric two-particle integrals $w_{\overline{ij}kl} = w_{ijkl} - w_{ijlk}$. Obviously, there are only terms with $i < j$ and $k < l$. Therefore, we end up with six different cases to be considered:

$$\begin{aligned}
\hat{W} = \; & \sum_{i=1} \sum_{j=i+1} w_{\overline{ij}ij}\, \hat{a}_j^\dagger \hat{a}_i^\dagger \hat{a}_i \hat{a}_j + \sum_{i=1} \sum_{j=i+1} \sum_{\substack{l=i+1 \\ l \neq j}} w_{\overline{ij}il}\, \hat{a}_j^\dagger \hat{a}_i^\dagger \hat{a}_i \hat{a}_l \\
& + \sum_{i=1} \sum_{j=i+1} \sum_{\substack{k=1 \\ k \neq i}}^{j-1} w_{\overline{ij}kj}\, \hat{a}_j^\dagger \hat{a}_i^\dagger \hat{a}_k \hat{a}_j + \sum_{i=1} \sum_{j=i+1} \sum_{k=1}^{i-1} w_{\overline{ij}ki}\, \hat{a}_j^\dagger \hat{a}_i^\dagger \hat{a}_k \hat{a}_i \\
& + \sum_{i=1} \sum_{j=i+1} \sum_{l=j+1} w_{\overline{ij}jl}\, \hat{a}_j^\dagger \hat{a}_i^\dagger \hat{a}_j \hat{a}_l + \sum_{i=1} \sum_{j=i+1} \sum_{\substack{k=1 \\ k \neq i,j}} \sum_{\substack{l=k+1 \\ l \neq i,j}} w_{\overline{ij}kl}\, \hat{a}_j^\dagger \hat{a}_i^\dagger \hat{a}_k \hat{a}_l.
\end{aligned} \tag{5.68}$$

For each of these cases, one readily computes the matrix elements $\langle \{n\}|\hat{a}_j^\dagger \hat{a}_i^\dagger \hat{a}_k \hat{a}_l|\{\bar{n}\}\rangle$. Rearranging of the terms finally yields

$$
\begin{aligned}
\langle\{n\}|\hat{W}|\{\bar{n}\}\rangle = {} & \delta_{\{n\},\{\bar{n}\}} \sum_{i=1}^{\infty} \sum_{j=i+1}^{\infty} w_{ijij}^{-} n_i n_j \\
& + \sum_{p=1}^{\infty} \sum_{q=1}^{\infty} \delta_{\{n\},\{\bar{n}\}}^{pq} \delta_{n_p,1} \delta_{\bar{n}_p,0} \delta_{n_q,0} \delta_{\bar{n}_q,1} \\
& \cdot \sum_{i=1}^{\infty} w_{ipiq}^{-} (-1)^{\alpha_{\{n\},i}+\alpha_{\{\bar{n}\},i}+\alpha_{\{n\},p}+\alpha_{\{\bar{n}\},q}} \Theta(i,p,q) n_i \\
& + \sum_{p=1}^{\infty} \sum_{q=p+1}^{\infty} \sum_{r=1}^{\infty} \sum_{s=r+1}^{\infty} \delta_{\{n\},\{\bar{n}\}}^{pqrs} \delta_{n_p,1} \delta_{\bar{n}_p,0} \delta_{n_q,1} \delta_{\bar{n}_q,0} \delta_{n_r,0} \delta_{\bar{n}_r,1} \delta_{n_s,0} \delta_{\bar{n}_s,1} \\
& \cdot w_{pqrs}^{-} (-1)^{\alpha_{\{n\},p}+\alpha_{\{n\},q}+\alpha_{\{\bar{n}\},r}+\alpha_{\{\bar{n}\},s}} ,
\end{aligned}
\tag{5.69}
$$

with

$$
\Theta(i,p,q) = \begin{cases} -1, & p < i < q, \text{ or } q < i < p \\ 0, & i = p \text{ or } i = q \\ 1, & \text{else} \end{cases} .
\tag{5.70}
$$

Thus, there are three different contributions to the matrix elements. These are the *Slater-Condon-rules* [32], which can be rewritten as follows:

$$
\begin{aligned}
\hat{W}_{\{n\},\{\bar{n}\}} &:= \langle\{n\}|\hat{W}|\{\bar{n}\}\rangle = \hat{W}^{\mathrm{I}}_{\{n\},\{\bar{n}\}} + \hat{W}^{\mathrm{II}}_{\{n\},\{\bar{n}\}} + \hat{W}^{\mathrm{III}}_{\{n\},\{\bar{n}\}} , \\
\hat{W}^{\mathrm{I}}_{\{n\},\{\bar{n}\}} &= \delta_{\{n\},\{\bar{n}\}} \sum_{i=0}^{\infty} \sum_{j=i+1}^{\infty} w_{ijij}^{-} n_i n_j , \\
\hat{W}^{\mathrm{II}}_{\{n\},\{\bar{n}\}} &= \sum_{p,q=1}^{\infty} \delta_{\{n\},\{\bar{n}\}}^{pq} \delta_{n_p,1} \delta_{\bar{n}_p,0} \delta_{n_q,0} \delta_{\bar{n}_q,1} \sum_{\substack{i=0 \\ i \neq p,q}} w_{ipiq}^{-} (-1)^{\sum_{l=\min(p,q)+1}^{\max(p,q)-1} n_l} n_i , \\
\hat{W}^{\mathrm{III}}_{\{n\},\{\bar{n}\}} &= \sum_{p=1}^{\infty} \sum_{q=p+1}^{\infty} \sum_{r=1}^{\infty} \sum_{s=r+1}^{\infty} \delta_{\{n\},\{\bar{n}\}}^{pqrs} \delta_{n_p,1} \delta_{\bar{n}_p,0} \delta_{n_q,1} \delta_{\bar{n}_q,0} \delta_{n_r,0} \delta_{\bar{n}_r,1} \delta_{n_s,0} \delta_{\bar{n}_s,1} \\
& \quad \cdot w_{pqrs}^{-} (-1)^{\sum_{l=p}^{q-1} n_l + \sum_{l=r}^{s-1} \bar{n}_l} .
\end{aligned}
\tag{5.71}
$$

The first term only gives a contribution if both states are equal and the second if left and right state differ in exactly two orbitals. In contrast, the third term is not vanishing only if both states differ in exactly four orbitals. In all cases, the left and right state must have the same particle number. All other matrix elements vanish. Collecting the obtained results, we finally end up with

$$\langle\{n\}|\hat{W}|\{\bar{n}\}\rangle = \begin{cases} \sum_{i=0}^{\infty} \sum_{j=i+1}^{\infty} w_{ijij}^{-} n_i n_j, & \{n\} = \{\bar{n}\} \\ \sum_{\substack{i=0 \\ i\neq p,q}} w_{ipiq}^{-} (-1)^{\sum_{l=\min(p,q)+1}^{\max(p,q)-1} n_l} n_i, & \{n\} = \{\bar{n}\}_q^p \\ w_{pqrs}^{-} (-1)^{\sum_{l=p}^{q-1} n_l + \sum_{l=r}^{s-1} \bar{n}_l}, & \{n\} = \{\bar{n}\}_{r<s}^{p<q} \\ 0, & \text{else} \end{cases} . \tag{5.72}$$

Thereby, $|\{n\}_{r<s}^{p<q}\rangle$ is the $N-$particle state that one obtains from $|\{n\}\rangle$ if a particle is added to the orbitals $|p\rangle$ and $|q\rangle$ each, with $p < q$, and removed from the orbitals $|r\rangle$ and $|s\rangle$, with $r < s$.

5.4 The Density Operator

The density operator $\hat{\rho}$ is of extreme importance for the description of thermodynamic properties of quantum many-particle systems. While in standard quantum mechanics a quantum system is described by a certain state $|\Psi\rangle$, this does not apply to a quantum system at a finite temperature (e.g. a system in a thermostate). In this case, in principle, all possible states may be observed with a certain probability, i.e. the system is in an ensemble of states or in a "mixed state". For an ensemble of possible many-particle states $\{|\Psi_i\rangle\}$, the density operator (which replaces the wave function in the previous case of a "pure" state) is defined as

$$\hat{\rho} = \sum_i P(|\Psi_i\rangle)\,|\Psi_i\rangle\,\langle\Psi_i|\,. \tag{5.73}$$

The sum goes over all states of the ensemble, and $P(|\Psi_i\rangle)$ is the real, positive probability to observe the state $|\Psi_i\rangle$. The matrix representation of $\hat{\rho}$ in an arbitrary N-particle basis $\{|\Phi_i\rangle\}$ is called the density matrix $\langle\Phi|\hat{\rho}|\Phi\rangle$. Further, it is hermitian, i.e. $\hat{\rho}^\dagger = \hat{\rho}$, and positive semi-definite, i.e. $\langle\Phi_i|\hat{\rho}|\Phi_i\rangle \geq 0$ for arbitrary states $|\Phi_i\rangle$. The probabilities obey the normalization $\sum_i P(|\Psi_i\rangle) = 1$, which corresponds to

$$\mathrm{Tr}\hat{\rho} = \sum_i \langle\Phi_i|\hat{\rho}|\Phi_i\rangle = 1, \tag{5.74}$$

where the trace can be evaluated in an arbitrary N-particle basis. The most important feature of the density matrix is that we can compute the expectation value of any observable via

$$\langle\hat{O}\rangle = \mathrm{Tr}(\hat{O}\hat{\rho}). \tag{5.75}$$

It is always $\mathrm{Tr}\hat{\rho}^2 \leq 1$, and for a mixed ensemble this expression is strictly smaller than 1. For a pure state $|\Psi\rangle$, the density operator becomes a projection operator with $\hat{\rho}^2 = \hat{\rho}$ and (5.75) turns into the standard quantum mechanical average $\langle\hat{O}\rangle = \langle\Psi|\hat{O}|\Psi\rangle$. The entropy of the system can be defined as

$$S = -\langle \ln \hat{\rho} \rangle . \tag{5.76}$$

Hence, the density operator contains the whole thermodynamic information of the system.

In the canonical ensemble, characterized by the particle number N, the temperature T and the volume V of the system, the density operator takes the form

$$\hat{\rho}(N, \beta, V) = \frac{1}{Z(N, \beta, V)} e^{-\beta \hat{H}}, \tag{5.77}$$

with the inverse temperature $\beta = \frac{1}{T}$ (we use $k_B = 1$). The normalization factor Z is the partition function

$$Z(N, \beta, V) = \mathrm{Tr} e^{-\beta \hat{H}}. \tag{5.78}$$

For most of the physical quantities, the whole information about the N-particle density operator is not required. Instead, it is sufficient to know the partition function. Another crucial quantity of the canonical ensemble is the free energy

$$F(N, \beta, V) = -T \ln Z. \tag{5.79}$$

Most of the physical quantities can also be derived from the free energy, e.g. for the total energy and the heat capacity at constant volume we have

$$\langle \hat{H} \rangle = -T^2 \frac{\partial}{\partial T} \frac{F}{T} \bigg|_{N,V} = -\frac{\partial}{\partial \beta} \ln Z \bigg|_{N,V}, \tag{5.80}$$

and

$$C_V = \frac{\partial}{\partial T} \langle \hat{H} \rangle \bigg|_V . \tag{5.81}$$

For a system of constant temperature T, chemical potential μ and volume V, we use the density operator of the *grand canonical ensemble*

$$\hat{\rho}(\beta, \mu, V) = \frac{1}{Z(\beta, \mu, V)} e^{-\beta(\hat{H} - \mu \hat{N})}, \tag{5.82}$$

with the grand canonical partition function

$$Z(\beta, \mu, V) = \mathrm{Tr} e^{-\beta(\hat{H} - \mu \hat{N})}, \tag{5.83}$$

and the grand canonical potential

$$\Omega(\beta, \mu, V) = -T \ln Z. \tag{5.84}$$

For example, the average particle number and the total energy can be computed according to

$$\langle \hat{N} \rangle = -\left.\frac{\partial}{\partial \mu}\Omega\right|_{\beta,V}, \qquad \langle \hat{H} \rangle = \left.\frac{\partial}{\partial \beta}\beta\Omega\right|_{\beta\mu,V}. \tag{5.85}$$

In second quantization, the one-particle density matrix is of particular importance:

$$d_{ij} := \langle \hat{a}_i^\dagger \hat{a}_j \rangle = \mathrm{Tr}\hat{a}_i^\dagger \hat{a}_j \hat{\rho}. \tag{5.86}$$

With the representation of one-particle operators in second quantization (5.56), we can apparently calculate all one-particle quantities from the knowledge of the one-particle density matrix:

$$\langle \hat{O} \rangle = \sum_{i,j=0}^{\infty} o_{ij} \mathrm{Tr}\hat{a}_i^\dagger \hat{a}_j \hat{\rho} = \sum_{i,j=0}^{\infty} o_{ij} d_{ij}. \tag{5.87}$$

In addition, the diagonal elements of the one-particle density matrix are the average occupation numbers of the orbitals, i.e. $\langle \hat{n}_i \rangle = d_{ii}$. For both the canonical and the grand canonical ensembles, the elements of the one-particle density matrix can be obtained from the partition function. In the canonical ensemble it is

$$\begin{aligned} d_{ij} &= \frac{1}{Z}\mathrm{Tr}\hat{a}_i^\dagger \hat{a}_j \; e^{-\beta\left(\sum\limits_{pq} h_{pq}\hat{a}_p^\dagger \hat{a}_q + \sum\limits_{p<q r<s} w^-_{pqrs}\hat{a}_p^\dagger \hat{a}_q^\dagger \hat{a}_s \hat{a}_r\right)} \\ &= -\frac{1}{\beta}\frac{\partial}{\partial h_{ij}} \ln Z. \end{aligned} \tag{5.88}$$

This relation also holds in the grand canonical ensemble. Moreover, using the definition of the field operators (5.53), we can compute the average particle density via

$$n(\mathbf{r}) = \sum_\sigma \langle \hat{\Psi}^\dagger(x)\hat{\Psi}(x)\rangle = \sum_\sigma \sum_{i,j=0}^{\infty} \phi_i(x)^* \phi_j(x) d_{ij}. \tag{5.89}$$

5.5 Configuration Path Integral Monte Carlo in the Interaction Representation

In this section, first, an expansion of the partition function suitable for a CPIMC algorithm shall be derived. Next, we will show how to obtain the estimators of physical quantities and explain the used Monte Carlo steps in detail. Finally, some results will be shown for a one-dimensional system of Coulomb-interacting electrons in a harmonic trap.

5.5.1 Expansion of the Partition Function

From now on, we will only consider fermions in the canonical ensemble. Hence, the partition function is given by (5.78). There are many different ways to expand this expression. First, we have to choose a certain basis in which we evaluate the trace, and second, there are quite a few possibilities to factorize the density operator. Each combination of basis and factorization will lead to a different, more or less efficient,[2] Monte Carlo algorithm. For example, we could choose coordinate space and use the *Trotter formula* [33, 34]:

$$e^{\beta \hat{H}} = \lim_{n\to\infty} \left(e^{\frac{\beta}{n}\hat{H}_0} e^{\frac{\beta}{n}\hat{W}} \right)^n . \tag{5.90}$$

In that case, we have to antisymmetrize the density operator which yields the DPIMC method [12]. We could also perform the trace in occupation number representation, which will automatically ensure the correct symmetrization. In combination with the standard expansion of the exponential function

$$e^x = \sum_{k=0}^{\infty} \frac{x^k}{k!}, \tag{5.91}$$

we would end up with the method called *Stochastic Series Expansion* [35].

Here, we follow a different approach based on the ideas of Prokof'ev et al. [25] and [36]. First, we split the Hamiltonian $\hat{H}$ into a diagonal part $\hat{D}$ and an off-diagonal part $\hat{Y}$, i.e. for an arbitrary basis $\{|\alpha\rangle\}$ it is

$$\langle\alpha|\hat{H}|\alpha'\rangle = \begin{cases} \langle\alpha|\hat{D}|\alpha\rangle = D_{\alpha,\alpha}\,, & \text{if} \quad \alpha = \alpha' \\ \langle\alpha|\hat{Y}|\alpha'\rangle = Y_{\alpha,\alpha'}\,, & \text{if} \quad \alpha \neq \alpha' \end{cases} . \tag{5.92}$$

Second, we switch to the interaction picture in imaginary time with respect to $\hat{D}$ and make use of the well-known identity[3]

$$e^{-\beta\hat{H}} \equiv e^{-\beta\hat{D}} \hat{T}_\tau \, e^{-\int_0^\beta \hat{Y}(\tau)\mathrm{d}\tau} , \quad \text{with} \quad \hat{Y}(\tau) = e^{\tau\hat{D}} \hat{Y} e^{-\tau\hat{D}} , \; \tau \in (0, \beta), \tag{5.93}$$

where the action of the time-ordering operator $\hat{T}_\tau$ on two arbitrary operators $\hat{A}(\tau_1)$ and $\hat{B}(\tau_2)$ is defined by

$$\hat{T}_\tau[\hat{A}(\tau_1)\hat{B}(\tau_2)] = \begin{cases} \hat{A}(\tau_1)\hat{B}(\tau_2)\,, & \text{if} \quad \tau_1 > \tau_2 \\ \hat{A}(\tau_2)\hat{B}(\tau_1)\,, & \text{if} \quad \tau_2 > \tau_1 \end{cases} . \tag{5.94}$$

[2] Efficient with respect to a certain system with fixed parameters to be simulated, i.e. currently, there is no method suitable for all systems.

[3] In this notation, operators in the interaction and the Schrödinger picture differ only concerning the presence of a time argument.

Next, we expand the exponential function of the off-diagonal operator:

$$\hat{T}_\tau e^{-\int_0^\beta \hat{Y}(\tau)\mathrm{d}\tau} = \sum_{K=0}^{\infty} \int_0^\beta \mathrm{d}\tau_1 \ldots \int_0^\beta \mathrm{d}\tau_K \frac{(-1)^K}{K!} \hat{T}_\tau[\hat{Y}(\tau_1)\hat{Y}(\tau_2) \cdot \ldots \cdot \hat{Y}(\tau_K)] \tag{5.95}$$
$$= \sum_{K=0}^{\infty} \int_0^\beta \mathrm{d}\tau_1 \int_{\tau_1}^\beta \mathrm{d}\tau_2 \ldots \int_{\tau_{K-1}}^\beta \mathrm{d}\tau_K (-1)^K \hat{Y}(\tau_K)\hat{Y}(\tau_{K-1}) \cdot \ldots \cdot \hat{Y}(\tau_1) \,.$$

In the first line, each integral goes from 0 to β. Since the time-ordering operator always arranges those off-diagonal operators with the latest times to the left, for each K, there are $K!$ equal terms. Therefore, we can eliminate the time-ordering operator by canceling the $K!$, modifying the integral borders and arranging the operator product in such a way that it is already time-ordered.

Combining (5.78), (5.93) and (5.95) yields

$$Z = \sum_{K=0}^{\infty} \sum_{\{n\}} \int_0^\beta \mathrm{d}\tau_1 \int_{\tau_1}^\beta \mathrm{d}\tau_2 \ldots \int_{\tau_{K-1}}^\beta \mathrm{d}\tau_K$$
$$(-1)^K e^{-\beta \hat{D}_{\{n\}}} \langle\{n\}|\hat{Y}(\tau_K)\hat{Y}(\tau_{K-1}) \cdot \ldots \cdot \hat{Y}(\tau_1)|\{n\}\rangle \,, \tag{5.96}$$

where we chose the occupation number representation to perform the trace, i.e. in (5.92) it is $|\alpha\rangle = |\{n\}\rangle$. Now we insert $K-1$ unit operators of the form $\hat{1} = \sum_{\{n^{(i)}\}} |\{n^{(i)}\}\rangle \langle\{n^{(i)}\}|$ in between the off-diagonal operators. By definition of the interaction picture, cf. (5.93), the resulting matrix elements are of the form

$$\langle\{n^{(i)}\}|\hat{Y}(\tau_K)|\{n^{(j)}\}\rangle = \langle\{n^{(i)}\}|e^{\tau_K \hat{D}}\hat{Y}e^{-\tau_K \hat{D}}|\{n^{(j)}\}\rangle$$
$$= e^{\tau_K D_{\{n^{(i)}\}}} Y_{\{n^{(i)}\},\{n^{(j)}\}} e^{-\tau_K D_{\{n^{(j)}\}}} . \tag{5.97}$$

After rearranging all factors and taking into account that $Y_{\{n^{(i)}\},\{n^{(i)}\}} \equiv 0$, we finally end up with

$$Z(N,V,\beta) = \sum_{\substack{K=0,\\ K\neq 1}}^{\infty} \sum_{\substack{\{n\},\\ =\{n^{(0)}\}=\{n^{(K)}\}}} \sum_{\substack{\{n^{(1)}\},\\ \neq\{n^{(0)}\}}} \cdots \sum_{\substack{\{n^{(K-1)}\},\\ \neq\{n^{(K-2)}\}\neq\{n^{(K)}\}}} \int_0^\beta \mathrm{d}\tau_1 \int_{\tau_1}^\beta \mathrm{d}\tau_2 \ldots \int_{\tau_{K-1}}^\beta \mathrm{d}\tau_K$$
$$(-1)^K \exp\left\{-\sum_{i=0}^{K} D_{\{n^{(i)}\}}(\tau_{i+1}-\tau_i)\right\} \prod_{i=0}^{K-1} Y_{\{n^{(i)}\},\{n^{(i+1)}\}} =: \sum_{C} W(C), \tag{5.98}$$

where the boundary conditions $\{n\} = \{n^{(0)}\} = \{n^{(K)}\}$, $\tau_0 = 0$ and $\tau_{k+1} = \beta$ hold. For the CPIMC Monte Carlo algorithm, this representationof the partition function

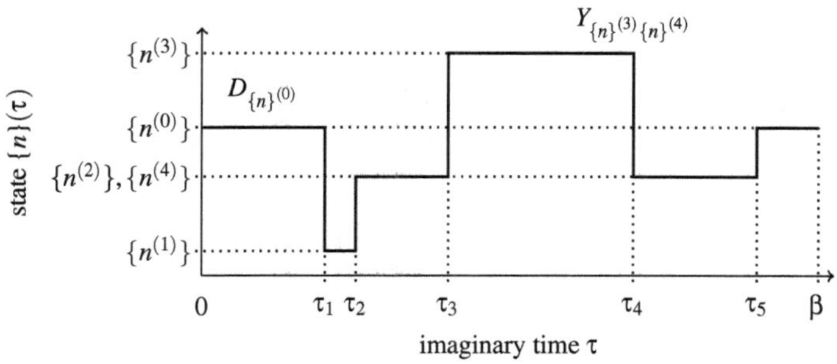

Fig. 5.1 Possible path in imaginary time. Horizontal lines correspond to diagonal matrix elements, whereas vertical lines correspond to off-diagonal elements

is used. Hence, we can identify the introduced weights W and configurations C of Sect. 5.2 in the following way. A contribution to the partition function is uniquely defined by the number of kinks K, K kink times τ_i and K *Slater determinants* which are realized in between the kinks:

$$C = \{K, \tau_1, \tau_2, \ldots, \tau_K, \{n^{(0)}\}, \{n^{(1)}\}, \ldots, \{n^{(K-1)}\}\}. \tag{5.99}$$

Thus, the weights defined by (5.98) are given by

$$W\Big(C = \{K, \tau_1, \tau_2, \ldots, \tau_K, \{n^{(0)}\}, \{n^{(1)}\}, \ldots, \{n^{(K-1)}\}\}\Big) \tag{5.100}$$
$$= (-1)^K \exp\left\{-\sum_{i=0}^{K} D_{\{n^{(i)}\}}(\tau_{i+1} - \tau_i)\right\} \prod_{i=0}^{K-1} Y_{\{n^{(i)}\},\{n^{(i+1)}\}}.$$

Such configurations can also be visualized as β−periodic paths in imaginary time. As an example, Fig. 5.1 shows a possible path. Each horizontal line corresponds to a diagonal matrix element in the weight of that configuration [cf. (5.100)], whereas vertical lines correspond to off-diagonal elements.

In addition to the correct statistics included in the occupation number representation, this expansion of the partition function yields paths that are continuous in imaginary time. This is different to DPIMC, where the corresponding paths are discrete and one has to ensure the convergence of the discretization by performing several simulations for the same system parameters but different discretizations (different numbers of "time slices"). For this reason, the presented CPIMC method belongs to the so-called continuous time QMC methods. A different, independent derivation of (5.98), that is based on the *Trotter Formula*, was given in [23] where the notion configuration PIMC was introduced.

Further, since we have only expanded the exponential function of the off-diagonal operator [cf. (5.96)], this representation of the partition function is nothing but a perturbation expansion with respect to the off-diagonal matrix elements. Hence, if we choose the ideal basis for the quantization, it is indeed a perturbation expansion with respect to the interaction.

Combining the *Slater-Condon-rules* (5.72) and (5.62), the matrix elements of the diagonal and off-diagonal operator are calculated according to[4]

$$D_{\{n^{(k)}\}} = \sum_{i=1}^{\infty} h_{ii} n_i^{(k)} + \sum_{i=0}^{\infty} \sum_{j=i+1}^{\infty} w_{\bar{ijij}} n_i^{(k)} n_j^{(k)}, \tag{5.101}$$

and

$$Y_{\{n^{(k)}\}\{n^{(l)}\}} = \begin{cases} \left(h_{pq} + \sum\limits_{\substack{i=0 \\ i \neq p,q}}^{\infty} w_{\bar{ipiq}} n_i^{(k)} \right) (-1)^{\sum_{m=\min(p,q)+1}^{\max(p,q)-1} n_m^{(k)}}, & \{n^{(k)}\} = \{n^{(l)}\}_q^p \\ w_{\bar{pqrs}} (-1)^{\sum_{m=p}^{q-1} n_m^{(k)} + \sum_{m=r}^{s-1} n_m^{(l)}}, & \{n^{(k)}\} = \{n^{(l)}\}_{r<s}^{p<q} \\ 0, & \text{else.} \end{cases} \tag{5.102}$$

Obviously, only those configurations (paths) contribute to the partition function (5.98) in which consecutive *Slater determinants* differ in either two or one orbitals, since otherwise the off-diagonal matrix elements vanish. Moreover, there are three sources of sign changes in the weights meaning that the fermion *sign problem* (see Sect. 5.2.2) survives. First, an increment of the number of kinks changes the sign due to the factor $(-1)^K$. Second, we have a phase factor in the off-diagonal matrix elements (5.102) and third, the one and two particle integrals can have positive and negative signs.

5.5.2 Estimators for Many-Particle Observables

The simplest way to obtain the estimators (cf. Sect. 5.2) is to use the thermodynamic relations of Sect. 5.4 between the partition function and the observables. Sometimes those *trivial estimators* can be numerically unstable, and one has to use different estimators, e.g. in DPIMC the *virial estimator* for the energy [37]. In CPIMC, the *trivial estimators* are numerically stable.

Starting with the energy, we have to compute $\langle \hat{H} \rangle = -\frac{1}{Z}\frac{\partial Z}{\partial \beta}$ with the partition function (5.98). Since both the integrand and the integral borders in (5.98) depend on the inverse temperature β, we substitute $t_i = \frac{\tau_i}{\beta}$ and $d\tau_i = \beta dt_i$ yielding

[4] Due to the finite number of one-particle states in a practical simulation, the sums will not go up to infinity but to N_B, the number of orbitals used in the simulation.

$$Z(N,V,\beta) = \sum_{\substack{K=0,\\ K\neq 1}}^{\infty} \sum_{\substack{\{n\},\\ =\{n^{(0)}\}=\{n^{(K)}\}}} \sum_{\substack{\{n^{(1)}\},\\ \neq\{n^{(0)}\}}} \cdots \sum_{\substack{\{n^{(K-1)}\},\\ \neq\{n^{(K-2)}\}\neq\{n^{(K)}\}}} \int_0^1 dt_1 \int_{t_1}^1 dt_2 \ldots \int_{t_{K-1}}^1 dt_K$$

$$(-\beta)^K \exp\left\{-\beta \sum_{i=0}^{K} D_{\{n^{(i)}\}}(t_{i+1}-t_i)\right\} \prod_{i=0}^{K-1} Y_{\{n^{(i)}\},\{n^{(i+1)}\}}. \quad (5.103)$$

Now the integration limits are independent of β ($t_i \in [0,1]$) and we can easily perform the differentiation:

$$\langle \hat{H} \rangle = \frac{1}{Z} \sum_{\substack{K=0,\\ K\neq 1}}^{\infty} \sum_{\substack{\{n\},\\ =\{n^{(0)}\}=\{n^{(K)}\}}} \sum_{\substack{\{n^{(1)}\},\\ \neq\{n^{(0)}\}}} \cdots \sum_{\substack{\{n^{(K-1)}\},\\ \neq\{n^{(K-2)}\}\neq\{n^{(K)}\}}} \int_0^1 dt_1 \int_{t_1}^1 dt_2 \ldots \int_{t_{K-1}}^1 dt_K$$

$$\left[-K\beta^{K-1} + \beta^K \sum_{i=0}^{K} D_{\{n^{(i)}\}}(t_{i+1}-t_i)\right]$$

$$\times (-1)^K \exp\left\{-\beta \sum_{i=0}^{K} D_{\{n^{(i)}\}}(t_{i+1}-t_i)\right\} \prod_{i=0}^{K-1} Y_{\{n^{(i)}\},\{n^{(i+1)}\}}, \quad (5.104)$$

and, after substituting back, we obtain the estimator for the total (internal) energy

$$\langle \hat{H} \rangle = \frac{1}{Z} \sum_{\substack{K=0,\\ K\neq 1}}^{\infty} \sum_{\substack{\{n\},\\ =\{n^{(0)}\}=\{n^{(K)}\}}} \sum_{\substack{\{n^{(1)}\},\\ \neq\{n^{(0)}\}}} \cdots \sum_{\substack{\{n^{(K-1)}\},\\ \neq\{n^{(K-2)}\}\neq\{n^{(K)}\}}} \int_0^\beta d\tau_1 \int_{\tau_1}^\beta d\tau_2 \ldots \int_{\tau_{K-1}}^\beta d\tau_K$$

$$\left[\frac{-K}{\beta} + \sum_{i=0}^{K} D_{\{n^{(i)}\}} \frac{(\tau_{i+1}-\tau_i)}{\beta}\right] W(C). \quad (5.105)$$

Thus, the energy of a configuration $C = \{K, \tau_1, \tau_2, \ldots, \tau_K, \{n^{(0)}\}, \{n^{(1)}\}, \ldots, \{n^{(K-1)}\}\}$ is given by

$$E(C) = \left[\frac{-K}{\beta} + \sum_{i=0}^{K} D_{\{n^{(i)}\}} \frac{(\tau_{i+1}-\tau_i)}{\beta}\right]. \quad (5.106)$$

It is worth mentioning that the contribution of the off-diagonal elements of the Hamiltonian to the total energy is given only by the number of kinks multiplied by the temperature. On the other hand, the form of the diagonal contribution is reasonable as well. It is the sum of all diagonal elements of all realized *Slater determinants* weighted with the relative length.

According to (5.81), the estimator for the heat capacity takes the following form:

$$C_V = \frac{1}{T^2}\frac{1}{Z} \sum_{\substack{K=0\\K\neq 1}}^{\infty} \sum_{\{n\}=\{n^{(0)}\}=\{n^{(K)}\}} \sum_{\substack{\{n^{(1)}\}\\ \neq\{n^{(0)}\}}} \cdots \sum_{\substack{\{n^{(K-1)}\}\\ \neq\{n^{(K-2)}\},\neq\{n^{(K)}\}}} \int_0^{\beta} \mathrm{d}\tau_1 \int_{\tau_1}^{\beta} \mathrm{d}\tau_2 \ldots \int_{\tau_{K-1}}^{\beta} \mathrm{d}\tau_K$$
$$\left[\left(\frac{1}{\beta}\sum_{i=0}^{K} D_{\{n^{(i)}\}}(\tau_{i+1}-\tau_i) - \frac{K}{\beta} - \langle H\rangle\right)^2 - \frac{K}{\beta^2}\right] W(C). \tag{5.107}$$

Other important quantities are the one-particle density matrix and, in particular, its diagonal elements–the average occupation numbers. Using (5.88), one readily computes the corresponding estimators:

$$d_{pq} = \frac{1}{Z}\sum_{\substack{K=0\\K\neq 1}}^{\infty} \sum_{\{n\}=\{n^{(0)}\}=\{n^{(K)}\}} \sum_{\substack{\{n^{(1)}\}\\ \neq\{n^{(0)}\}}} \cdots \sum_{\substack{\{n^{(K-1)}\}\\ \neq\{n^{(K-2)}\},\neq\{n^{(K)}\}}} \int_0^{\beta} \mathrm{d}\tau_1 \int_{\tau_1}^{\beta} \mathrm{d}\tau_2 \ldots \int_{\tau_{K-1}}^{\beta} \mathrm{d}\tau_K$$
$$\left(-\frac{1}{\beta}\sum_{i=0}^{K-1} \frac{\langle\{n^{(i)}\}|\hat{a}_p^\dagger \hat{a}_q|\{n^{(i+1)}\}\rangle}{Y_{\{n^{(i)}\},\{n^{(i+1)}\}}}\right) W(C) + \langle \hat{n}_p\rangle\, \delta_{pq}, \tag{5.108}$$

$$\langle \hat{n}_p\rangle = \frac{1}{Z}\sum_{\substack{K=0\\K\neq 1}}^{\infty} \sum_{\{n\}=\{n^{(0)}\}=\{n^{(K)}\}} \sum_{\substack{\{n^{(1)}\}\\ \neq\{n^{(0)}\}}} \cdots \sum_{\substack{\{n^{(K-1)}\}\\ \neq\{n^{(K-2)}\},\neq\{n^{(K)}\}}} \int_0^{\beta} \mathrm{d}\tau_1 \int_{\tau_1}^{\beta} \mathrm{d}\tau_2 \ldots \int_{\tau_{K-1}}^{\beta} \mathrm{d}\tau_K$$
$$\left(\frac{1}{\beta}\sum_{i=0}^{K} n_p^{(i)}(\tau_{i+1}-\tau_i)\right) W(C). \tag{5.109}$$

The average occupation of the pth orbital in the configuration C

$$n_p(C) = \sum_{i=0}^{K} n_p^{(i)} \frac{(\tau_{i+1}-\tau_i)}{\beta}, \tag{5.110}$$

is nothing but the average of the occupation of the pth orbital over realized *Slater determinants*, again weighted with the relative length of the path on which the determinant is realized.

5.5.3 CPIMC Procedure: Monte Carlo Steps

When developing a Monte Carlo algorithm, the most difficult task lies in the construction of a good set of Monte Carlo steps that is most efficient while simultaneously ensuring the ergodicity. In the program, we start from an initial configuration with zero kinks. Thus, the path consists of only one *Slater determinant* realized from 0 to β, i.e. it is $C_0 = \{0, \{n^0\}\}$. If we simulate N particles, then we populate the first N one-particle orbitals of $|\{n^{(0)}\}\rangle$ while the remaining $(N_B - N)$ orbitals are initially unoccupied. Now we need a sufficient number of Monte Carlo steps that allow for sampling of the whole configuration space (all possible paths) in an acceptable amount of computation time. As a consequence of the summation constraint in the partition function (5.98), configurations with only a single kink are forbidden. Hence, to reach configurations with more than zero kinks, we will not only need a step in which a single kink is added but also a step in which two kinks are added at once.

Obviously, we also need a step that changes the occupation of the realized *Slater determinants*. One might expect that an implementation of a step that swaps the occupation of an occupied and an unoccupied orbital of a determinant (one-particle excitation) is sufficient. Actually, it can be shown that for a fixed number of kinks the configuration space separates into subspaces, and hence, ergodicity is not fulfilled [38]. Meaning, if we only use the one-particle excitation, we will be stuck in a certain subspace, which results in wrong expectation values. The number of separated subspaces for a fixed number of kinks can be reduced by the implementation of a two particle excitation.

Since the off-diagonal elements vanish if the two determinants differ in more than four orbitals (cf. (5.102)), a step in which three particles are excited will in most cases be rejected.

Since we cannot exclude that a separation of the configuration space occurs for a variable number of kinks (thus destroying ergodicity of our procedure), we assign a so-called *virtual weight* to certain forbidden paths. This is achieved by setting each off-diagonal matrix element $Y_{\{n\}\{\bar{n}\}} < Y_{\min} = 10^{-10}$, with $\{n\} = \{\bar{n}\}_{r<s}^{p<q}$ to a fixed *virtual weight* Y_V. Thereby, the subspaces of the configuration space become connected via the forbidden configurations. If all forbidden paths are neglected in the calculation of the expectation values, then these will still be correct. During the equilibration time of the *Markov process*, the virtual weight will be determined in such a way that between 70 to 90 % off all configurations in the *Markov chain* are allowed. As a consequence, the runtime of the program increases by up to 30 % by the use of *virtual weights*. On the other hand, the autocorrelation time of those configurations contributing to the measurements is reduced, and so the extended runtime is partially compensated.

A particular step to remove a kink is not needed. Whenever a determinant, after a one- or two-particle excitation has been applied to it, equals its left or right neighboring determinant (in imaginary time), the corresponding kink will be removed. Besides, a particular step to remove kinks would have a factor in its acceptance probability that decreases with the basis size N_B.

Finally, we need a step to shift the position of a kink (in imaginary time). In fact, this step might not be necessary for the ergodicity, since we can remove a certain kink and add it at a different time. However, a particular step in which a kink is shifted helps reducing the autocorrelation time.

In the following, the Monte Carlo steps shall be explained in detail following [38] and [23]. As an illustration, Fig. 5.2 shows the changes in the path for each step. Nevertheless, the $\beta-$periodicity has not been taken into account explicitly. Instead

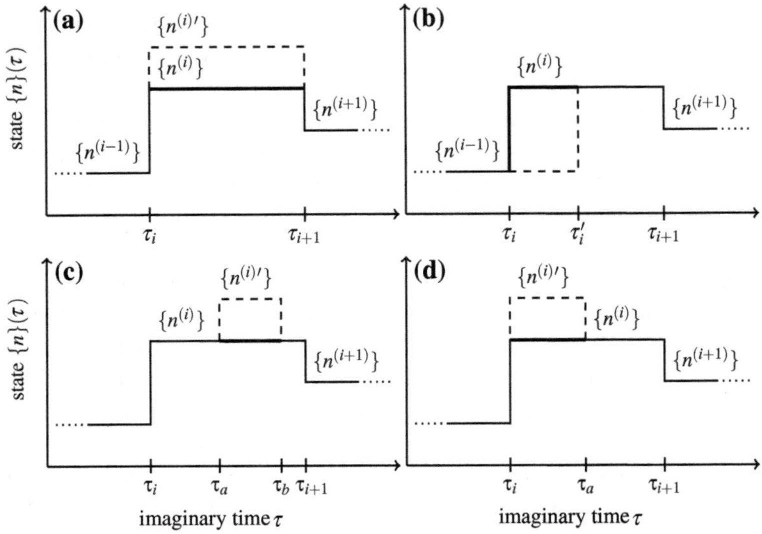

Fig. 5.2 Visualization of the Monte Carlo steps. All pictures only show the relevant part of the path in continuous imaginary time. Dashed lines indicate the new path after the step has been performed. Thick lines mark the part of the path that will be removed. **a** Step 1: Change state (determinant). The selected state is at $|\{n^{(i)}\}\rangle$. **b** Step 2: Shift kink. The selected kink is at τ_i. Possible new kink positions are in between the kinks τ_{i-1} (not shown) and τ_{i+1}. **c** Step 3: Add pair of kinks. The selected interval is $[\tau_i, \tau_{i+1}]$. **d** Step 4: Add single kink. The added kink is at τ_a, and the selected interval is $[\tau_i, \tau_{i+1}]$

there are two intervals $[0, \tau_1]$ and $[\tau_K, \beta]$ which are both assigned to the same *Slater determinant*:

(i) Change state:

i. Randomly select one of the N_S states $|\{n^{(i)}\}\rangle$ of the path. For a configuration without kinks ($K = 0$) it is $N_S = 1$, else $N_S = K$.
ii. Select randomly either (a) a one-particle or (b) a two-particle excitation
   (a) Randomly select one of the N occupied and one of the $N_B - N$ unoccupied one-particle orbitals. Invert the occupation of both orbitals.
   (b) Randomly select two of the N occupied and two of the $N_B - N$ unoccupied one-particle orbitals. Invert the occupation of selected orbitals.
iii. If $K \geq 2$, check the following two cases:
   (A) If for a new many-particle state $|\{n^{(i)\prime}\}\rangle$ $|\{n^{(i)\prime}\}\rangle = |\{n^{(i-1)}\}\rangle$ <u>or</u> $|\{n^{(i)\prime}\}\rangle = |\{n^{(i+1)}\}\rangle$
   $\Longrightarrow$ Remove the corresponding kink.
   (B) If for a new many-particle state $|\{n^{(i)\prime}\}\rangle$ $|\{n^{(i)\prime}\}\rangle = |\{n^{(i-1)}\}\rangle$ <u>and</u> $|\{n^{(i)\prime}\}\rangle = |\{n^{(i+1)}\}\rangle$
   $\Longrightarrow$ Remove both kinks.

If none of the above cases applies or if $K = 0$, then

(C) the number of kinks remains unchanged.

(ii) Shift kink

i. Randomly select one of the K kinks with time τ_i.
ii. Randomly select a new kink position τ_i' from the interval $[\tau_{i-1}, \tau_{i+1}]$, where τ_{i-1} and τ_{i+1} are the times of the adjacent kinks. Hence, the ordering of the kinks remains unchanged.

(iii) Add pair of kinks

i. Randomly select one of the $K + 1$ intervals $[\tau_i, \tau_{i+1}]$.
ii. Randomly select two new kink positions within the interval $\tau_a < \tau_b \in [\tau_i, \tau_{i+1}]$.
iii. Change the state on the interval $[\tau_a, \tau_b]$ as for MC move 1:
Select randomly either (a) a one-particle or (b) a two-particle excitation
(a) Randomly select one of the N occupied and one of the $N_B - N$ unoccupied one-particle orbitals. Invert the occupation of both orbitals.
(b) Randomly select two of the N occupied and two of the $N_B - N$ unoccupied one-particle orbitals. Invert the occupation of the selected orbitals.

(iv) Add a kink

i. Randomly select one of the K kinks with time τ_i.
ii. Randomly choose the interval $[\tau_{i-1}, \tau_i]$ or $[\tau_i, \tau_{i+1}]$ for the new kink to be added.
iii. Randomly select the new kink position τ_a in the chosen interval.
iv. Depending on the interval, change the state on the interval $[\tau_a, \tau_i]$ or $[\tau_i, \tau_a]$ as for step 1:
Select randomly either (a) a one-particle or (b) a two particle excitation
(a) Randomly select one of the N occupied and one of the $N_B - N$ unoccupied one-particle orbitals. Invert the occupation of both orbitals.
(b) Randomly select two of the N occupied and two of the $N_B - N$ unoccupied one-particle orbitals. Invert the occupation of selected orbitals.

These four steps are selected with a kink-dependent probability. If $K = 0$, only steps 1 and 3 are selected with the probabilities p_{cs} and p_{add2}, respectively. In all other cases also the steps 2 and 4 are possible with the probabilities p_{sk} and p_{add1}. Then, the probabilities p_{cs} and p_{add2} are adjusted. In the implementation which was used to produce the results of Sect. 5.6 all currently possible steps are equally selected.

5.5.4 Acceptance Probabilities of the Monte Carlo Steps

When determining the acceptance probabilities from the *detailed balance* (cf. (5.3) and (5.6)), one only has solve the *detailed balance* for pairs of Monte Carlo steps that yield the original configuration when performed consecutively. In addition, the weight (5.100) divided by the partition function is a true probability only with respect to the number of kinks and *Slater determinants*, but with respect to the kink times, it is a probability density. Only after multiplying with the infinitesimal kink times $d\tau_i$, the latter becomes a true probability with respect to the kink times. For steps that change the number of kinks, this would result in an infinitely large or vanishing acceptance probability, as some of the $d\tau_i$ do not cancel in the fraction of the weights. Fortunately, it was shown that one can simply treat the weights divided by the partition function as probabilities with respect to the kink times. This leads to finite acceptance probabilities while still being mathematically correct, see e.g. [36].

Instead of giving the detailed balance for all steps, we will rather explain how to set up and the solve the *detailed balance* for one specific step, and simplify the obtained result of the acceptance probability so that it can actually be implemented. We choose step 3 of Sect. 5.5.3 (add pair of kinks). The *detailed balance* for this step takes the following form:

$$p_{\text{add2}}(C)\frac{1}{K+1}\frac{2}{(\tau_{i+1}-\tau_i)^2}E|W(C)|A(C\to C') = p_{\text{cs}}(C')\frac{1}{K'}E|W(C')|A(C'\to C), \tag{5.111}$$

where the weights $W(C)$ are defined by (5.100) and $A(C \to C')$ is the acceptance probability of the step that converts the configuration C with K kinks into the configuration C' with $K' = K + 2$ kinks. The sampling probability to add two specific kinks consists of the product of the probabilities to

- select step 3 ($p_{\text{add}} = \frac{1}{2}$, if $K = 0$ and $\frac{1}{4}$ else): p_{add2} .
- select one of the $K+1$ intervals of the configuration C, where the two kinks shall be added: $\frac{1}{K+1}$.
- select two new kink times τ_a and τ_b from the chosen interval $[\tau_i, \tau_{i+1}]$. Since there are two possibilities of the time ordering of the two randomly selected times, and since we will name the smaller one τ_a, we have to add an additional factor 2: $\frac{2}{(\tau_{i+1}-\tau_i)^2}$
- choose one of the two possible changes of the determinant $|\{n^{(i)}\}\rangle$: E. For one-particle excitation it is

$$E = E_1 = \frac{1}{2}\frac{1}{N}\frac{1}{(N_B - N)}, \tag{5.112}$$

  where the factor $\frac{1}{2}$ is the probability to choose the one-particle excitation and $\frac{1}{N}\frac{1}{(N_B-N)}$, the probability to select one of the occupied and one of the unoccupied orbitals. Analogously, for the two-particle excitations, it is:

$$E = E_2 = \frac{1}{2}\frac{1}{N}\frac{1}{(N-1)}\frac{1}{(N_B - N)}\frac{1}{(N_B - N - 1)}. \tag{5.113}$$

The sampling probability to propose exactly that change which transfers the configuration C' back to the configuration C consists of the product of the probabilities to

- select step 1 (change state): p_{cs}, where $p_{\text{cs}} = \frac{1}{2}$ if $K' = 0$ and $\frac{1}{4}$ else. Since $K' = K + 2 \neq 0$, it is $p_{\text{cs}}(C') \equiv \frac{1}{4}$.
- select one of the $K' = K + 2$ Slater determinants of the configuration C': $\frac{1}{K'}$
- choose exactly that excitation which transfers $|\{n^{(i)'}\}\rangle$ into $|\{n^{(i)}\}\rangle$: E

By inserting $K' = K + 2$ and $p_{\text{cs}(C')} = \frac{1}{4}$ into (5.111), we obtain the acceptance probability for step 3:

$$A_{\text{add2}}(C \to C') = \min\left[1, \frac{(K+1)}{(K+2)} \frac{(\tau_{i+1} - \tau_i)^2}{8 p_{\text{add2}}(C)} \frac{|W(C')|}{|W(C)|}\right]. \tag{5.114}$$

It is highly inefficient to calculate the whole fraction of weights for each Monte Carlo step, since this would very much increase the computation time. Therefore, the fraction of the weights has to be simplified as far as possible. For that purpose, we rewrite the weights as follows:

$$\begin{aligned} W(C) =& (-1)^K e^{-\sum_{\substack{j=0,\\ j\neq i}}^{K} D_{\{n^{(j)}\}}(\tau_{j+1}-\tau_j)} e^{D_{\{n^{(i)}\}}(\tau_{i+1}-\tau_i)} \prod_{j=0}^{K-1} Y_{\{n^{(j)}\},\{n^{(j+1)}\}} , \\ W(C') =& (-1)^{K'} e^{-\sum_{\substack{j=0,\\ j\neq i}}^{K} D_{\{n^{(j)}\}}(\tau_{j+1}-\tau_j)} e^{D_{\{n^{(i)}\}}(\tau_a-\tau_i)} e^{D_{\{n^{(i')}\}}(\tau_b-\tau_a)} e^{D_{\{n^{(i)}\}}(\tau_{i+1}-\tau_b)} \\ & \times \prod_{j=0}^{K-1} Y_{\{n^{(j)}\},\{n^{(j+1)}\}} \times Y^2_{\{n^{(i)}\},\{n^{(i')}\}} . \end{aligned} \tag{5.115}$$

When keeping in mind (Fig. 5.2c), rewriting it in this particular way seems reasonable. Now most of the factors cancel in the weight fraction and we end up with

$$A(C \to C') = \min\left[1, \frac{(K+1)}{(K+2)} \frac{(\tau_{i+1} - \tau_i)^2}{8 p_{\text{add2}}(C)} e^{-\left(\left(D_{\{n^{(i)}\}} - D_{\{n^{(i')}\}}\right)(\tau_b - \tau_a)\right)}\right]. \tag{5.116}$$

This is the implemented acceptance probability to add a kink pair in the interval $[\tau_i, \tau_{i+1}]$ at the times τ_a and τ_b. The acceptance probabilities for the remaining three steps can be derived with similar considerations.

5.6 Results for Interacting Fermions in a One-Dimensional Harmonic Trap

As an example for the applicability of the introduced *Configuration Path Integral Monte Carlo* (CPIMC) algorithm, we calculate thermodynamic properties of a one-dimensional system of spin-polarized, Coulomb-interacting fermions in a harmonic trap. Despite its simplicity, this system is general enough to draw conclusions about the behavior of the method when simulating more complex systems. In particular, the Hamiltonian contains a pair interaction of the most general form $\hat{W} = \sum_{ijkl} w_{\bar{i}\bar{j}kl} \hat{a}_i^\dagger \hat{a}_j^\dagger \hat{a}_l \hat{a}_k$. In contrast to earlier applications of the *Continuous Time Path Integral Monte Carlo* method, which usually consider lattice models, we investigate a spatially continuous system. At first, the test system will be characterized, and the used units will be introduced. In the following section, results for expectation values of several thermodynamic observables will be presented and compared to results obtained from the exact *Configuration Interaction* (CI) method, see e.g. [32], and from the *Direct Path Integral Monte Carlo* (DPIMC) method [37], respectively. Finally, the dependence of the efficiency of the CPIMC and DPIMC methods on different parameters will be investigated.

5.6.1 The System

The Hamiltonian of the considered system in natural units is given by

$$\hat{H} = \sum_{\alpha=1}^{N} \left(\frac{\hat{p}_\alpha^2}{2m} + \frac{m\omega^2 \hat{r}_\alpha^2}{2} \right) + \sum_{1 \le \alpha < \beta}^{N} \frac{e^2}{\sqrt{\left|\hat{r}_\alpha - \hat{r}_\beta\right|^2 + \kappa^2}}. \tag{5.117}$$

Here, m is the mass of the fermions and ω the frequency of the trap. The parameter κ is necessary in one dimension to prevent the divergence of the two-particle integrals of the Coulomb-interaction, cf. (5.64). Introducing the coupling parameter $\lambda = E_C/\omega$ which denotes the ratio of the Coulomb-energy $E_C = e^2/r_0$ with $r_0^2 = 1/(m\omega)$ and the potential energy of the trap ω, the Hamiltonian can be written in dimensionless units:

$$\hat{H} = \frac{1}{2} \sum_{\alpha=1}^{N} (\hat{p}_\alpha^2 + \hat{r}_\alpha^2) + \sum_{1 \le \alpha < \beta}^{N} \frac{\lambda}{\sqrt{\left|\hat{r}_\alpha - \hat{r}_\beta\right|^2 + \kappa^2}}. \tag{5.118}$$

In these units, the energy is given by E/ω, the chemical potential by μ/ω and the temperature by T/ω and $\beta = \omega/T$, respectively.

The eigenfunctions of the one-particle Hamiltonian are the well known eigenfunctions of the quantum mechanical harmonic oscillator. In this basis, the one- and two-particle integrals can be calculated in advance with high precision.[5] Furthermore the one- and two-particle integrals in the Hartree-Fock (HF) basis will be used in the following. They have been calculated for each specific value of the temperature, the particle number and the coupling strength.[6]

[5] The one- and two-particle integrals used in these calculations have been calculated by K. Balzer with Mathematica [39].

[6] These integrals have been calculated with a program by K. Balzer.

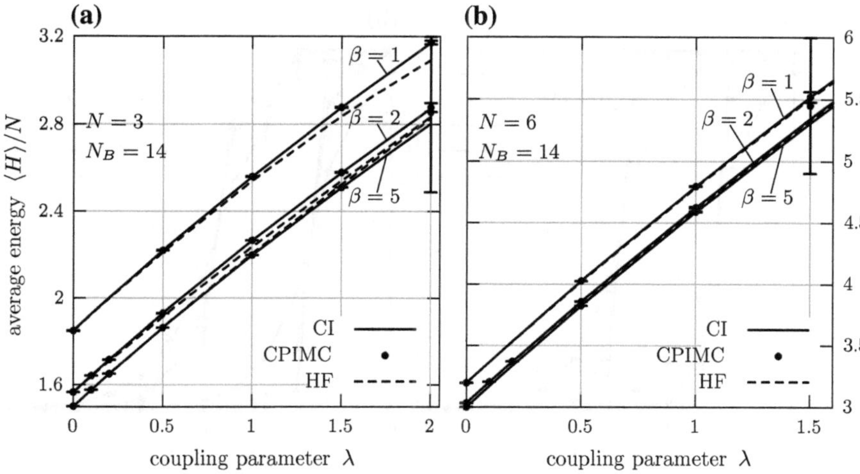

Fig. 5.3 Average energy per particle versus coupling parameter for **a** $N = 3$ and **b** $N = 6$ particles and different temperatures. Comparison of CPIMC results (*dots*), CI results (*solid lines*) and HF results (*dashed lines*). Temperatures are given by labels and *small arrows* within the figure. *Error bars* show the statistical error of the CPIMC simulation

The results in this work have been obtained under the additional assumption of spin-polarized fermions, i.e. all particles have the same spin projection remaining fixed in the simulation. Such a system is realized e.g. in the presence of a strong magnetic field. It should be noted that this was chosen for simplicity and is not a restriction of the method at hand. Generalizations to systems with arbitrary spin configurations or even with a spin dependent Hamiltonian are straightforward.

5.6.2 Numerical Results

The results in this section were obtained with the CPIMC method using 5×10^9 Monte Carlo steps per data point, if not noted otherwise. Samples for the thermodynamic averages were taken every 5,000 steps. This large number of steps is necessary because of the large auto-correlation time. In total, this results in 5×10^6 samples per observable. The equilibration time was chosen to 5×10^7 steps. The runtimes for these calculations were in between $3h$ and $12h$ on a single-core CPU depending on the particle number, temperature and coupling parameter.

First, we will consider the energy per particle, for it is a central and simple observable. Figure 5.3 shows the results for $N = 3$ and $N = 6$ particles obtained with 3 different methods: CPIMC, HF and CI.[7] For HF and CI, the first $N_B = 14$ eigenfunctions of the ideal one-particle Hamiltonian were used as basis functions. For

[7] The finite temperature HF and CI programs were written by D. Hochstuhl.

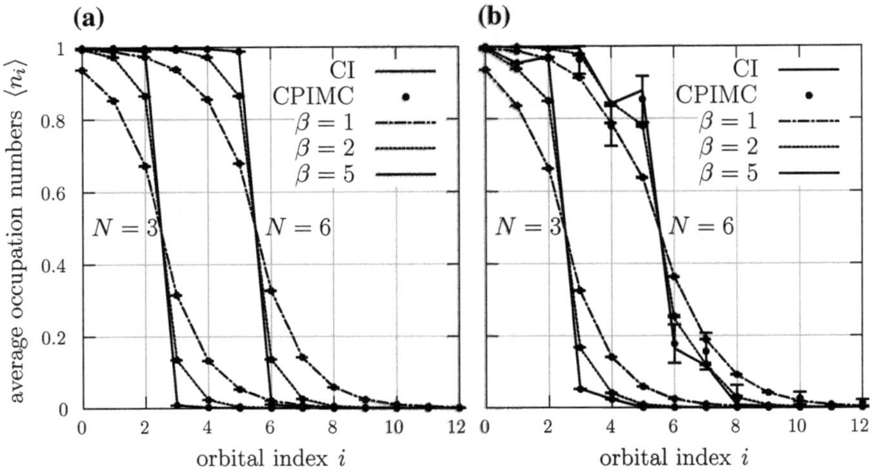

Fig. 5.4 Average occupation numbers $\langle \hat{n}_i \rangle$ of the one-particle orbitals $|i\rangle$. **a** in the HF basis and **b** in the basis of eigenfunctions of the ideal Hamiltonian. Shown are results obtained from CPIMC calculations (points with *error bars*) and CI results (*lines*) for the coupling strength $\lambda = 1$ and the temperatures $\beta = 1$ (*dashed dotted lines*), $\beta = 2$ (*dotted lines*) and $\beta = 5$ (*solid lines*). Adapted from [23]

CPIMC, the basis functions from the HF calculation were used. The CI results can be considered as reference data, because the only approximation of CI, the usage a finite basis set, is the same for all methods. For all shown coupling strengths and temperatures, the energies from the CI calculation match the energies from the MC calculation within the statistical errors.[8] For weak and moderate coupling, $\lambda \leq 1$, the statistical error is small and the agreement is very good. For stronger couplings, the statistical error strongly depends on the temperature and on the particle number. While the precision for $N = 3$ and $\beta = 2$ is high enough to distinguish the CPIMC result from the HF result up to $\lambda = 2$, this is not the case for $\beta = 5$ or $N = 6$. For $\lambda = 1.5$, $N = 6$ and $\beta = 5$, the relative error exceeds 100 % and the corresponding data point is not shown.

A quantity that is much more sensitive to systematical errors is the average occupation number $\langle \hat{n}_i \rangle$ of each one-particle orbital $|i\rangle$. The occupation numbers of the orbitals in the HF and the ideal basis for different temperatures and particle numbers are shown in Fig. 5.4. In the case of HF orbitals the shape of the curves is similar to the Fermi-Dirac distribution of non-interacting fermions in the grand-canonical ensemble. It should be noted that it is not possible to relate a specific energy to each orbital as in the case of basis functions that diagonalize the Hamiltonian, e.g. for ideal systems. Moreover the orbital energies of the HF orbitals have a meaning only within the HF approximation. Additionally, the HF orbitals with the same index differ from each other for different values of the temperature. For small temperatures, the

[8] Shown is the standard error, see (5.15). Within two times the standard error the results can be considered to coincide.

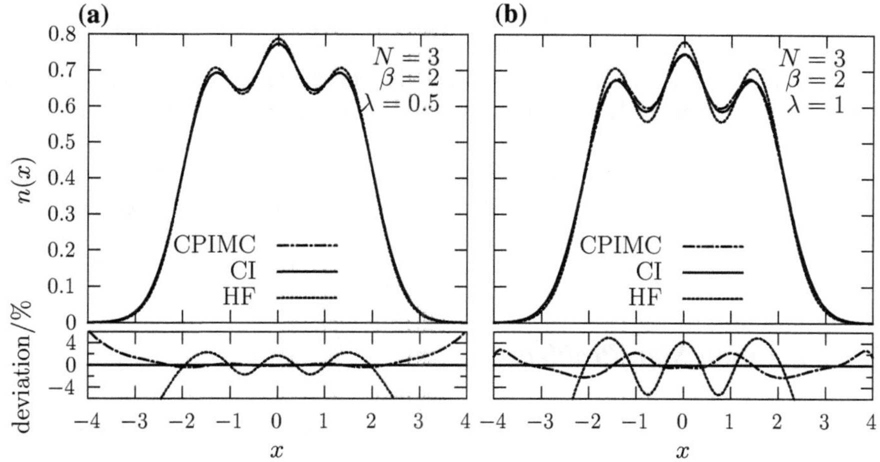

Fig. 5.5 One-particle density $n(x)$ for $N = 3$ particles, $\beta = 2$ and **a** $\lambda = 0.5$ and **b** $\lambda = 1$. *Solid lines* indicate CI results, which are considered to be exact. CPIMC and HF results are indicated by *dashed dotted* and *dotted lines*, respectively. Below, the relative deviation $(n(x) - n_{\text{CI}}(x))/n_{\text{CI}}(x)$ is shown

shape of the curves approaches a step functions, but as long as the exact ground state does not equal the HF ground state (which is the general case due to the interaction), there will be non-zero occupation numbers above the step even in the ground state at $T = 0$. There is an excellent agreement between the CPIMC and CI results for all shown parameters.

Considering the results for the occupation numbers of the ideal orbitals, the non-monotonic behavior is obvious. This behavior is confirmed by the CI results, which agree well within the statistical error. However, the statistical errors in these calculations are much larger than in the calculations using the HF orbitals, which will be discussed later.

The final observable, for which we will compare the results between the different methods in this chapter, is the one-particle density $n(x)$ where x denotes the spatial coordinate. As can be seen from (5.89), all matrix elements of the one-particle density matrix contribute to this quantity. Compared to the diagonal elements $d_{ii} = \langle \hat{n}_i \rangle$ shown before, the off-diagonal elements d_{ij} with $i \neq j$ converge much slower. The one-particle density contains a large statistical error since each of the off-diagonal elements can be determined only in certain configurations of the Markov-chain, see (5.108). Correspondingly, a larger number of MC steps would be necessary for a precise calculation of the whole one-particle density matrix. Nevertheless, Fig. 5.5 shows that for $N = 3$ particles, a temperature of $\beta = 2$, and a coupling strength of $\lambda = 0.5$ and $\lambda = 1$ the agreement of the CPIMC results with the exact density is better than in the case of the HF results. The relative deviation amounts to only a few per cent. This can be explained by the diagonal elements (which can be calculated efficiently, see Fig. 5.4) dominating the contribution to the density. Additionally, the statistical errors in the off-diagonal elements may partly cancel each other in the calculation of the one-particle density.

For all thermodynamic quantities presented in this chapter, we could demonstrate the very good agreement within the statistical error between the results obtained from the CPIMC method and the reference results obtained from the exact diagonalization of the Hamiltonian. For weak and moderate coupling strengths $\lambda \leq 1$ and moderate temperatures, i.e. $\beta = 1$ and $\beta = 2$, results can be obtained with high precision within several hours of runtime on a single CPU.

5.6.3 Fermion Sign Problem

In the last section, the focus lay on the exact calculation of thermodynamic averages of physical observables. In this section, we will investigate the main factor that determines the efficiency of QMC methods: the fermion sign problem. Taking the modulus of the weights to allow for the MC sampling of configurations with negative weights introduces the average sign $\langle s \rangle'$ in (5.9). For small average sign, the statistical error of MC calculations is approximately inverse proportional to it. For path integral Monte Carlo methods in coordinate space, e.g. DPIMC, the average sign decreases exponentially with the particle number N and the inverse temperature β. This leads to exponentially growing statistical errors. Alternatively, if results with a fixed precision are targeted, this leads to exponentially growing runtimes, see (5.18).

It turns out that the dependence of the average sign on temperature and particle number for CPIMC is essentially the same as for DPIMC, namely exponential. This is as expected since the general behavior is governed by (5.18). The dependence of the average sign on the coupling strength is shown in Fig. 5.6 for $N = 3$ and $N = 6$ particles and different temperatures. Additionally, the figure shows the average sign of DPIMC calculations for the same parameters.[9] For large coupling strength, $\lambda \gg 1$, the average sign of DPIMC is close to 1 and therefore, calculations with DPIMC are highly efficient. For decreasing λ, the sign decreases and calculations with DPIMC become more demanding. The simulation of the ideal system with $\lambda = 0$ is infeasible. The reason for this behavior is the following: For strong repulsive interactions, the overlap of the wave functions of the particles is small and the statistics resemble those of a classical system of point-like particles. The weaker the Coulomb repulsion compared to the strength of the trap, the larger is the overlap of the wave functions. With larger overlap, particle exchange, which is the source of the alternating sign in DPIMC, becomes more important.

CPIMC calculations, on the other hand, show a behavior completely opposite to that of DPIMC. While for the ideal system there exists no sign problem at all, the average sign decreases with increasing coupling parameter. This decrease becomes very fast at $\lambda \approx 1$, and calculations for larger coupling parameters become inefficient. The value of λ for which this rapid decrease of the average sign starts depends on

[9] The data was produced by A. Filinov.

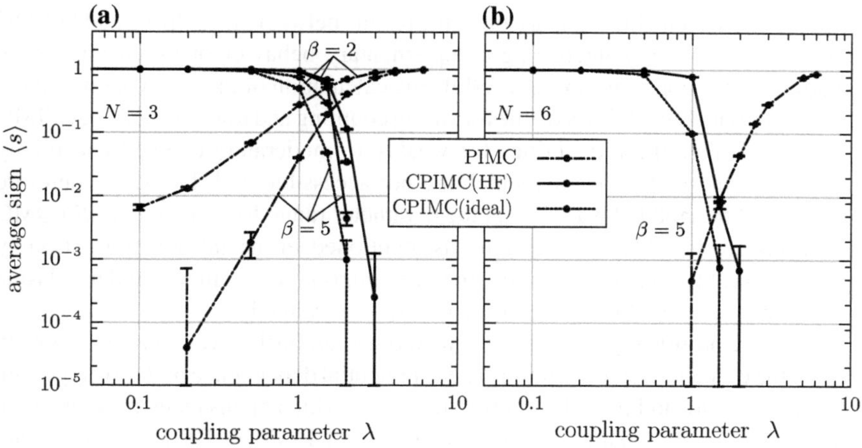

Fig. 5.6 Average sign versus coupling parameter for different temperatures and **a** $N = 3$ and **b** $N = 6$ particles. Shown are results obtained from CPIMC calculations with the HF basis (*solid lines*) and with the ideal basis (*dotted lines*). *Dashed dotted lines* indicate results from DPIMC calculations. Temperatures are given by labels and small arrows within the figure. *Error bars* show the statistical errors. Note the log scale. Adapted from [23]

the temperature and the particle number. For lower temperature and larger particle number, the average sign decreases earlier. Figure 5.6 also shows the influence of the choice of the basis on the average sign. Using HF basis functions, the average sign and therefore the efficiency of CPIMC can be increased by several magnitudes.

5.7 Summary and Outlook

In this chapter, we presented a path integral Monte Carlo method in the Slater determinant space for the exact calculation of thermodynamic properties of correlated fermionic many-particle systems. We introduced the appropriate description of the physical system in terms of second quantization and described the transition from the coordinate representation to the occupation number representation of the path integral in detail. In contrast to the widely used application of the underlying formulation to lattice models with onsite (or short-range) interaction, such as the Hubbard model, the method at hand allows for ab initio calculations of spatially continuous systems with arbitrary interaction. We gave a detailed description of the Monte Carlo steps including the acceptance probabilities for the exmple of one step. At the end we demonstrated the application of this method to a test system of Coulomb-interacting fermions in a one-dimensional harmonic trap. While we have presented only a small selection of thermodynamic quantities, our simulations are able to yield all other thermodynamic quantities. Moreover, it was recently demonstrated that also dynamic quantities, including the energy spectrum and collective excitation spectrum, can be efficiently computed within quantum Monte Carlo simulations [40].

For the test system, the comparison of the results between the CPIMC and DPIMC method revealed a very interesting complementary behavior of the average sign, depending on the coupling strength. While the calculation of thermodynamic expectation values with DPIMC is possible for strong coupling and low degeneracy, CPIMC allows to calculate these quantities for weakly to moderately coupled and highly degenerate many-body systems, which are not accessible with DPIMC. Therefore the presented method helps to extend the parameter range for which *ab initio* QMC simulations are feasible. Furthermore, as mentioned in the introduction, strongly degenerate weakly coupled electrons are of key importance for many modern plasma applications, so here CPIMC is expected to be very valuable.

Despite the promising results shown in this chapter, further work has to be done to improve the efficiency of the method. The exponential dependence of the average sign on the temperature and particle number still exists, which opens an increasing gap in the ranges of coupling parameters for which either CPIMC or DPIMC methods are efficient. This gap can, at the moment, be bridged only by additional approximations. Approaches to this problem can be assigned to two classes. The first class contains improvements to the algorithm that increase the effective number of MC samples per time, e.g. by optimizing the acceptance probability of the steps and by decreasing the auto-correlation time. The so-called heat bath idea is such an approach. By choosing the sampling probability to be proportional to the transition probability, the acceptance probability becomes unity. This is possible if the normalization for the transition can be calculated efficiently, e.g. for choosing the time of the kinks in the MC steps. Another promising approach in this class is the so-called worm algorithm [25] which also allows for more efficient sampling of off-diagonal quantities related to the Matsubara Green function, including the one-particle density. Other approaches aim at reducing the sign problem directly. One method in this class is the so-called blocking, which tries to analytically combine configurations of opposite sign into groups with a much lower modulus of the weight than the single configurations.

To achieve even higher efficiency additional approximations can be used. The similarities of the CPIMC method to MC on a lattice (such as the Hubbard model) as well as to CI methods, give access to a large variety of approximations developed for these latter methods. As an example we mention the restricted active space (RAS) approach well known in quantum chemistry, see [41]. Special cases are the "CI-Singles" and "CI-Doubles" approximations. The main idea is to restrict the occupation number of the orbitals in different energy (or spatial) regions. For example, one can require that the n orbitals within the lowest energy interval must be occupied by at least $n - 1$ particles, the m orbitals within the next-higher energy interval be occupied by at least $m - 2$ particles and so on, until the orbitals with the highest energy are occupied at most by one particle. These restrictions are arbitrary but can be motivated by physical insights. This approach strongly decreases the size of the configuration space and consequently the sign problem. The advantage of this approximation is that it is easy to control and to fine-tune. Calculations with gradually reduced restrictions allow one to estimate the introduced systematical error. With these improvements applications of CPIMC to larger particle numbers and, eventually, to macroscopic quantum plasmas seem to be possible.

Acknowledgments We are grateful to K. Balzer, A. Filinov and D. Hochstuhl for providing supplementary data shown in this chapter. We acknowledge financial support by the Deutsche Forschungsgemeinschaft via grant BO1366-10 and a grant for CPU time at the HLRN.

References

1. Proceedings of the conference. Strongly Coupled Coulomb systems. J. Phys. A 42 No. 21 (2009)
2. M. Bonitz, C. Henning, D. Block, Complex plasmas: a laboratory for strong correlations. Rep. Prog. Phys. **73**(6), 066501 (2010)
3. H. Haberland, M. Bonitz, D. Kremp, Harmonics generation in electron-ion collisions in a short laser pulse. Phys. Rev. E **64**(2), 026405 (2001)
4. R. Redmer, G. Röpke, Progress in the theory of dense strongly coupled plasmas. Contrib. Plasma Phys. **50**(10), 970 (2010)
5. B. Holst, M. French, R. Redmer, Electronic transport coefficients from ab initio simulations and application to dense liquid hydrogen. Phys. Rev. B **83**(23), 235120 (2011)
6. M. Bonitz, A. Filinov, J. Böning, J.W. Dufty, in *Introduction to Complex Plasmas*, vol. 59, ed. by M. Bonitz, N. Horing, P. Ludwig (Springer, Berlin, Heidelberg, 2010), pp. 41–77
7. J.S. Tse, Ab initio molecular dynamics with density functional theory. Annu. Rev. Phys. Chem. **53**(1), 249 (2002)
8. D. Kremp, M. Schlanges, W. Kraeft, T. Bornath, *Quantum Statistics of Nonideal Plasmas*, 1st edn. (Springer, Berlin, Heidelberg, 2005)
9. M. Bonitz, *Quantum Kinetic Theory* (B G Teubner, Stuttgart, Leipzig, 1998)
10. K. Balzer, M. Bonitz, *Nonequilibrium Green's Functions Approach to Inhomogeneous Systems* (Springer, Heidelberg, New York, 2013)
11. R. Feynman, *Quantum Mechanics and Path Integrals* (McGraw-Hill, New York, 1965)
12. D.M. Ceperley, Path integrals in the theory of condensed helium. Rev. Mod. Phys. **67**(2), 279 (1995)
13. A. Filinov, N.V. Prokof'ev, M. Bonitz, Berezinskii-Kosterlitz-Thouless transition in two-dimensional dipole systems. Phys. Rev. Lett. **105**(7), 070401 (2010) NULL
14. D. Ceperley,*in Monte Carlo and Molecular Dynamics of Condensed Matter Systems* (Editrice Compositori, Bologna, Italy, 1996)
15. M. Troyer, U. Wiese, Computational complexity and fundamental limitations to fermionic quantum Monte Carlo simulations. Phys. Rev. Lett. **94**(17), 170201 (2005)
16. D.M. Ceperley, Path-integral calculations of normal liquid ^{3}He. Phys. Rev. Lett. **69**(2), 331 (1992)
17. K.P. Driver, B. Militzer, All-Electron path integral Monte Carlo simulations of warm dense matter: application to water and carbon plasmas. Phys. Rev. Lett. **108**(11), 115502 (2012)
18. C. Pierleoni, D. Ceperley, in Computer Simulations, in *Condensed Matter Systems: From Materials to Chemical Biology Volume 1*, vol. 703, ed. by M. Ferrario, G. Ciccotti, K. Binder (Springer, Berlin, Heidelberg, 2006), pp. 641–683
19. M.A. Morales, C. Pierleoni, D.M. Ceperley, Equation of state of metallic hydrogen from coupled electron-ion Monte Carlo simulations. Phys. Rev. E **81**(2), 021202 (2010)
20. R. Egger, W. Häusler, C.H. Mak, H. Grabert, Crossover from fermi liquid to wigner molecule behavior in quantum dots. Phys. Rev. Lett. **82**(16), 3320 (1999)
21. V.S. Filinov, M. Bonitz, W. Ebeling, V.E. Fortov, Thermodynamics of hot dense H-plasmas: path integral Monte Carlo simulations and analytical approximations. Plasma Phys. Control. Fusion **43**(6), 743 (2001)
22. A.V. Filinov, M. Bonitz, Y.E. Lozovik, Wigner crystallization in mesoscopic 2D electron systems. Phys. Rev. Lett. **86**(17), 3851 (2001)

23. T. Schoof, M. Bonitz, A. Filinov, D. Hochstuhl, J.W. Dufty, Configuration path integral Monte Carlo. Contrib. Plasma Phys. **51**(8), 687 (2011)
24. N.V. Prokofev, B.V. Svistunov, I.S. Tupitsyn, Exact quantum Monte Carlo process for the statistics of discrete systems, J. Exp. Theor. Phys. Lett. **64**(12), 911 (1996)
25. N.V. Prokofev, B.V. Svistunov, I.S. Tupitsyn, Exact, complete, and universal continuous-time worldline Monte Carlo approach to the statistics of discrete quantum systems, J. Exp. Theor. Phys. **87**(2), 310 (1998)
26. M. Troyer, F. Alet, S. Trebst, S. Wessel, Non-local updates for quantum Monte Carlo simulations. AIP Conf. Proc. **690**, 156 (2003)
27. E. Gull, A.J. Millis, A.I. Lichtenstein, A.N. Rubtsov, M. Troyer, P. Werner, Continuous-time Monte Carlo methods for quantum impurity models. Rev. Mod. Phys. **83**, 349 (2011)
28. N. Metropolis, A.W. Rosenbluth, M.N. Rosenbluth, A.H. Teller, E. Teller, Equation of state calculations by fast computing machines. J. Chem. Phys. **21**(6), 1087 (1953)
29. W. Janke, in *Quantum Simulations of Complex Many-Body Systems: From Theory to Algorithms*, ed by J. Grotendorst, D. Marx, A. Muramatsu (John von Neumann Institute for Computing, NIC, Jülich, 2002)
30. T. Nakamura, Vanishing of the negative-sign problem of quantum Monte Carlo simulations in one-dimensional frustrated spin systems. Phys. Rev. B **57**(6), R3197 (1998)
31. A.P. Lyubartsev, Simulation of excited states and the sign problem in the path integral Monte Carlo method. J. Phys. A **38**(30), 6659 (2005)
32. T. Helgaker, P. Jorgensen, J. Olsen, *Molecular Electronic-Structure Theory*, 1st edn. (Wiley, Chichester, Hoboken, 2000)
33. H.F. Trotter, On the product of semi-groups of operators. Proc. Amer. Math. Soc. **10**(4), 545 (1959)
34. M. Suzuki, Generalized Trotter's formula and systematic approximants of exponential operators and inner derivations with applications to many-body problems. Commun. Math. Phys. **51**(2), 183 (1976)
35. A.W. Sandvik, J. Kurkijrvi, Quantum Monte Carlo simulation method for spin systems. Phys. Rev. B **43**(5950) (1991)
36. N.V. Prokof'ev, Worm algorithm for classical and quantum statistical models, Lecture notes at "Advanced School on Quantum Monte Carlo Methods in Physics and Chemistry", 2008, Trieste, http://mcwa.csi.cuny.edu/umass/lectures.html
37. M. Bonitz, D. Semkat (eds.), *Introduction to Computational Methods in Many Body Physics* (Rinton Press Inc, Paramus, 2006)
38. T. Schoof, Thermodynamische Eigenschaften entarteter, korrelierte Fermionen, Diplomarbeit (in German). (Kiel University, 2011)
39. Wolfram Research Inc, Mathematica. Version 7.0 (Wolfram Research Inc, Champaign, Illinois, 2008)
40. A. Filinov, M. Bonitz, Collective and single-particle excitations in two-dimensional dipolar Bose gases. Phys. Rev. A **86**(4), 043628 (2012)
41. D. Hochstuhl, M. Bonitz, Time-dependent restricted-active-space configuration-interaction method for the photoionization of many-electron atoms. Phys. Rev. A **86**, 053424 (2012)

Part III
Low-Temperature Plasmas

Chapter 6
Calorimetric Probes for Energy Flux Measurements in Process Plasmas

Sven Bornholdt, Maik Fröhlich and Holger Kersten

Abstract This chapter gives an overview of the method of calorimetric probes which are used for characterizing the interaction between low-temperature plasmas and substrates in materials processing. Although the focus is on low-temperature non-equilibrium plasmas most of the concepts can also be transferred to thermal plasmas or are in fact adopted from fusion research. An introductory section showing the importance and complexity of plasma wall interactions is followed by a section providing an overview and comparison of various probe concepts, which have been developed in the last decades. Special focus is on the type of probes which are similar to the probes used by J.A. THORNTON (In fact, THORNTON was not the first, to use this type of probe. From his work from 1978 [1], one can follow the citations back to the work of JACKSON in 1969, who used equation (6.7) for the determination of the power dissipated by a copper block located at substrate position in a sputtering discharge [2]) who was one of the pioneers connecting plasma characteristics with resulting surface properties. Thereafter, a short section gives a basic overview of the different contributions to the total energy influx and which are of importance for the plasma wall interaction. The last section shows different examples of applications of calorimetric probes. It demonstrates the applicability and flexibility of these types of probes for characterization of different low-temperature plasmas. The examples

Sections of this chapter are reprinted from: Sven Bornholdt: PARTICLE AND ENERGY FLUXES DURING PLASMA BASED DEPOSITION OF NANOSTRUCTURED MATERIALS. PhD thesis, Kiel University (2013).

S. Bornholdt (✉) · M. Fröhlich · H. Kersten
Institut für Experimentelle und Angewandte Physik, Christian-Albrechts-Universität zu Kiel, Leibnizstr. 19, 24098 Kiel, Germany
e-mail: bornholdt@physik.uni-kiel.de

M. Fröhlich
e-mail: froehlich@physik.uni-kiel.de

H. Kersten
e-mail: kersten@physik.uni-kiel.de

M. Bonitz et al. (eds.), *Complex Plasmas*, Springer Series on Atomic, Optical, and Plasma Physics 82, DOI: 10.1007/978-3-319-05437-7_6,

mainly focus on complex plasmas, where the variety of the involved processes also causes a specific plasma wall interaction. Such complex situations can be found in systems where reactive gases or gas mixtures, nano or micro particles (dust), magnetic fields or large surface to volume ratios are involved.

6.1 Introduction

Since all man made plasmas ranging from smallest micro plasmas up to cubic meter sized plasmas in Tokamak or Stellerator devices for fusion research are in contact with confining and surrounding walls or substrate materials the characterization of plasma wall interactions has been and still is a hot topic in all fields of plasma research and technology. The various processes at walls, electrodes or substrates lead to an inseparable connection and a reciprocative feedback between plasma and surface. The interaction is of quite complex nature, since all plasma constituents or so called species (electrons, ions, neutrals and radiation), their energy or velocity distribution and density as well as all the different surface properties like chemical composition, secondary electron emission coefficient, electrical and thermal conductivity, binding energies etc. have an important impact and contribute in different ways. A deeper understanding of these processes is of great importance for the improvement of fusion devices (high-temperature plasmas) as well as for the modification and coating of substrate surfaces by using non-thermal plasmas in technology. Again, looking at different types of plasmas one can identify different roles of the walls and, therefore, also of the interaction between plasma and surface:

Atmospheric pressure plasmas have a relatively large surface to volume ratio, due to their usually small dimensions. This leads to a more pronounced influence of the surface and its properties. Working at atmospheric pressure they are also suited for treatment of living cells or tissues in biomedical applications. Especially those plasmas, which are used for in vivo processes (sterilization, wound healing, etc.) the knowledge of the amount and origin of transferred energy is of essential relevance. Too high temperature load of the living tissue by the plasma might lead to its destruction or other unwanted results.

Fusion plasmas open a wide field of plasma surface interactions. In general, the types of interactions are similar but the total energy fluxes are much higher compared to low temperature plasmas [3, 4]. The high energy fluxes are caused by much higher energies of the different species. Especially particle formation from eroded wall materials leads to safety and operational issues due to radioactive content, tritium retention or chemical activity [5, 6].

Low pressure plasmas are widely used in various fields of materials processing. The energy balance of the substrate not only plays an important role during the treatment of heat sensitive materials, which might be damaged or destroyed due to a high thermal load, but also for all kinds of substrates and modifications. The substrate temperature influences the growth and structure of deposited films [7, 8] and also

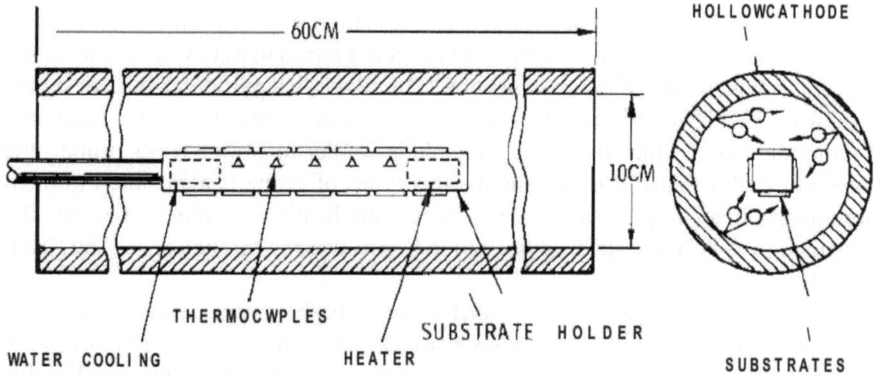

Fig. 6.1 Schematic of hollow cathode sputtering apparatus with temperature gradient substrate holder used by THORNTON [7] (Reprinted with permission from [7], ©1974 American Vacuum Society)

surface reactions during plasma etching [9] and other applications like cleaning or functionalization.

When Thornton investigated the "influence of apparatus geometry and deposition conditions on the structure and topography of thick sputtered coatings", he recognized that the substrate temperature, which is determined by heating or cooling by the substrate holder and by the influence of the plasma itself, has an important impact on the structure and properties of deposited films [7].

By using a substrate holder, which allows mounting many substrates at different temperatures simultaneously (see Fig. 6.1), he was able to develop a zone model which describes the dependence of different film structures and morphologies on the energetic conditions at the surface [7, 10].

For the investigation of the influence of the plasma on the thin film properties THORNTON developed a so called heat transfer probe [1] (see Sect. 6.2.1 and Fig. 6.4c), which is one of various types of calorimetric probes used today for characterization of plasma wall interaction. This probe usually consists of a thin metal plate. The backside of the metal plate is connected to a thermocouple which allows the continuous recording of the plate's temperature. The sensor area is embedded into a massive shielding which prevents the energy flux from reaching the sensor except the front area. Only small adjustments have been made on that initial setup, but the concept has been extended from the investigation of heat or energy transfer, respectively, during growth of thin films on substrates to many complex processes which take place at the plasma boundary during treatment of various substrate materials. The simple and robust design of the probe enables a flexible integration into existing experimental setups. This has led to applications under a large variety of conditions in nearly all kinds of low temperature plasmas ranging from small atmospheric plasmas to extended low pressure plasmas. Calorimetric probes can be used for spatial characterization of the plasma in general or as a dummy for the determination of the energy influx and the resulting surface temperature at substrate position.

A second technical approach for the determination of energy fluxes is based on the radiometer developed by GARDON [11] (often referred to as GARDON sensor, see Sect. 6.2.2) which was an improvement of "The silver disk pyrheliometer" designed by ABBOT [12]. This type of probe was intentionally developed for the measurement of radiation fluxes. Since the basic principle is similar to THORNTON's probe - the temperature of a thin foil is measured - this types of probe is also suited to measure energy fluxes in plasmas. One of the first applications of this probes was the investigation of "Charge states of heavy-ion beams passing through gases" by STIER et al. [13].

Besides these two methods also other types of probes were developed in order to measure the energy influx towards substrates. The common similarity is usually the temperature measurement of a "substrate dummy" exposed to a plasma. For this temperature measurement different technical methods are used such as thermopile sensors [14] (a stack of thermocouples, usually operated in parallel), or the optical measurement of the temperature dependent band gap of an GaAs-crystal [15]. Also an active calorimetric probe is on the market [16]. In contrast to the mentioned passive probes, which record the temperature slope during processing, the active probe is heated to a certain temperature and the dissipated power is measured, which is needed to keep the probe at a constant temperature. During the plasma exposure the heating power needed to sustain this temperature is lowered due to the power delivered from the plasma. By monitoring the heating power the additional energy flux from the plasma can be determined directly.

The different probe types and designs have been used for characterization of various plasmas which are used in materials processing, such as different types of magnetron discharges (dc [23, 1, 17–22], rf [20, 22, 24], HiPIMS [25–28]) and hollow cathode arcs [29–31] for thin film deposition purposes. Also rf [24, 32–36] or microwave [24] plasmas and ion beams [37] for surface modification have been investigated. Further examples are hollow cathode plasma jets [21] and plasma downstream reactors (PDR) [15, 38, 39] or atmospheric pressure plasmas like plasma needle [40] and non-thermal atmospheric pressure plasma jets [41]. Since dusty plasmas received growing interest also the stationary temperature of micro particles ($\approx$10–200 μm) which are confined in the sheath region are used to calculate energy fluxes towards their surfaces in order to obtain information about the surrounding plasma [42, 43].

6.2 Calorimetric Probes

Since heat flux measurement is a well established technique in several fields of engineering and science many different technical approaches are available. It is not possible to cover all different devices. Some selected examples are given here which were found to be suitable in the field of plasma diagnostics. Also the mathematical description of the underlying physical mechanisms involved, would be beyond the scope of this work. The interested reader is referred to the countless textbooks and

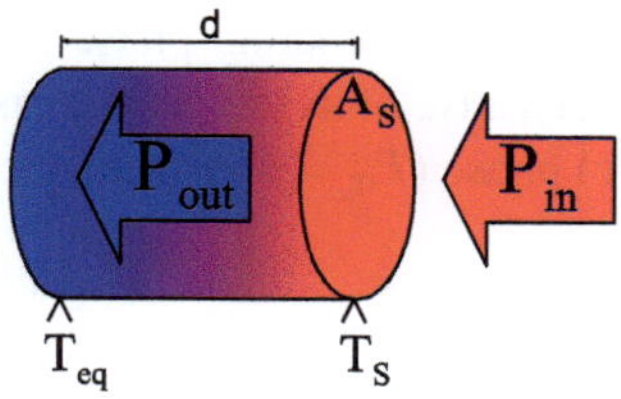

Fig. 6.2 A schematic drawing of the heat transfer through a *cylinder* (probe) with the length d and the cross section area A_S

papers available covering the variety of techniques and the physics of heat fluxes, e.g. see [44].

This section describes some different technical approaches, the experimental setups and their mathematical description. Since heat fluxes are always related to a temperature T, which is a measure for the transferred energy $\dot{Q}$ connected via the heat capacity C

$$\dot{Q} = P = C\dot{T} \tag{6.1}$$

the methods use commonly the measurement of temperatures, spatial temperature distributions or temporal temperature characteristics.

6.2.1 THORNTON*'s "Heat Flux Probe"*

As already mentioned many of the various calorimetric probes are very similar to the one developed by THORNTON, so they can be described by the same model. The temperature T_S of a surface area A_S which is connected to a reservoir at constant temperature T_{eq} depends on the incoming power P_{in} and the outgoing power P_{out}. P_{in} is the sum of all processes caused by impinging particles, processes taking place at surface as well as radiation. The loss P_{out} is the sum of conduction along connections, convection by the background gas and radiative losses. In the case of low pressure plasmas losses due to convection and radiation are often negligible and heat conduction is the dominant loss process.

The principal situation is illustrated in Fig. 6.2 and the general case can be written as [29]:

$$C_S\dot{T}_{S,h} = P_{in} - P_{out} = P_{in} - \underbrace{a(T_S - T_{eq})}_{\text{conduction}} \tag{6.2}$$

Here C_S is the heat capacity of the probe and $a = \lambda\frac{A_S}{d}$ (λ: heat conduction coefficient and d: length of the connection).

The conventional method for determining P_{in} is to expose the area A_S to a constant power for a certain time and then switch off the source of power or shield the probe with a shutter.

During the heating phase ($P_{\text{in}} > 0$)) the surface is heated (index h) and the time derivative of its temperature is described by equation (6.2). During the cooling (index c) phase ($P_{\text{in}} = 0$) this reads:

$$C_S \dot{T}_{\text{S,c}} = -P_{\text{out}} = -a(T_{\text{S}} - T_{\text{eq}}) \tag{6.3}$$

This two first-order linear differential equations have the solution

$$T_{\text{S,h}}(t) = \left(T_{\text{eq}} + \frac{P_{\text{in}}}{a}\right) - \left(\frac{P_{\text{in}}}{a}\right) \exp(-\frac{a}{C_{\text{S}}}t) \tag{6.4}$$

for the temperature $T_{\text{S,h}}$ during the heating phase and the solution

$$T_{\text{S,c}}(t) = T_{\text{eq}} + (T_{\text{st}} - T_{\text{eq}}) \exp(-\frac{a}{C_{\text{S}}}t) \tag{6.5}$$

for the cooling phase. The term T_{st} denotes the starting temperature at the beginning of the cooling phase. The solutions results in typical temporal temperature profiles obtained with calorimetric probes.(see Fig. 6.3a).
Figure 6.3a shows that the maximum temperature $T_{\max}$ to which the probe can be heated does not depend on the heat capacity of the probe. Only the time, which is needed to reach the equilibrium temperature is longer for larger heat capacities. This temperature can be obtained by setting $t \to \infty$ into equation (6.4), which than yields:

$$T_{\text{S,h}}(t \to \infty) = T_{\text{eq}} + \frac{P_{\text{in}}}{a} =: T_{\max} \tag{6.6}$$

$T_{\max}$ is indeed independent from C_{S} and depends only on the heat conductivity of the material and its geometrical dimensions. Different approaches for the determination of P_{in} from the temporal temperature behavior have been used:

1. One can determine the slope at the beginning of the temperature rise [1, 45]. Near the equilibrium temperature the losses can be neglected and equation (6.2) gives $C_{\text{S}} \dot{T}_{\text{S,h}} = P_{\text{in}}$. This can also be seen in Fig. 6.3b, where all the $\dot{T}_{\text{S,c}}$ lines meet at the point (T_{eq}, 0). Hence, the slope of the temperature rise is directly proportional to the incoming power. The quantitative determination of P_{in} by this method requires the knowledge of C_{S}. The advantage of this method is that only very short heating times are needed and, therefore, the shielding of the probe (see Sect. 6.2.1.1) usually remains near or even at equilibrium temperature (T(t=0)).
2. A method which is assumed to be more accurate uses the combination of equations (6.2) and (6.3) to yield [2]:

$$P_{\text{in}} = C_{\text{S}} \left[\dot{T}_{\text{S,h}} - \dot{T}_{\text{S,c}}\right] \tag{6.7}$$

For the calculation of the time derivatives it is often convenient to fit the heating and cooling curves by equations (6.4) and (6.5) or by polynomial functions. This

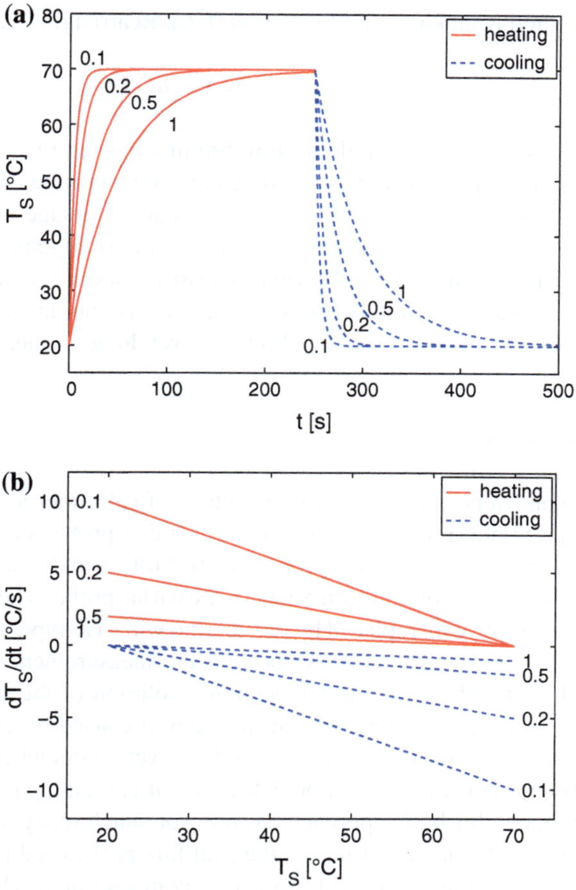

Fig. 6.3 **a** Temporal temperature characteristics of probes with different heat capacities during heating and subsequent cooling. **b** Time derivatives of the surface temperature versus the surface temperature for the curves shown in (**a**). Typical data obtained by calorimetric probes with different heat capacities. The power delivered to the probe (P_{in}) equals 1 W for all curves. The unit of the different heat capacity values is JK^{-1}

should be especially done if the resulting substrate heating and the subsequent signal to noise ratios are low. However, the accuracy of this method depends on the number of data pairs, which can be enhanced by increasing the heating time, which leads also to a temperature rise of the shielding. However, the heat capacity of the substrate dummy is usually much lower than the heat capacity of the surrounding shielding and this effect can usually be neglected. The time derivatives of the probe temperature from Fig. 6.3a are shown in Fig. 6.3b, which shows a typical $\dot{T}_S$ versus T_S plot for different heat capacities of the probe.

3. The third method is making use of equation (6.6). Rearranging yields

$$P_{in} = a\left(T_{max} - T_{eq}\right) = a\Delta T \qquad (6.8)$$

The idea is to heat the probe until the equilibrium temperature is reached. From this equilibrium temperature the energy influx can be determined without knowledge of the heat capacity of the probe, which is an advantage of this method. The drawback is that the probe has to reach the equilibrium temperature, which elongates the measurement time. This may lead to a situation where the shielding ceramic or the plasma source is essentially heated. As an unwanted result the equilibrium cannot be reached within hours or even longer time.

6.2.1.1 Probe Layouts

In the following an overview of the different setups of calorimetric probes, which have a working principle similar to THORNTON's heat flux probe, are provided.

The most simple construction is the probe used by GOLAN and AXELEVITCH [46]. It is a further development of an electrostatic LANGMUIR probe. The probing part, which is usually a simple wire replaced by two thermo wires (K-type) twisted around each other (see Fig. 6.4a). This allows not only the measurement of typical I-V characteristics, but also the temporal temperature evolution of the probe tip. This probe was mainly used to calculate the transferred heat caused by electron energy flux. Due to the large volume to surface ratio it takes several minutes for the probe to reach its equilibrium temperature. Although the idea of combining two probe types is very interesting, but this basic approach has not won much recognition.

A simple probe with a larger surface is the heat flux probe used by MAKRINICH et al. for the characterization of the heat flux and its composition in a helicon plasma source [33, 47]. This probe (see Fig. 6.4b) consists of a molybdenum plate with the area of 17.5 × 19 mm and 2.5 mm thickness, which is the probing surface. The plate is glued to a thermocouple for measuring the plate's temperature via a digital voltmeter and a PC. This probe was mainly used for estimating the energy influx by impinging ions at different (negative) probe potentials. The results show that the amount of energy transferred per ion is only in the order of 50 % up to 90 % of the maximum kinetic energy, depending on the ion mass. One drawback of this unshielded type of probes might be that the probing area is not well defined since the wires are also exposed to the plasma and, therefore, also collect electrical current and energy flux. Another point is that the energy flux, which can be highly anisotropic inside the plasma can hardly be determined with an angular resolution. However, this method gives a simple access to energy fluxes and plasma composition.

The probe invented by THORNTON [1] (see Fig. 6.4c) already makes use of a steel shielding, which surrounds the backside of the plate and the connection to the thermocouples. It overcomes the drawback of the probe by MAKRINICH because the probe allows energy flux measurements from distinct directions. The probe was intentionally constructed for the determination of substrate temperatures during sput-

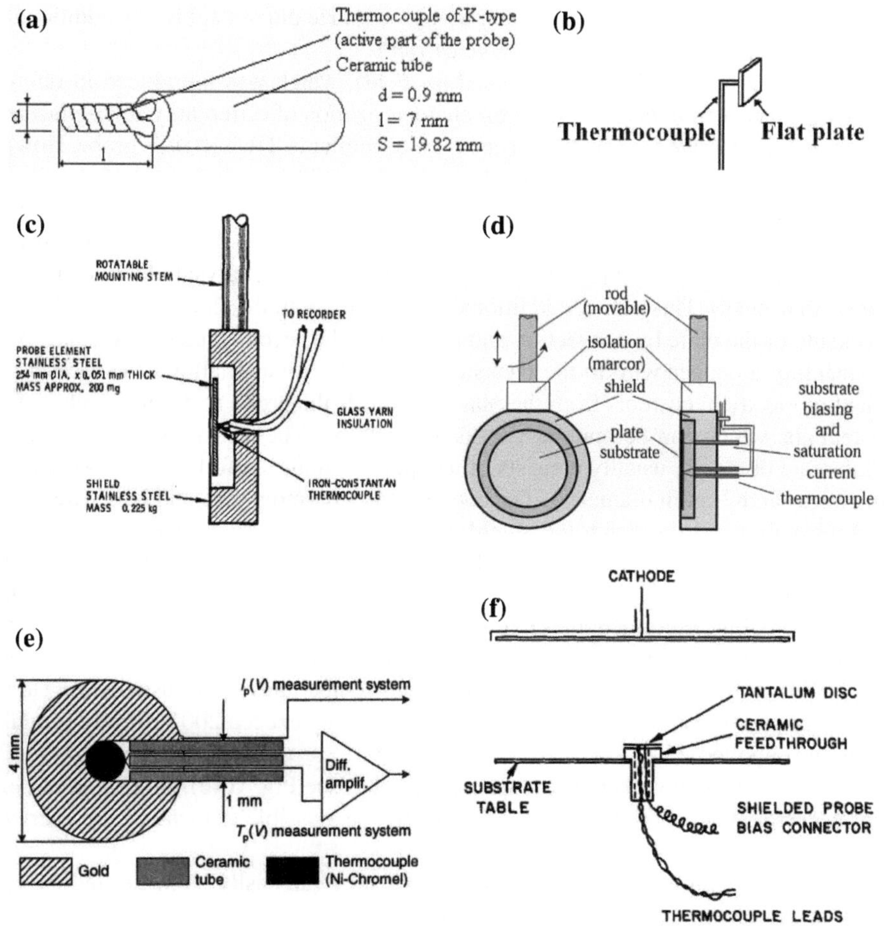

Fig. 6.4 **a** Modified Langmuir probe used by GOLAN and AXELEVITCH [46]. (Reprinted with permission from [46], ©2003 JOAM/OAM-RC). **b** Heat flux probe used by MAKRINICH and FRUCHTMAN [47]. (Reprinted with permission from [47], ©2006 American Institute of Physics). **c** THORNTON's heat flux probe [1]. (Reprinted with permission from [1], ©1978 Elsevier) **d** Calorimetric probe design applied by KERSTEN et al. [24]. (Reprinted with permission from H. KERSTEN et al. [24], ©2000 Elsevier). **e** Spherical calorimetric probe by STAMATE et al. [48]. (Reprinted with permission from E. STAMATE et al. [48], ©2002 American Institute of Physics) **f** Original probe by BALL [45]. (Reprinted with permission from D.J. BALL [45], ©1972 American Institute of Physics). Calorimetric probe setups similar to THORNTON's heat flux probe, used for investigations of plasma-wall-interaction

ter deposition of different metallic films. With the help of these studies the structure zone model was developed describing the internal film structure and morphology in dependence on the energetic conditions at the substrate surface influenced by external heating and plasma exposure [7, 10, 51]. THORNTON's investigations were mainly focused on the impact of neutral sputtered particles and the release of heat by condensation during the subsequent film growth. He did not apply any additional bias voltage for the separation of charge carriers.

The probe used by KERSTEN et al. (Fig. 6.4d), which was introduced in reference [24] has been intensively used for characterization of different types of plasma sources. This passive calorimetric probe is quite similar to THORNTON's probe. However, two essential changes were made in the setup. At first the shield of stainless steel was replaced by a ceramic one consisting of Macor®. This ceramic shielding has a much lower temperature rise during the heating phase due to its lower heat conductivity (10 times smaller than steel) and the larger heat capacity (about 2 times larger than steel). This reduces additional heat fluxes from the shielding towards the backside of the plate by convection and radiation. The second change was made by connecting a copper wire to the backside of the plate, which allows the measurement of electrical currents from the plasma towards the probe at different substrate potentials, which can be applied via this connection. The sensor area, also called "substrate dummy", usually consists of a copper or tungsten foil (about 100 μm in thickness and 2 cm in diameter). The connections (thermocouple and bias wire) are fed through a rod on which the shielding is mounted to protect the wires and to allow easy positioning or rotation of the probe for spatially and angularly resolved measurements [35, 36]. Drawback of the latter three probes is the relatively weak connection of the thermocouples to the probe area. The welding spot or the glue can usually not sustain temperatures much higher than 400 °C, which makes them less applicable in plasmas with elevated neutral gas temperatures. But usually the substrate temperature in low temperature plasmas does not exceed 200 °C if no external heating is applied.

STAMATE and coworkers use a spherical probe (see Fig. 6.4e), so a shielding is not necessary. This probe can be biased. Thus, it is possible to measure the energy influx at different probe potentials [48].

Due to the simplicity of calorimetric probes they can easily be integrated into a substrate holder as performed by BALL [45] or CADA et al. [49] (see Figs. 6.4f and 6.5a).

For plasma sources which operate at high power densities also the energy flux to the probe can be very high and the temperature might overcome the probes limitations. The easiest way to avoid such problems is to use a probe with larger heat capacity. For example, a simple cylindric metal body can be used facing the plasma with the top area (see Fig. 6.5b). In steady state condition, there will be a linear temperature profile along the cylinder, which can be measured by two or more thermocouples attached [50, 52].

So far, thermocouples have been used for the temperature measurement at the probing area. ROTH et al. [15, 38, 39] used an optical method for monitoring the probe temperature (see Fig. 6.5c, d). The sensor is a GaAs-crystal, which shows a temperature dependent band edge displacement. This method has a slightly lower time and temperature resolution than a thermocouple, but it is not sensitive to intense electro magnetic fields and can, therefore, be used where thermocouples can not be applied. Unfortunately, a disadvantage is that this probe can not be biased. It is always at floating potential.

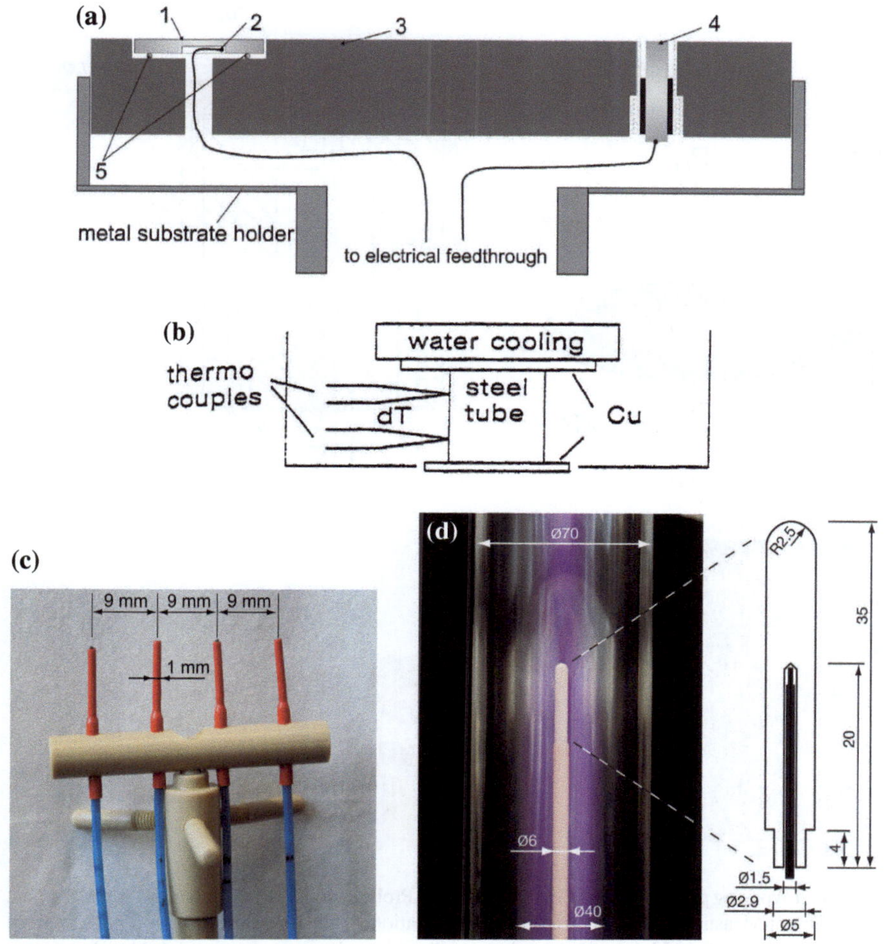

Fig. 6.5 **a** Calorimetric probe by CADA et al. inserted into a substrate holder: *1* The copper calorimeter probe, *2* the K-type thermo couple, *3* the Macor® block, *4* the planar Langmuir probe, and *5* ceramic tubes [49]. (Reprinted with permission from M. CADA et al. [49], ©2007 American Institute of Physics). **b** Schematic of the probe used by STEFFEN et al. [50]. (Reprinted with permission from H. STEFFEN et al. [50], ©1994 American Vacuum Society). **c** Four-headed probe by ROTH et al. using a GaAs crystal [38]. (Reprinted with permission from C. ROTH et al. [38], ©2011 Institute of Physics). **d** Encapsulated probe by ROTH et al. also using a GaAs crystal [39]. (Reprinted with permission from C. ROTH et al. [39], ©2012 Institute of Physics). Additional calorimetric probe setups similar to THORNTON's heat flux probe

6.2.2 GARDON's Radiometer

When GARDON constructed his simple radiometer as "An Instrument for the Direct Measurement of Intense Thermal Radiation" [11] he developed a probe (see Fig. 6.6b) which is also suitable for the characterization of energy fluxes in plasmas. This probe

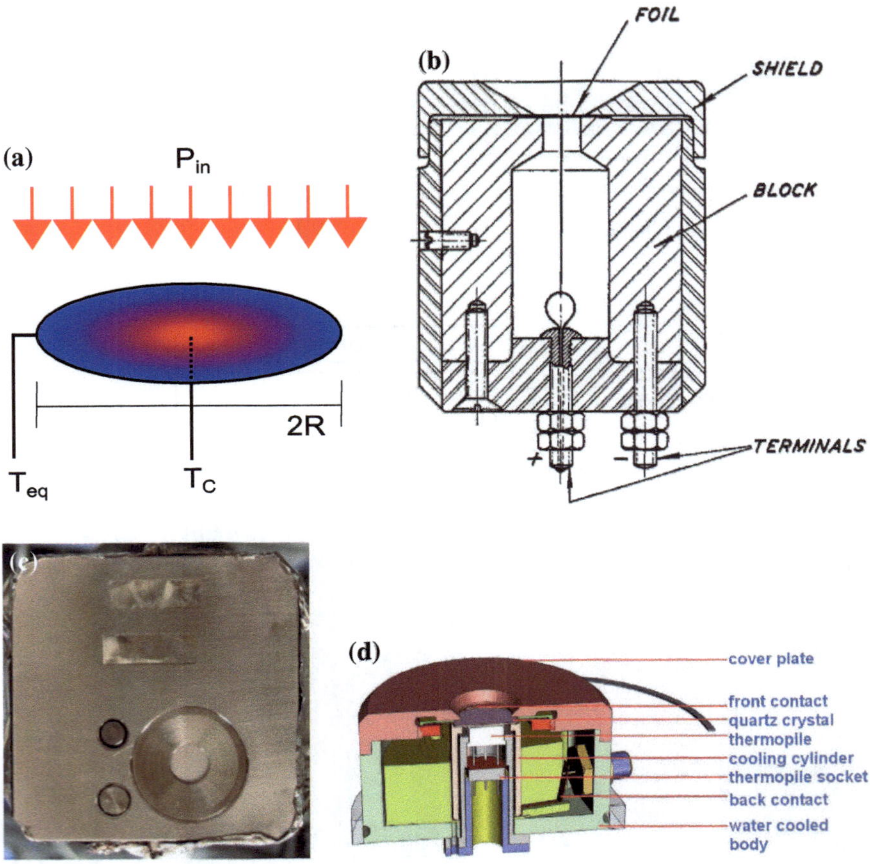

Fig. 6.6 **a** Working principle of the GARDON probe. **b** Probe designed by GARDON as "An Instrument for the Direct Measurement of Intense Thermal Radiation" [11]. (Reprinted with permission from R. GARDON [11], ©1953 American Institute of Physics). **c** Integrated sensor head by HARBAUER et al. consisting of a LANGMUIR probe (LP), a quartz crystal microbalance (QCM) and a GARDON probe (GP) [54]. (Reprinted with permission from K. HARBAUER et al. [54], ©2012 Elsevier) **d** Sensor head by WELZEL et al. combining a LANGMUIR probe, a quartz crystal microbalance and a GARDON probe [55]. (Courtesy of T. Welzel et al.) Working principle and illustration of different calorimetric probe setups based on GARDON's radiometer

is not only capable to measure radiation by photons, but also heating due to impinging particles and surface reactions [13, 53].

The theory describes the radial temperature distribution $T(r)$ across a circular plate (radius R and thickness d) exposed to a spatially homogeneous energy flux P_{in}, while the edge of the plate is kept at a constant temperature T_{eq} due to a connection to a large temperature bath (see Fig. 6.6b). The energy influx will lead to a steady-state temperature distribution across the foil which can be expressed as [11]

$$T\left(1+\frac{\alpha}{2}T\right) = P_{\text{in}}\left(\frac{R^2-r^2}{4\lambda_0 d}\right) \quad (6.9)$$

Here α and λ_0 are described by the temperature dependent heat conductivity $\lambda = \lambda_0(1+\alpha T)$ of the foil. The relation of the temperature difference $\Delta T = T_C - T_{\text{eq}}$ between the sink T_{eq} and the temperature $T_C = T(r=0)$ at the center of the foil is given by

$$\Delta T\left(1+\frac{\alpha}{2}\Delta T\right) = P_{\text{in}}\left(\frac{R^2}{4\lambda_0 d}\right), \quad (6.10)$$

thus, it is possible to determine the energy influx by measuring the temperature at the center and the temperature of the heat sink. Until the steady-state temperature is reached the temperature follows an exponential law [11] and, thus, shows the same time dependence as THORNTON's probe [53]. For a complete derivation of the formulas and the basic model we refer to [11, 53].

The principal setup used for GARDON's probe is shown in Fig. 6.6b. The sensor area consists of a thin constantan foil, which is clamped to a copper block at the edge. The shield made from aluminum protects the outer part of the foil leading to a well defined probing area. It is electrically insulated from the foil by paper washers. The copper block, which acts as heat sink, is embedded into insulating housing to protect the instrument from stray electromagnetic fields. By measuring the thermo electric voltage between the terminals (thermocouple) one can obtain the energy flux towards the surface according to equation (6.10). Constantan is chosen as foil material because it acts as one part of the thermocouple.

ELLMER and coworkers designed a sensor array (see Fig. 6.6c) that combines a flat LANGMUIR probe (LP) for the determination of plasma parameters (n_e, n_{i}, T_{e} and Φ_{pl}), a quartz crystal microbalance (QCM) in order to calculate the deposition rate R from the mass change Δm and a GARDON probe for the characterization of the total energy flux P_{in} [54]. A further improvement was made due to the combination of this three probes into one sensing area [55] (see Fig. 6.6d). Since it is desirable to build very small heat flux sensors there are also approaches available on the basis of microelectronics, e.g. see [56–58].

6.2.2.1 Calibration of Passive Probes

For a quantitative determination of the total energy influx the heat capacity C_S or the factors describing the heat conduction of the probe have to be known for THORNTON or GARDON like passive probes (see equations (6.7) and (6.9)). The general procedure is to expose the probe to a source of known power. From the measured temperature evolution during heating and subsequent cooling and by the known power influx one obtains the heat capacity by transposing equation (6.7). Various methods for the calibration of calorimetric probes to obtain the heat capacity have been used so far [59], using lasers [35], plasma ions [59] or plasma electrons [49, 60]. Since all these methods show several drawbacks a simple, less error prone method was developed using accelerated electrons under high vacuum conditions [37]. Applying this procedure, a thin tungsten wire is heated by resistive heating with an electrical current of about 1 A. The filament and the calorimetric probe are placed in a vacuum chamber pumped down to a pressure of 10^{-5} Pa to ensure a sufficiently long mean free path of the electrons. The electrons are accelerated due to the positive probe potential between 500 and 1000 V. Electron current I_e and acceleration voltage U_{acc} are monitored and give the total power of the electrons ($P_{in,e} = U_{acc} \cdot I_e$) which is delivered to the probe.

Also GARDON probes need to be calibrated for a quantitative determination of the energy influx. Since they cannot be biased, only non-electrostatic calibration methods are suited for this purpose. Usually, a black-body radiator is used as a source of known power [61–64].

6.2.3 Other Examples

In addition to the two basic probe designs described above, there are several other approaches given in the literature. They are only briefly mentioned here since a complete description would be far beyond the scope of this work.

6.2.3.1 Thermopile Sensor

A well characterized probe using a commercially available heat flux probe [14, 28, 65, 66] based on thermopile (a stack of thermo couples) sensors has been used by THOMANN, DUSSARD and coworkers for measurements in different plasma environments.

This probe type is not described here in detail because recently there is a review article written by DUSSARD et al. [67]. It gives an overview of the working principle and measurements in different plasma environments. This probe has been compared with a calorimetric probe used by KERSTEN and coworkers and the results obtained by both probes have shown quite good agreement [65].

6.2.3.2 Active Calorimetric Probe

An active calorimetric probe [16, 68] was developed to overcome the drawback of calibration procedures and to monitor the energy flux continuously. The working principle is as follows:

The probing area is heated up by ohmic heating to a certain temperature level T_S. The heating power P_{heat} can easily be monitored via a resistance thermometer. After a source of power e.g., plasma, ion beam, etc. is switched on, the temperature of the probe will rise (see Fig. 6.7). A feedback loop reduces the applied heating

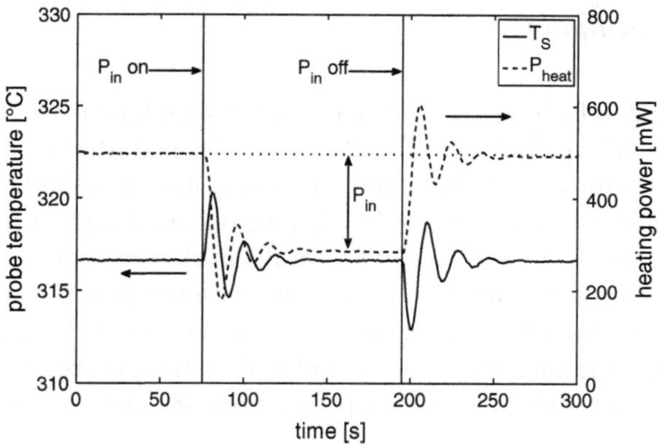

Fig. 6.7 Typical data obtained by an active calorimetric probe showing the probe temperature T_S (*full line*) and the heating power P_{heat} (*dashed line*)

power in order to cool down the probe to the original temperature level. Hence, the reduction of heating power is a direct measure for the energy influx applied from the source. After the source is switched off the probe temperature will decrease and the feed back loop will increase the heating power until the previous temperature level is reached again (see Fig. 6.7).

Since the energy flux from the plasma is monitored directly and independently from the heat capacity there is no need for a calibration procedure for this type of probe. Depending on the speed of response of the feedback loop it is also possible to measure transient energy fluxes. This makes it also possible to use this probe for process control, because there is no need to measure any cooling process like for the conventional/passive calorimetric probe. Thus, the source can remain switched on.

6.3 Contributions to the Total Energy Influx

Most process plasmas are of complex nature, because they contain different species which are involved in various physical and chemical reactions. Therefore, the total/integral energy influx P_{in} towards surfaces is the sum of different contributions caused by kinetic impact of different species and various surface reactions. However, the quantity which is measured by calorimetric probes is the total energy influx. There are recipes to distinguish between the different contributions. This section gives a brief overview of the different contributions. They can be roughly divided into the impact of kinetic particle bombardment and processes taking place at the surface itself.

6.3.1 Particle Impact

Energetic particles which hit a surface may transfer their kinetic energy to the bombarded material. The particles can be electrons, ions or fast neutrals. The kinetic impact of heavy particles (ions, neutrals) is known to have an important influence on film growth and structure [7, 51, 69–71]. On the one hand high energetic particles lead to a densification of the films, which results in changed optical, electrical and mechanical properties. On the other hand high energetic particle bombardment can also have a destructive effect or can even lead to an ablation of the film, e.g. by sputtering. Hence, the knowledge and control of the kinetic energy is very important for most applications. The next paragraphs describe the energy transfer by kinetic particle impact.

6.3.1.1 Electrons

The most important species in a process plasma are the free electrons, which determine the basic plasma properties like plasma potential (Φ_{pl}) or ionization degree. They also have a strong impact on the chemical reactions like e.g. molecule fragmentation via electron impact dissociation in the plasma. The impact of electrons at surfaces leads to emission of secondary electrons, which carry away kinetic energy. There are two populations of secondary electrons caused by electron impact at surfaces:

Reflected electrons usually have approximately the same energy as the impinging electrons, since they undergo an elastic collision. Since they take away the same amount of energy and electrical charge they do not contribute to the electrical current and the energy influx.

True secondary electrons usually have much lower energy than the impinging electrons. So the energy taken away is very small, but the current measurement is falsified. This might lead to the conclusion that the energy per electron is larger than expected. However, since the secondary electron emission yield is usually very small at low electron energies which are typically in the range of some eV for the electrons reaching a surface in low temperature plasmas, this contributions seem to be negligible.

The general term for the energy flux J_{e} transferred by the kinetic energy ($\bar{E}_{\text{kin,e}}$: mean kinetic energy) of electrons ($\Gamma_{\text{e}} = j_{\text{e}}/e_0$: electron flux density) can be written as

$$J_{\text{e}} = \Gamma_{\text{e}} \bar{E}_{\text{kin,e}}. \tag{6.11}$$

Depending on the electron energy distribution function (EEDF) it can be expressed as

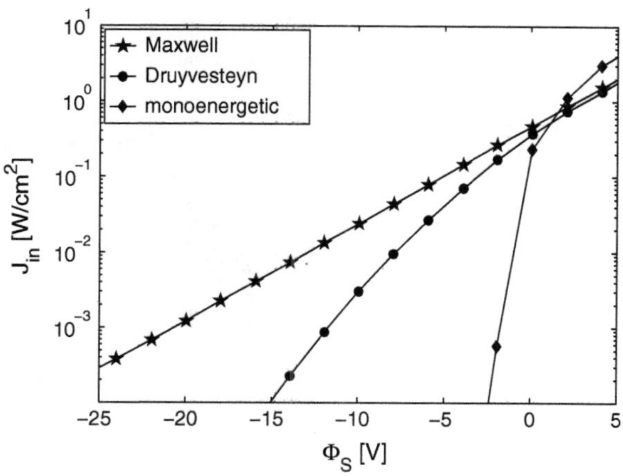

Fig. 6.8 Calculated energy flux by electrons in dependence on the substrate voltage Φ_S for various EEDFs ($n_e = 3.10^9/\text{cm}^3$, $k_B T_e = 4.2\,\text{eV}$, $\Phi_{pl} = 1$ V). Values taken from [17]

$$J_e = n_e \sqrt{\frac{k_B T_e}{2\pi m_e}} \exp(-\xi) 2 k_B T_e = \frac{j_e}{e_0} 2 k_B T_e \tag{6.12}$$

for a MAXWELLIAN EEDF [31] with ($\xi = E/k_B T_e$). For a DRUYVESTEYN EEDF J_e yields

$$J_e = n_e \sqrt{\frac{k_B T_e}{2\pi m_e}} k_B T_e \left(-\xi \exp(-0.25\xi) + \left(1.79 + 0.89\,(\xi)^2\right) \text{erfc}\,(0.5\xi)\right) \tag{6.13}$$

For a monoenergetic electron beam equation (6.11) becomes

$$J_e = n_e \sqrt{\frac{k_B T_e}{2\pi m_e}} 2\pi\,(1 - \xi)^2\, k_B T_e \tag{6.14}$$

The energy influx by electrons for different EEDF is shown in Fig. 6.8. Again, no secondary electron emission is included here. This effect is considered in Sect. 6.3.4.2.

6.3.1.2 Ions

The second group of impinging energetic particles are ions, which can be either ionized atoms or ionized molecules. They are formed by various processes like electron impact ionization, Penning ionization or for negatively charged ions like e.g. O^- by electron attachment. Depending on the sign of their charge they play a

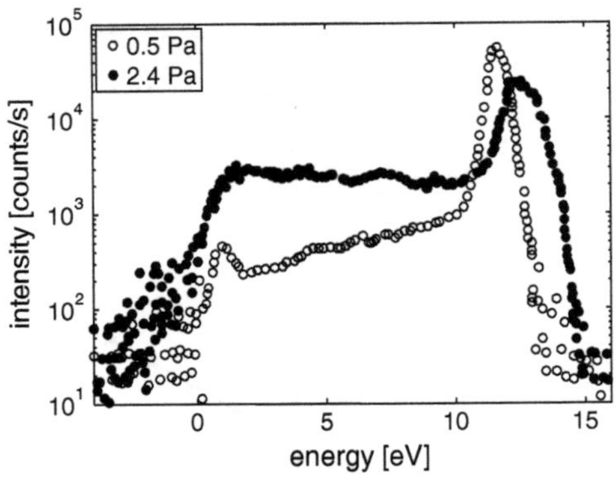

Fig. 6.9 IEDF measured for two different Ar pressures in a capacitively coupled rf plasma (values taken from [72]). The ions are extracted from $\Phi_{\text{pl}} \approx$ 15 eV to a grounded substrate. Mean kinetic energies are 11.4 eV for 0.5 Pa and 9.9 eV for 2.4 Pa, respectively

different role in plasma surface interactions. However, this paragraph just covers the transfer of kinetic energy of the impinging ions. Surface processes like film formation or sputtering are considered in Sects. 6.3.2 and 6.3.4, respectively.

In general, the kinetic energy transferred by ions is determined by the product of the ion flux density $\Gamma_{\text{i}} = j_{\text{i}}/e_0$ and the mean kinetic energy of the ions $\bar{E}_{\text{kin,i}}$:

$$J_{\text{i}} = \Gamma_{\text{i}} \bar{E}_{\text{kin,i}} \tag{6.15}$$

The current density j_{i} of the ions depends on the discharge regime and can be described by the following expressions [72]:

$$j_{\text{i}} = \begin{cases} n_{\text{e}}\sqrt{\frac{k_{\text{B}}T_{\text{e}}}{m_{\text{i}}}}\exp(-0.5) & \text{BOHM – flux of ions} \\ \frac{4}{9}\frac{\varepsilon_0}{e_0}\sqrt{\frac{e_0}{m_{\text{i}}}}\frac{\Phi_{\text{sh}}^{3/2}}{d_{\text{sh}}^2} & \text{ion flux without collisions in the sheath (CHILD)} \\ \frac{1}{4}n_{\text{i}}v_{\text{i}} & \text{thermal flux of ions} \\ n_{\text{i}}v_{\text{amb}} & \text{ambipolar diffusion of ions} \end{cases} \tag{6.16}$$

Also the (mean) kinetic energy of the ions strongly depends on the type of discharge and the geometry of the experimental setup as well as on the used gas mixture and pressure (see Fig. 6.9).

Positive ions are accelerated in the sheath in front of a substrate. Thus, their kinetic energy is mainly determined by the voltage drop Φ_{sh} in that sheath. In a DC discharge at low pressures (without collisions) the contribution by ions reads:

$$J_{\text{i}} = j_{\text{i}}\Phi_{\text{sh}} \tag{6.17}$$

This is the most simple case. Unfortunately, there are several reasons, why this simple formula cannot be always applied.

- At higher pressures, when the mean free path length λ_{mfp} of the ions is in the order of the sheath thickness d_{sh} or even shorter, the ions undergo collisions and, thus, their kinetic energy is lowered and not monoenergetic (see Fig. 6.9).
- In radio frequency (rf) discharges the electrical asymmetry effect [73] occurs and leads to a more complex shaped ion energy distribution function (IEDF) [60, 72, 73]. Of course the maximum energy of the ions is also determined by the sheath potential, but the mean energy is well below that value. It has recently been shown, that the mean kinetic energy is in the range of $0.3 - 0.5 \cdot \Phi_{sh}$ [74, 75].

Negative ions such as O^-, O_2^- or SF_6^- can have different kinetic energies due to their origin:

- When they are formed in the bulk plasma they are attracted to the most dense region where the plasma potential has its most positive value. Therefore, they stay in the plasma for a long time, because their kinetic (thermal) energy is too low to overcome the sheath potential barrier.
- Negative ions can also be produced at electrodes, e.g. at the target of a magnetron or in the sheath in front of that electrode. Once they are created in a strong electric field (several $100 \mathrm{V cm}^{-1}$) they gain high kinetic energy, sufficiently enough to overcome the potential barrier of the sheath at the substrate. Subsequently, they impinge at the substrate with very high kinetic energy. Such high energetic negative ions usually have a small current density, but their high kinetic energy can be a crucial factor for film properties [69, 76].

6.3.1.3 Neutrals

The description of the contribution of neutral particles is very similar to that of the ions. In general, their contribution is described by the product of flux density (Γ_n) and their mean kinetic energy $\bar{E}_{kin,n}$:

$$J_n = \Gamma_n \bar{E}_{kin,n} \tag{6.18}$$

The kinetic energy of neutrals also strongly depends on their origin:

- Since low temperature plasmas are characterized by low ion and neutral gas temperatures (T_i, T_n) between room temperature and 1000 °C the kinetic energy of neutral atoms is usually quite low (30 up to 100 meV). Since they are not accelerated by electric fields they do not gain high energies. As long as they do not contribute to film growth they will desorb from the surface thermalized. The resulting energy transfer depends only on the gas temperature T_{gas} and the temperature of the surface T_S:

$$J_n = \Gamma_n k_B (T_{gas} - T_S) \tag{6.19}$$

Subsequently, this contribution is very low as long as the energy and flux densities are small. However, for some plasmas, especially atmospheric pressure discharges at elevated gas temperatures and high gas throughput substrate heating due to neutral gas can play an important role or can even be the most dominant heating mechanism.

- In sputtering processes, the kinetic energy of the sputtered neutral atoms is much higher than the neutral background gas temperature. Depending on the sputtered material the sputtering gas, the energy of the impinging ions or neutrals and the angle between the surface normal and the trajectory of the sputtered species the kinetic energy of the sputtered neutrals can be in the range of several 10 eV for typical process plasmas.
- Neutrals with even higher kinetic energy can be produced e.g. in ion beams. In that case, for example, ions are created in a plasma and then extracted and accelerated via a grid system. After passing the grid holes they are often neutralized by a neutralizer or they are converted into neutrals by charge exchange collisions with the background gas. Depending on the acceleration voltage of the grid, the kinetic energy of the particles can easily exceed 1 keV [77].

6.3.2 *Surface Processes*

This section discusses reactions going on at the surface. The surface processes also influence the energy balance of the substrate during plasma treatment. Note that the role of metastable atoms or molecules is not mentioned in the following sections because measurements have shown that their contribution is about $3.10^{-4}\mathrm{mW/cm^2}$ and, thus, negligible in comparison to other contributions in typical low temperature low pressure plasmas [78]. However, they may play an important role at higher pressures, but up to now there are no results known to the authors.

6.3.2.1 Electron-Ion Recombination

The recombination of ions and electrons at the substrate surface is mainly influenced by the ion flux density $\Gamma_\mathrm{i} = j_\mathrm{i}/e_0$ since it is the limiting factor for this process. Ions usually can reach surfaces at potentials more negative than the plasma potential. Thus, this contribution only exists for $\Phi_\mathrm{S} < \Phi_\mathrm{plasma}$. For more positive potentials no ions are accessible for this process.

An ion reaching a surface always finds an electron for the recombination process. Grounded chamber walls and substrates provide electrons for recombination. Even at floating (dielectric) walls there are enough electrons present since the current densities of ions and electrons from the plasma balance each other ($j_\mathrm{i} = j_\mathrm{e}$). During the recombination the ion captures an electron from the surface, so the electron has to overcome the electron work function E_ewf, which is necessary to release an electron from the solid. In a simple theory the recombination process takes place directly at the surface, meaning that the total energy released by this mechanism (equal to the ionization energy of the ion E_{ion}) is transferred to the surface. This reads:

$$J_{rec} = \Gamma_i(E_{ion} - E_{ewf}) \quad (6.20)$$

An other theory claims that the recombination process takes place somewhat above the surface [79]. In that case the electron has to tunnel through the potential barrier finding the ion. For this situation the ionization energy is released in form of an photon. Assuming isotropic emission of photons leads to the case where only half of the photons can reach the surface. Depending on the reflectivity ($R(\nu)$) only a small fraction of the energy released by recombination is transferred to the surface. This yields:

$$J_{rec} = \Gamma_i(\frac{1}{2}R(\nu)E_{ion} - E_{ewf}) \quad (6.21)$$

It is not easy to prove one or the other assumption because the process cannot be isolated. There are always ions involved, which also transfer their kinetic energy and one will always measure the sum of both contributions. As mentioned above the contribution by ions is also quite complex since they are usually not monoenergetic.

6.3.2.2 Molecule Formation

In many complex plasmas gas mixtures of atomic and molecular gases are used for different applications. This can be "simple" molecules like N_2, O_2 or H_2 or complex molecules consisting of a variety of atoms, such as organic precursors (e.g. tetraethyl orthosilicate (TEOS), hexamethyldisiloxane (HMDSO), aluminium triisopropoxide (ATI), etc.). Depending on the electron temperature many different fragments are formed in the plasma due to electron impact dissociation. The radical concentration depends on the production and loss rates, leading to a dynamic equilibrium. Depending on the chemical structure and the charge of the fragments different reactions can follow like formation of polymeric chains or further destruction into smaller fragments. Due to momentum conservation, the formation processes are usually three body collisions that are often very unlikely in the bulk of low pressure plasmas. This molecule forming reactions can easily happen at chamber walls or at substrates since the surfaces act as the third collision partner. During the formation of a molecule, e.g. the formation of an hydrogen molecule, the binding energy of the molecule (E_b) will be released at the surface. For a two atomic molecule consisting of atoms of the same type (neutral flux density Γ_n) the transferred energy flux can be written as

$$J_{mol} = \frac{1}{2}\Gamma_n E_b \quad (6.22)$$

6.3.2.3 Film Formation

After striking the substrate (and transferring kinetic energy) sputtered particles (or precursor fragments) usually lead to film formation at the exposed surface. During this condensation the enthalpy of sublimation is released. Let us consider a simple example, where a metal film of density ρ is growing from neutral metal atoms with the growth rate R:

There will be a flux of neutral atoms Γ_n hitting the surface. The sticking coefficient γ_n describes the probability of the atoms to stick at the surface. The effective neutral particle flux is described by $\Gamma_{eff} = \Gamma_n \gamma_n$. If E_{sub} is the sublimation enthalpy per single atom the total energy influx by heat of condensation J_{sub} can be written as

$$J_{sub} = \Gamma_n \gamma_n E_{sub} = \Gamma_{eff} E_{sub} \tag{6.23}$$

In general, the effective neutral particle flux is not easy to determine, but it can be deduced from the growth rate, film density and mass m_n of the single atoms:

$$\Gamma_{eff} = \frac{\rho R}{m_n} \tag{6.24}$$

Growth rate and film density can easily be measured e.g. by using X-ray techniques. One should be aware of using Γ_{eff} instead of Γ_n for the calculation of J_n by using equation (6.24), because the sticking coefficient can be drastically lowered at elevated substrate temperatures leading to underestimation of the neutral particle flux.

6.3.3 *Radiation*

Besides particles also radiation is present in each plasma and, therefore, has an impact on the plasma wall interaction. Most of the light (UV, visible, IR) has its origin in the line emission of several plasma species. Also thermal radiation may be present inside a plasma chamber produced by hot electrodes, filaments or the chamber wall itself.

The simplest description of the contribution of heat radiation from hot surfaces is using BOLTZMANN's radiation law:

$$J_{rad,1} = \sigma \left(\varepsilon T_{rad}^4 - \varepsilon_S T_S^4 \right) \tag{6.25}$$

Here, $\sigma = 5.67 \times 10^{-8}\ \mathrm{W/(m^2K^4)}$ denotes the Stefan-BOLTZMANN constant, ε, ε_S the emissivity of the hot surface and the probe, respectively, and T_{rad} and T_S the temperature of the hot surface and the probe, respectively. Often a geometrical factor $f(A_{rad}, A_S)$ describing the geometrical configuration and the probe orientation has to be taken into account [72]. In common low temperature plasma reactors, this contribution is usually small, but for methods involving chemical vapor deposition (CVD) or arc evaporation, this contribution can make up to several tens of percent [50].

The second contribution by radiation, which is caused by the line radiation (photon flux Γ_{ph} and mean photon energy E_{ph}) from the plasma species itself can be described as follows:

$$J_{rad,2} = \xi_{ph} \Gamma_{ph} E_{ph} \tag{6.26}$$

The factor ξ_{ph} denotes the probability of a photon to be absorbed at the surface, similar to a sticking coefficient for atoms.

However, even if the description of the contribution by radiation is quite simple and intuitive, the measurement of this contribution is very complex. Also, the temperature of hot surfaces may depend on the operation time of the experiment. Thus, they do not have a strong influence directly after the beginning of the experiment, but they may get important after some hours of operation when e.g. electrodes or the vacuum vessel are heated up [43].

6.3.4 Particle Emission from Surfaces

Not only particles impinging on surfaces contribute to the energy balance, but also particles emitted from surfaces take away kinetic energy and, therefore, they have an effect on the energy balance of the surface. The emission of particles is usually caused by particle impact, such as electrons, ions and neutrals. Also chemical etching can lead to surface erosion. In general, the contributions of emitted particles are similar to impinging particles. Because the particles are leaving the surface, they are loss processes and, therefore, they have a negative sign.

6.3.4.1 Sputtering

In general, sputtering is the evaporation of solid materials by impact of fast ions or neutrals. This physical vapor deposition (PVD) is one of the most common plasma based technique for deposition of thin films. It overcomes the drawbacks of chemical vapor deposition (CVD) which are based on the evaporation of materials, e.g. by heating. Thus, the thermal load of substrates is drastically reduced, because the gas is usually not heated directly and no source of intense thermal radiation is present. Most common plasma sources used for sputtering are ion beams or magnetron sputter sources. In both cases neutrals and / or ions are accelerated towards a target material, which is subsequently sputtered. The incident particles transfer their energy and momentum to the target material and cause a collision cascade, which leads to the release of atoms, molecules or clusters from the target [80]. The released species, which are normally neutral, move across the chamber and hit walls or substrates to form a film.

The threshold energy for sputtering is usually in the range of some 10 eV, but at this low energies the sputtering yield Y, which is defined as

$$Y = \frac{\text{number of sputtered particles}}{\text{number of incident particles}} \tag{6.27}$$

is very low. Typically, energies of several 100 eV to some 1000 eV are used in applications. Since such high energies are required for an efficient process, sputtering is usually limited to low pressures, where the mean free path of the fast particles is long and they do not undergo a considerable number of collisions, which would lead to a reduction of their incident energy.

It is well known that the angular distributions of released particles and the energy of the sputtered material strongly depend on several parameters, such as [80]

- energy of impinging particles
- angle of impinging particles
- material (target) constants such as surface binding energy or lattice energy, etc.
- ratio of the masses of impinging and the target atoms

Sputtered particles have kinetic energies, which can reach several 10 eV to about 100 eV. If the kinetic energy of particles impinging at the substrate is sufficiently high, also sputtering of the substrate can occur, and the sputtered particles carry away energy. This fraction is usually small, because the energy taken away by sputtered material is much smaller than the energy delivered by impinging particles.

6.3.4.2 Secondary Electron Emission (SEE)

The generation of secondary electrons at electrodes in a plasma is an essential mechanism for plasma generation. Only if the generation of electrons at the cathode is efficient enough to overcome the losses of electrons at the anode and surroundings the generation of a self sustained discharge is possible.

In general, secondary electrons are generated by the impact of energetic ions (γ process) or by impinging electrons (δ process). The latter process produces different populations of secondary electrons: Those, which undergo an elastic or inelastic scattering and have nearly the same kinetic energy as the impinging electron and true secondary electrons which have a thermal energy distribution function [81].

Before an electron is released the electron work function E_{ewf} has to be overcome. Hence, the energy taken away by a secondary electron carrying the kinetic energy E_{kin} is expressed by

$$J_{\text{see}} = \Gamma_{\text{see}}(E_{\text{ewf}} + E_{\text{kin}}). \tag{6.28}$$

Here Γ_{see} denotes the flux of secondary electrons away from the surface. Since the initial energy of the secondary electrons is usually small in comparison to the energy of the impinging particle as it is also the case for sputtering, the contribution of the energy taken away by secondary electrons is almost small.

6.3.5 *Total Energy Influx*

In the previous paragraphs it has been shown, that the total energy influx consists of various contributions:

$$J_{\mathrm{total}} = J_{\mathrm{e}} + J_{\mathrm{i}} + J_{\mathrm{n}} + J_{\mathrm{rec}} + J_{\mathrm{mol}} + J_{\mathrm{sub}} + J_{\mathrm{rad}} - J_{\mathrm{loss}} \tag{6.29}$$

Here J_{loss} is summing up all loss processes like particle emission (sputtering, SEE), cooling by convection, conduction and radiation. Calorimetric investigations allow for measuring the integral or total energy influx J_{total}. The determination of the total energy influx by determining the different contributions separately and subsequently adding them up is very error prone, because a lot of unknown parameters and variables have to be taken into account.

6.4 Examples of Application

In this section some examples for the application of calorimetric probes in process plasmas are discussed. It will be demonstrated how they can be used in order to understand plasma surface interactions in more detail and how they can support different diagnostic methods to draw a quantitative picture of the involved processes. The examples were chosen from the field of complex plasmas, where micro particles, reactive gases or large surface to volume ratios are involved. The experimental details and scientific results cannot be covered exhaustively here. The interested reader is referred to the references given for the particular example.

6.4.1 *Ion Beam*

Ion beams have high potentials in various fields of science and technology, since they offer various applications such as coating deposition and surface modification or surface patterning. They are also commonly used for space applications as thrusters for propulsion or positioning of satellites.

In the latter case an exact knowledge of the thrust is of essential interest, because for the exact positioning of satellites the forces needed are in the range of μN. For the investigation of such small forces an experiment (VIBEx) was designed (see Fig. 6.10a), which allows the measurement of forces that act on micro particles exposed to an ion beam [82]. The idea is explained in the following:

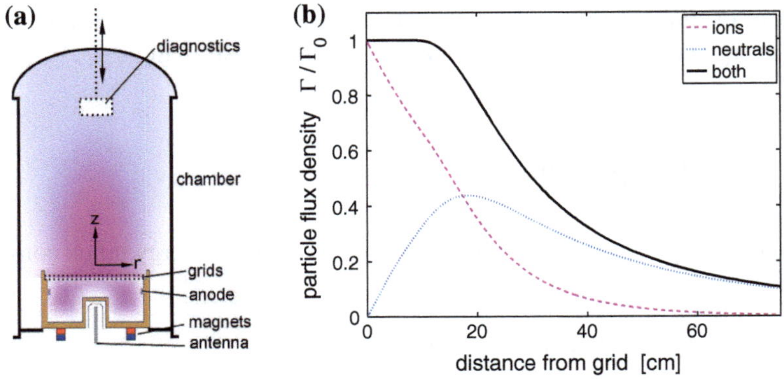

Fig. 6.10 **a** Schematic drawing of the ion beam experiment. **b** Ion, neutral and total particle fluxes in the ion beam chamber in dependence on the distance from the grid system. Data taken from [83]. Experimental setup and particle fluxes in the vertical ion beam experiment (VIBEx)

Micro particles (glass hollow spheres, $2r = 100 \pm 13\,\mu\text{m}$) are injected via a dust dispenser in a vertically upwards directed ion beam source. The falling particles are stroboscopically illuminated (500 Hz) by a laser fan and observed with a charge-coupled device (CCD) camera. The particle trajectory $s(t)$ in the field of view has been deduced from one or more frames of the camera. Subsequently, the trajectory is used to determine the resulting acceleration caused by the ion beam.

The obtained results show the expected trend: The higher the ion energy determined by the voltage of the acceleration grid (U_{acc}), the higher the force acting on the micro particles. At higher ion energies, the acting force can even overcome the gravitational force and the particles are not only slowed down during their free fall, but accelerated upwards following a parabolic trajectory. For a comparison between measurements and theory FARADAY cup measurements were performed to measure the ion current densities j_{i}, but a calculation of the acting force from the mean ion energy $\bar{E}_{\text{i}} = ZeU_{\text{acc}}$ (here $Z = 1$) and current density can only explain a part of the force measured. The other part is due to fast neutral atoms, created by charge exchange collisions of the ions with the background gas leading to an additional momentum transfer. The fast neutrals can not be detected with a FARADAY cup, since it is only sensitive to charged particles. However, calorimetric probes are also sensitive to neutral particles. Thus, they allow to measure the effect of these species.

The total particle flux $\Gamma_{\text{i+n}}$ of the ions and neutrals for different distances z from the grid can be written as [77]:

$$\Gamma_{\text{i+n}}(z) = \Gamma_0 \left[1 - \exp\left(-\frac{R^2}{2z^2 \tan^2 \vartheta} \right) \right] \tag{6.30}$$

Here R denotes the distance from the center axis and ϑ the divergence half-angle of the GAUSSian shaped "beamlet" (= grid hole) profiles. Γ_0 is the initial intensity at the grids.

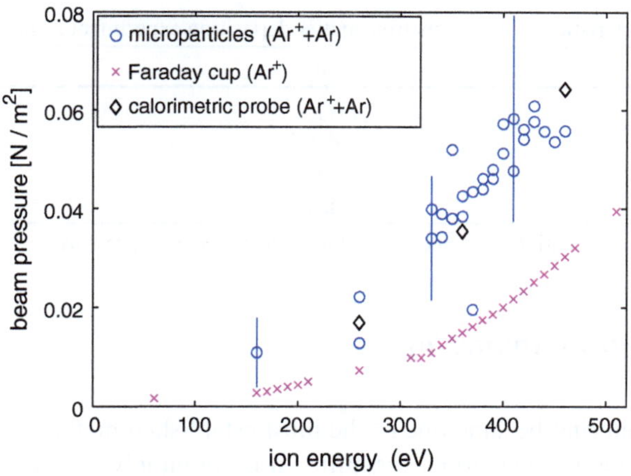

Fig. 6.11 Comparison between three different measurement methods for the determination of the beam pressure. Micro particles and the calorimetric probe, both measure the contributions from ions and neutrals, while the FARADAY cup just measures the ions. Data taken from [77]

The total particle flux is put together [82] from the ion flux

$$\Gamma_{\mathrm{i}}(z) = \Gamma_{\mathrm{i+n}}(z)\ \exp\left(-\frac{z}{\lambda_{\mathrm{cx}}}\right) \tag{6.31}$$

and the flux of the fast neutrals

$$\Gamma_{\mathrm{n}}(z) = \Gamma_{\mathrm{i+n}}(z)\left[1 - \exp\left(-\frac{z}{\lambda_{\mathrm{cx}}}\right)\right], \tag{6.32}$$

where λ_{cx} is the mean free path length for charge exchange collisions, which is in the range of approximately 25 cm for the given pressure of about $5 \cdot 10^{-2}$ Pa. The different contributions as well as the total particle flux are shown in Fig. 6.10b.

The results obtained by the three different measurement methods (micro particles, FARADAY cup and calorimetric probe) at a distance of 30 cm from the grid are shown in Fig. 6.11. All results show the same trend, an increasing beam pressure with increasing ion energies. Since the FARADAY cup is just sensitive to ions, the total values are lower than for the other two methods, which are both sensitive to ions and neutrals. At the given distance from the grid, a large fraction of the ions has already undergone a charge exchange collision (see Fig. 6.10b) and is converted to fast neutrals.

Summarizing, calorimetric probe measurements helped to understand the results obtained by falling micro particles, that could not be explained quantitatively by electrostatic methods in detail. Thus, calorimetric probes are also an easy and robust method for the investigation of (fast) neutrals.

Table 6.1 Characteristics of the magnetron sputter deposition experiment discussed here

Quantity	Unit	Value
P_{rf}	[W]	3×100
f_{rf}	[MHz]	13.56
p	[Pa]	0.3
d_t	[inch]	2

P_{rf} = power, f_{rf} = excitation frequency, d_t = target diameter, p = pressure

6.4.2 Magnetron Sputtering

Magnetron sputtering became one of the most established method of plasma based deposition techniques. Due to its versatility, it is commonly used for the deposition of various films ranging from optical applications to wear resistant and catalytically active layers as well as thermal barrier coatings. There are only few restrictions for the choice of target materials, thus, it is possible to deposit films consisting of various metals, ceramics and even some polymers or complex mixtures of them.

Beneath the different geometrical configurations, target materials and excitation mechanisms, also various gas mixtures can be used for magnetron sputtering. In general, e.g. for deposition of metal films Ar is used as sputter gas, because it is the best compromise between process costs and sputter yield. However, various applications need an addition of molecular gases like O_2 or N_2 for reactive sputtering and deposition of alloys and ceramics.

These admixtures do not only influence the chemical composition, but also the structure of the films, because they have an important impact on the plasma wall interaction and the resulting substrate temperature.

The following examples show how the admixture of different diatomic gases have an impact on the energy influx towards the substrate in magnetron sputter deposition of zinc oxide (ZnO) films, from a ZnO target.

Since the target material is a ceramic (non-conductive), dc-sputtering cannot be applied because the target surface will charge up negatively and the discharge is interrupted. For this reason rf excitation was used for sputtering (operational parameters listed in Table 6.1). Calorimetric probe measurements were carried out in a typical substrate position 5 cm above the center of the targets.

In the first case nitrogen was added to the argon sputtering gas. The motivation for adding nitrogen to the sputtering gas is the improvement of the crystal structure, which is important for the electrical properties of the ZnO films. From the currents to the probe (I-V-characteristics), which were recorded during energy flux measurements, the plasma parameters have been calculated. The trend of the results show a quite good agreement with comparable measurements [74]. The influence of the nitrogen admixture on different plasma parameters is presented in Fig. 6.12.

The obtained plasma parameters are used to calculate the contributions from charge carriers and recombination to the total energy influx according to equations (6.11), (6.17) and (6.20). From growth rate and film density measurements the con-

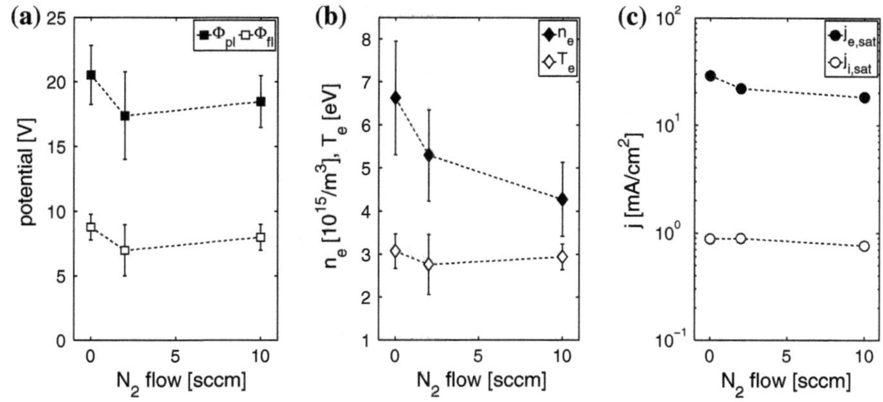

Fig. 6.12 **a** Plasma and floating potentials. **b** Electron density and temperature. **c** Electron and ion saturation currents. Plasma parameters at substrate position in dependence on the nitrogen admixture for magnetron sputter deposition of ZnO films deposited at 0.3 Pa

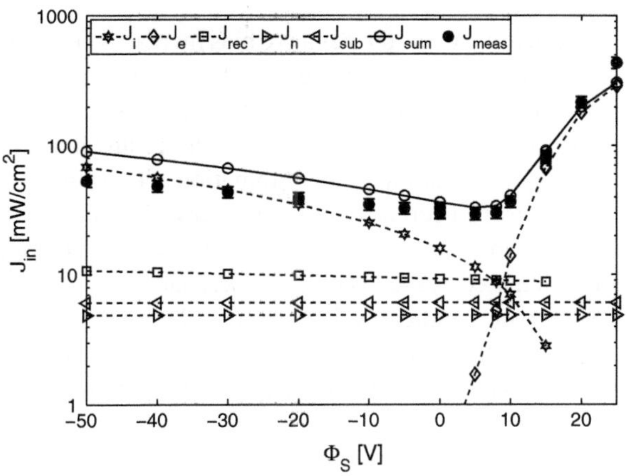

Fig. 6.13 Calculated contributions from electrons, ions, recombination, neutrals and film formation as well as the sum of all calculated contributions in comparison with the measured total energy influx in dependence on the probe bias. The flux ratio of Ar/N2 was 16 sccm/10 sccm

tribution of the film formation was calculated using equation (6.23). In this case the part of the neutral sputtered particles can not be calculated, because the mean kinetic energy of the film forming particles is not known for this particular setup. Values from the literature [60] were taken to get a rough estimation of the contribution J_n.

Figure 6.13 exemplarily shows the energy influx in dependence on the probe bias for an Ar-N_2 flux ratio of 16sccm/10sccm (sccm=standard cubic centimeter per minute). Also the different calculated contributions are plotted. The results show that the contributions of neutral particles and film formation are low in comparison to the total energy influx. They are not depending on the substrate bias. The contributions by

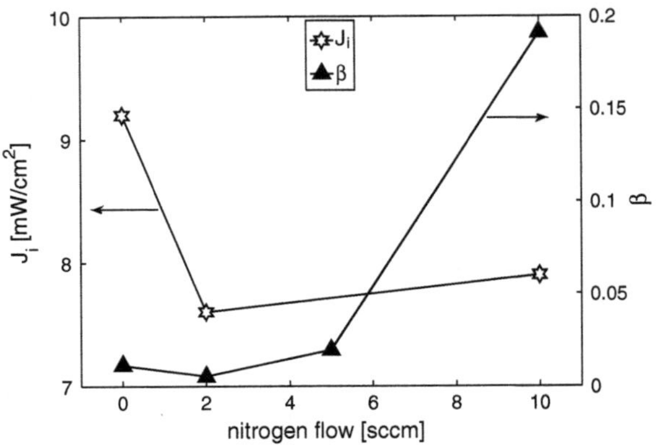

Fig. 6.14 Influence of the nitrogen admixture on the energy influx caused by ions and the crystallinity of the ZnO films obtained from the integral width β of the X-ray reflectometry pattern (Reprinted with permission from [75], ©2012 Institute of Physics)

ions and recombination increase as the negative bias increases. The ionic contribution does not increase linearly, as it is expected from the linear slope of the potential difference $\Phi_{\text{pl}} - \Phi_{\text{S}}$, because also the ion current increases due to a sheath expansion and an increasing effective collection area [84]. The increasing ion current also causes the slightly increasing contribution of recombination. For values of $\Phi_{\text{S}} \geq \Phi_{\text{fl}}$ the contribution of the electrons becomes dominant. Lowest values for both, measured and calculated values, respectively, are obtained at $\Phi_{\text{S}} \approx \Phi_{\text{fl}}$. A comparison between the measured and calculated values shows a discrepancy for more negative probe potentials. This discrepancy indicates that the ions do not carry the kinetic energy $E_{\text{kin}} = (\Phi_{\text{pl}} - \Phi_{\text{S}})/e_o$, but a lower value, which is caused by energy losses due to collisions. This reduction factor has been estimated to be about 0.5 for this particular condition.

It is known from the literature that especially the energy influx of ions has a drastic impact on the resulting film properties. Figure 6.14 shows, that lowest β-values ($\beta =$ integral intensity divided by total intensity of the X-ray reflectometry pattern), which is a measure for the crystallinity, are found when also the contribution of ions has its minimum value. This observation indicates that high ion energies can lead to a destruction of the crystalline structure during film growth.

In a second experiment, hydrogen was added into the argon sputtering atmosphere. It is known from the literature, that hydrogen admixture leads to an increased surface heating caused by the formation of H_2 molecules from hydrogen atoms at the surface [85]. For the process described above a subsequent annealing is a typical process step for the further improvement of the crystallinity. Often common annealing processes require temperatures, which are too high for temperature sensitive substrate materials, like e.g. polymers. The motivation for the admixture of hydrogen was to perform a local annealing during the deposition process.

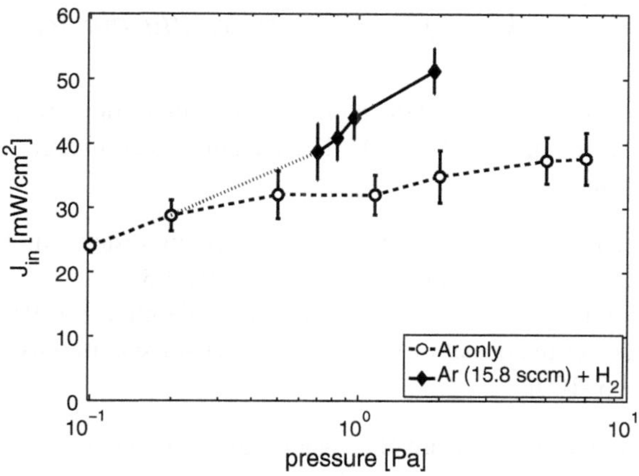

Fig. 6.15 Comparison between the total energy influx in a pure Ar atmosphere and a Ar/H_2 mixture at different pressures. (Reprinted with permission from [75], ©2012 Institute of Physics)

Figure 6.15 shows the total energy influx in a pure Ar atmosphere at different pressures as well as the total energy influx to the probe at different Ar/H_2 mixtures. The Ar pressure was kept constant at 0.2 Pa and different amounts of H_2 were added. A certain amount of Ar (15.8sccm) was needed, otherwise the discharge did not ignite. Both, the pure argon and the mixture show an increasing energy influx for increasing pressure. There are two competitive processes taking place as the pressure is raised: An increasing plasma density leading to a higher energy influx and a reduction of mean free path length due to more frequent collisions leading to a lower energy influx per impinging particle. This two mechanisms lead to a maximum energy influx which is expected to be around 10 Pa. However, the effect of additional local heating due to H_2 formation can clearly be observed.

ZnO films deposited at small hydrogen admixtures show an increased charge carrier density that leads to a decreased resistivity. For higher H_2 admixtures the increased carrier density is balanced by a decreasing carrier mobility. Thus, the resistivity remains constant. However, the major drawback of adding H_2 is increased etching and reduction of the ZnO films during plasma treatment. More detailed informations about the influence of H_2 addition is given in the literature [86].

It should be mentioned that the contribution caused by molecule formation strongly depends on the electron temperature and the molecule binding energy. Low electron temperatures might not be sufficient to initiate dissociation of the molecules. In this particular case, nitrogen admixture does not lead to an increase of the total energy influx, because the electrons ($T_e \approx 3\,\text{eV}$) can hardly break the strong bonds between the nitrogen atoms in the N_2-molecule ($E_b = 9.8\,\text{eV}$ [87]). In case of H_2 addition at least the high energetic tail of the electron energy distribution leads to electron impact dissociation of the hydrogen molecule, which has a binding energy of $E_b = 4.5\,\text{eV}$ [87]. This phenomenon has also been observed for the comparison between nitrogen and oxygen admixtures [74].

6.4.3 Non-Thermal Atmospheric Pressure Plasma Jet

During the last years the number of different plasma sources operating at atmospheric pressure increased remarkably [88–90]. The advantages compared to low pressure plasmas are obvious:

- Usually no need for time and cost intensive vacuum systems and pumps makes them less expensive and maintenance-intensive [91–93].
- Easy integration into existing process lines and scalability [94, 95].
- Many of them offer the possibility to treat heat sensitive materials, tissue or living cells [40, 93, 95–100].

Especially atmospheric pressure plasma jets have a great potential for applications in various fields of plasma processing due to their flexibility and scalability. They can be used for treatment of inner walls of trenches and cavities. It is also possible to arrange them in an array to cover large surface areas or to mount them on a robotic arm to process complex 3D structures. Especially in the field of plasma medicine a high potential is claimed. They are not only suited for cleaning and sterilization of medical equipment (catheters or implants), but also for the treatment of living tissues for disinfection and wound healing purposes. Especially for the latter applications but also for the treatment of temperature sensitive materials like polymers, where the thermal load should be as low as possible to prevent the treated material from getting melted or burned, the knowledge of the energy fluxes and the resulting surface temperatures is of important interest. Even though the number of different atmospheric pressure plasma sources increased dramatically, only few applications of calorimetric probes for energy flux measurements can be found in the literature [40, 41]. Due to the small size and strong spatial inhomogeneities, the interpretation of the measurements is quite difficult. However, if a quantitative description cannot be given, at least the maximum substrate temperature [41], the temperature distribution across the substrate [101, 102] or the ratios of the different contributions [40, 41] are interesting and important characteristics.

The plasma source investigated here (see Fig. 6.16a) is a commercially available non-thermal atmospheric pressure plasma jet (nt-APPJ) [103] (model kINPen 08). The jet consists of a quartz capillary, through which the feed gas (in this case Argon only) is flowing (flow rate $\approx 5\,\mathrm{L\,min^{-1}} = 0.08\,\mathrm{L\,s^{-1}}$).

The electrode configuration is arranged in such a way, that the rf-driven ($f_{\mathrm{rf}} = 1.7\,\mathrm{MHz}$) needle electrode is located in the center of the capillary and surrounded by the flowing gas. A ring shaped grounded counter electrode is mounted around the capillary. Other modifications work with two outer ring electrodes (without a inner needle) or with an additional inner capillary for feeding a second gas into the plasma zone. The volume of the plasma generation is located near the electrodes in the inner capillary of the jet, while the effluent consisting of different plasma species is blown outwards by the gas flow. The visible part of developing plasma plume expands up to 17 mm into the surrounding atmosphere, depending on the tuning of the internal capacities. Optical emission spectroscopy (OES) measurements in

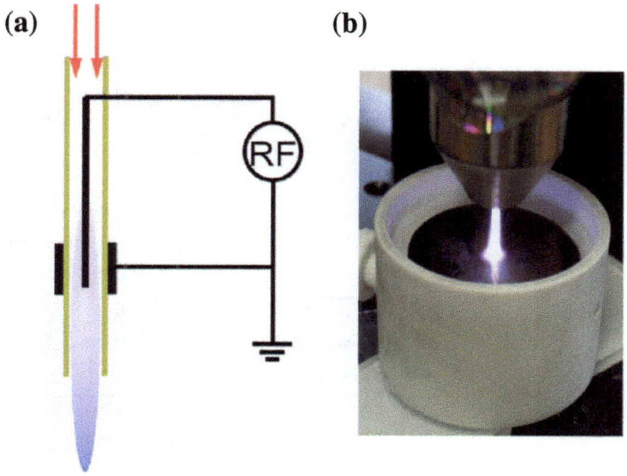

Fig. 6.16 **a** Schematic setup of the nt-APPJ. **b** Calorimetric probe exposed to a non thermal atmospheric plasma jet. Schematic and application of the non thermal atmospheric pressure plasma jet

the range between 300 and 1000nm have shown that the effluent mainly consists of excited neutrals (Ar, O, N_2 and OH) [41, 104], which originate from the feed gas and from the constituents of the surrounding air, that diffuses inside the gas stream [105]. The jet shows a pronounced non-equilibrium character. OES measurements have also shown that the excitation temperature T_{exc}, which is a rough estimate for the electron temperature [93, 106, 107], reaches values up to 3150 K ($\approx$0.27 eV), while the gas temperature remains slightly above room temperature [41].

In the following we discuss the results of the energy flux measurements performed with a calorimetric probe (see Fig. 6.16b). In low pressure plasmas, which usually have a much bigger volume than atmospheric pressure plasmas, the energy flux is assumed to be homogeneously distributed across the probe surface, since spatial inhomogeneities of plasma parameters are small in the plasma bulk. In case of the investigated plasma jet this is not the case. The different contributions from the different species and processes show strong spatial variations. Radiation, for example, is emitted into all directions, so one can measure it also perpendicular to the jet. Molecules, their fragments or metastable species follow the gas flow. Depending on the lifetime of the species they can be detected quite far away from the visible effluent. Also the interaction of the jet with the surface affect the gas flow, leading to different sizes of the "footprint" of the jet and, thus, to different coverage of the surface with the effluent. However, this case is treated like a homogeneously distributed plasma that covers the whole probing area to have comparable results. The used probe was calibrated using a laser, because this method can also be applied at ambient pressure. The calibration at atmospheric pressure is desired, because at higher pressures the cooling of the probe due to convection has a remarkable influence.

Main focus of the results presented here is on the energy influx in dependence of the distance between the exit nozzle and the probe as well as on the influence of probe bias. Subsequently, for the investigation of the spatial distribution of the energy influx, the distance between probe and orifice was varied in a range up to 24 mm. For gaps smaller than 4 mm the plasma did not ignite, probably caused by

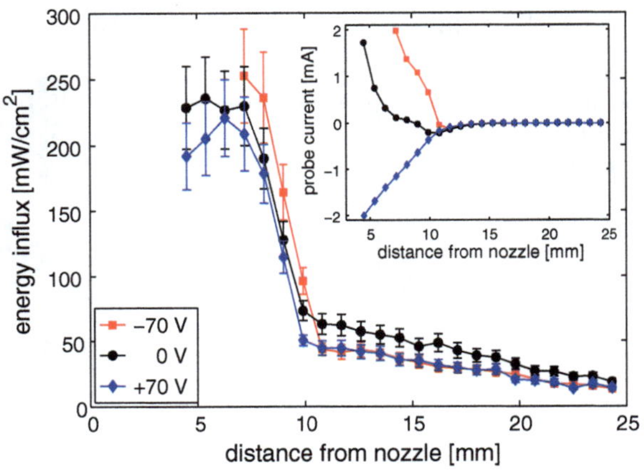

Fig. 6.17 Energy influx and current (*inset*) to the calorimetric probe for three different probe bias voltages (−70, 0 and +70 V) at different distances from the nozzle. Argon was used as process gas

the disturbance of the gas flow. For the measured distances the probe bias was set to −70, 0 and +70 V, respectively. The results of the energy flux measurements are shown in Fig. 6.17.

Starting at ≈24 mm away from the nozzle and going inwards the total energy influx increases slightly for all probe potentials until the probe "touches" the visible part of the effluent at ≈11 mm. For the same distance range the electrical current to the probe is close to 0 mA. For distances between ≈15 and 11 mm the current becomes slightly negative. Coming closer to the orifice, the energy influx increases rapidly and gets saturated for distances smaller than 7 mm. Depending on the probe bias the current changes in the visible region of the jet. For ground potential and even more pronounced for negative bias it becomes positive. For positive probe bias the current shows a linear slope towards negative values.

For the remote part of the effluent, which is not visible for the naked eye it is assumed that radiation (UV and visible), as well as slightly heated neutral species and formation of molecules from long living species are responsible for the measured energy influx. For the visible region it is assumed that most of the energy influx is caused by charge carriers as indicated by the current measurements. The results indicate that the energy influx in the visible region is higher for more negative probe potentials. This can be understood if the contribution from recombination is considered. No ions can reach the probe surface at positive probe potential, thus, no recombination can take place.

Although the measured energy influx is quite high in comparison to low pressure plasmas, the resulting surface temperature does not exceed 55 °C, because the cooling by surrounding gas is much more effective at atmospheric pressure and even enhanced by the gas flow. This makes the characterized plasma jet suitable for treatment of heat sensitive materials or even living cells.

6.5 Summary and Conclusion

Basic methods of calorimetric probes, different probe setups and working principles are described and discussed. Main focus is on the description of probes which are similar to THORNTON's heat flux probe and on radiometer like probes (GARDON probes). Moreover, different contributions of the total energy influx towards the probe originating from particle impact and radiation as well as surface processes are described. Three examples from the field of complex plasmas show how different plasma sources can be characterized my means of energy flux measurements.

It can be stated that calorimetric measurements are a versatile and robust method for the determination of total energy fluxes in plasmas, which are not easy to obtain by calculation from plasma parameters. Even if these parameters can be obtained by standard diagnostic methods a lot of parameters (number of collisions, secondary electron emission coefficients, sticking coefficients, etc.) are not known, but desired for a quantitative calculation.

Acknowledgments The authors would like to thank M. Stahl, T. Trottenberg, V. Schneider, N. Bergemann, J. Teliban, N. Itagaki, K. Koga, G. Uchida, K. Kuwahara, K. Nishiyama, M. Shiratani, H. Wulff, M. Häckel and K.D. Weltmann for their kind contributions. Also the fruitful discussions with M. Wolter, J. Ye, S. Ulrich, A. Winter, T. Peter. V. Zaporojchenko, T. Strunskus, F. Faupel, P.A. Cormier, A.L. Thomann, D. Lundin, U. Helmerson, C. Roth, V. Zuber, A. Sonnenfeld, P. Rudolph von Rohr and K. Ostrikov are greatfully acknowledged.

This work were financially supported under the DFG via SFB/TR 24 project B13.

References

1. J. Thornton, Thin Solid Films **54**(1), 23 (1978)
2. G. Jackson, Vacuum **19**(11), 493 (1969)
3. P. Stangeby, G. McCracken, Nucl. Fusion **30**(7), 1225 (1990)
4. K. Nishiyama, Y. Morita, G. Uchida, D. Yamashita, K. Kamataki, H. Seo, N. Itagaki, K. Koga, M. Shiratani, N. Ashikawa, S. Masuzaki, K. Nishimura, A. Sagara, the LHD Experimental Group, S. Bornholdt, H. Kersten, J. Nucl. Mater. doi:10.1016/j.jnucmat.2013.01.169. Accepted.
5. J. Girard, P. Garin, N. Taylor, J. Uzan-Elbez, L. Rodríguez-Rodrigo, W. Gulden, Fusion Eng. Des. **82**(5), 506 (2007)
6. N. Taylor, D. Baker, S. Ciattaglia, P. Cortes, J. Elbez-Uzan, M. Iseli, S. Reyes, L. Rodriguez-Rodrigo, S. Rosanvallon, L. Topilski, Fusion Eng. Des. **86**(6), 619 (2011)
7. J. Thornton, J. Vac. Sci. Technol. **11**(4), 666 (1974)
8. H. Deutsch, H. Kersten, S. Klagge, A. Rutscher, Contrib. Plasma. Phys. **28**(2), 149 (1988)
9. H. Deutsch, H. Kersten, A. Rutscher, Contrib. Plasma. Phys. **29**(3), 263 (1989)
10. J. Thornton, Annu. Rev. Mater. Sci. **7**(1), 239 (1977)
11. R. Gardon, Rev. Sci. Instrum. **24**(5), 366 (1953)
12. C. Abbot, Smithsonian Misc. Coll. **56**(19) (1911)
13. P. Stier, C. Barnett, G. Evans, Phys. Rev. **96**(4), 973 (1954)
14. A. Thomann, N. Semmar, R. Dussart, J. Mathias, V. Lang, Rev. Sci. Instrum. **77**(3), 033501 (2006)
15. C. Roth, A. Spillmann, A. Sonnenfeld, P. von Rohr, Plasma Processes Polym. **6**(S1), S566 (2009)

16. R. Wiese, H. Kersten, Galvanotechnik **99**(6) (2008)
17. H. Kersten, G. Kroesen, R. Hippler, Thin Solid Films **332**(1–2), 282 (1998)
18. S. Ekpe, S. Dew, J. Vac. Sci. Technol. A **22**, 1420 (2004)
19. C. Paturaud, G. Farges, M. Sainte Catherine, J. Machet, Surf. Coat. Technol. **98**(1–3), 1257 (1998)
20. T. Drüsedau, T. Bock, T. John, F. Klabunde, W. Eckstein, J. Vac. Sci. Technol. A **17**, 2896 (1999)
21. M. Čada, P. Virostko, Š. Kment, Z. Hubička, Plasma Processes Polym. **6**(S1), S247 (2009)
22. T. Drüsedau, K. Koppenhagen, Surf. Coat. Technol. **153**(2–3), 155 (2002)
23. S. Bornholdt, T. Peter, T. Strunskus, V. Zaporojtchenko, F. Faupel, H. Kersten, Surf. Coat. Technol. **205**, 388 (2011)
24. H. Kersten, D. Rohde, J. Berndt, H. Deutsch, R. Hippler, Thin Solid Films **377**, 585 (2000)
25. D. Lundin, M. Stahl, H. Kersten, U. Helmersson, J. Phys. D: Appl. Phys. **42**, 185202 (2009)
26. V. Stranak, M. Cada, Z. Hubicka, M. Tichy, R. Hippler, J. Appl. Phys. **108**(4), 043305 (2010)
27. W. Leroy, S. Konstantinidis, S. Mahieu, R. Snyders, D. Depla, J. Phys. D: Appl. Phys. **44**(11), 115201 (2011)
28. P. Cormier, A. Balhamri, A. Thomann, R. Dussart, N. Semmar, J. Mathias, R. Snyders, S. Konstantinidis, J. Appl. Phys. **113**(1), 013305 (2013)
29. H. Kersten, G. Kroesen, Contrib. Plasma. Phys. **30**(6), 725 (1990)
30. H. Kersten, D. Steffen, D. Vender, H. Wagner, Vacuum **46**(3), 305 (1995)
31. H. Kersten, D. Rohde, H. Steffen, H. Deutsch, R. Hippler, G. Swinkels, G. Kroesen, Appl. Phys. A **72**(5), 531 (2001)
32. H. Kersten, E. Stoffels, W. Stoffels, M. Otte, C. Csambal, H. Deutsch, R. Hippler, J. Appl. Phys. **87**(8), 3637 (2000)
33. G. Makrinich, A. Fruchtman, S. Shitrit, G. Appelbaum, J. Ashkenazy, in *IEPC Proceedings* (2005), pp. 1–6
34. R. Dussart, A. Thomann, L. Pichon, L. Bedra, N. Semmar, P. Lefaucheux, J. Mathias, Y. Tessier, Appl. Phys. Lett. **93**, 131502 (2008)
35. M. Wolter, M. Stahl, H. Kersten, Vacuum **83**(4), 768 (2008)
36. M. Wolter, M. Stahl, H. Kersten, Plasma Processes Polym. **6**(S1), S626 (2009)
37. M. Stahl, T. Trottenberg, H. Kersten, Rev. Sci. Instrum. **81**, 023504 (2010)
38. C. Roth, S. Bornholdt, V. Zuber, A. Sonnenfeld, H. Kersten, P. Rudolf von Rohr, J. Appl. Phys. **44**, 095201 (2011)
39. C. Roth, G. Oberbossel, P. von Rohr, J. Phys. D: Appl. Phys. **45**(35), 355202 (2012)
40. E. Stoffels, R. Sladek, I. Kieft, H. Kersten, R. Wiese, Plasma Phys. Contr. F. **46**, B167 (2004)
41. S. Bornholdt, M. Wolter, H. Kersten, Eur. Phys J. D **60**(3), 653 (2010)
42. S. Khrapak, G. Morfill, Phys. Plasmas **13**, 104506 (2006)
43. H. Maurer, R. Basner, H. Kersten, Rev. Sci. Instrum. **79**, 093508 (2008)
44. D. Kondepudi, Introduction to Modern Thermodynamics (Wiley, Chichester, 2008)
45. D. Ball, J. Appl. Phys. **43**(7), 3047 (1972)
46. G. Golan, A. Axelevitch, J. Optoelectron. Adv. M. **5**(5), 1417 (2003)
47. G. Makrinich, A. Fruchtman, J. Appl. Phys. **100**, 093302 (2006)
48. E. Stamate, H. Sugai, K. Ohe, Appl. Phys. Lett. **80**(17), 3066 (2002)
49. M. Čada, J. Bradley, G. Clarke, P. Kelly, J. Appl. Phys. **102**, 063301 (2007)
50. H. Steffen, H. Kersten, H. Wulff, J. Vac. Sci. Technol. A **12**(5), 2780 (1994)
51. J. Thornton, J. Vac. Sci. Technol. **12**(4), 830 (1975)
52. H. Jouhara, S. Saloum, M. Alsous, Vacuum **86**, 1898 (2012)
53. K. Ellmer, R. Mientus, Surf. Coat. Technol. **116**, 1102 (1999)
54. K. Harbauer, T. Welzel, K. Ellmer, Thin Solid Films **520**, 6429 (2012)
55. T. Welzel, K. Harbauer, F. Kreis, M. Kellermeier, K. Ellmer, Characterisation of a combined plasma sensor and its application to a magnetron sputtering discharge. *Poster Contribution at the 13th International Conference on Plasma Surface Engineering - PSE* (2012)
56. C. Cho, G. Fralick, H. Bhatt, Meas. Sci. Technol. **8**(7), 721 (1999)
57. S. Oh, K. Lee, J. Chun, M. Kim, S. Lee, J. Micromech. Microeng. **11**(3), 221 (2001)

58. M. Freed, M. Krüger, K. Poolla, C. Spanos, IEEE T. Semiconduct. M **18**(1), 148 (2005)
59. M. Stahl, Energiestrommessungen in Prozessplasmen. Diploma Thesis, Kiel University, 2009
60. R. Wendt, K. Ellmer, K. Wiesemann, J. Appl. Phys. **82**(5), 2115 (1997)
61. R. Ash, W. Weller, R. Wright, Rev. Sci. Instrum. **41**(7), 1112 (1970)
62. A. Murthy, B. Tsai, R. Saunders, Metrologia **35**(4), 501 (2003)
63. J. Ballestrin, C. Estrada, M. Rodríguez-Alonso, C. Pérez-Rábago, L. Langley, A. Barnes, Metrologia **41**(4), 314 (2004)
64. J. Ballestrin, C. Estrada, M. Rodríguez-Alonso, C. Perez-Rabago, L. Langley, A. Barnes, Sol. Energy **80**(10), 1314 (2006)
65. P. Cormier, M. Stahl, A. Thomann, R. Dussart, M. Wolter, N. Semmar, J. Mathias, H. Kersten, J. Phys. D: Appl. Phys. **43**(46), 465201 (2010)
66. L. Bedra, A. Thomann, N. Semmar, R. Dussart, J. Mathias, J. Phys. D: Appl. Phys. **43**(6), 065202 (2010)
67. R. Dussard, A.L. Thomann, N. Semmar, *Microsensors (InTech, 2011), Chap. A Heat Flux Microsensor for Direct Measurements in Plasma Surface Interactions* (2011)
68. R. Wiese, H. Kersten, G. Wiese, M. Häckel, Vakuum in Forschung und Praxis **23**(3), 20 (2011)
69. K. Ellmer, J. Phys. D: Appl. Phys. **33**(4), 17 (2000)
70. R. Wiese, H. Kersten, M. Hannemann, V. Sittinger, F. Ruske, R. Menner, Plasma Processes Polym. **4**(1), 527 (2007)
71. S. Schlichtherle, G. Strauss, H. Pulker, Vakuum in Forschung und Praxis **23**(5), 27 (2011)
72. H. Kersten, H. Deutsch, H. Steffen, G. Kroesen, R. Hippler, Vacuum **63**(3), 385 (2001)
73. J. Schulze, E. Schüngel, U. Czarnetzki, J. Phys. D: Appl. Phys. **42**(9), 092005 (2009)
74. S. Bornholdt, J. Ye, S. Ulrich, H. Kersten, J. Appl. Phys. **112**(12), 123301 (2012)
75. S. Bornholdt, N. Itagki, K. Kuwahara, H. Wulff, M. Shiratani, H. Kersten (To be submitted)
76. K. Ellmer, T. Welzel, J. Mater. Res. **1**(1), 1 (2012)
77. T. Trottenberg, A. Spethmann, V. Schneider, M. Stahl, M. Giesenhagen, H. Kersten, Contrib. Plasma. Phys. **52**(7), 584 (2012)
78. H. Do, H. Kersten, R. Hippler, New J. Phys. **10**(5), 053010 (2008)
79. F. Aumayr, H. Winter, Philosophical transactions of the royal society of London. Ser. A: Math. Phys. Eng. Sci. **362**(1814), 77 (2004)
80. G. Betz, K. Wien, Int. J. Mass Spectrom. **140**(1), 1 (1994)
81. B. Chapman, *Glow Discharge Processes: Sputtering and Plasma Etching* (Wiley New York, 1980)
82. V. Schneider, T. Trottenberg, I. Teliban, H. Kersten, Rev. Sci. Instrum. **81**(1), 013503 (2010)
83. T. Trottenberg, A. Spethmann, J. Rutscher, H. Kersten, Plasma Phys. Contr. F. 54 (2012)
84. I. Hutchinson, *Principles of Plasma Diagnostics* (Cambridge University Press, 2002)
85. M. Wolter, I. Levchenko, H. Kersten, K. Ostrikov, Appl. Phys. Lett. **96**(13), 133105 (2010)
86. I. Suhariadi, K. Matsushim, K. Kuwahara, K. Oshikawa, D. Yamashita, H. Seo, G. Uchida, K. Kamtaki, K. Koga, M. Shiratani, S. Bornholdt, H. Kersten, H. Wulff, N. Itagaki, Jpn. J. Appl. Phys. **52**, 01AC08 (2013)
87. R. Weast, *CRC Handbook of Chemistry and Physics*, 48th edn. (The Chemical Rubber Company, 1968)
88. A. Napartovich, Plasmas Polym. **6**(1), 1 (2001)
89. R. Foest, E. Kindel, A. Ohl, M. Stieber, K. Weltmann, Plasma Phys. Contr. Fusion **47**(12B), B525 (2005)
90. R. Foest, M. Schmidt, K. Becker, Int. J. Mass Spectrom. **248**(3), 87 (2006)
91. A. Schütze, J. Jeong, S. Babayan, J. Park, G. Selwyn, R. Hicks, IEEE T. Plasma. Sci. **26**(6), 1685 (1998)
92. G. Selwyn, H. Herrmann, J. Park, I. Henins, Contrib. Plasma. Phys. **41**(6), 610 (2001)
93. E. Stoffels, A. Flikweert, W. Stoffels, G. Kroesen, Plasma Sources Sci. T. **11**(4), 383 (2002)
94. Z. Cao, J. Walsh, M. Kong, Appl. Phys. Lett. **94**(2), 021501 (2009)
95. K. Weltmann, R. Brandenburg, T. Von Woedtke, J. Ehlbeck, R. Foest, M. Stieber, E. Kindel, J. Phys. D: Appl. Phys. **41**(19), 194008 (2008)

96. M. Noeske, J. Degenhardt, S. Strudthoff, U. Lommatzsch, Int. J. Adhes. Adhes. **24**(2), 171 (2004)
97. M. Laroussi, C. Tendero, X. Lu, S. Alla, W. Hynes, Plasma Processes Polym. **3**(6–7), 470 (2006)
98. R. Brandenburg, J. Ehlbeck, M. Stieber, J. Zeymer, O. Schlüter, K. Weltmann et al., Contrib. Plasma. Phys. **47**(1–2), 72 (2007)
99. J. Walsh, M. Kong, Appl. Phys. Lett. **91**(22), 221502 (2007)
100. M. Laroussi, W. Hynes, T. Akan, X. Lu, C. Tendero, IEEE T. Plasma. Sci. **36**(4), 1298 (2008)
101. J. Meister, G. Böhm, I. Eichentopf, T. Arnold, Plasma Processes Polym. **6**(S1), S209 (2009)
102. J. Meister, T. Arnold, Plasma Chem. Plasma Process. **31**(1), 91 (2011)
103. NEOPLAS GmbH, www.neoplas.eu (2009)
104. K. Fricke, H. Tresp, R. Bussiahn, K. Schröder, T. von Woedtke, K. Weltmann, Plasma Chem. Plasma Process. **32**, 801 (2012)
105. R. Foest, T. Bindemann, R. Brandenburg, E. Kindel, H. Lange, M. Stieber, K. Weltmann, Plasma Processes Polym. **4**(S1), S460 (2007)
106. Q. Nie, C. Ren, D. Wang, J. Zhang, Appl. Phys. Lett. **93**(1), 011503 (2008)
107. J. Walsh, M. Kong, Appl. Phys. Lett. **93**(11), 111501 (2008)

Chapter 7
Fundamental and Applied Studies of Molecular Plasmas Using Infrared Absorption Techniques

J. Röpcke, P. B. Davies, J. H. van Helden, M. Hübner, N. Lang and S. Welzel

Abstract Over the past few years mid infrared absorption spectroscopy (MIR-AS) over the region from 3 to 20 ?m has progressed considerably as a powerful diagnostic technique for in situ studies of the fundamental physics and chemistry of molecular plasmas. The increasing interest in processing plasmas containing hydrocarbons, fluorocarbons, nitrogen oxides and organo-silicon compounds has led to further applications of MIR-AS because most of these compounds and their decomposition products are infrared active. MIR-AS provides a means of determining the absolute concentrations of the ground states of stable and transient molecular species at time resolutions below a micro second, which is of particular importance for the investigation of reaction kinetics and dynamics. Information about gas temperature and population densities can also be derived from MIR-AS measurements. Since plasmas with molecular feed gases are used in many applications such as thin film deposition, semiconductor processing, surface activation and cleaning, and materials and waste treatment, this has stimulated the adaptation of MIR-AS techniques to industrial requirements including the development of new diagnostic equipment. The aim of the present chapter is fourfold: (i) to briefly summarize the basic principles of infrared absorption spectroscopy and related instrumentation, (ii) to report on recent achievements in our understanding of molecular phenomena in plasmas using different types of MIR-AS techniques, (iii) to describe examples of industrial process monitoring in the mid infrared and (iv) to discuss the potential of advanced instrumentation based on quantum cascade lasers (QCLs) for plasma diagnostics.

J. Röpcke (✉) · J. H. van Helden · M. Hübner · N. Lang
INP Greifswald, Felix-Hausdorff-Str. 2, 17489 Greifswald, Germany
e-mail: roepcke@inp-greifswald.de

P. B. Davies
Department of Chemistry, University of Cambridge, Lensfield Road, Cambridge CB2 1EW, UK

S. Welzel
Eindhoven University of Technology, 513, 5600 MB Eindhoven, The Netherlands

M. Bonitz et al. (eds.), *Complex Plasmas*, Springer Series on Atomic, Optical, and Plasma Physics 82, DOI: 10.1007/978-3-319-05437-7_7,

7.1 Introduction

Molecular plasmas are increasingly being used not only for basic research but also, due to their favorable properties, for materials processing technology. While offering a wide field of potential applications however a number of processes and properties are far from being fully understood for this type of plasma. This situation concerns plasmas generated under low-pressure conditions, in particular, with growing numbers of applications involving gas discharges at atmospheric pressures, including phenomena of plasma surface interactions. Therefore, the analysis of important plasma parameters, for example, the measurement of molecular precursor concentrations, the monitoring of reaction products and the determination of the gas temperature can lead to a better understanding of plasma chemistry and reaction kinetics. In addition, the detection of low molecular weight transient species, such as radicals and molecular ions, is of key importance, because they play a central role in most of the chemical and kinetic processes in plasmas.

Over the last 2 decades chemical sensing using mid infrared absorption spectroscopy (MIR-AS) in the molecular fingerprint region from 3 to 20 μm, which contains strong rovibrational absorption features of a large variety of gaseous species, has been established as a powerful insitu diagnostic tool for molecular plasmas [1–5]. The methods of MIR-AS provide a means of detecting stable and transient molecular species in ground and excited states and of measuring the concentrations and temperatures of reactive species in plasmas. Compared to "pure" trace gas detection, e.g. in environmental and combustion studies, high sensitivity is not the only issue for plasma diagnostic applications. Since kinetic processes are inherent to discharges excited in molecular gases, high time resolution is frequently desired for fundamental studies as well as for process monitoring in applied research and industry. In addition to high sensitivity and good time resolution the capacity for multi-component detection including molecules with broader absorption structures can further widen the field of use of MIR-AS techniques.

When applying absorption spectroscopic techniques two principle cases need to be distinguished: (I) the measurement of the absorption of light emitted by the plasma itself (*self absorption* or *reabsorption*), or (II) absorption of light generated by an external light source. In the case of self absorption measurements the light from the plasma is measured with and without a retro-reflector or compared with the direct light emitted from an identical plasma. This ensures identical line profiles for the emitted and the absorbed light. This method is particularly important for checking the optical thickness of the plasma. The probing light intensity of an external light source in case (II) has to be low to avoid saturation effects of the exited states of the species under investigation.

The change in light intensity dI_ν when a beam passes through a transparent homogeneous plasma zone of thickness dl is given by the net balance of the intensity due to absorption and emission within the layer dl:

$$dI_\nu = (\varepsilon(\nu) - \kappa(\nu) I_\nu)\, dl \tag{7.1}$$

where $\varepsilon(\nu)$ is the emission coefficient and $\kappa(\nu)$ the absorption coefficient per unit length. The absorption coefficient $\kappa(\nu)$, describing the light absorption in an infinitesimally thin layer of a gas or plasma, is given by

$$\kappa(\nu) = \sum_i N_i \sigma_i(\nu) \tag{7.2}$$

where the summation is over all absorbing species and states. N_i are the species population densities and $\sigma_i(\nu)$ their absorption cross sections at frequency ν. The line profile P_ν is normalised to

$$\int_{line} P_\nu d\nu = 1. \tag{7.3}$$

Assuming the upper level k is not populated, then the absolute value of $k(\nu)$ results:

$$\int_{line} \kappa(\nu)\, d\nu = \frac{h\nu}{c_0} N_i B_{ik}. \tag{7.4}$$

The line absorption coefficient is then given by

$$\kappa(\nu) = \frac{h\nu}{c_0} N_i(\alpha, \beta, \gamma, \ldots) B_{ik} P_\nu \tag{7.5}$$

where $N_i(\alpha, \beta, \gamma, \ldots)$ is the population density of the ith level, which depends on different plasma parameters $(\alpha, \beta, \gamma, \ldots)$, and B_{ik} is the Einstein coefficient for the transition between the levels i and k.

In the case where (i) the external light source has much higher intensity than that of the plasma itself and (ii) the absorption is continuous over the whole distance $l(\kappa(\nu) = \text{constant})$, the absorption of radiation can be described by the Beer-Lambert law:

$$I_\nu(l) = I_\nu(0) \exp(-\kappa(\nu)\, l). \tag{7.6}$$

$I_\nu(0)$ and $I_\nu(l)$ are the fluxes of the radiation entering and leaving the plasma and l is the length of the absorbing (homogeneous) plasma zone, see Fig. 7.1 for illustration. It should be noted that the definition of the effective absorption length of the plasma zone and the assumption of a constant absorption coefficient over the complete distance l are two non-trivial questions in plasma spectroscopy. Inhomogeneity, anisotropy and temperature gradients caused by local changes in the plasma can lead to serious errors [6].

A wide variety of light sources, dispersive elements, detectors and data acquisition methods can be used for absorption spectroscopy [7]. The classic dispersion experiment for measuring the density of atomic or molecular states in a plasma by absorption spectroscopy uses a continuous light source in combination with a narrow bandwidth frequency filter, e.g. a spectrograph, with a detector suited for the spectral range under investigation. The Fourier Transform Infrared (FTIR) technique uses

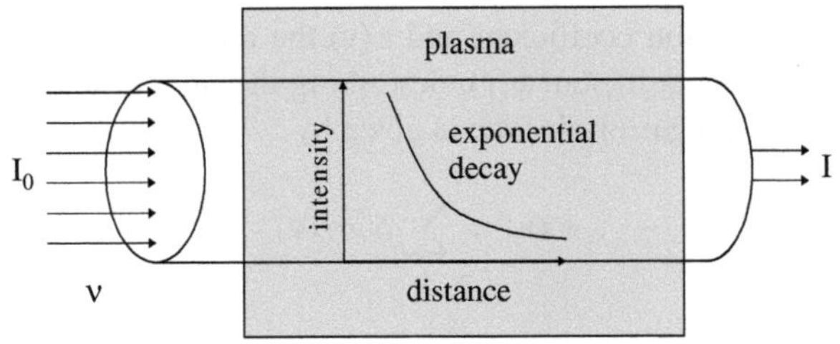

Fig. 7.1 Absorption of external radiation in a plasma according to the Beer-Lambert Law [1]

a Michelson interferometer setup. The light of a continuous infrared light source interferes with the light transmitted through the media under investigation, e.g. the plasma. The intensity depends on the (variable) optical path difference between the mirrors in the two arms of the interferometer. While one mirror is fixed the other is moved continuously between identical path length (maximum signal) and a shifted path length of $\lambda/2$ (zero signal). The inverse transform yields an absorption spectrum in the frequency domain. In contrast to dispersion techniques the FTIR spectrometer records the whole spectrum simultaneously (multiplex or Fellgett advantage). In principle the resolution of FTIR can be as high as $0.001\,\text{cm}^{-1}$, determined by the distance scanned by the moveable mirror, but at the expense of time resolution. Fractional absorptions as small as 10^{-4} can be measured [6].

The current development of tuneable, narrow band light sources such as tuneable dye lasers, infrared diode lasers (TDLs) or quantum cascade lasers (QCLs), has led to their substitution as continuous light sources in AS experiments. These narrow band laser sources have the advantage of high spectral intensity, narrow bandwidth, and continuous tuneability over the absorption profile. Alternatively tuneable infrared radiation can be generated using difference frequency generation (DFG) or Optical Parametric Oscillators (OPO). Although in the past these systems had the disadvantage of rather low radiation power and were restricted to specific wavelength regions [8, 9] new technical developments have led to solutions which provide up to a mW of single mode power and tunability of up to $100\,\text{cm}^{-1}$ [10, 11].

Cavity enhanced techniques like cavity ring-down spectroscopy (CRDS) or cavity enhanced absorption spectroscopy (CEAS) are other high-sensitive laser based absorption techniques [12]. CRDS is based on the measurement of the intensity decay rate of a short laser pulse injected into an optical cavity formed by two very highly reflective mirrors which may also enclose the plasma. For each reflection of the laser pulse, a small fraction (depending on the reflectivity of the mirrors forming the cavity) leaks through one of the mirrors. Therefore, the light intensity detected outside the cavity decays with time dependent on the absorption inside the cavity. In contrast, in CEAS—sometimes also called integrated cavity output spectroscopy (ICOS)—the steady-state transmission or integrated transmitted intensity through the cavity is observed as a function of the laser frequency. Since the laser radiation is trapped in the cavity for many thousands of round trips, absorption path lengths in the kilometres range can be achieved easily, and noise equivalent absorptions as low as $10^{-9}\,cm^{-1}\,Hz^{-1/2}$ can be measured with an acceptable signal to noise ratio [13–16].

The present chapter is intended to give an overview of recent achievements which have led to an improved understanding of phenomena in non-equilibrium molecular plasmas based on the application of MIR-AS techniques. In Chap. 2 special attention is devoted to recent achievements in our understanding of molecular phenomena in plasmas using FTIR, methods of frequency mixing, and TDL techniques. Chapter 3 concerns an overview of some selected new applications of QCLs in the mid infrared including the current status of advanced spectroscopic instrumentation.

7.2 Studies of Molecular Plasmas Using Frequency Mixing, FTIR and TDL Techniques

7.2.1 On Ethylene Destruction in a Dielectric Packed-Bed Plasma Reactor Using FTIR Spectroscopy

Pollution abatement from gas exhaust has been the focus of a wealth of studies during the last few decades. Volatile Organic Compounds (VOCs) belong to an important group of environmentally harmful species, which are widely used in industry. Their emission into the ambient air, even at low levels, should be essentially reduced or even better completely eradicated. Apart from conventional techniques such as catalysis, scrubbers and active charcoal, the use of electric discharges is a promising technique for VOC removal, especially when these gases are present at low concentrations [23, 24]. In the field of pollution destruction, the combination of plasmas with dielectric materials including catalysts has been a technical as well as a scientific challenge. In particular non-thermal discharges at ambient pressure, such as microwave plasmas, DC or AC corona discharges, and dielectric barrier discharges (DBDs) have demonstrated their ability as promising technologies for VOC destruction. Ideally the main products of VOC decomposition should be carbon dioxide and water. The robustness and life time of the discharge configurations and the energy efficiency are the other main issues.

A new promising approach has been the combination of several identical discharge cells to a multistage plasma reactor with the motivation to achieve an essential higher level of conversion of VOCs and other pollution gases. Using a multistage dielectric packed-bed plasma reactor at atmospheric pressure the destruction of ethylene, a typical VOC, and the generation of major by-products have been studied by means of Fourier Transform Infrared Spectroscopy (FTIR). As test gas mixture humidified air with 0.1 % ethylene admixture was used. The gas flow was 1 slm.

Figure 7.2 shows the experimental arrangement of the multistage atmospheric packed-bed plasma reactor combined with gas handling system and FTIR analysis unit. The application of three discharge cells in series and operated at a frequency of $f = 4\,kHz$ and voltage values between 9 and 12 kV led to the nearly complete

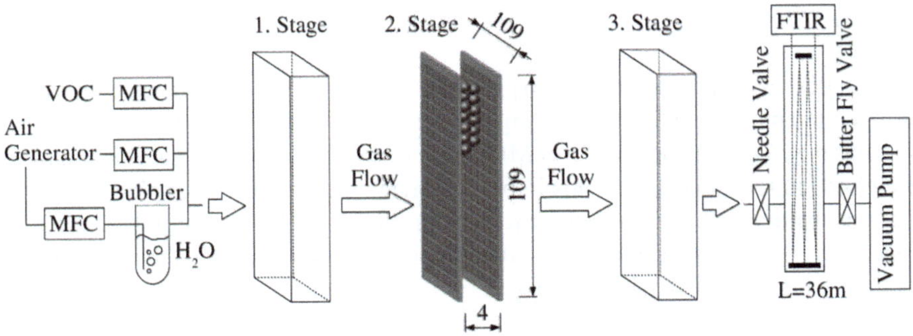

Fig. 7.2 Experimental arrangement of the multistage admospheric pressure-bed plasma reactor combined with gas handling system and FTIR analysis unit. Lengths are given in millimeter [25]

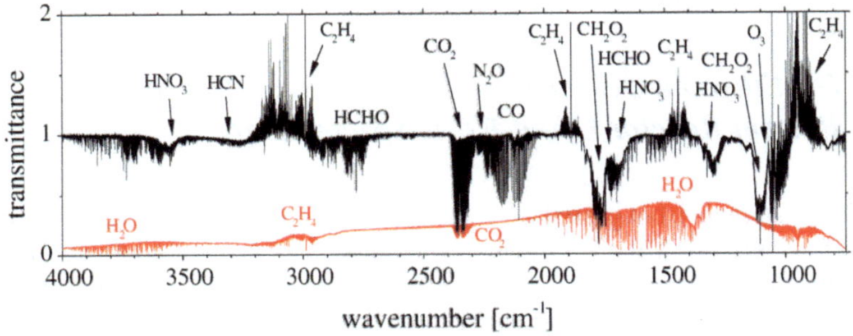

Fig. 7.3 Example of a transmittance spectrum measured by the FTIR spectrometer analyzing the gas composition after treatment in three active stages (U = 12 kV). A spectrum of the initial gas mixture is presented in the lower part of the figure [25]. Absorption features showing transmittance values <1 correspond to decomposition products while transmittances >1 embody the destruction of C_2H_4

decomposition of C_2H_4. The maximum specific energy of the destruction process was found to be about $900\,Jl^{-1}$. In addition to the fragmentation of the precursor gas, the evolution of the concentration of ten stable reaction products, CO, CO_2, O_3, NO_2, N_2O, HCN, H_2O, HNO_3, CH_2O and CH_2O_2 has been monitored. In Fig. 7.3, an example is given of a transmittance spectrum measured by the FTIR spectrometer analyzing the gas composition after treatment through three active stages.
The concentrations of the by-products range between 5 ppm, in the case of NO_2, and 1200 ppm, for H_2O. Based on this multi-component detection the carbon balance of the plasma chemical conversion of ethylene was analyzed.

Fragmentation efficiencies of ethylene ($R_F(C_2H_4) = 5.5 \times 10^{19}$ molecules J^{-1}) and conversion efficiencies to the molecular species produced ($R_C = (0.1–3) \times 10^{16}$ molecules J^{-1}) could be estimated in dependence on the discharge conditions in the multistage plasma reactor. As a typical example, Fig. 7.4 presents the level of destruction of 0.1 % ethylene admixed to an air stream with a flow of 1 slm as a function of the power consumed in the reactor [25].

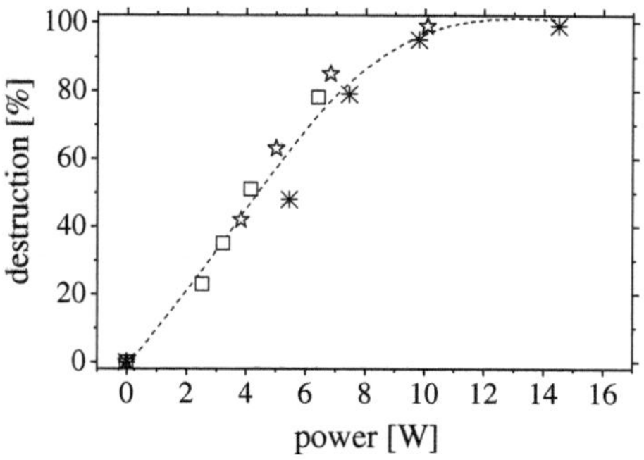

Fig. 7.4 Destruction of 0.1 % ethylene admixed to an air stream with a flow of 1 slm; as a function of the power consumed in the reactor. Destruction values are given for 1 stage (□), 2 stages (⋆) and 3 stages (∗) working simultaneously [25]

7.2.2 Studies of Hydrocarbon RF Plasmas Using a Difference Frequency Laser Source

Difference frequency generation (DFG) is an alternative method for the production of MIR radiation via nonlinear optical processes utilizing widely available room temperature sources in the near-infrared (NIR). These lasers have the same advantageous features as the lead salt diode lasers but without the necessity for cryogenic operation making their use much simpler and because of their importance to telecommunications are based on mature technology. The NIR lasers are mixed in a nonlinear crystal to produce a beam in the MIR with optical qualities directly related to those of the source lasers. The advent of reliable production of periodically poled materials for applications in nonlinear optics in combination with widely tunable sources, such as external cavity diode lasers (ECDL), means that transitions can be accessed over a relatively large spectral range at high resolution (<20 MHz). Primarily, periodically poled lithium niobate (PPLN) has been used either in a DFG or optical parametric oscillator arrangement to generate radiation in the MIR (∼3 μm) for gas detection applications [17–21].

Recently, the practicality of a compact solid-state laser-based difference frequency generation (DFG) system is demonstrated as a tool for probing hydrocarbon-based plasmas [22]. The laser light of a cw Nd :YAG operating at 1064 nm and one of two distributed feedback diode lasers operating at 1560 and 1620 nm were mixed in a periodically poled lithium niobate crystal producing mid-infrared radiation at 3.35 and 3.1 μm for the detection of CH_4 and C_2H_6 at the first wavelength and C_2H_2 and C_2H_4 at the latter.

The radiation with a power of about 18 μW was used to probe a (radio frequency) RF capacitively coupled CH_4 plasma for a matrix of conditions, varying power (<180 W) and pressure (<1 Torr) in both direct absorption spectroscopy and wavelength modulation spectroscopy (WMS) experiments to achieve relatively high sensitivities while retaining accurate spectral information in the form of linewidths. Figure 7.5 shows the schematic of the DFG experimental setup [22].

A minimum detectable absorption coefficient α_{min} of $2 \times 10^{-5}\,cm^{-1}$ was achieved for direct absorption spectroscopy which corresponds to a minimum

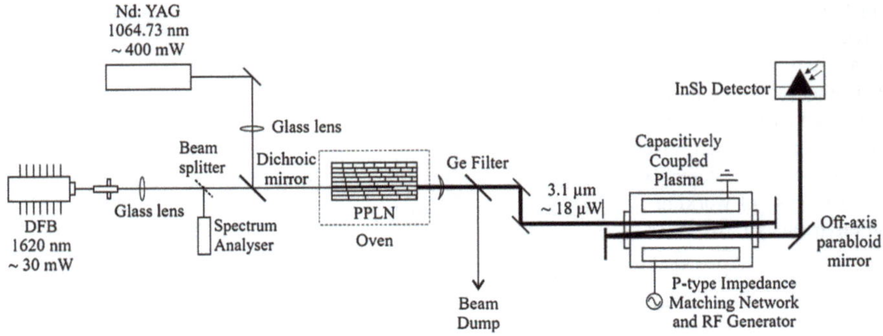

Fig. 7.5 Schematic of the DFG experimental setup [22]

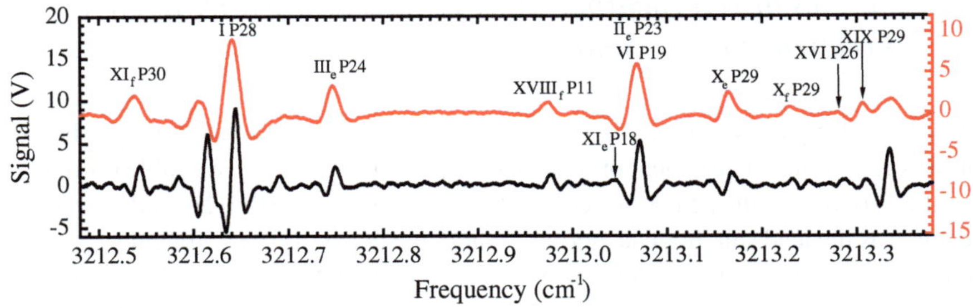

Fig. 7.6 WMS spectra of C_2H_2 in the region 3212.5–3213.4 cm^{-1} as measured in a reference cell of 115 Torr of C_2H_2 (*top*) and in the CH_4 plasma (*bottom*) using $2f$- harmonic detection; the operating power of the plasma was 180 W and the pressure was 1.57 Torr. Marked in the spectra are transitions which could be assigned in both the 115 Torr cell and in the plasma and the vibrational bands they belong to. The other lines observed are not assigned, but part of them are from C_2H_2 transitions as they are observed in both spectra. The other unassigned transitions could be from CH_xH_y molecules, or radicals such as CH_3 and C_2H [22]

detectable density for CH_4 of 1.7×10^{12} cm^{-3} and an α_{min} value of 2×10^{-6} cm^{-1} was achieved for WMS measurements on C_2H_2. Figure 7.6 shows WMS spectra of C_2H_2 in the region 3212.5–3213.4 cm^{-1} as measured in a reference cell of 115 Torr of C_2H_2 (top) and in the CH_4 plasma (bottom) using $2f$-harmonic detection. The depletion of CH_4 was measured to increase from 23 % at 40 W to 40 % at 180 W at a pressure of 0.84 Torr, shown in Fig. 7.7, and the depletion increased to 56 % as the pressure decreased to 0.45 Torr.

7.2.3 Chemical and Kinetic Studies in Dust Producing RF Plasmas Using TDLAS

In reactive low temperature molecular plasmas the formation of dust is a well-known phenomenon. Numerous properties of dusty plasmas and dust particles have been

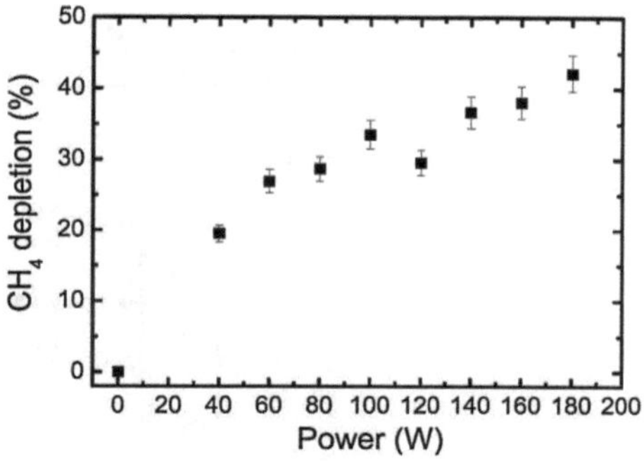

Fig. 7.7 The depletion of CH_4 as a function of plasma power at a pressure of 0.84 Torr [22]

extensively scrutinized in recent decades. This included dust formation, particle growth, astrophysical importance of dust, build-up of strongly coupled systems, formation of voids and many other aspects [26–28]. However, the basic plasma chemical phenomena are still not fully analysed. For a better understanding of the plasma chemistry in dust-producing discharges information about the time-dependent absolute ground state concentration of precursor molecules and of transient and stable molecular products in relation to the dust formation would be very helpful. The hydrocarbon chemistry has been studied in Ar and He RF plasmas with admixtures of C_2H_2 and CH_4 in relation to dust particle formation by means of infrared tuneable diode laser absorption spectroscopy (TDLAS) combined with Fourier Transform Infrared (FTIR) spectroscopy. The experiments were performed in a RF capacitively coupled parallel plate reactor operated at a frequency of $f = 13.56$ MHz and a pressure of $p = 0.1$ mbar. The power applied to the gas mixture composed of a flow of 8 sccm of Ar or He with admixtures of 0.5 sccm C_2H_2 or 1 sccm CH_4 was $P = 15$ W. Using TDLAS, the temporal evolution of the concentrations of the methyl radical and of four stable molecules, C_2H_2, CH_4, C_2H_4 and CO was monitored in the plasma.

Figures 7.8 and 7.9 show the time dependence of concentrations of molecular species measured by TDLAS in Ar and He plasmas with admixtures of C_2H_2 or CH_4, the degree of dissociation of the precursor gases combined with the in situ absorbance measurement using the FTIR system at 5000 cm^{-1}, monitoring the formation of dust in the reactor chamber. The left columns of Figs. 7.8 and 7.9 present the experimental data measured in the Ar plasma case, while in both Figures the right columns represent the conditions found in the He plasma case. The experiments have been structured in time in different phases, which characterize the change of important plasma and particle parameters, see Table 7.1.

The growth process of the dust particles was analysed by FTIR spectroscopy, Fig. 7.9a, b. The degree of dissociation of the acetylene precursor was found to be nearly constant in the range of 96 % under stabilized conditions for both the Ar and He plasmas. In contrast, the degree of dissociation of the methane precursor varied between 45 and 90 % depending (i) on the appearance of dust particles in the reactor volume and (ii) on the Ar or He plasma conditions, Fig. 7.9c, d. The methyl radical concentration was found to be in the range of 10^{11} molecules cm^{-3}.

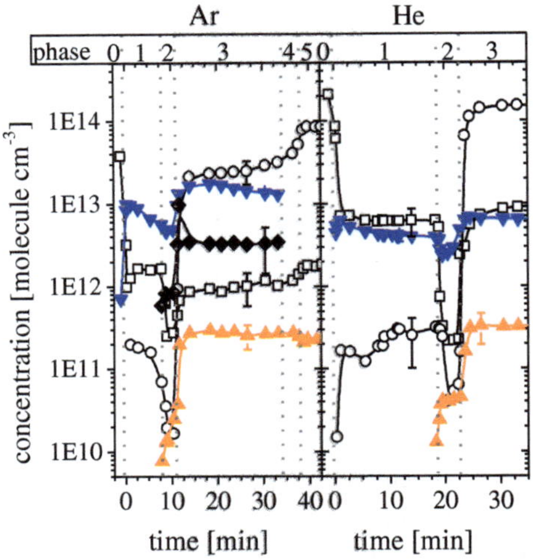

Fig. 7.8 Molecular concentrations in Ar, *left panel*, and He RF plasmas, *right panel*, with admixtures of C_2H_2 or CH_4 in different experimental phases, see Table 7.1, as a function of the time (□ - C_2H_2, ○ - CH_4, ♦ - C_2H_4, △ - CH_3, ▼ - CO) [30]

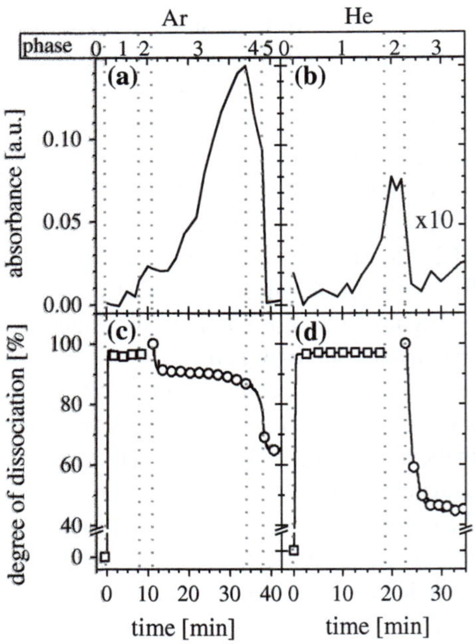

Fig. 7.9 Temporal evolution of (i) the absorbance at 5000 cm^{-1}, measured by FTIR, monitoring the dust production in the reactor (subdiagrams (**a**) and (**b**)) and (ii) of the degree of dissociation of the precursor gases C_2H_2(□) and CH_4 (○)(subdiagrams (**c**) and (**d**)) in Ar, *left panel*, and He RF plasmas, *right panel*, with admixtures of C_2H_2 or CH_4 in different experimental phases, see Table 7.1 [30]

The concentrations of all hydrocarbon species were strongly correlated with the dynamics of the dust formation.

In the Ar plasma case several interesting features of the temporal behavior of the molecular concentrations related to the disappearance of the particles have been

Table 7.1 Experimental phases of the measurements in the Ar and He plasma cases [30]

Phase	Plasma on/off	Hydrocarbon precursor gas	Status of particle formation
Ar 0	Off	C_2H_2	No growth
Ar 1	On	C_2H_2	Initiation of nucleation
Ar 2	On	None	No further growth
Ar 3	On	CH_4	Growth of particle diameter up to maximum
Ar 4	On	CH_4	Particles start to disappear from reactor volume
Ar 5	On	CH_4	Reactor volume free of particles
He 0	Off	C_2H_2	No growth
He 1	On	C_2H_2	Slight initiation of nucleation
He 2	On	None	Reaching of a distinct maximum of number and size
He 3	On	CH_4	Small particle number present

detected, shown in Fig. 7.10. As soon as phase Ar 4 begins the particles start to move out of the volume of observation, see Fig. 7.9a. This process changes the plasma conditions, that is reflected (i) in a reduced degree of dissociation of the precursor CH_4, (ii) in an increasing concentration of the product C_2H_2 and (iii) in a slight decrease of the methyl concentration. Directly at the transition between the phases Ar 4 and 5 the remaining particles tend to vanish on a very short time scale. The changing plasma conditions, in particular the lower T_e, can be seen immediately in a steep decrease of the methyl radical concentration, which is reduced by about 25 % within a time of $T_1 = 5.4$ s. After reaching this minimum the methyl concentration increases again within about $T_2 = 7.2$ s, and shows a relatively constant value. The specific dynamics of the methyl radical concentration, displayed in Fig. 7.10b, can be interpreted as proof of a new start of the nucleation process of the dust production after about 12.6 s, i.e. after $T_1 + T_2$ from the disappearance of nearly all particles out of the volume of observation in the reactor. Further, it is interesting to note, that the CH_3 radical, which has a life time of about 40 ms, is a fast and sensitive probe of changing plasma conditions [29].

Fragmentation efficiencies of acetylene ($R_F(C_2H_2) = 3.2 \text{ x } 10^{16}$ molecules J^{-1}) and of methane ($R_F(CH_4) = 0.16$–$2.5 \text{ x } 10^{16}$ molecules J^{-1}) and conversion efficiencies to the hydrocarbons produced ($R_C = 0.23 - 8.5 \text{ x } 10^{14}$ molecules J^{-1}) could be estimated in dependence on the discharge conditions in the RF plasma.

In summary, it was possible to determine the absolute ground state concentration of the methyl radical and of four stable species under a wide range of experimental conditions by TDLAS providing a link with the dynamic behaviour of the dust formation in a capacitively coupled RF discharge, the ultimate objective being a better understanding of chemical and reaction kinetic processes occurring in dust producing plasmas [30].

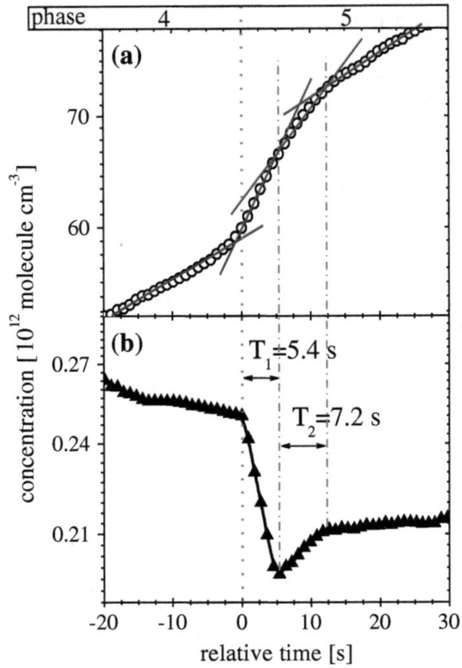

Fig. 7.10 Detailed view of the temporal behavior of the concentrations of CH_4, sub-diagram (**a**), and CH_3, sub-diagram (**b**), in an Ar RF plasma as a function of time in the transition between the experimental phases Ar 4-5 and, see also Table 7.1 [30]

7.3 Quantum Cascade Laser Absorption Spectroscopy for Plasma Diagnostics and Control

7.3.1 General Considerations

The recent development and commercial availability of pulsed and continuous wave quantum cascade lasers (QCLs) for wavelengths longer than 3.4 μm and inter-band cascade lasers (ICLs) for shorter wavelengths offer an attractive new option for MIR-AS for plasma diagnostic purposes. The first QCL devices could only be used at room temperature in pulsed mode at low duty cycles. Room temperature continuous wave (cw) lasers have become available following the successful solution of heat dissipation problems [31, 32]. Compared to lead salt lasers distributed feedback (DFB) QCLs provide (i) continuous mode-hop free wavelength tuning, (ii) increasingly high output powers, (iii) near room temperature operation, and in the case of cw-DFB-QCL (iv) narrow linewidth radiation. In conjunction with integrated DFB gratings, the emission wavelength of this new class of thermoelectrically cooled semiconductor laser can be tailored over a wide wavelength range throughout the infrared molecular fingerprint region. Their total emission range is typically limited to less than 7 cm^{-1} (see [2] and references therein). Therefore multi-component detection to study infrared active compounds in reactive plasmas requires, as in the case of TDLAS, the combination of several lasers in a spectrometer [33]. Meanwhile

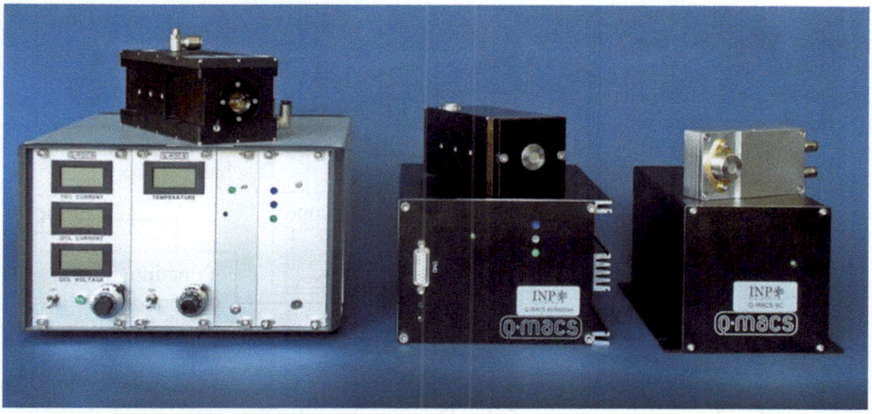

Fig. 7.11 Photo of 3 generations of driver electronics for QCLs, Q-MACS Basic (2005–2011)

the variety of QCLs and ICLs operating at room temperature provide alternatives to cryogenically cooled lead salt lasers. Based on an ongoing technical development focused on compactness and robustness, (Fig. 7.11), this has led to rapid development of MIR-LAS from a niche position to a standard diagnostic technique [2, 3, 16].

Recently, tunability over much broader spectral ranges than with a typical lead salt laser or DFB-QCL has been achieved using external cavity (EC) configurations. Nowadays, EC-QCLs, which work in pulsed or continuous wave (cw) modes, can be tuned over more than $100\,\text{cm}^{-1}$. Modern cw EC-QCLs provide mode-hop free (MHF) tuning ranges as large as $80\,\text{cm}^{-1}$, with narrow line widths of typically 20–60 MHz ($6\text{–}18 \times 10^{-4}\,\text{cm}^{-1}$) and powers of up to 350 mW [2]. Even higher tuning ranges and powers have been achieved in specific cases [34]. At present EC-QCLs are used for an increasing number of applications including high resolution isotope analysis, explosive detection and trace gas monitoring [34–39, 70].

7.3.2 Application of Pulsed QCLs

7.3.2.1 Time Resolved Gas Temperature Measurements Using Pulsed QCLs

In plasma physics, plasma chemistry and also in plasma technology the gas temperature is a very important parameter, since the rate coefficients of chemical reactions in the gas volume, the diffusion of gaseous species and also heterogeneous reactions on surfaces usually show a strong dependence on the temperature [40]. Even in a monoatomic gas, the plasma density distribution cannot be calculated without knowledge of the temperature distribution. The determination of the gas temperature by emission and absorption spectroscopy is quite common in plasma diagnostics.

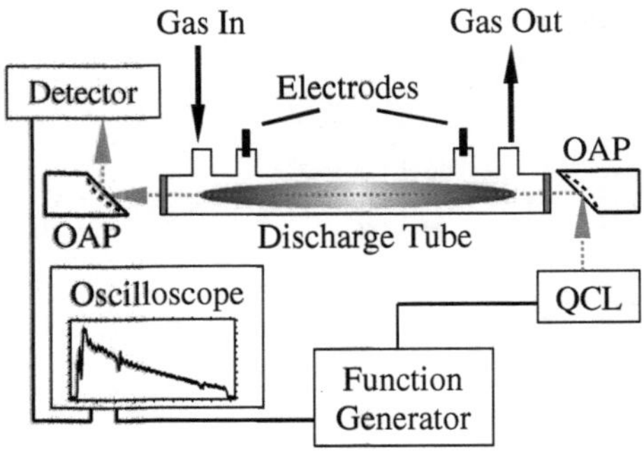

Fig. 7.12 Principal sketch of the optical setup. The tube was 60 cm in length with an inner diameter of 20 mm. The experiments were performed under static gas conditions, $p_{initial}$ = 1.33 mbar [44]

For example, the measurement of the Doppler broadening of spectral lines is a well known and widely used method of temperature determination. Nevertheless, the implementation of the method is not that simple in practice, because the main problems are: (i) the determination of the instrumental function of the spectrometer, (ii) taking into account the multiplet and fine structure of lines and (iii) extraction of the Doppler broadening contribution from the measured line profile while (iv) considering possible pressure broadening and line profile distortion effects.

An alternative spectroscopic approach is the analysis of rotational temperatures of molecules in excited vibronic states. Here, the rotational temperature has definite physical meaning if the time of rotational relaxation in a certain vibronic state is much smaller than the mean lifetime of the rotational levels. In this case, the rotational distribution of the populations is close to a Boltzmann distribution, with the temperature equal to the translational temperature of the colliding particles; see ref. [41] and references therein.

In particular, methods of mid infrared laser absorption spectroscopy, using lead salt lasers or quantum cascade laser absorption spectroscopy, are useful for the study of plasmas with infrared active molecules. Already in 1998, Allen described a TDLAS method to determine rotational temperature values, which were shown to be in good agreement with the gas temperature, employing the intensity ratios of measured absorption lines [42]. Both methods, TDLAS and QCLAS, can be used as a non-invasive temperature probe. A time resolution approaching the sub-microsecond time scale can be achieved when pulsed quantum cascade lasers (p-QCL) are applied. This high temporal resolution will improve the understanding of kinetic processes in e.g. pulsed plasmas *on line* and in situ.

A pulsed DC air plasma admixed with 0.8 % NO was studied in a Pyrex tube reactor. A time resolution of 33 μs was provided by the QCLAS system. The line ratio method was applied to determine the gas temperature. Figure 7.12 presents the principal sketch of the optical setup. The tube was 60 cm in length with an inner diameter of 20 mm. The experiments were performed under static gas conditions, $p_{initial} = 1.33$ mbar. For this purpose, the temperature dependent intensity ratio of two absorption structures of NO at $1900\,\text{cm}^{-1}$ (5.26 μm) was used. Figure 7.13 shows the NO spectrum detected in the time domain. The QCLAS system was operated in

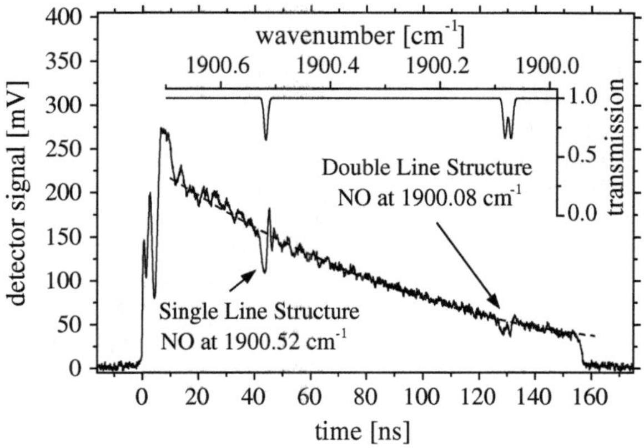

Fig. 7.13 Detected NO spectrum in the time domain measured by the QCLAS system. The spectrum (*solid line*) corresponds to 0.8 % NO in air at a pressure of p = 1.33 mbar with an absorption length $L = 180$ cm. The *dashed curve* shows the baseline. The embedded diagram shows a simulated NO spectrum using the HITRAN database [43, 44]

Fig. 7.14 Calculated temporal evolution of the temperature of NO diluted in air (symbols) for a 150 mA plasma pulse. Exponential functions have been fitted to the temperature rise during the pulse, t_2, as well as to the cooling period after the plasma pulse, t_3 [44]

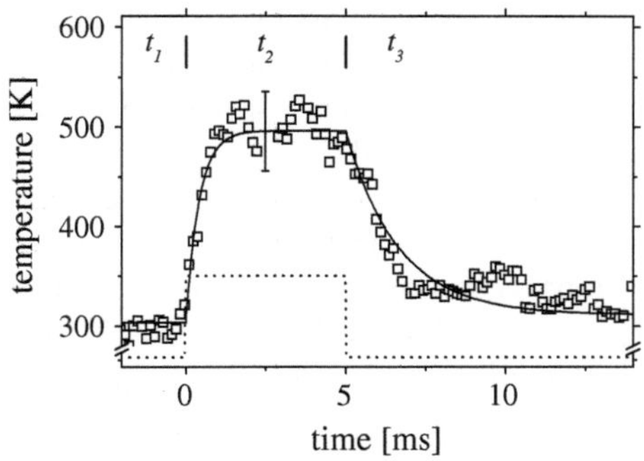

the *Intra Pulse Mode* with a pulse repetition frequency of 30 kHz leading to an entire spectrum each 33 μs [3]. The influence of non-linear absorption phenomena, which are usually observed under low pressure conditions, and that cause strongly distorted absorption structures of NO were taken into account by a calibration routine based on tabulated line strengths. As a typical result for a 150 mA plasma pulse the calculated temporal evolution of the temperature of NO diluted in air is shown in Fig. 7.14. Different mean plasma currents have been applied to the discharge leading to gas temperature values ranging from about 300 K up to about 500 K, (Fig. 7.15) [44].

7.3.2.2 Kinetic of Species in Pulsed CF_4/H_2 RF Plasmas

Fluorocarbon plasmas are applied in industry and fundamental research for etching or deposition purposes and as model systems. Pulsing of the discharge provides an

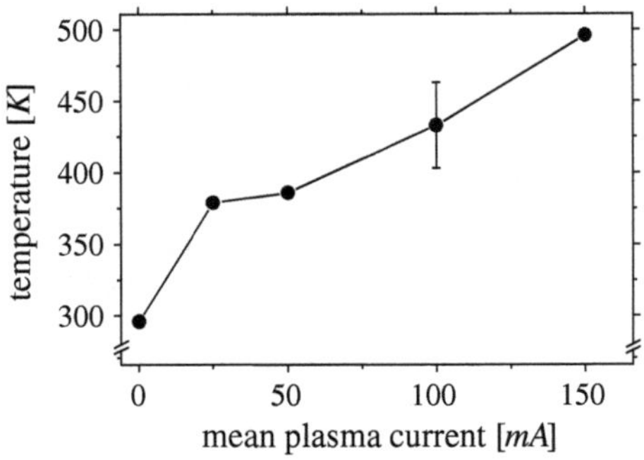

Fig. 7.15 Calculated gas temperatures during the plasma pulse as a function of the mean plasma current. The power was determined from the electric measurement. The total pressure of the gas mixture containing 0.8 % NO in air was 1.33 mbar [44]

additional means of tuning plasma parameters, e.g. densities and temperatures, which become time dependent.

However, a better understanding of the plasma chemistry and kinetics in these complex discharges is still required. Thus time resolved information on absolute number densities of transient and stable species is desirable. The kinetics of stable species has been studied in situ in pulsed CF_4/H_2 RF discharges by means of time resolved QCLAS. The absorption spectra were recorded with a time resolution of 5ms and required a multi-species analysis, because of interfering complex absorption features of CF_4 and C_3F_8.

For this reason, measurements were carried out at two different spectral positions. High resolution spectroscopic data were established by calibrating effective absorption cross sections and their relative temperature dependences for the relevant low pressure conditions (10 Pa). Figure 7.16 shows the experimental arrangement used for diagnostic studies in pulsed CF_4/H_2 RF plasmas. During the discharge a decrease in the CF_4 density by ~12 % was observed. The off-phase was characterized mainly by the gas exchange. The C_3F_8 density in the off-phase was found to be of the order of the detection limit ($3 \times 10^{13}\,\mathrm{cm}^{-3}$). Spectra acquired during the plasma-on phase showed a rapid temperature-induced increase in the absorption signal and, additionally, suggested the influence of a short-lived broadband absorbing species. In Fig. 7.17, as an example, the time resolved absolute number densities of species CF_4 and C_3F_8 with a rapid change of the gas temperature between 296 K during the off-phase and 390 K during the plasma pulse is shown [45].

7.3.2.3 Industrial Process Monitoring

In order to achieve enhanced efficiency, increased stability and higher product quality, a deeper insight and more direct control of plasma-assisted chemical vapor deposition (PACVD) applications becomes an important goal, as well as a challenging subject, for plasma research and technology. Boron trichloride (BCl_3) is used as a source gas in various industrial plasma applications, e.g. for the surface modification of steel components in the mechanical engineering and automotive industries as well

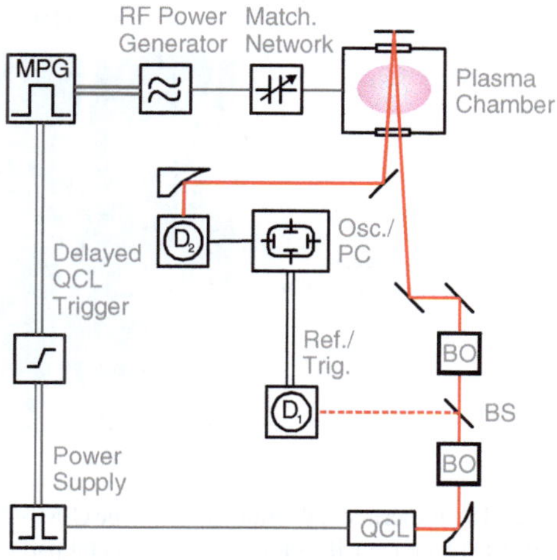

Fig. 7.16 Experimental arrangement used for diagnostic studies in pulsed CF_4/H_2 RF plasmas. (*MPG* master pulse generator, *D* detector, *BO* beam shaping optics, *BS* beam splitter, *Osc./PC* oscilloscope/personal computer) [45]

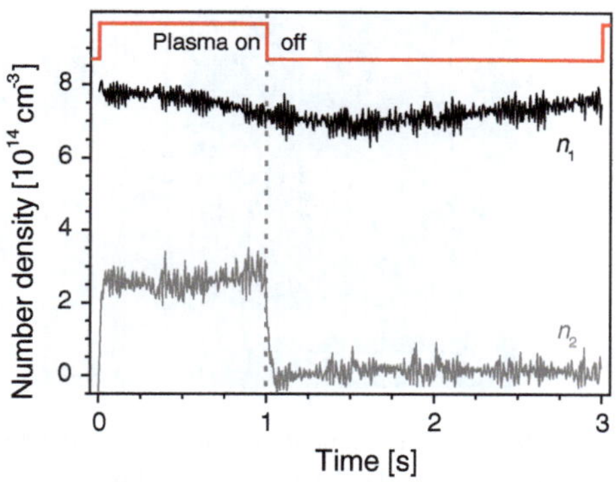

Fig. 7.17 Time resolved absolute number densities of species n_1 (CF_4) and n_2 (C_3F_8) with a rapid change of the gas temperature between 296 K during the off-phase and 390 K during the plasma pulse (10 Pa, 7 sccm CF_4 + 3 sccm H_2, 100 W) [45]

as for plasma etching of semiconductor devices. The goal of plasma-assisted boron treatment of steel substrates is the optimization of surface properties for various loading conditions. In particular iron boride layers on iron and steel surfaces enhance the surface properties of components such as hardness, wear resistance and adhesion. It is known that by varying the plasma parameters, coatings with a wide range of favorable chemical and physical properties can be produced.

Several investigations on gas and plasma treatment using mixtures of BCl_3 with Ar and H_2 have been reported with varying degrees of success. In particular, the use of heated-wall pulsed glow discharges showed promising results for plasma-assisted boron treatment. Boron nitride coatings have been used to increase the hardness, chemical stability and resistance to oxidation of metal substrate surfaces (see [46] and references therein). In order to understand the basic plasma chemistry processes, different diagnostic approaches have been applied and integrated into plasma systems. In the past this also included the application of TDLAS methods [47]. The feasibility of QCLAS has been proven using the quantum cascade laser measurement

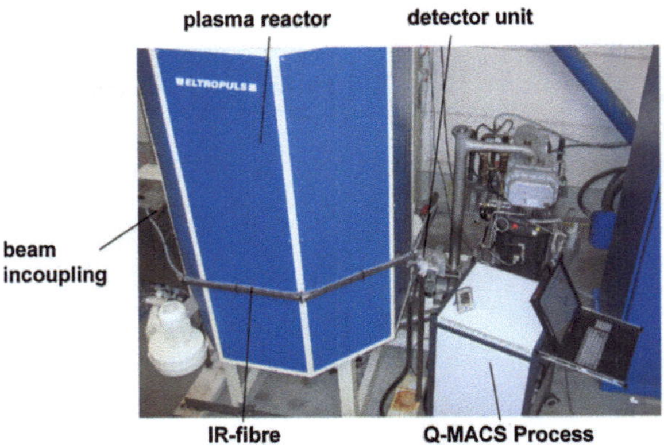

Fig. 7.18 Experimental arrangement of the Q-MACS Process with IR fibre coupling at the industrial PACVD reactor for Boron treatment of metal parts [46]

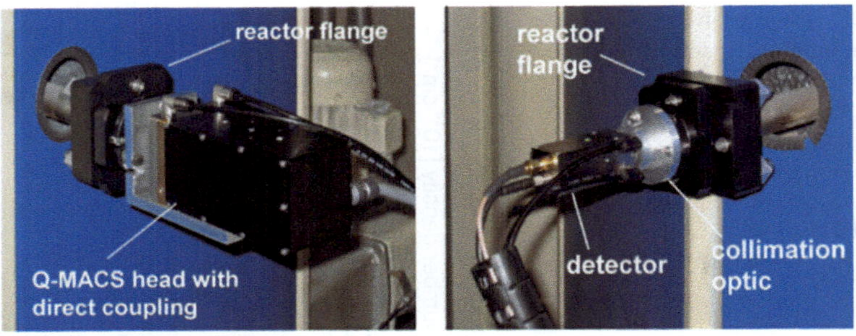

Fig. 7.19 Direct implementation of the Q-MACS Process at the PACVD reactor: *left*—the Q-MACS laser head, *right*—the detector with collimation optics [46]

and control system Q-MACS for on-line process monitoring in research and industry [47–49].

Recently time-resolved infrared absorption spectroscopy investigations of the concentration behavior of the precursor BCl_3 in a pulsed DC plasma with heated reactor walls have been performed. A mixture of $Ar/H_2/N_2/BCl_3$ was used as the boron treatment feed gas. Figure 7.18 shows the experimental arrangement of the Q-MACS Process system with IR fibre coupling at the industrial PACVD reactor for boron treatment of metal parts. The direct implementation of the Q-MACS Process at the PACVD reactor is shown in Fig. 7.19. With reference to the conditions in an industrial environment, the Q-MACS Process system, described in ref. [50], was further modified and optimized to enable time-resolved and in situ QCLAS measurements of the BCl_3 concentrations. The Q-MACS instrumentation was improved using specially developed TO3-QCLs with an integrated lens to allow for direct and dust sealed (modular) optical coupling at the PACVD reactor.

The structure of the BCl_3 molecule did not permit the measurement of rotationally resolved lines, mainly because of overlapping spectra from isotopic species.

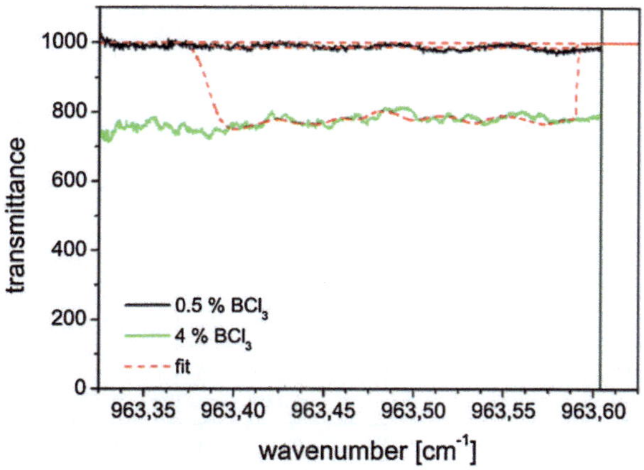

Fig. 7.20 QCLAS spectrum for different partial pressures (*solid line*, *black* and *green*) of BCl_3 at $963.5\,cm^{-1}$ with corresponding multi peak fit (*dashed line*, *red*) [46, 50]

Therefore, special routines had to be used to determine the line strength data experimentally for quantitative measurements. Stancu et al. have proposed an approach for unresolved spectra in the mid-infrared region [51]. Instead of determining line strengths an effective absorption coefficient for the spectral range as well as for the temperature and pressure of interest was used for the measurement. Following this, the effective absorption coefficient for BCl_3 had to be determined experimentally by measuring the averaged absorbance across the spectral range for a given set of BCl_3 concentrations at known absorption length. Figure 7.20 shows a QCLAS spectrum for different partial pressures of BCl_3 at $963.5\,cm^{-1}$ with a corresponding multi peak fit [46, 50]. Following this approach, the determination of the BCl_3 ground state concentration was performed using effective absorption cross sections leading to lower computational times without loss of sensitivity.

A sensitivity of $10^{-6}\,cm^{-1}\,Hz^{-1/2}$ was achieved in a single pass measurement. Synchronization of the Q-MACS with the process control unit provided an insight into individual process cycles with a time resolution of 0.2 s. The measurements gave an insight into the dynamics of an industrial PACVD process. It was found that the fragmentation of the precursor molecule BCl_3 exhibits a non-linear behavior during an individual process cycle resulting in variable fragmentation rates. The BCl_3 concentrations detected were on the order of 10^{14} molecules cm^{-3}. The time resolved changes of the BCl_3 precursor concentration on an absolute scale during several plasma cycles are shown in Fig. 7.21. The results of in situ monitoring using QCLAS demonstrate the potential of this type of diagnostics for effective optimization procedures in industrial PACVD processes hence improving productivity, reliability and reproducibility [46].

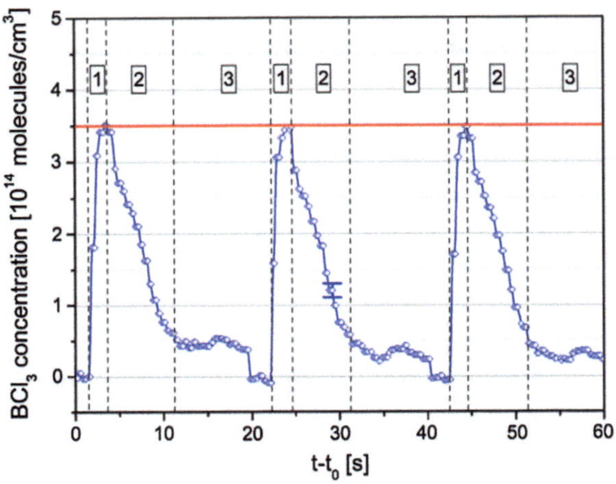

Fig. 7.21 Change of the BCl_3 precursor concentration on an absolute scale during several plasma cycles: phase *1*: injection of BCl_3, phase *2*: plasma, phase *3*: pump down of the reactor. The horizontal red line denotes the injected BCl_3 concentration corresponding to a mixture of $N_2/H_2/Ar/BCl_3$ = 30/45/20/2 slh at a total pressure of 180 Pa and a wall temperature of 873 K. The error bar represents the typical uncertainty in the measured BCl_3 concentration of about 8 % [46]

7.3.3 Application of cw QCLs

7.3.3.1 Monitoring of Dissociation of Molecular Precursors in Direct Absorption Experiments

Within the last few years DFB–cw–QCLs have become increasingly available and possess the advantages of near room temperature operation and continuously tuneable single mode emission. DFB–cw–QCLs are characterized by convenient tuning options comparable to the older TDLAS.

The application of cw QCLs enables higher signal to noise ratios to be achieved due to the higher output power levels of QCLs compared with TDLs. Additionally, intensity fluctuations, which are inherent in pulsed QCLs, are considerably less. The earliest plasma related experiment using cw-QCLs was reported in 2009. The dissociation of a silane precursor in the exhaust line of an industrial PACVD reactor was studied [52]. The silane absorbance in the 3.7 m long exhaust line of the reactor was monitored during the low pressure ($p < 5$ mbar) plasma deposition process. The deposition rate of microcrystalline silicon films was estimated from the depletion of the precursor concentration in a SiH_4 plasmas diluted with H_2 and was in good agreement with complementary ex situ profilometric measurements. The experiments were carried out by employing direct absorption spectroscopy in a single pass configuration. The recorded QCLAS spectra were compared with high-resolution FTIR spectra measured earlier.

Recently, a cw-QCL has been used for in situ plasma measurement to quantify the sensitivity of cw-QCLAS compared to traditional TDLAS. In this experiment a planar MW discharge reactor was employed that was specifically designed for high resolution IRLAS studies on stable and transient molecular species and therefore equipped with a White type multiple pass cell with a 1.50 m mirror separation [3, 53]. A relatively homogeneous plasma can usually be achieved across the entire absorption volume. The QCL spectrometer was mounted perpendicular to the long axis of the reactor to measure the precursor concentration parallel to a cross sectional plane of 0.20 m length in a direct single pass absorption configuration. Figure 7.22a is a schematic diagram of the experimental arrangement.

The cw-QCL emitting around 1303 cm^{-1} was swept in frequency by adding a milliseconds ramp on the operating current (known as the sweep integration method used in TDLAS [50]) to achieve a 0.8 cm^{-1} spectral scan. Examples of cw-QCLAS transmission spectra for CH_4 measured in an Ar(9 %) CH_4 and an Ar(33 %) O_2 plasma are presented in Fig. 7.22b. The absorption of the strongest CH_4 line covered by the spectral scan was changed from 57 % without plasma to some 3 % (Fig. 7.22b) with the discharge on. This suggests efficient precursor dissociation and emphazised the good SNR achieved in the QCLAS measurement, in fact comparable to the TDLAS results recorded under the same conditions.

The peak absorption of the strongest CH_4 feature was followed with time to determine the sensitivity of the spectrometer by means of the Allan variance σ_a, see Fig. 7.22c. The sensitivity was established to be 10^{-3} cm^{-1} $Hz^{-1/2}$, which corresponds to a CH_4 number density of 7×10^{-11} cm^{-3} or 19 ppm for the conditions under study. This detection limit was achieved in a single pass configuration without reducing the noise by excessive spectral averaging, post-processing of data or applying modulation techniques [3].

7.3.3.2 Molecular Trace Gas Detection Using Optical Cavities

Increasing the sensitivity of (MIR) plasma diagnostic methods is desirable either for lowering the limit of detection to levels where important transient species might be expected to be produced in the plasma or for obtaining better spatial resolution, i.e., for reducing the line-of-sight to obtain more precise local molecular number densities. Increasing the effective absorption length by employing high finesse optical cavities [55, 56] can provide increased sensitivities at inherently small base lengths (see e.g. [54] and references therein) and would therefore be ideally suitable for localized measurements in discharge environments. The application of cwQCLs to CEAS methods benefits from their milliwatt output power without being hindered by the frequency chirp reported for pulsed QCLs [54]. In the absence of a rapid spectral chirp the effective line width is then only limited by the performance of the power supply and is typically of the order of ~ 0.001 cm^{-1} (30 MHz) [57–59]. The number of excited cavity modes is considerably reduced and the intensity gain in the developing modes is not limited by the (fast) sweep rate α. Furthermore, CEAS is

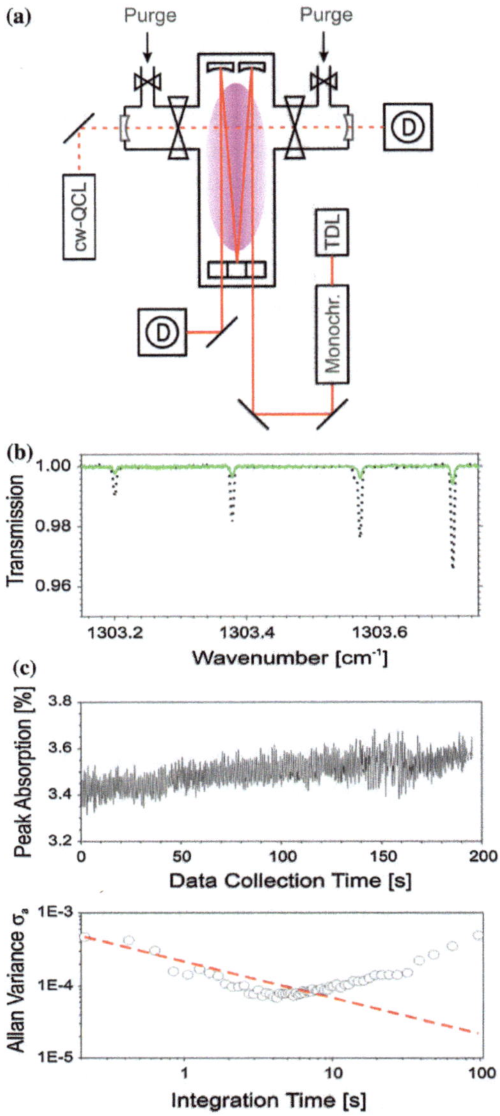

Fig. 7.22 **a** Schematic diagram of the experimental setup used for IRLAS with a planar MW discharge reactor. A 7.67 μm cw-QCL was guided through purged viewing apertures perpendicular to a lead salt TDL spectrometer aligned to achieve 24 passes within an in situ White cell with a 1.5 m mirror separation ($L_{TDL} = 36\,\text{m}$). The single pass QCL absorption path (reactor clearance) was $L_{QCL} = 0.2\,\text{m}$ [3]. Discharge active region shown by red shading. **b** CH_4 transmission spectra measured with the cw-QCL in a CH_4-Ar plasma [25 sccm + (225 sccm + 20 sccm purge gas), 1.5 mbar, 0.7 kW input power, *dotted line*] and an O_2-Ar plasma [20 sccm + (15 sccm + 25 sccm purge gas), 1.5 mbar, 0.7 kW input power, *solid line*] [3]. **c** *Upper*: CH_4 peak absorption at 1303.71 cm^{-1} as a function of time in the CH_4-Ar discharge shown in (**b**). *Lower*: Allan plot retrieved from the data stream plotted in the *upper panel* (symbols). The trend in white noise is indicated (*dashed line*) [3]

relatively straightforward to implement if sophisticated cavity stabilization schemes or laser locking by optical feedback are omitted [3].

In a recent proof of principle study CEAS using cw-QCLs was applied to detect molecular species in low-temperature microwave plasmas. The measurements, focused on the monitoring of HCN and NO, were performed in Ar-CH_4-N_2 or Ar-O_2 plasmas, respectively. The operating conditions used provided a link to earlier TDLAS studies [60].

The impressive performance of cw-QCLAS and the absence of a rapid frequency chirp with these devices facilitate the extension of CEAS methods to the mid-IR spectral range. The CEAS approach offers the advantage of reduced absorption volumes and the potential for in situ diagnostics compared to the application of long optical path cells. This is of special interest for in situ detection of plasma produced molecular species, specifically transient molecules, which cannot be extracted into (ex situ) high sensitivity trace gas spectrometers. It may also provide better spatial resolution along the line-of-sight and thus more accurate local densities compared to in situ multi pass optics [3].

7.3.4 Applications of EC-QCLs

7.3.4.1 Specific Issues Using EC-QCLs for Plasma Spectroscopy

Based on their intrinsic properties, in particular (i) wide spectral tunability, (ii) intrinsic narrow line width, and (iii) high output powers, EC-QCLs have high potential for applications in plasma diagnostics, in particular for the monitoring of multiple species including larger molecules characterized by broader spectral features [2].

Recently several cw EC-QCLs operating between 1305 and 2260 cm^{-1} (4.42–7.66 μm) have been tested as radiation sources for an absorption spectrometer focused on the analysis of physical and chemical phenomena in molecular plasmas.

It should be noted that stable operation of cw MHF EC-QCLs at short pulse lengths, e.g. on the ms time scale, is not feasible due to the very narrow mode spacing [61]. A reliable concentration determination using the Beer-Lambert law requires measuring the relative radiation fluxes entering and leaving the absorbing gas or plasma column. This also requires monitoring the detector signal without any radiation, i.e. in the laser-off phase. Therefore, a periodic interruption of the laser radiation is essential. Different methods of chopping the laser radiation can be used for this purpose, e.g. by mechanical choppers or shutters or by acousto-optic modulators (AOM).

An example of a spectrum using an AOM for chopping the laser radiation at a frequency of 5 kHz while tuning the laser over its entire wavenumber range is given in Fig. 7.23. The radiation passed through a reference gas cell containing CO at a pressure of 0.05 mbar. The relative intensities of the CO lines are given in the lower part of the figure. The inset shows a single line as representative of the whole spectrum. Optical fringes are clearly visible in the spectrum. Additionally,

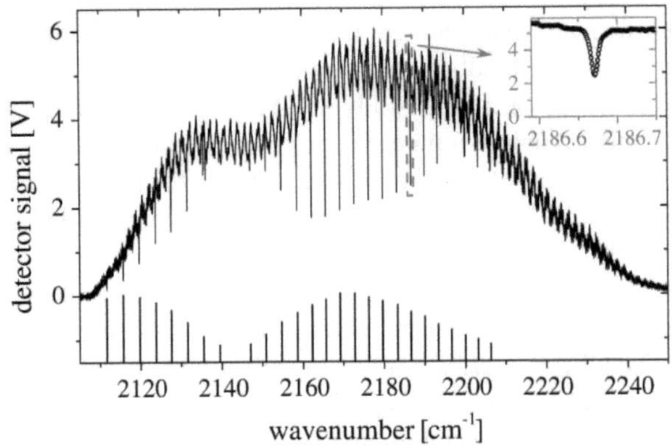

Fig. 7.23 Spectrum of an EC-QCL using stepper motor tuning and an AOM for chopping the laser radiation at a frequency of 5 kHz. The radiation passed through a reference gas cell (l = 10 cm) containing CO at a pressure of 0.05 mbar. Line positions and relative intensities of the CO lines are shown in the lower part of the diagram [16, 43, 63]. *Inset* representative single CO line

in this configuration the use of an optical isolator is very important to avoid back reflections of the laser radiation into the EC-QCL from any of the optical elements, which can lead to serious instabilities, in particular in the appearance of mode hops. The applicability of EC-QCLs for plasma diagnostic purposes has been tested using a three channel absorption spectrometer combined with a planar microwave (MW) plasma source (f = 2.45 GHz). An optical multi pass White cell arrangement has been installed inside the MW plasma source providing absorption lengths of up to 30 m for high sensitivity detection [62]. The discharge configuration in the planar MW plasma is well suited for spectroscopic observations, because plasma homogeneity can be achieved to a large extent for these long plasmas [53]. The schematic setup of the 3-channel EC-QCL spectrometer is shown in Fig. 7.25.

Multiple species detection has been feasible in low pressure Ar/N_2 microwave plasmas containing methane as hydrocarbon precursor. Using the direct absorption technique the evolution of the concentrations of CH_4, C_2H_2, HCN and H_2O have been monitored under different discharge conditions. An example of an absorption spectrum of an Ar/N_2/CH_4 MW plasma is presented in Fig. 7.24 showing absorption lines of the CH_4 precursor and the plasma reaction products C_2H_2 and HCN. Furthermore, in a pure He/Ar microwave plasma the wavelength modulation spectroscopy technique (WMS) has been applied for high sensitivity detection of transient plasma species. A typical example of WMS in a He MW discharge with a 3 % admixture of Ar at a pressure of p = 0.265 mbar and a power value of P = 2.7 kW is shown in Fig. 7.26. The signals of the measurement, reference and etalon channels are detected simultaneously. In Fig. 7.26a a clear absorption feature of a transient species, has been found at a spectral position of 1410.639 cm^{-1}. The sensitivity for detecting small fractional absorptions is estimated to be about 10^{-5}. In addition, the neutral gas temperature T_g has been calculated as a function of the discharge power based on the analysis of the profile of selected absorption lines.

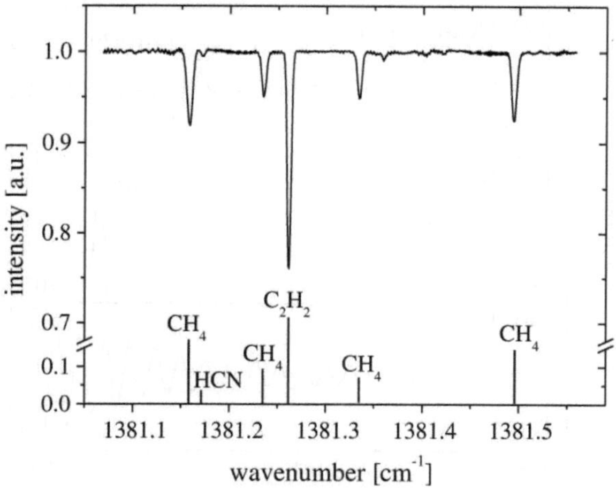

Fig. 7.24 Example of an absorption spectrum in an $Ar/N_2/CH_4$ MW plasma at p = 0.5 mbar and P = 0.2 kW showing absorption lines of CH_4, C_2H_2 and HCN at an absorption length of l = 30 m in the MW reactor shown in Fig. 7.25 [16, 63]

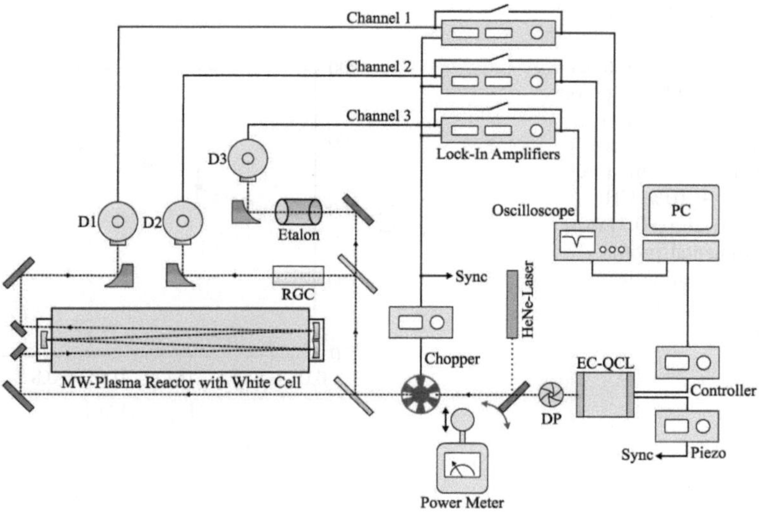

Fig. 7.25 Schematic arrangement of the 3-channel EC-QCL spectrometer in combination with a planar microwave (MW) plasma reactor equipped with an optical multi pass configuration of the White cell type [53, 62]. (*D1, D2, D3* HgCdTe detectors, *RGC* reference gas cell, *DP* iris) [16, 63]

The typical spectral line widths of the radiation of the EC-QCLs in this study were found to be in the range of 24–38 MHz. The increase of the gas temperature with the applied power in the MW discharge from about 400–700 K is depicted in Fig. 7.27. Another objective of this study was the analysis of possible saturation effects under low pressure conditions caused by the high power of the lasers. In

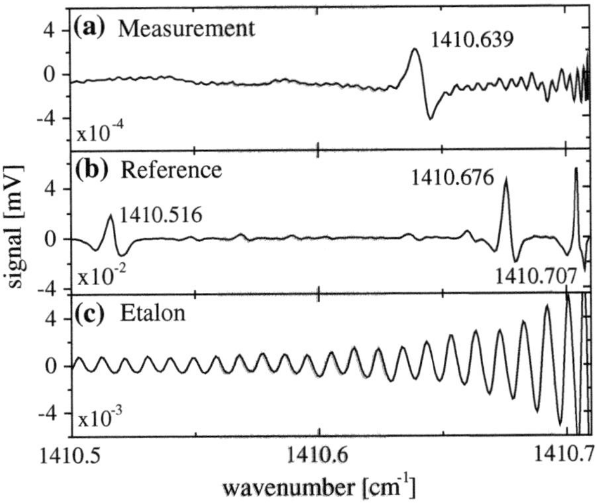

Fig. 7.26 Example of WMS in a He MW discharge with 3 % Ar admixture at $p = 0.265$ mbar and $P = 2.7$ kW, sub-diagram (**a**). The calibration of the wavenumber is based on absorption features of a C_2H_4 reference gas cell, sub-diagram (**b**) and an air spaced etalon (FSR: 0.01 cm^{-1}), sub-diagram, (**c**) see reference [16, 63]

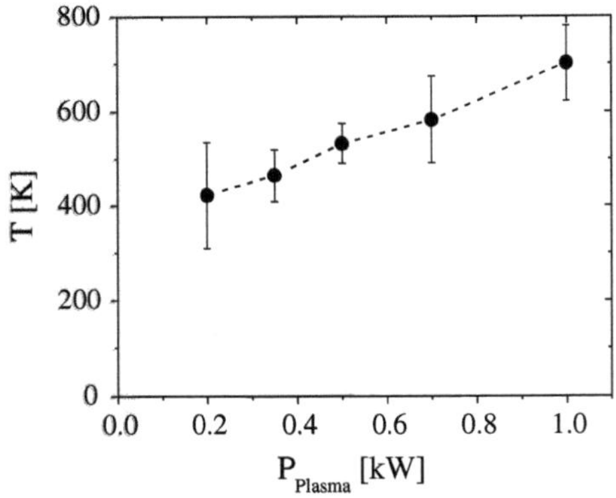

Fig. 7.27 Gas temperatures as a function of plasma power calculated from line profiles. Measurements were performed in a gas mixture of 10 % N_2 + 10 % CH_4 diluted in Ar at $p = 0.5$ mbar. The instrumental profile used for calculation was 0.00127 cm^{-1} (38.1 MHz) [16, 63]

direct absorption measurements under low pressure conditions no saturation effects were found while determining the concentrations of methane, acetylene and carbon monoxide. The absorption lines of these molecules covered line strengths between 10^{-19} to 10^{-22} cm molecule^{-1} and the output power of the EC-QCL was varied over several orders of magnitude [63].

In summary, the EC-QCLs used are advantageous for multiple species detection, in cases where time resolution is not important. When tuning the laser simply by the stepper motor over the entire spectral range the sensitivity suffers from relatively big fringes even in the mode hop free regions. Hence using a PZT for tuning through small spectral windows, where the influence of these fringes is less perturbing, is one of the preferred working protocols for plasma diagnostics. The high power values of up to several hundred mW on the one hand eliminate several problems in earlier spectroscopic approaches in plasma diagnostics, but reflection of light back into the EC-QCL causes mode instabilities. Possible non-linear effects should also be considered at such high power levels.

7.3.4.2 Plasma Chemical Study of a RF Discharge Containing Aluminum Tri-isopropoxide (ATI)

Metal organic precursors are frequently used in plasma processes for the deposition of metal oxide layers. The advantage of these precursors is that they have relatively low boiling points and can thus easily be introduced into a plasma. Typical precursors for PACVD include, for example, silane [52], hexamethyldisiloxane (HMDSO) [64, 65] and aluminium tri-isopropoxide (ATI). In contrast to silicon-based precursors, which have been widely studied in plasmas, relatively few publications exist about the use of aluminium-based precursors [66]. An important use of ATI is for the deposition of Al_xO_y layers in CVD processes [67]. Although ATI containing plasmas are also employed for deposition in PECVD processes [68] the associated chemical phenomena in these plasmas are far from being completely understood.

Recently the chemical processes in Ar and Ar/N_2 RF discharges containing ATI as the metal-organic precursor have been the focus of interest. Using MIR-LAS based on a cw EC-QCL arrangement the evolution of the concentrations of six stable molecules, CH_4, C_2H_2, C_2H_4, H_2O, HCN and HNO_3, has been monitored in the plasma. One of the objectives was to determine the main fragmentation pathways of the ATI molecule.

The investigations were performed in an asymmetric capacitively coupled RF discharge at a frequency of f = 13.56 MHz. The cylindrical vacuum chamber has a volume of about 50 dm^3 with a plain electrode and several ports for diagnostic interrogations. Argon and nitrogen were used as carrier gases and ATI as the metal-organic precursor. The ATI vapor was provided by a specially designed heating system. By variation of one of the parameters, (i) discharge power, (ii) total pressure or (iii) gas mixture, molecular concentrations of stable molecular products, CH_4, C_2H_2, C_2H_4, H_2O, HCN and HNO_3, were measured with an EC-QCL absorption spectrometer. The configuration of the spectrometer is shown in Fig. 7.28.

The set-up consists of (i) the radiation source, an EC-QCL, with a full tuning range of 1305–1445 cm^{-1}, providing a mode hop free range between 1345 and 1400 cm^{-1}at a maximum power of about 150 mW, (ii) a chopper, (iii) two standard HgCdTe detectors for measurement and reference paths and (iv) opto-mechanical elements for beam manipulation. The identification of lines and the determination of their absolute positions was carried out using well-documented reference gas spectra and an etalon with known free spectral range for interpolation. Absolute species concentrations are produced from the nonlinear least-squares fits of the measured absorption features so that external calibration is not required. A gas temperature of 300 K was used for concentration calculations since this is a suitable average value

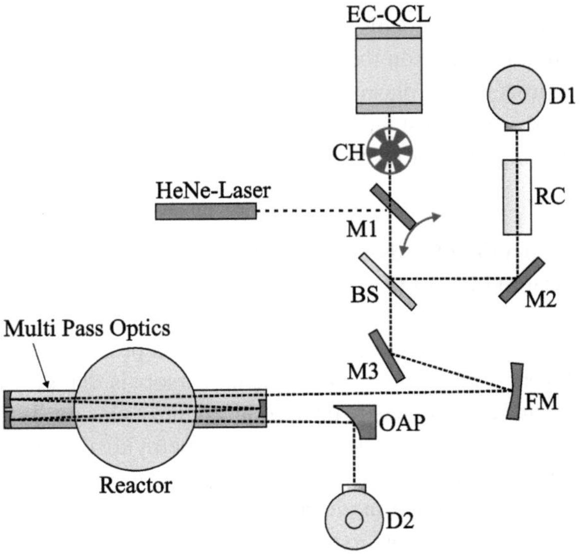

Fig. 7.28 Schematic of the EC-QCLAS spectrometer in combination with the RF plasma reactor. (*CH* chopper, *M1-M3* mirrors, *BS* beam splitter, *RC* reference gas cell, *D1, D2* HgCdTe detectors, *FM* focal mirror, *OAP* off axis paraboloid) [16, 69]

Fig. 7.29 Example of a spectrum of CH_4, C_2H_2 and HNO_3 absorption lines in an Ar/N_2/ATI plasma [69]

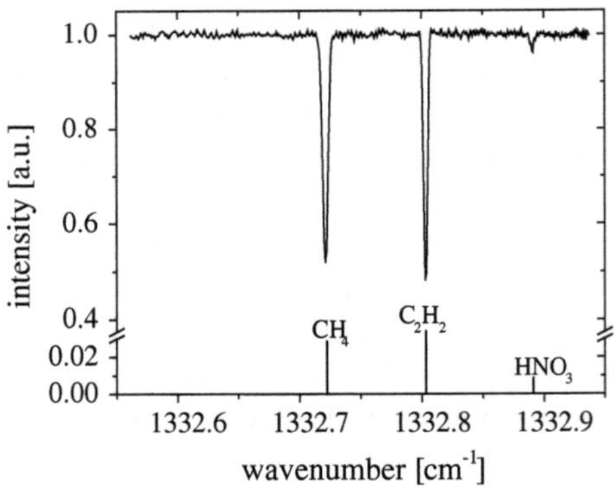

for the gas temperature over the line of sight, comprising the active zone over the discharge electrode and the afterglow surrounding it, and over the time period of the measurement.

Figure 7.29 shows a typical spectrum of CH_4, C_2H_2 and HNO_3 absorption lines recorded from an Ar/N_2/ATI plasma. The concentration of CH_4, C_2H_2, C_2H_4, H_2O, HCN and HNO_3, and their dependence on the discharge power in an Ar/N_2/ATI plasma with a mixing ratio of 1/1/1 at a total pressure of p = 6 Pa are shown in Fig. 7.30. It is interesting to note, that the concentrations of all molecules, except water, show a maximum at medium power values between 50 and 90 W. The decrease of concentration above 90 W may be explained by more effective colli-

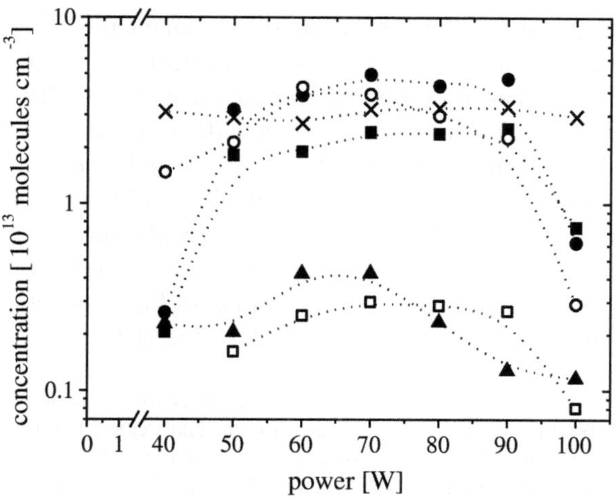

Fig. 7.30 Concentration dependence of CH_4 – ■, C_2H_2 – •, C_2H_4 – ▲, H_2O - ×, HCN - ○ and HNO_3 – ◇ on the discharge power in an $Ar/N_2/ATI$ plasma, p = 6 Pa, ratio $Ar/N_2/ATI = 1/1/1$ [69]

sions of electrons with nitrogen molecules, leading eventually to a limited fragmentation of ATI. The maximum concentration for $C_2H_2$was found to be about 5 x 10^{13} molecules cm^{-3}, followed by smaller concentrations of HCN, H_2O and CH_4. A second group of molecular species is formed by C_2H_4 and HNO_3 with concentrations of about one order of magnitude lower than the species of the former group.

In this study the absolute ground state concentrations of six stable species under a wide range of experimental conditions were determined by infrared laser absorption spectroscopy providing a link with the fragmentation behaviour of ATI in a capacitively coupled RF discharge. The final objective, to gain a better understanding of chemical processes occurring in plasmas containing metal-organic precursors for metal oxide PECVD deposition, can therefore be achieved by applying the EC-QCLs [16, 69].

7.4 Summary and Conclusions

A variety of phenomena in molecular non-equilibrium plasmas in which many short lived and stable species are produced have been successfully studied with absorption spectroscopy in the mid-infrared spectral range, with which the present chapter is concerned. It has been possible to determine absolute concentrations of ground states using spectroscopy thereby providing a link with chemical modeling of the plasma, the ultimate objective being a better understanding of the chemical and reaction kinetic processes occurring in the plasma and on surfaces. The other essential component needed to reach this objective was to determine physical parameters of the plasma, as for example, temperatures, degrees of dissociation, dynamics of reaction processes and phenomena involving plasma–surface interactions. The need for a better scientific understanding of plasma physics and chemistry, including the kinetics of transient molecules, has stimulated the application of several MIR-AS techniques. QCLAS has proven to be one of the most versatile techniques for studying molecular plasmas. More than 30 QCLAS studies in research and industry have been reported in the last few years. The recent availability of EC-QCL offers a promising new avenue for multi-component detection and for plasma diagnostics generally. Based on the recent development of advanced instrumentation including compact and robust QCLAS systems the spread of this method of high-resolution mid-infrared spectroscopy to industrial applications has rapidly materialized.

Acknowledgments This work was partly supported by (a) the German Research Foundation (DFG) within the framework of the Collaborative Research Centre Transregio 24 "Fundamentals of Complex Plasmas", (b) the Federal Ministry of Education and Research (BMBF), FKZ 13N7451/8 and by the Federal Ministry of Economics and Technology (BMWi), FKZ KF 0086503FK6, (c) the German Academic Exchange Service (DAAD) and EGIDE as part of the French–German PROCOPE Collaboration Program, and (d) the EPSRC with an award of an Advanced Research Fellowship.

The authors give sincere thanks to all present and former members of the laboratories involved in Greifswald, Palaiseau, Eindhoven, Oxford and Cambridge for permanent support and a stimulating working climate. In particular, the authors are indebted to all co-authors of former papers whose contributions made the present chapter possible.

Finally, the authors acknowledge the support given by Neoplas Control, Alpes Lasers, Nanoplus, Daylight Solutions and MGO Optical Solutions.

References

1. J. Röpcke, G. Lombardi, A. Rousseau, P.B. Davies, Plasma Sources Sci. Technol. **15**, S148 (2006)
2. R.F. Curl, F. Capasso, C. Gmachl, A.A. Kostorev, B. McManus, R. Lewicki, M. Pusharsky, G. Wysocki, F.K. Tittel, Chem. Phys. Lett. **487**, 1 (2010)
3. S. Welzel, F. Hempel, M. Hübner, N. Lang, P.B. Davies, J. Röpcke, Sensors **10**, 6861 (2010)
4. N. Lang, S. Zimmermann, B. Uhlig, M. Schaller, J. Röpcke, S.E Schulz, in *proceedings of the 30th ICPIG conference*, vol. B6 (University of Belfast, Belfast, 2011), p. 141
5. O. Guaitella, M. Hübner, S. Welzel, D. Marinov, J. Röpcke, A. Rousseau, Plasma Sources Sci. Technol. **19**, 045026 (2010)
6. J. Röpcke, P.B. Davies, F. Hempel, B.P. Lavrov, in *Low Temperature Plasmas—Fundamentals, Technologies and Techniques*, ed. by R. Hippler, H. Kersten, M. Schmidt, K.H. Schönbach. Emission and Absoption Spectroscopy, vol 1 (Wiley, Berlin, 2008), pp. 215–242
7. W. Demtröder, *Laserspektroskopie* (Springer-Verlag, Berlin, 1991)
8. G. Duxbury, in *Infrared Vibration-Rotation Spectroscopy; From Free radicals to the Infrared Sky* (Wiley, Chichester, 2000)
9. F.K. Tittel, D. Richte, A. Freed, in *Mid-Infrared Laser Applications in Spectroscopy*, ed. by I.T. Sorokina, K.L. Vodopyanov. Solid State Infrared Sources, (Topics Appl. Phys. **89**, 445) (Springer, Berlin, Heidelberg, 2003)
10. D.G. Lancaster, D. Richter, F.K. Tittel, Appl. Phys. B. **69**, 459 (1999)
11. D. Rehle, D. Leleux, M. Erdely, F.K. Tittel, M. Fraser, S. Friedfeld, Appl. Phys. B **72**, 947 (2001)
12. G. Berden, R. Engeln (eds.), in *Cavity Ring-Down Spectroscopy: Techniques and Applications*, ISBN 978-1-4051-7688-0, (Wiley, Berlin, 2009)

13. M.D. Wheeler, S.M. Newman, A.J. Orr-Ewing, M.N.R. Ashfold, J. Chem. Soc. Faraday Trans. **94**(3), 337 (1998)
14. M. Mazurenka, A.J. Orr-Ewing, R. Peverall, G.A.D. Ritchie, Annu. Rep. Prog. Chem., Sect. C **101**, 100 (2005)
15. S. Welzel, G. Lombardi, P.B. Davies, R. Engeln, D.C. Schram, J. Röpcke, J. Appl. Phys. **104**, 093115 (2008)
16. J. Röpcke, P.B. Davies, N. Lang, A. Rousseau, S. Welzel, J. Phys. D: Appl. Phys. **45**, 423001 (2012)
17. H.Y. Clark, L. Corner, W. Denzer, G. Hancock, A. Hutchinson, M. Islam, R. Peverall, G.A.D. Ritchie, Chem. Phys. Lett. **399**, 102 (2004)
18. W. Chen, J. Cousin, E. Poullet, J. Burie, D. Boucher, X. Gao, M.W. Sigrist, F.K. Tittel, C. R. Phys. **8**, 1129 (2007)
19. O. Tadanaga, T. Yanagawa, Y. Nishida, H. Miyazawa, K. Magari, M. Asobe, H. Suzuki, Appl. Phys. Lett. **88**, 061101 (2006)
20. L.H. Deng, X.M. Gao, Z.S. Cao, W.D. Chen, Y.Q. Yuan, W.J. Zhang, Z.B. Gong, Opt. Commun. **281**, 1686 (2008)
21. G. D'Amico, G. Pesce, G. Rusciano, A. Sasso, Opt. Laser Eng. **37**, 481 (2002)
22. J.H. van Helden, G. Hancock, R. Peverall, G.A.D. Ritchie, J. Phys. D: Appl. Phys. **44**, 125202 (2011)
23. P. Hunter, S.T. Oyama, in *Control of Volatile Organic Compound Emissions : Conventional and Emerging Technologies* (Wiley, New York, 2000)
24. L.A. Rosocha, in *Removal of Volatile Organic Compounds (VOCs) by Atmospheric-Pressure Dielectric-Barrier and Pulsed-Corona Electrical Discharges* ed. by E.M. Velduizen. Electrical Discharges for Environmental Purposes. (NOVA Science Publishers, New York, 1999)
25. M. Hübner, O. Guaitella, A. Rousseau, J. Röpcke, J. Appl. Phys. **144**, 033301 (2013)
26. J. Perrin, in *Sources and Growth of Particles Dusty Plasmas*, ed. by A. Bouchoule (Wiley, New York, 1999)
27. U. Kortshagen, U. Bhandarkar, Phys. Rev. E **60**, 887 (1999)
28. E. Kovacevic, I. Stefanovic, J. Berndt, J. Winter, J. Appl. Phys. **93**, 2924 (2003)
29. G.D. Stancu, J. Röpcke, P.B. Davies, J. Chem. Phys. **122**, 014306 (2005)
30. F. Hempel, D. Lopatik, B. Sikimic, I. Stefanovic, J. Winter, J. Röpcke, Plasma Sources Sci. Technol. **21**, 055001 (2012)
31. R.F. Kazarinov, R.A. Suris, Sov. Phys. Semicond. **5**, 707 (1971)
32. J. Faist, F. Capasso, D.L. Sivco, C. Sirtori, A.L. Hutvhinson, A.Y. Cho, Science **264**, 553 (1994)
33. M. Hübner, S. Welzel, D. Marinov, O. Guaitella, S. Glitsch, A. Rousseau, J. Röpcke, Rev. Sci. Instr. **82**, 093102 (2011)
34. A. Hugi, R. Maulini, J. Faist, Semicond. Sci. Technol. **25**, 083001 (2010)
35. R.J. Walker, R.J. Van Helden, G.A.D. Richie, Chem. Phys. Lett. **501**, 20 (2010)
36. A. Karpf, G.N. Rao, Appl. Optics **49**, 1406 (2010)
37. R. Furstenberg, C.A. Kendziora, J. Stepnowski, S.V. Stepnowski, M. Rake, M.R. Papantonakis, V. Nguyen, G.K. Hubler, R.A. McGill, Appl. Phys. Lett. **93**, 224103 (2008)
38. A. Karpf, G.N. Rao, Appl. Optics **48**, 408 (2009)
39. R. Lewicki, J.H. Doty III, R.F. Curl, F.K. Tittel, G. Wysocki, PNAS **106**, 12587 (2009)
40. M. Capitelli, C.M. Ferreira, B.F. Gordiets, A.I. Osipov, *Plasma Kinetics in Atmospheric gases. Springer Series on Atomic, Optical, and Plasma Physics*, vol. 31 (Springer, Berlin, 2000)
41. J. Röpcke, P.B. Davies, F. Hempel, B.P. Lavrov, *Emission and Absorption Spectroscopy in Low Temperature Plasmas—Fundamentals, Technologies and Techniques*, vol. 1 (Wiley-VCH, Berlin, 2008)
42. M.G. Allen, Meas. Sci. Technol. **9**, 545 (1998)
43. L.S. Rothman, D. Jacquemart, A. Barbe, D.C. Benner, M. Birk, L.R. Brown, M.R. Carleer, C. Chackerian, K. Chance, L.H. Coudert, V. Dana, V.M. Devi, J.M. Flaud, R.R. Gamache, A. Goldman, J.M. Hartmann, K.W. Jucks, A.G. Maki, J.Y. Mandin, S.T. Massie, J. Orphal, A. Perrin, C.P. Rinsland, M.A.H. Smith, J. Tennyson, R.N. Tolchenov, R.A. Toth, P. Varanasi, G. Wagner, J. Quant. Spectrosc. Radiat. Transf. **96**, 139 (2005)

44. M. Hübner, D. Marinov, O. Guaitella, A. Rousseau, J. Röpcke, Meas. Sci. Technol. **23**, 115602 (2012)
45. S. Welzel, S. Stepanov, J. Meichsner, J. Röpcke, J. Phys. D: Appl. Phys. **43**, 124014 (2010)
46. N. Lang, F. Hempel, S. Strämke, J. Röpcke, Jpn. J. Appl. Phys. **50**, 08JB04 (2011)
47. F. Hempel, N. Lang, H. Zimmermann, S. Strämke, J. Röpcke, Meas. Sci. Technol. **21**, 085703 (2010)
48. N. Lang, J. Röpcke, A. Steinbach, S. Wege, IEEE Trans. Plasma Sci. **37**, 2335 (2009)
49. N. Lang, J. Röpcke, S. Wege, A. Steinbach, Eur. Phys. J.: Appl. Phys. **49**, 13110 (2010)
50. F. Hempel, V. Artyushenko, F. Weichbrodt, J. Röpcke, J. Phys.: Conf. Ser. **157**, 12003 (2009)
51. G.D. Stancu, N. Lang, J. Röpcke, M. Reinicke, A. Steinbach, S. Wege, Chem. Vap. Deposition **13**, 351 (2007)
52. R. Bartlome, A. Feltrin, C. Ballif, Appl. Phys. Lett. **94**, 201501 (2009)
53. F. Hempel, P.B. Davies, D. Loffhagen, L. Mechold, J. Röpcke, Plasma Sources Sci. Technol. **12**, S98 (2003)
54. M.T. McCulloch, E.L. Normand, N. Langford, G. Duxbury, D.A. Newnham, J. Opt. Soc. Am. B **20**, 1761 (2003)
55. D.Z. Anderson, J.C. Frisch, C.S. Masser, Mirror reflectometer based on optical cavity decay time. Appl. Optics **23**, 1238–1245 (1984)
56. A. O'Keefe, D.A,G. Deacon, Rev. Sci. Instrum. **59**, 2544 (1988)
57. D.D. Nelson, J.B. McManus, S.C. Herndon, J.H. Shorter, M.S. Zahniser, S. Blaser, L. Hvozdara, A. Muller, M. Giovannini, J. Faist, Opt. Lett. **31**, 2012 (2006)
58. S.W. Sharpe, J.F. Kelly, J.S. Hartman, C. Gmachl, F. Capasso, D.L. Sivco, J.N. Baillargeon, A.Y. Cho, Opt. Lett. **23**, 1396 (1998)
59. J.B. McManus, D.D. Nelson, S.C. Herndon, J.H. Shorter, M.S. Zahniser, S. Blaser, L. Hvozdara, A. Muller, M. Giovannini, J. Faist, Appl. Phys. B-Lasers Opt. **85**, 235 (2006)
60. R.A.B. Zijlmans, O. Gabriel, S. Welzel, F. Hempel, J. Röpcke, R. Engeln, D.C. Schram, Plasma Sources Sci. Technol. **15**, 564 (2006)
61. Daylight Solutions, User Manual, Rev. A (2009). http://www.daylightsolutions.com
62. J.U. White, J. Opt. Soc. Am. **32**, 285 (1942)
63. D. Lopatik, N. Lang, U. Macherius, H. Zimmermann, J. Röpcke, Meas. Sci. Technol. **23**, 115501 (2012)
64. M. Hähnel, V. Brüser, H. Kersten, Plasma Process. Polym. **6**, 629 (2007)
65. S. Bornholdt, M. Wolter, H. Kersten, Eur. Phys. J. D **60**, 653 (2010)
66. H. Kersten, G. Thieme, M. Fröhlich, D. Bojic, D.H. Tung, M. Quaas, H. Wulff, R. Hippler, Pure Appl. Chem. **77**, 415 (2005)
67. S. Blittersdorf, N. Bahlawane, K. Kohse-Höinghaus, B. Atakan, J. Müller, Chem. Vap. Deposition **9**, 194 (2003)
68. H. Kersten, R. Wiese, G. Thieme, M. Fröhlich, A. Kopitov, D. Bojic, F. Scholze, H. Neumann, M. Quaas, H. Wulff, R. Hippler, New J. Phys. **5**(93), 1 (2003)
69. D. Lopatik, S. Niemietz, M. Fröhlich, J. Röpcke, H. Kersten, Contrib. Plasma Phys. **52**, 864 (2012)
70. G.N. Rao, A. Karpf, Appl. Opt. **50**, A100 (2011)

Chapter 8
Surface Electrons at Plasma Walls

Rafael Leslie Heinisch, Franz Xaver Bronold and Holger Fehske

Abstract In this chapter we introduce a microscopic modelling of the surplus electrons on the plasma wall which complements the classical description of the plasma sheath. First we introduce a model for the electron surface layer to study the quasi-stationary electron distribution and the potential at an unbiased plasma wall. Then we calculate sticking coefficients and desorption times for electron trapping in the image states. Finally we study how surplus electrons affect light scattering and how charge signatures offer the possibility of a novel charge measurement for dust grains.

8.1 Introduction

When a macroscopic object is brought into contact with an ionised gas its surface is exposed to a fast influx of electrons. Due to the collection of electrons it acquires a negative charge. The wall electrons give rise to a repulsive Coulomb potential which reduces the further influx of electrons until—in quasi-stationarity—electron and ion flux onto the surface balance each other. This coincides with the formation of the plasma sheath, an electron depleted region adjacent to the wall.

The traditional modelling of the plasma wall focuses mainly on the plasma sheath which is a macroscopic manifestation of the charge transfer at the plasma boundary. Electrons and ions are usually assumed to annihilate instantly at the wall. Thus the wall is considered to be a perfect absorber for the charged species of the plasma. This boundary condition implicitly assumes that the charge transfer at the plasma wall

R. L. Heinisch (✉) · F. X. Bronold · H. Fehske
Institut für Physik, Universität Greifswald, 17487 Greifswald, Germany
e-mail: heinisch@physik.uni-greifswald.de

F. X. Bronold
e-mail: bronold@physik.uni-greifswald.de

H. Fehske
e-mail: fehske@physik.uni-greifswald.de

M. Bonitz et al. (eds.), *Complex Plasmas*, Springer Series on Atomic, Optical, and Plasma Physics 82, DOI: 10.1007/978-3-319-05437-7_8,

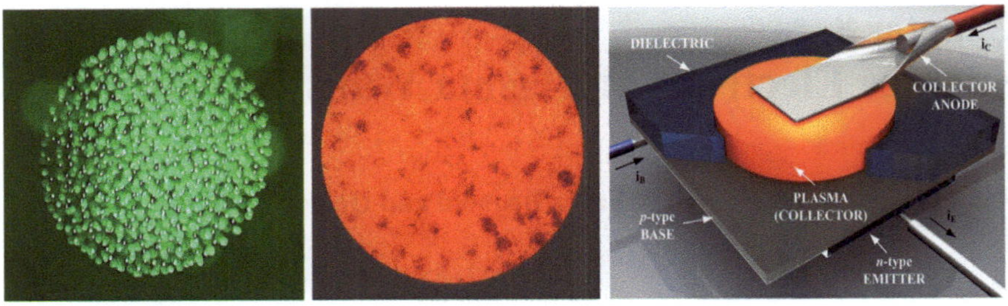

Fig. 8.1 Three novel bounded plasmas where surface charges matter. *Left* The particle charge controls interactions between dust grains forming a spherical dust ball. *Middle* Surface charges (negative charges in *blue* over neutral background in *red*) pin the location of filaments for a filamentary dielectric barrier discharge. *Right* The plasma serves as a collector in a plasma transistor (Reprinted with permission from [2]. Copyright [2010], AIP Publishing LLC)

occurs on time and length scales too short to be of significance for the development of the discharge. It is thus not necessary to track them [1]. Improvements to the perfect absorber condition characterise the wall by an electron-ion recombination coefficient and secondary electron emission coefficients for various impacting species. Still, replacing the electron kinetics by an ad-hoc boundary condition does not allow for a description of the charge transfer at the surface.

While this approach may be sufficient for large discharges with small surface to volume ratios it becomes questionable for the modelling of small discharges with large surface to volume ratios. Indeed, the nature and build-up of surface charges have received increasing attention in various novel set-ups of bounded plasmas (illustrated in Fig. 8.1). In dusty plasmas [3–5], for instance, the charge soaked up by the dust particles controls the overall charge balance of the discharge [6]. Moreover, the particle charge controls the interaction between dust particles and between dust particles and external electromagnetic fields. The particle charge is thus a central quantity of interest which should be known as precisely as possible. In dielectric barrier discharges [7–12] or microplasmas [13, 14] surface charges determine the spatio-temporal structure of the discharge [15]. Surface charges, for instance, may pin the filaments in the filamentary mode of the dielectric barrier discharge and give the discharge a memory across several cycles. Lastly, in solid-state based microdischarges [16, 17], the miniaturisation of the discharge increases the surface to volume ratio to such an extent that the (biased) plasma wall becomes an integral part of the discharge. The charge transfer from the plasma into the solid and vice versa may even be used explicitly, as for instance in the plasma transistor [2].

While the most prominent effect of charge transfer at the plasma wall is the plasma sheath the microscopic processes responsible for it cannot be captured in any sheath model which is solely based on the long range Coulomb potential and classical mechanics. Inspired by Emeleus and Coulter [18] as well as Behnke and coworkers [7, 19] who envisaged a surface plasma of ions and electrons coupled to the bulk plasma by phenomenological rate equations characterised by sticking coefficients, residence times, and recombination coefficients we consider the build-up, distribution

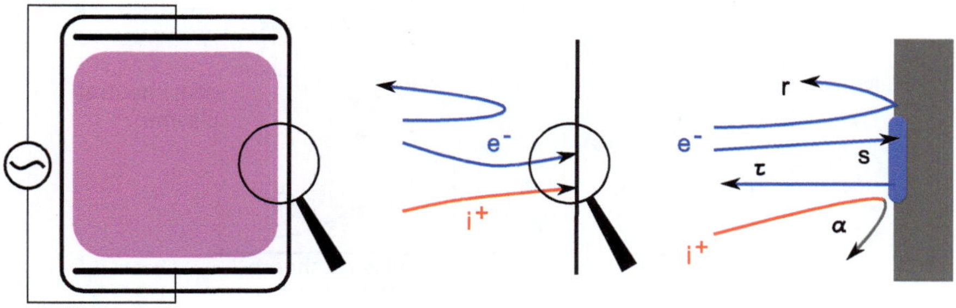

Fig. 8.2 Illustration of the plasma and its boundary. The bulk plasma (*left*) is characterised by quasi-neutrality. In the sheath (*middle*) ions outweigh electrons, quasi-neutrality does not hold and a potential gradient is formed. Note that the bulk as well as the sheath have macroscopic extent and can be described classically. The microscopic processes at the surface (*right*), for instance electron reflection, sticking or desorption as well as electron ion recombination occur on a length scale smaller than the shortest collision length of the plasma and have to be described by quantum mechanics

and release of surface electrons from a surface physics perspective. A sketch of the wall bound-processes which cannot be described in a sheath model is given by Fig. 8.2. The description of the plasma boundary has thus to be augmented by an interface region—the electron surface layer (ESL)—which encompasses not only the Coulomb potential but also short ranged surface potentials. This gives a framework for identifying electron trapping sites, determining the electron distribution and the potential across the interface in quasi-stationarity and calculating electron sticking coefficients. Moreover it serves us to discuss the optical properties of the electron adsorbate on a dust grain.

In the following section we will introduce our model for the ESL for an unbiased plasma wall. In Sect. 8.3 we will calculate sticking coefficients and desorption times, that are average residence times at the surface, for the case where electron trapping takes place in the image states in front of the surface. Finally in Sect. 8.4 we apply our microscopic model to the electrons on a dust particle in order to study charge signatures in light scattering by a spherical dust particle and propose thereby a novel optical charge measurement.

8.2 Electron Surface Layer

After the initial charge-up is completed the plasma wall carries a negative charge. In this section we extend the modelling of the plasma boundary to the region inside the solid and calculate the plasma-induced modifications of the potential and charge distribution of the surface within a quasi-static model. Although knowing the potential and charge distribution across the interface may not be of particular importance for present day technological plasmas, it is of fundamental interest from an interface

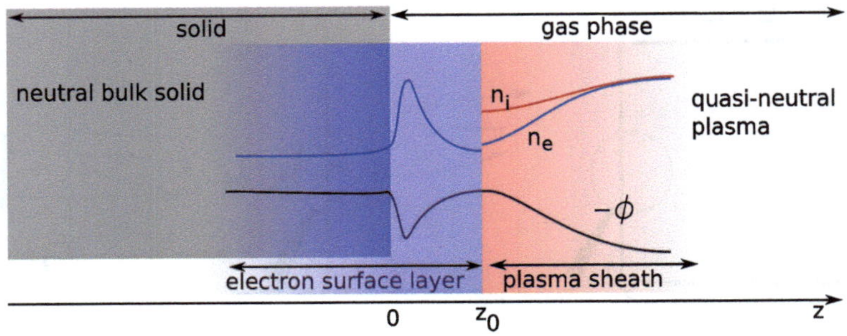

Fig. 8.3 Qualitative sketch of the charge double layer formed at a plasma wall. The electrons depleted from the plasma sheath are accumulated in the ESL. Note that the boundary between sheath and ESL is located in front of the crystallographic interface

physics point of view. In addition, considering the plasma wall as an integral part of the plasma sheath may become critical when the miniaturisation of solid-state-based plasma devices continues.

The current modelling of the plasma wall focuses on the plasma sheath and gives the potential as well as the electron and ion densities across that region. Our model for the ESL extends this into the solid [20]. Specifically, we calculate the potential as well as the distribution of wall bound electrons. Moreover, we give the spatial position, width and chemical potential of the surplus electrons. For this we use a graded interface potential [21] to interpolate between the sheath and the wall potential and distribute the surplus electrons making up the wall charge in this potential under the assumption that at quasi-stationarity they are thermalised with the wall [22].

As depicted in Fig. 8.3, we consider an ideal, planar interface at $z = 0$ with the dielectric occupying the half-space $z < 0$ and the discharge occupying the half-space $z > 0$. At the moment we focus on the physical principles controlling the electronic properties of the plasma-wall interface. Chemical contamination and structural damage due to the burning gas discharge are discarded. In the model we propose the wall charge to be treated as an ESL, which is an interface specific electron distribution thermalised with the solid stretching from the plasma sheath over the crystallographic interface into the bulk of the dielectric.

The boundary between the ESL and the plasma sheath at z_0 is located in front of the surface. It is the position where the attractive force due to the surface potential ϕ_{surf} and the repulsive force due to the sheath potential ϕ_{sheath} balance each other. Thus, z_0 is given by

$$\phi'_{\text{surf}}(z_0) + \phi'_{\text{sheath}}(z_0) = 0. \tag{8.1}$$

For $z < z_0$ an electron is attracted to the surface and thus contained in the ESL while for $z > z_0$ it is repelled back into the plasma. The position z_0 is an effective wall for plasma electrons and ions at which the flux balance between electron flux j_e and ion flux j_i has to be fulfilled. In the following we use for simplicity the perfect absorber model $j_e = j_i$. The effective wall position also gives the distance to the surface at which the description of the plasma sheath based on the long-range potential breaks down as the short-range surface potential becomes dominant. On the solid side, for $z < 0$, the ESL is bounded because of the shallow potential well formed by the restoring force from the positive charge in the plasma sheath.

Within this quasi-static model the electrons missing in the sheath are accumulated on the surface. For the construction of our one-dimensional interface model we need the total number per unit area of missing sheath electrons (that is, the total surface density of missing sheath electrons) as a function of the wall potential because it is this number of electrons which can be distributed across the ESL. Hence, we require a model for the plasma sheath.

For simplicity, we use a collisionless sheath model [1], more realistic sheath models [1, 23, 24] make no difference in principle. In the collisionless sheath electrons are thermalised, that is, the electron density $n_e = n_0 \exp(e\phi/k_B T_e)$, with ϕ the potential, n_0 the plasma density and T_e the electron temperature. The ions enter the sheath with a directed velocity v_{i0} and satisfy a source-free continuity equation, $\mathrm{d}(n_i v_i)/\mathrm{d}z = 0$, implying $n_i v_i = n_0 v_{i0}$, and an equation of motion $M v_i \mathrm{d}v_i/\mathrm{d}z = -e\mathrm{d}\phi/\mathrm{d}z$, with n_i the ion density, and M the ion mass. The potential ϕ satisfies Poisson's equation $\mathrm{d}^2\phi/\mathrm{d}z^2 = -4\pi e(n_i - n_e)$. Thus, the governing equations for the collisionless plasma sheath are [1]

$$v_i \frac{\mathrm{d}v_i}{\mathrm{d}z} = -\frac{e}{M}\frac{\mathrm{d}\phi}{\mathrm{d}z} \tag{8.2}$$

and

$$\frac{\mathrm{d}^2}{\mathrm{d}z^2}\phi = -4\pi e n_0 \left[\frac{v_0}{v_i} - \exp\left(\frac{e\phi}{k_B T_e}\right)\right]. \tag{8.3}$$

Introducing dimensionless variables $\eta = -e\phi/(k_B T_e)$, $\xi = z/\lambda_D$ and $u = v_i/c_s$, where $\lambda_D = \sqrt{k_B T_e/4\pi n_0 e^2}$ and $c_s = \sqrt{k_B T_e/M}$, equations (8.2) and (8.3) become

$$uu' = \eta' \quad \text{and} \quad \eta'' = \frac{u_0}{u} - \exp(-\eta). \tag{8.4}$$

In the ESL model the plasma occupies not the whole half space $z > 0$ but only the portion $z > z_0$ (see Fig. 8.3). Integrating the first equation gives $\left(u = -\sqrt{2\eta + u_0^2}\right)$, where $u_0 = v_{i0}/c_s$ is the reduced velocity of ions entering the sheath, so that the second equation becomes

$$\eta'' = -\frac{u_0}{\sqrt{2\eta + u_0^2}} - \exp(-\eta). \tag{8.5}$$

Using the boundary condition that the potential and the field vanish far inside the plasma, that is, $\eta \to 0$ and $\eta' \to 0$ for $\xi \to \infty$, (8.5) can be integrated once and we obtain

$$\eta' = -\sqrt{-2u_0\sqrt{2\eta + u_0^2} + 2\exp(-\eta) + 2u_0\sqrt{u_0^2} - 2}. \tag{8.6}$$

For ions entering the sheath with the Bohm velocity $u_0 = -1$. The field at the effective wall as a function of the wall potential $\eta_w = \eta(\xi_0)$ is then given by

$$\eta_w' = -\sqrt{2\sqrt{2\eta_w + 1} + 2\exp(-\eta_w) - 4}. \tag{8.7}$$

The total surface density of ions in the sheath equals the total surface density of electrons in the ESL N. It can be related by Poisson's equation to the field at the wall and reads

$$N = \int_{z_0}^{\infty} \mathrm{d}z(n_i - n_e) = -\frac{1}{4\pi e}\int_{z_0}^{\infty} \mathrm{d}z \frac{\mathrm{d}^2\phi}{\mathrm{d}z^2} = \frac{1}{4\pi e}\frac{\mathrm{d}\phi}{\mathrm{d}z}(z_0) = -n_0\lambda_D\eta_w'. \tag{8.8}$$

Combining (8.8) and (8.7) gives the total surface density of electrons to be inserted into the ESL as a function of the wall potential.

The wall potential itself is determined by the flux balance condition, $j_e = j_i$, which, in the ESL model, is assumed to be fulfilled at $z = z_0$. Using the Bohm flux for the ions and the thermal flux for the electrons,

$$j_i = n_0\sqrt{\frac{k_B T_e}{M}}, \qquad j_e = \frac{1}{4}n_0\sqrt{\frac{8k_B T}{\pi m_e}}\exp\left(\frac{e\phi}{k_B T_e}\right), \tag{8.9}$$

the wall potential is given by [1]

$$\eta_w = \frac{1}{2}\ln\left(\frac{M}{2\pi m_e}\right), \tag{8.10}$$

that is,

$$\phi_w = -\frac{k_B T_e}{2e}\ln\left(\frac{M}{2\pi m_e}\right). \tag{8.11}$$

In the collisionless sheath model the wall potential depends only on the electron temperature and the ion to electron mass ratio.

We can now move on to the distribution of surplus electrons in the surface layer. It is primarily determined by the potential in the interface region surrounding an ideal dielectric interface. We first single out the relevant microscopic potentials at the interface (illustrated in Fig. 8.4) and then describe an effective model for them (Fig. 8.5).

Far away from the surface the potential is given by the repulsive sheath potential. Closer to the surface an attractive surface potential, the image potential, which stems from the dielectric mismatch at the interface, becomes dominant. Classically the image potential is given by the expression [25]

$$\phi_{\mathrm{im}}(z) = \frac{\varepsilon_0 - 1}{4(\varepsilon_0 + 1)}\frac{e}{z} \tag{8.12}$$

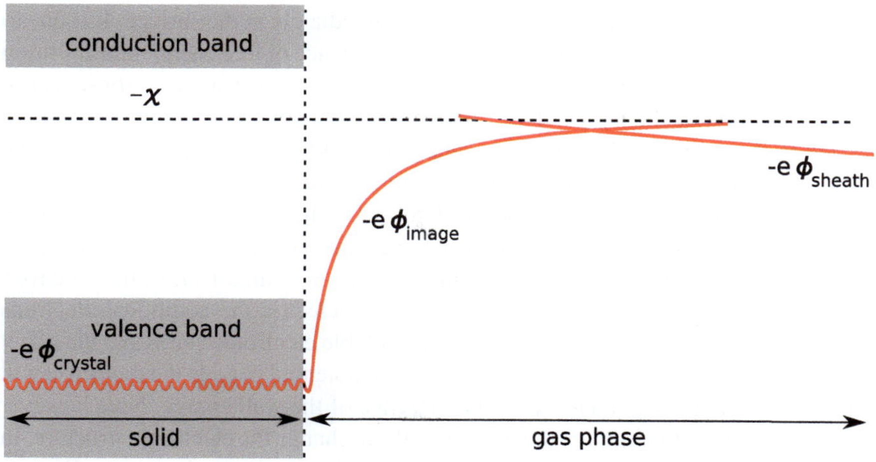

Fig. 8.4 Microscopic potentials at the interface between a plasma and a dielectric wall. Shown is the microscopic crystal potential merging with the image potential and the sheath potential. Inside the dielectric the crystal potential gives rise to a band structure. Reproduced from [20]. Copyright (2012) by the American Physical Society

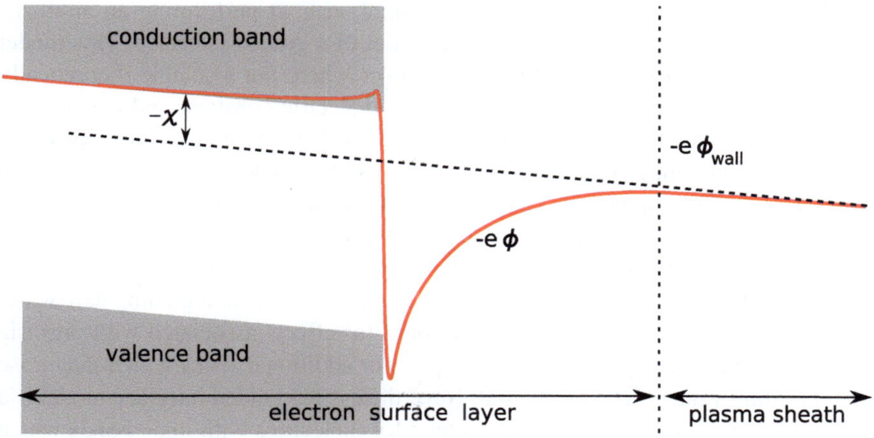

Fig. 8.5 Effective potential for the graded interface on which the model of an ESL is based. It encompasses the image potential as well as the potential offset at the interface. Reproduced from [20]. Copyright (2012) by the American Physical Society

where ε_0 is the static dielectric constant. This expression is valid far away from the wall. Closer to the wall the $1/z$ shape is modified as the potential merges continuously with the crystal potential in the bulk of the dielectric. Even more important inside the dielectric is the band structure. For a dielectric the valence band is full while the conduction band is completely empty. Thus surplus electrons can populate the conduction band and the conduction band minimum acts as a long range potential for the electrons. In addition to the image potential, the surface potential comprises an offset due to a charge double layer formed immediately at the surface. It is due to electron density leaking out of the solid as a consequence of free energy minimisation or to relaxation and polarisation of the atomic bonds at the surface as a consequence of the truncation of the lattice. This offset of the conduction band minimum at the surface to the potential in front of it is characterised by the electron affinity of the solid χ. For $\chi < 0$ the conduction band minimum lies above the potential just in front of the solid, for $\chi > 0$ it lies below it. So far we have considered an uncharged surface. This raises the question whether the electronic structure of the surface is changed by the surplus electrons from the plasma. In comparison to the electrons responsible for the chemical binding within the dielectric the additional electrons coming from the plasma are only a few. The available electronic states and the offset of the energy bands in the bulk with respect to the potential outside the dielectric will thus not be changed significantly by the presence of the wall charge. Note however that a chemical modification of the plasma wall can change the electronic structure. In particular the electron affinity is susceptible to surface coating and adatoms. Surface termination by elements with small electronegativity, for instance hydrogen, may induce a negative electron affinity [26] while elements with larger electronegativity, for instance oxygen, can lead to a positive electron affinity [27].

From these microscopic potentials we construct an effective surface potential. To obtain a realistic image and offset potential without performing an atomistically accurate calculation we employ the model of a graded interface. This model is parametrised by experimentally measured (or where not available theoretically calculated) values for the electron affinity, the dielectric constant, and the conduction band effective mass. First proposed by Stern [21] to remove the unphysical singularity of the image potential at $z = 0$ and later extended to potential offsets at semiconductor heterojunctions [28] it assumes the smooth variation of parameters that change abruptly at the interface over a length on the order of the lattice constant. For our purposes we assume a sinusoidal interpolation of the dielectric constant, offset of the long range potential and effective electron mass with a grading half-width of $d = 5$Å [20]. The model does not account for effects associated with intrinsic surface states (Shockley or Tamm states [29]) or additional states which may arise from the short-range surface potential. Nevertheless the graded interface model is a reasonable description of the surface, suited for dielectrics with ionic bonds which typically have no intrinsic surface states.

The total surface potential

$$\phi_{\text{surf}}(z) = \phi_{\text{im}}(z) + \phi_{\text{offset}}(z) \tag{8.13}$$

comprises the graded image and offset potential. It is continuous across the crystallographic interface at $z = 0$ and enables us thereby to also calculate a smoothly varying electron distribution in the ESL.

Using (8.1) we can now determine the position z_0 of the effective wall, that is, the maximum extent of the ESL on the plasma side. The derivative of the surface potential is $\phi'_{\text{surf}} = \phi'_{\text{offset}} + \phi'_{\text{im}}$. Due to the relatively weak field in the sheath compared to the image force, the boundary z_0 will be so far away from the interface that ϕ'_{offset}

vanishes and the image potential obeys (8.12). Thus, the boundary between the ESL and the plasma sheath is given by

$$z_0 = \sqrt{\frac{(\varepsilon_0 - 1)e}{4(\varepsilon_0 + 1)\phi'_w}} \tag{8.14}$$

with $\phi'_w = -(k_B T_e \eta'_w)/(e\lambda_D)$, and η'_w given by (8.7).

We can now turn to the distribution of the plasma-supplied excess electrons in the ESL. For this quasi-static model we assume that the wall-bound surplus electrons are in thermal equilibrium with the wall. Thus, their distribution minimises the grand canonical potential and satisfies Poisson's equation. Inspired by Tkharev and Danilyuk [22] we apply density functional theory [30, 31] to the graded interface. For the purpose of this exploratory calculation we will use density functional theory in the local approximation. Quite generally, the grand canonical potential of an electron system in an external potential $V(\mathbf{r})$ is given in the local approximation by

$$\Omega = \int V(\mathbf{r})n(\mathbf{r})\mathrm{d}\mathbf{r} - \frac{e}{2}\int \phi_C(\mathbf{r})n(\mathbf{r})\mathrm{d}\mathbf{r} + G[n] - \mu \int n(\mathbf{r})\mathrm{d}\mathbf{r}, \tag{8.15}$$

where $G[n]$ is the grand canonical potential of the homogeneous system with density $n(\mathbf{r})$ and the Coulomb potential is determined by

$$\nabla\left(\varepsilon(\mathbf{r})\nabla\phi_C(\mathbf{r})\right) = 4\pi e n(\mathbf{r}). \tag{8.16}$$

The ground state electron density minimises Ω, that is, it satisfies

$$V(\mathbf{r}) - e\phi_C(\mathbf{r}) + \mu^h(n) - \mu = 0\,, \tag{8.17}$$

where $\mu^h(n) = \delta G[n]/\delta n$ is the chemical potential for the homogeneous system.

In our one-dimensional graded interface model (8.17) reduces to

$$-e\phi(z) + \mu^h(z) - \mu = 0, \tag{8.18}$$

where $\mu^h(z) \equiv \mu^h(n(z), T)$, and the electrostatic potential

$$\phi(z) = \phi_{\mathrm{surf}}(z) + \phi_C(z) \tag{8.19}$$

consists of the potential of the bare surface given by (8.13) and the internal Coulomb potential which satisfies Poisson's equation

$$\frac{\mathrm{d}}{\mathrm{d}z}\left(\varepsilon(z)\frac{\mathrm{d}}{\mathrm{d}z}\phi_C(z)\right) = 4\pi e n(z) \tag{8.20}$$

with the graded dielectric constant $\varepsilon(z)$. The boundary conditions $\phi_C(z_0) = \phi_w$ and $\phi'_C(z_0) = \phi'_w$ guarantee continuity of the potential at z_0 and include the restoring force from the positive charge in the sheath. Note that the Coulomb potential derived from this equation already includes the attraction of an electron to the image of the charge distribution. The image potential contains only the self-interaction of an electron to its own image.

For the functional relation $\mu^h(z) \equiv \mu^h(n(z), T)$ we take the expression adequate for a homogeneous, non-interacting, non-degenerate electron gas

$$n(z) = \frac{1}{\sqrt{2}} \left(\frac{m(z) k_B T}{\pi \hbar^2} \right)^{\frac{3}{2}} e^{\frac{\mu^h(z)}{k_B T}} \tag{8.21}$$

where $m(z)$ is the graded effective electron mass. This is justified because the density of the excess electrons is rather low and the temperature of the surface is rather high, typically a few hundred Kelvins.

For the calculation of the quasi-stationary distribution of surplus electrons (8.18) and (8.20) have to be solved iteratively (until μ is stationary) subject to the additional constraint $\int_{z_s}^{z_0} \mathrm{d}z\, n(z) = N$ which guarantees charge neutrality between the ESL and the plasma sheath. The position $z_s < 0$ is a cut-off which has to be chosen large enough in order not to affect the numerical results. In a more refined model for the ESL the crossover of the wall charge to the neutral bulk of the dielectric can be taken into account by splitting the ESL into an interface specific region and a space charge region. This approach which turns z_s from an ad-hoc cutt-off into the boundary between the two regions is described in [20]. For the charge distribution at the interface the simple model is however sufficient.

We now use the ESL model to calculate for a helium discharge in contact with a LiF and Al_2O_3 surface the electron density and potential across the plasma wall. Our main focus lies in the identification of generic types of electron distributions in the ESL.

Depending on the electron affinity χ, one of two scenarios for the distribution of surplus electrons in the ESL may be realised. For negative electron affintiy $\chi < 0$, the conduction band minimum lies above the long-range potential just in front of the surface. It is thus energetically unfavourable for surplus electrons to populate the conduction band. Instead they are located in the image potential in front of the surface. The energy of an electron in the image potential reaches its minimum just in front of the surface at the beginning of the graded interface. This is confirmed by the electron density and the potential for LiF ($\chi < 0$) shown in the left panel of Fig. 8.6. For $\chi < 0$, the excess electrons coming from the plasma form therefore an external surface charge in the image potential in front of the crystallographic interface. As the electrons do not penetrate into the solid the band bending associated with it is negligible. The external surface charge is very narrow. It can thus be considered as a quasi-two-dimensional electron gas, similar to the surface plasma anticipated by Emeleus and Coulter [18, 32].

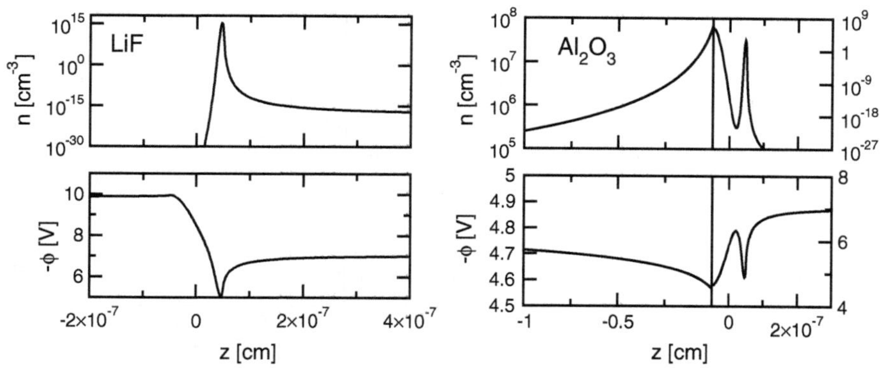

Fig. 8.6 Plasma-supplied excess electron density n (*upper panel*) and the potential $-\phi$ (*lower panel*) it gives rise to for an LiF (*left*) and Al_2O_3 (*right*) surface in contact with a helium discharge with $n_0 = 10^7 \text{cm}^{-3}$ and $k_B T_e = 2\text{eV}$

For positive electron affinity $\chi > 0$ the situation is dramatically different. In this case the conduction band minimum is below the potential just outside. It is thus energetically favourable for electrons to accumulate inside the dielectric. This can be seen in the right panel of Fig. 8.6 which shows the electron density and the potential for an Al_2O_3 surface ($\chi > 0$). Although the image potential still induces an attractive well in front of the surface the minimum of the potential energy for electrons $-e\phi$ is reached inside the dielectric. Excess electrons coming from the plasma are thus mostly located inside the wall where they form a space charge in the conduction band. The electron distribution extends deep into the bulk and entails bending of the energy bands near the surface. Note the different scales of the axes for the left and right panels of Fig. 8.6. On the scale where variations in the space charge for Al_2O_3 are noticeable the electron distribution for LiF is basically a vertical line.

So far we have shown the potential and the electron distribution in the ESL. Now, we will compare the potential and the charge distribution in the ESL with the ones in the plasma sheath. The wall bound electrons in the surface layer constitute an electron distribution in thermal equilibrium with the wall which balances electron and ion influx at the sheath-ELS boundary z_0. On top of this quasistationary electron distribution there are fluxes of electrons and ions, coming from the plasma. These currents, which begin at z_0 and persist to the location in the ESL where either electron trapping or electron-ion recombination takes place are not described by our simple ESL model. The electron and ion densities in this model are thus discontinuous at z_0. The potential, however, which has been obtained from the integration of Poisson's equation is continuous and differentiable everywhere. Between the crystallographic interface and z_0 the electron and ion flux from the plasma would be important. The neglect of the charge densities associated with these fluxes does however not affect the potential because they are too small to have a significant effect.

Figure 8.7 compares the ESL and the plasma sheath for LiF ($\chi < 0$, left) and Al_2O_3 ($\chi > 0$, right). We plot the electron and ion densities (upper panel) as well

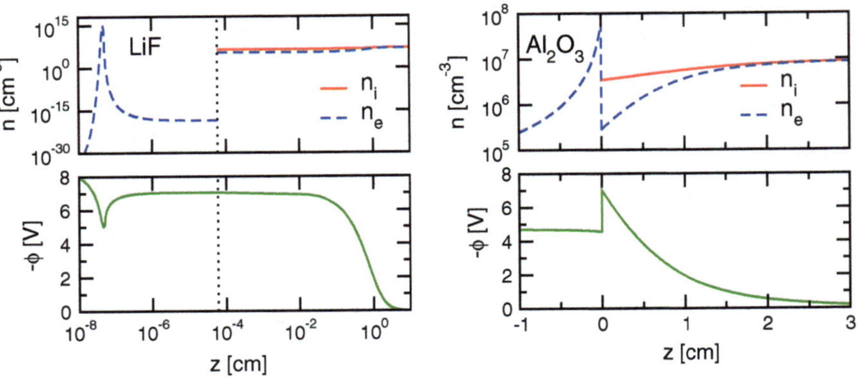

Fig. 8.7 Density of plasma-supplied surplus electrons trapped in the ESL, electron and ion density in the plasma sheath, and electric potential for a LiF (*left*) and an Al_2O_3 surface (*right*) in contact with a helium discharge with plasma parameters $n_0 = 10^7 \text{cm}^{-1}$ and $k_B T_e = 2eV$. The crystallographic interface is at $z = 0$. Note the different scales of the two panels. The deep penetration of the Al_2O_3 wall charge is due to the neglect of defect states and other collision centres. Reproduced from [33]. Copyright (2012) by Wiley-VCH Verlag

as the electric potential (lower panel). Note the logarithmic scale in the left panel and the linear scale in the right panel. For LiF the surface electrons are bound in the image states in front of the surface due to the negative electron affinity. Far from the surface, the potential approaches the bulk plasma value chosen to be zero. In the sheath the potential develops a Coulomb barrier and reaches the wall potential ϕ_w at z_0, the distance where the sheath merges with the surface layer (vertical dotted line). The wall potential is the potential just outside to which the energies of the bulk states are referenced. Closer to the surface the potential follows the attractive image potential while at the surface the repulsive potential due to the negative electron affinity prevents the electron from entering the dielectric (only scarcely seen on the scale of the figure).

For Al_2O_3 the excess electrons constituting the wall charge penetrate deep into the dielectric due to the positive electron affinity. Compared to the variation of the electric potential in the sheath the band bending in the dielectric induced by the wall charge is rather small as indicated by the variation of ϕ inside the dielectric. This is because ε_0 is large and most surplus electrons are bound close to the crystallographic interface. The interface specific region where the short ranged surface potential varies strongly is very narrow compared to the extent of the two space charge regions.

The ESL should be regarded as the ultimate boundary layer of a bounded gas discharge. Our model for the ESL captures the coupling of the neutral bulk plasma to an electron plasma at the boundary separated from the bulk plasma by the plasma sheath. Depending on the electron affinity this boundary plasma can be either a quasi-two-dimensional electron plasma or an electron(-hole) plasma inside the wall.

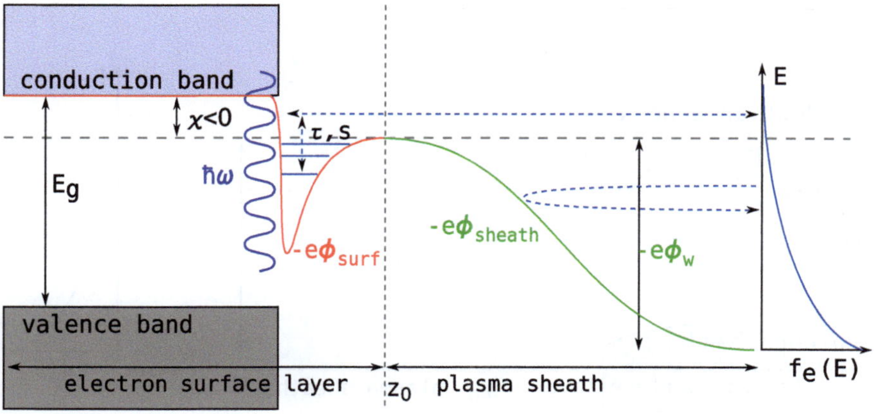

Fig. 8.8 Electron trapping in the ESL in front of a dielectric with $\chi < 0$

8.3 Electron Physisorption

We can now turn to the electron kinetics in the ESL. Depending on the electron affinity two distinct scenarios for electron trapping emerge which are schematically shown in Figs. 8.8 and 8.9. For negative electron affinity (see Fig. 8.8) electrons that overcome the sheath potential do not penetrate the solid as their energy falls in the band gap where no internal states are available. Instead, they are trapped in the image potential in front of the surface. Energy-relaxation is in this case due to surface vibrations which trigger transitions to the bound image states. As the image potential is relatively deep compared to the energies of the longitudinal acoustic phonon responsible for electron trapping and the wavefunctions of the unbound states are suppressed close to the surface the probability for sticking is small. This scenario, which we have investigated in detail [34–36], will be presented below.

For positive electron affinity (see Fig. 8.9) electrons that overcome the sheath potential penetrate into the solid. There they initially occupy high-lying states of the conduction band. Subsequent electron energy relaxation is due to scattering with bulk phonons of the dielectric and the plasma-supplied electrons are eventually trapped at the bottom of the conduction band in the potential well created by the restoring force from the plasma. Due to the high-density of states and an efficient scattering mechanism we expect the sticking coefficient to be much larger in this case. However, a detailed analysis of this scenario still remains to be done.

We will now consider the physisorption of an electron in the image potential in front of a dielectric with $\chi < 0$ in detail. We describe the time evolution of the occupancy of the bound surface states with a quantum-kinetic rate equation [37, 38]. It captures all three characteristic stages of physisorption: initial trapping, subsequent relaxation and desorption [39, 40].

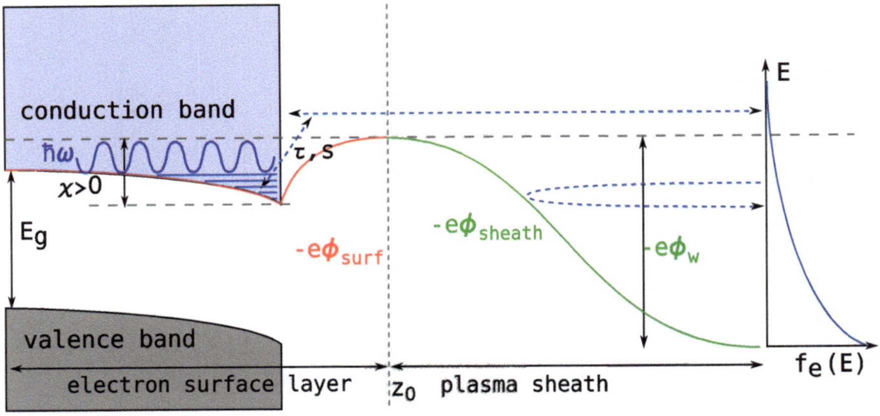

Fig. 8.9 Electron trapping in the ESL in a dielectric with $\chi > 0$

The time dependence of the occupancies of the bound states is given by [37, 38]

$$\frac{\mathrm{d}}{\mathrm{d}t}n_n(t) = \sum_{n'} [W_{nn'}n_{n'}(t) - W_{n'n}n_n(t)] - \sum_k W_{kn}n_n(t) + \sum_k \tau_t W_{nk}j_k(t) \quad (8.22)$$

which can be rewritten as

$$\frac{\mathrm{d}}{\mathrm{d}t}n_n(t) = \sum_{n'} T_{nn'}n_{n'}(t) + \sum_k \tau_t W_{nk}j_k(t), \quad (8.23)$$

where $W_{n'n}$ is the probability per unit time for a transition from a bound state n to another bound state n', W_{kn} and W_{nk} are the probabilities per unit time for a transition from the bound state n to the continuum state k and vice versa, and $\tau_t = 2L/v_z$ is the transit time through the surface potential of width L, which, in the limit $L \to \infty$, can be absorbed into the transition probability. The matrix $T_{nn'}$ is defined implicitly by the above equation. The last term in (8.22) and (8.23), respectively, gives the increase in the bound state occupancy due to the coupling between bound and unbound surface states.

The probability for an approaching electron in the continuum state k to make a transition to any of the bound states is given by the prompt energy-resolved sticking coefficient

$$s_{e,k}^{\text{prompt}} = \tau_t \sum_n W_{nk}. \quad (8.24)$$

Treating the incident electron flux as an externally specified parameter, the solution of (8.22) describes the subsequent relaxation and desorption. It is given by

$$n_n(t) = \sum_\kappa e^{-\lambda_\kappa t} \int_{-\infty}^{t} \mathrm{d}t' e^{\lambda_\kappa t'} e_n^{(\kappa)} \sum_{kl} \tilde{e}_l^{(\kappa)} \tau_t W_{lk} j_k(t') \quad (8.25)$$

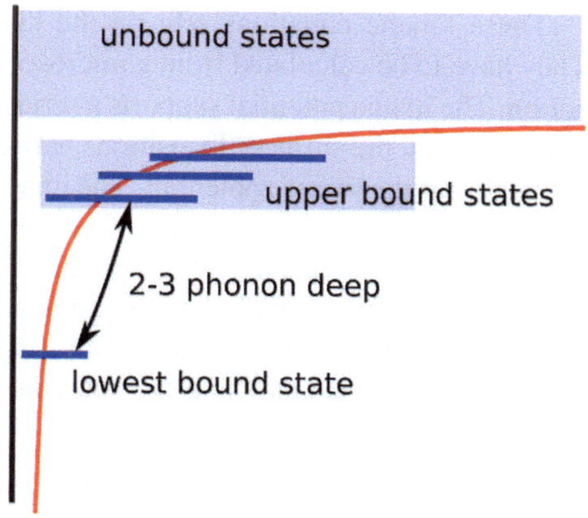

Fig. 8.10 Schematic representation of the series of image states in the image potential. The lowest bound states lies very deep and is connected to a group of upper bound states only by a two or three-phonon processes. The upper bound states are connected by one-phonon processes to the continuum

where $e_n^{(\kappa)}$ and $\tilde{e}_n^{(\kappa)}$ are the right and left eigenvectors to the eigenvalue $-\lambda_\kappa$ of the matrix $T_{nn'}$.

If the modulus of one eigenvalue, λ_0, is considerably smaller than the moduli of the other eigenvalues, λ_κ, a unique desorption time and a unique sticking coefficient can be identified [38]. In this case λ_0 governs the long time behaviour of the equilibrium occupation of the bound states, $n_n^{\mathrm{eq}} \sim e^{-E_n/k_B T_s}$, and its inverse can be identified with the desorption time, $\lambda_0^{-1} = \tau_e$. The bound state occupancy $n_n(t)$ splits into a slowly varying part $n_n^0(t)$ given by the $\kappa = 0$ summand in (8.25) and a quickly varying part $n_n^f(t)$ given by the sum over $\kappa \neq 0$ in (8.25).

The electron adsorbate, i.e. the fraction of the trapped electron remaining in the surface states for times on the order of the desorption time, is given by the slowly varying part only, $n^0(t) = \sum_n n_n^0(t)$. Differentiating $n^0(t)$ with respect to the time,

$$\frac{d}{dt} n^0(t) = \sum_k s_{e,k}^{\mathrm{kinetic}} j_k(t) - \lambda_0 n^0(t) \tag{8.26}$$

we can, following Brenig [40], identify the kinetic energy-resolved sticking coefficient

$$s_{e,k}^{\mathrm{kinetic}} = \tau_t \sum_{n,n'} e_{n'}^{(0)} \tilde{e}_n^{(0)} W_{nk} \tag{8.27}$$

giving the probability for both, initial trapping and subsequent relaxation.

If the incident unit electron flux corresponds to an electron with Boltzmann distributed kinetic energies, the prompt or kinetic energy-averaged sticking coefficient is given by

$$s_e^{\cdots} = \frac{\sum_k s_{e,k}^{\cdots} k e^{-\beta_e E_k}}{\sum_k k e^{-\beta_e E_k}}, \tag{8.28}$$

where $\beta_e^{-1} = k_B T_e$ is the mean electron energy.

These kinetic equations rely on the knowledge of the transition probabilities. They have to be calculated from a microscopic model for the electron-surface interaction. The image potential supports a series of bound states which can trap electrons temporarily at the surface. Transitions between image states are due to dynamic perturbations of the surface potential. The image potential is very steep near the surface. A particularly strong perturbation arises therefore from the surface vibrations induced by the longitudinal acoustic bulk phonon perpendicular to the surface. Describing them with a Debye model, the maximum phonon energy of this mode is the Debye energy $\hbar\omega_D$. Measuring energies in units of the Debye energy, important dimensionless parameters characterising the potential depth are

$$\varepsilon_n = \frac{E_n}{\hbar\omega_D} \quad \text{and} \quad \Delta_{nn'} = \frac{E_n - E_{n'}}{\hbar\omega_D}, \tag{8.29}$$

where $E_n < 0$ is the energy of the n-th bound state.

The depth of the image potential can be classified with respect to the Debye energy. If $-n+1 > \Delta_{12} > -n$, we call the potential n-phonon deep. If $\varepsilon_1 > -1$ we call it shallow. For a shallow potential the lowest bound state is coupled by a one-phonon transition to the continuum. If the potential is n-phonon deep an n-phonon process links the lowest bound state to the second bound states, which is linked by one-phonon transitions either directly or via intermediates to the continuum. For the dielectrics which we consider the potential is two-phonon (MgO) or three-phonon deep (LiF). A schematic representation of the image states is given in Fig. 8.10.

For the calculation of the transition probabilities used in the rate equation we need a microscopic model for the image potential and the electron-surface vibration interaction. Classically the image potential ensues from the dielectric mismatch at the interface. On a microscopic level it arises from the coupling of the electron to a polarisable surface mode of the solid. For a dielectric this is a surface phonon. For LiF and MgO the low frequency dielectric function is dominated by a transverse optical (TO) phonon with frequency ω_{TO}. It can be approximated by [41]

$$\varepsilon(\omega) = 1 + (\varepsilon_0 - 1)\frac{\omega_{TO}^2}{\omega_{TO}^2 - \omega^2}, \tag{8.30}$$

where ε_0 is the static dielectric constant. The bulk TO-phonon gives rise to a surface phonon. Its frequency $\omega_s = \omega_{TO}\sqrt{(1+\varepsilon_0)/2}$ is determined by the condition $\varepsilon(\omega_s) = -1$. The electron couples to this surface phonon according to [41, 42]

$$H = -\frac{\hbar^2}{2m}\Delta + \sum_{\mathbf{Q}_\parallel} \hbar\omega_s a^\dagger_{\mathbf{Q}_\parallel} a_{\mathbf{Q}_\parallel} + \sum_{\mathbf{Q}_\parallel} \left(\phi_{\mathbf{Q}_\parallel}(\mathbf{x}_\parallel) a_{\mathbf{Q}_\parallel} + \phi^*_{\mathbf{Q}_\parallel}(\mathbf{x}_\parallel) a^\dagger_{\mathbf{Q}_\parallel} \right) \tag{8.31}$$

with

$$\phi_{\mathbf{Q}_\parallel}(\mathbf{x}_\parallel) = \frac{2\sqrt{\pi\omega_s\Lambda_0\hbar}e}{Q_\parallel^{1/2}A^{1/2}} e^{-Q_\parallel|z|} e^{i\mathbf{Q}_\parallel\mathbf{x}_\parallel} \tag{8.32}$$

where $\Lambda_0 = (\varepsilon_0 - 1)/(4(\varepsilon_0 + 1))$ and the subscript $\|$ denotes vectors parallel to the surface and $a^{(\dagger)}_{\mathbf{Q}_\|}$ are annihilation (creation) operators for surface phonons.

Applying a unitary transformation [42] separates the coupling into a static part which takes the classical form of the image potential $\sim 1/z$ and a dynamic part of the electron-surface phonon coupling, which encodes recoil effects and encompasses momentum relaxation parallel to the surface. While the classical image potential allows a simple description of the image states which captures their properties fairly well it is not sufficient for the calculation of probabilities for surface vibration induced transitions, as they perturb the electronic states most strongly close to the surface where the image potential is steepest. Unfortunately, in this region the classical image potential has an unphysical divergence. To remedy this we use a variational procedure to extract the static image potential from [41]. Thereby we keep some recoil effects which make the recoil-corrected image potential $\sim 1/(z + z_c)$ with a cut-off parameter $z_c = \sqrt{\hbar/2m\omega_s\pi^2}$ [34] divergence free.

Transitions between the eigenstates of the recoil-corrected image potential are due to the longitudinal-acoustic bulk phonon perpendicular to the surface. The Hamiltonian from which we calculate the transition probabilities is given by

$$H = H_e^{\text{static}} + H_{ph} + H_{e-ph}^{\text{dyn}} \tag{8.33}$$

where

$$H_e^{\text{static}} = \sum_q E_q c_q^\dagger c_q \tag{8.34}$$

describes the electron in the recoil-corrected image potential,

$$H_{ph} = \sum_Q \hbar\omega_Q b_Q^\dagger b_Q \tag{8.35}$$

describes the free dynamics of the bulk longitudinal acoustic phonon responsible for transitions between surface states, and

$$H_{e-ph}^{\text{dyn}} = \sum_{q,q'} \langle q'|V_p(u,z)|q\rangle c_{q'}^\dagger c_q \tag{8.36}$$

denotes the dynamic coupling of the electron to the bulk phonon.

The perturbation $V_p(u,z)$ can be identified as the difference between the displaced surface potential and the static surface potential. It reads, after the transformation $z \to z - z_c$,

$$V_p(u,z) = -\frac{e^2\Lambda_0}{z+u} + \frac{e^2\Lambda_0}{z}. \tag{8.37}$$

In general, multi-phonon processes can arise both from the nonlinearity of the electron-phonon coupling H_{e-ph}^{dyn} as well as from the successive actions of H_{e-ph}^{dyn}

encoded in the T matrix equation,

$$T = H_{e-ph}^{\text{dyn}} + H_{e-ph}^{\text{dyn}} G_0 T, \tag{8.38}$$

where G_0 is given by

$$G_0 = \left(E - H_e^{\text{static}} - H_{ph} + i0^+\right)^{-1}. \tag{8.39}$$

The transition rate from an electronic state q to an electronic state q' encompassing both types of processes is given by [43]

$$W_{q'q} = \frac{2\pi}{\hbar} \sum_{s,s'} \frac{e^{-\beta_s E_s}}{\sum_{s''} e^{-\beta_s E_{s''}}} \left|\langle s', q'|T|s, q\rangle\right|^2 \delta(E_s - E_{s'} + E_q - E_{q'}), \tag{8.40}$$

where $\beta_s = (k_B T_s)^{-1}$, with T_s the surface temperature and $|s\rangle$ and $|s'\rangle$ the initial and final phonon states. We are only interested in the transitions between electronic states. It is thus natural to average in (8.40) over all phonon states. The delta function guarantees energy conservation.

In principle, multiphonon transition rates can be obtained by iterating the T-matrix and evaluating (8.40). Up to $\mathscr{O}(u^3)$, for instance, the T-matrix reads

$$T = V_1 + V_2 + V_3 + V_1 G_0 V_1 + V_2 G_0 V_1 + V_1 G_0 V_2 + V_1 G_0 V_1 G_0 V_1, \tag{8.41}$$

where the $V_i \sim u^i$ originate from expanding (8.37) in the displacement u. The T-matrix enters as $\langle T\rangle\langle T^*\rangle$ into the transition probability. The term $\langle V_1\rangle\langle V_1^*\rangle$ can be identified as the Golden Rule transition probability. Proportional to u^2 it is a one-phonon process. Two-phonon processes, proportional to u^4, are represented by the terms

$$\langle V_1\rangle\langle V_3^*\rangle, \langle V_3\rangle\langle V_1^*\rangle, \langle V_1\rangle\langle V_2^* G_0^* V_1^*\rangle, \langle V_1 G_0 V_2\rangle\langle V_1^*\rangle, \langle V_1\rangle\langle V_1^* G_0^* V_2^*\rangle,$$
$$\langle V_2 G_0 V_1\rangle\langle V_1^*\rangle, \langle V_1\rangle\langle V_1^* G_0^* V_1^* G_0^* V_1^*\rangle, \langle V_1 G_0 V_1 G_0 V_1\rangle\langle V_1^*\rangle, \tag{8.42}$$

and

$$\langle V_2\rangle\langle V_2^*\rangle, \langle V_1 G_0 V_1\rangle\langle V_1^* G_0^* V_1^*\rangle, \langle V_2\rangle\langle V_1^* G_0^* V_1^*\rangle, \langle V_1 G_0 V_1\rangle\langle V_2^*\rangle. \tag{8.43}$$

A complete two-phonon calculation would take all these processes into account as they stand. This is however not always necessary. A closer analysis (see [34]) of the first group of terms reveals that they are two-phonon corrections to transitions already enabled by a one-phonon process. We assume that theses corrections are small and evaluate only the second group for transitions where they enable two-phonon transitions which are not merely corrections to one-phonon transitions. The

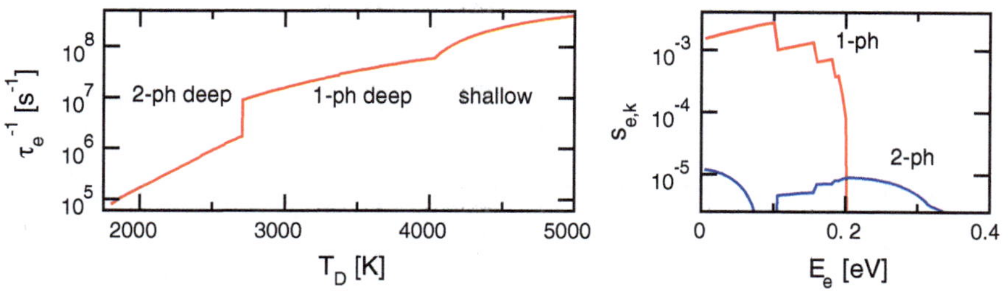

Fig. 8.11 *Left* Inverse desorption time τ_e^{-1} in the two-phonon approximation for $T_D/T = 5$ as a function of the Debye temperature T_D. Data for $T_D = 2500K$ apply to graphite. *Right* Prompt energy-resolved sticking coefficient as a function of the electron energy for a two phonon deep potential ($T_D = 2500K$ and $T_s = 357K$). The contribution of one-phonon processes (*red*) far outweighs the contribution from two phonon processes (*blue*)

details of the evaluation of the transition probabilities, including a regularisation of divergences by taking a finite phonon lifetime into account, can be found in [34, 35].

The expansion of the T-matrix allows the calculation of transition probabilities for the two-phonon deep potentials of graphite and MgO. However, for a three-phonon deep potential, for instance for LiF and CaO, this approach is no longer feasible. From Refs. [34, 35] we qualitatively know the relevance of different types of multi-phonon processes. For continuum to bound state transitions, for instance, one phonon processes are sufficient at low electron energies. We will therefore compute the transition probabilities between bound and continuum states in the one-phonon approximation. For transitions between bound states, we found that multi-phonon processes due to the nonlinearity of the electron-phonon coupling tend to be more important than the multi-phonon processes due to the iteration of the T-matrix. Hence we expect that an approximation which takes only the nonlinearity of the electron-phonon interaction nonperturbatively into account to be sufficient for the identification of the generic behaviour of multi-phonon-mediated adsorption and desorption. This approach is described in [36].

We now turn to results for the desorption time and the sticking coefficient for physisorption in the image states. Before showing results for MgO and LiF, we consider the dependence of the electron kinetics on the potential depth and the relevance of one or two-phonon processes. For this we show in the left panel of Fig. 8.11 the inverse of the desorption time as a function of the Debye temperature $T_D = \hbar\omega_D/k_B$ which sets the energy scale of the acoustic phonons. While the absolute depth of the potential remains constant, varying the phonon energy tunes the effective potential depth. Figure 8.11 reveals that for a shallow potential desorption is most efficient while for a one-phonon deep potential it gradually becomes less efficient. When the potential becomes two-phonon deep desorption suddenly becomes even slower which reflects the small magnitude of two-phonon transitions compared to one-phonon transitions. This justifies our approximation of neglecting two-phonon corrections to transitions already enabled by a one-phonon process. For a Debye temperature $T_D = 2500K$ the results apply to graphite.

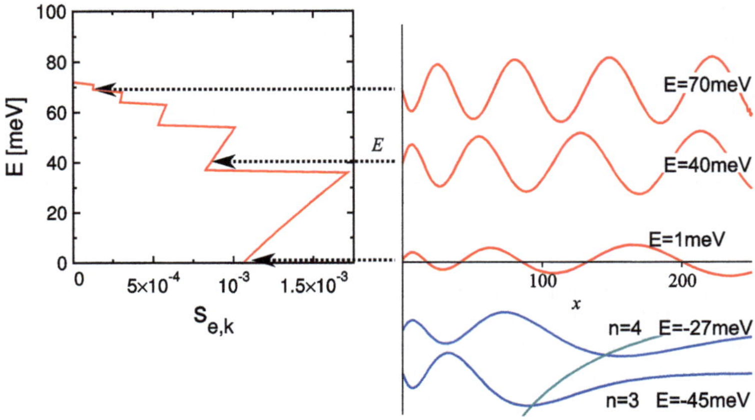

Fig. 8.12 Energy resolved prompt sticking coefficient for MgO (*left*) and schematic representation of the bound and continuum wavefunctions (*right*). The suppression of the amplitude of the continuum wavefunction close to the surface for small electron energies causes the small sticking coefficient

The energy-resolved prompt sticking coefficient, plotted in the right panel of Fig. 8.11 for graphite, shows that also for bound state-continuum transitions one-phonon processes far outweigh two-phonon processes. Thus, sticking is mainly due to one-phonon processes. Moreover we find that the sticking coefficient drops sharply at specific energies. These accessibility thresholds occur whenever one bound state becomes no longer accessible from the continuum because the energy difference exceeds one Debye energy.

The sticking coefficient takes relatively small values—on the order of 10^{-3}. This is due to the small matrix elements between bound and continuum states. Figure 8.12 shows the wave functions of representative bound and continuum states for MgO. For large energy the continuum wave functions have sinusoidal shape whereas for small energies their amplitude is significantly suppressed close to the surface. These two behaviours correspond to two limits for the wavefunction. For simplicity we discuss this for a $1/x$ potential without cut-off. The continuum wave-functions read $\phi_k(x) \sim \frac{1}{\sqrt{k}} M_{\frac{-i}{k},\frac{1}{2}}(2ik\Lambda_0 x)$ where $x = z/a_B$. For $x \to \infty$ we obtain $\phi_k(x) \sim \sin(\Lambda_0 kx)$ which also holds for large k while for $k \to 0$ we obtain $\phi_k(x) \sim \sqrt{k}\sqrt{2\Lambda_0 x} J_1(\sqrt{8\Lambda_0 x})$. The proportionality to $\sqrt{k}$ entails a strong suppression of the wavefunction for low energy.

Electron physisorption at a dielectric surface with negative electron affinity is an intriguing phenomenon due to the interplay of potential depth, magnitude of matrix elements and surface temperature [34–36]. Initial trapping of the electrons, characterised by the prompt sticking coefficient, occurs in the upper bound states by one-phonon transitions. Relaxation after initial trapping depends on the strength of transitions from the upper bound states to the lowest bound states. If the lowest bound state was linked to the second bound state by a one-phonon process the electron would relax for all surface temperatures. If they are linked by a multiphonon process

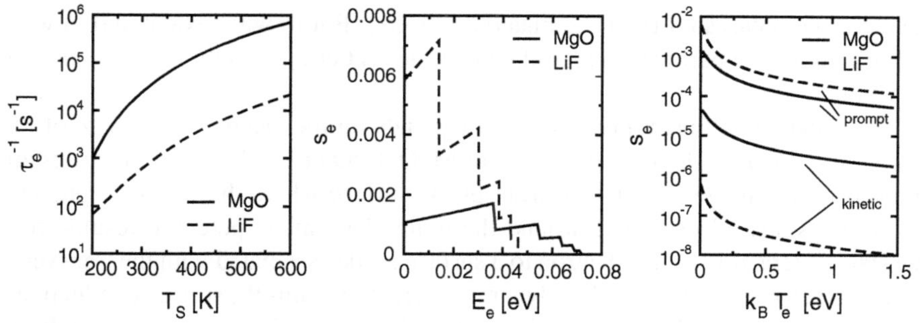

Fig. 8.13 Inverse desorption time τ_e^{-1} (*left*), prompt energy resolved sticking coefficient (*middle*) and prompt as as well as kinetic energy averaged sticking coefficient (*right*) for MgO and LiF

relaxation takes place only for low temperature, whereas for room temperature a relaxation bottleneck ensues as the electron desorbs from the upper bound states before it drops to the lowest bound state. This leads to a severe reduction of the kinetic sticking coefficient compared to the prompt sticking coefficient. The dominant desorption channel depends also on the depth of the potential. For a shallow potential desorption occurs directly from the lowest bound state to the continuum. For deeper potentials desorption proceeds via the upper bound states. Desorption occurs then via a cascade in systems without and as a one-way process in systems with relaxation bottleneck [36]. An overview of results for the desorption time as well as prompt and kinetic sticking coefficients for MgO and LiF is given by Fig. 8.13. Most important for the plasma context is that $s_e \ll 1$ and $\tau_e^{-1} \neq 0$ [34–36], implying that a dielectric surface with a negative electron affinity is not a perfect absorber for electrons.

8.4 Mie Scattering by a Charged Particle

In the previous section we have employed our model for the ESL to analyse the electron kinetics at the plasma boundary and to calculate desorption times and sticking coefficients for a dielectric with negative electron affinity. Another application of the model for the ESL is the study of the optical properties of the wall-bound electrons. In this section we will study the effect of surplus electrons on the scattering of light by a sphere (Mie scattering) with an eye on a charge measurement for dust particles. Most established methods for measuring the particle charge [44–46] require plasma parameters which are not precisely known. The particle charge, however, is the fundamental quantity of interest in the field of dusty plasmas. It determines the interactions between dust grains and between dust grains and external electromagnetic fields. While Mie scattering is already used in the field of dusty plasmas to determine the particle size [47], it has up to now not been used to determine the particle charge. Quite generally the Mie signal contains information about excess electrons as their electrical conductivity modifies either the boundary condition for electromagnetic fields or the polarizability of the material [48–51]. But to what extent and in what spectral range the particle charge influences the scattering of light remains an open question.

The location of the surplus electrons depends on the electron affinity χ of the dust grain. For $\chi < 0$, as it is the case for MgO, CaO or LiF [52, 53], electrons are trapped in the image potential in front of the surface where they form a spherical two-dimensional electron gas around the grain. The image potential results from the static interaction of the electron to a surface mode associated with the transverse optical (TO) phonon. The residual dynamic interaction limits the surface conductivity σ_s of the image bound electrons. The electrical field parallel to the surface induces a surface current $\mathbf{K} = \sigma_s \mathbf{E}_{\|}$ which is proportional to the surface conductivity σ_s. The surface current enters the boundary condition for the magnetic field. Thus, $\hat{\mathbf{e}}_r \times (\mathbf{H}_i + \mathbf{H}_r - \mathbf{H}_t) = \frac{4\pi}{c}\mathbf{K}$, where c is the speed of light [51].

For $\chi > 0$, as it is the case for Al_2O_3 or SiO_2, electrons form a space charge layer in the conduction band [20]. Its width, limited by the screening in the dielectric, is typically larger than a micron. For micron sized particles we can thus assume a homogeneous distribution of the excess electrons in the bulk. The effect on light scattering is now encoded in the bulk conductivity of the excess electrons σ_b, which is limited by scattering with a longitudinal optical (LO) bulk phonon. The bulk conductivity gives rise to a polarizability $\alpha = 4\pi i \sigma_b/\omega$, with ω the frequency, which enters the refractive index $N = \sqrt{\varepsilon + \alpha}$.

To solve the scattering problem we expand the incident plane wave,

$$\mathbf{E}_i = E_0 e^{ikz - i\omega t} \hat{\mathbf{e}}_x, \qquad \mathbf{H}_i = \frac{ck}{\omega} E_0 e^{ikz - i\omega t} \hat{\mathbf{e}}_y \tag{8.44}$$

in spherical vector harmonics $\mathbf{m}_{emn}$, $\mathbf{m}_{omn}$, $\mathbf{n}_{emn}$, and $\mathbf{n}_{omn}$ [54] which are solution of the vector wave equation $\nabla^2 \mathbf{C} + k^2 \mathbf{C} = 0$, in spherical coordinates

$$\mathbf{E}_i = E_0 e^{-i\omega t} \sum_{n=1}^{\infty} i^n \frac{2n+1}{n(n+1)} \left(\mathbf{m}_{o1n}^{(1)} - i \mathbf{n}_{e1n}^{(1)} \right), \tag{8.45}$$

$$\mathbf{H}_i = -\frac{ck}{\omega} E_0 e^{-i\omega t} \sum_{n=1}^{\infty} i^n \frac{2n+1}{n(n+1)} \left(\mathbf{m}_{e1n}^{(1)} + i \mathbf{n}_{o1n}^{(1)} \right) \tag{8.46}$$

and match the incident wave with the reflected wave,

$$\mathbf{E}_r = E_0 e^{-i\omega t} \sum_{n=1}^{\infty} i^n \frac{2n+1}{n(n+1)} \left(a_n^r \mathbf{m}_{o1n}^{(3)} - i b_n^r \mathbf{n}_{e1n}^{(3)} \right), \tag{8.47}$$

$$\mathbf{H}_r = -\frac{ck}{\omega} E_0 e^{-i\omega t} \sum_{n=1}^{\infty} i^n \frac{2n+1}{n(n+1)} \left(b_n^r \mathbf{m}_{e1n}^{(3)} + i a_n^r \mathbf{n}_{o1n}^{(3)} \right) \tag{8.48}$$

and the transmitted wave,

$$\mathbf{E}_t = E_0 e^{-i\omega t} \sum_{n=1}^{\infty} i^n \frac{2n+1}{n(n+1)} \left(a_n^t \mathbf{m}_{o1n}^{(1)} - i b_n^t \mathbf{n}_{e1n}^{(1)} \right), \tag{8.49}$$

$$\mathbf{H}_t = -\frac{cq}{\omega} E_0 e^{-i\omega t} \sum_{n=1}^{\infty} i^n \frac{2n+1}{n(n+1)} \left(b_n^t \mathbf{m}_{e1n}^{(1)} + i a_n^t \mathbf{n}_{o1n}^{(1)} \right) \tag{8.50}$$

(q is the wavenumber inside the particle) at the boundary of a dielectric sphere with radius a. The boundary conditions at the surface are given by

$$\hat{\mathbf{e}}_r \times \left(\mathbf{C}_i + \mathbf{C}_r - \mathbf{C}_t \right) = 0 \tag{8.51}$$

for $\mathbf{C} = \mathbf{E}, \mathbf{H}$ in the case of $\chi > 0$ and in the case of $\chi < 0$ by the above equation for $\mathbf{E}$ and

$$\hat{\mathbf{e}}_r \times (\mathbf{H}_i + \mathbf{H}_r - \mathbf{H}_t) = \frac{4\pi}{c} \sigma_s \mathbf{E}_\parallel . \tag{8.52}$$

This gives, following Bohren and Hunt [51], the scattering coefficients

$$a_n^r = \frac{\psi_n(N\rho)\psi_n'(\rho) - \left[N\psi_n'(N\rho) - i\tau\psi_n(N\rho) \right] \psi_n(\rho)}{\left[N\psi_n'(N\rho) - i\tau\psi_n(N\rho) \right] \xi_n(\rho) - \psi_n(N\rho)\xi_n'(\rho)} \tag{8.53}$$

and

$$b_n^r = \frac{\psi_n'(N\rho)\psi_n(\rho) - \left[N\psi_n(N\rho) + i\tau\psi_n'(N\rho) \right] \psi_n'(\rho)}{\left[N\psi_n(N\rho) + i\tau\psi_n'(N\rho) \right] \xi_n'(\rho) - \psi_n'(N\rho)\xi_n(\rho)} \tag{8.54}$$

where for $\chi < 0 \, (\chi > 0)$ the dimensionless surface conductivity $\tau(\omega) = 4\pi\sigma_s(\omega)/c$ ($\tau = 0$) and the refractive index $N = \sqrt{\varepsilon}$ ($N = \sqrt{\varepsilon + \alpha}$), $\rho = ka = 2\pi a/\lambda$ the size parameter, where k is the wavenumber, $\psi_n(\rho) = \sqrt{\pi\rho/2} J_{n+1/2}(\rho)$, $\xi_n(\rho) = \sqrt{\pi\rho/2} H_{n+1/2}^{(1)}(\rho)$ with $J_n(\rho)$ the Bessel and $H_n^{(1)}(\rho)$ the Hankel function of the first kind. As for uncharged particles the scattering efficiency is given by

$$Q_s = \frac{2}{\rho^2} \sum_{n=1}^{\infty} (2n+1) \left(|a_n^r|^2 + |b_n^r|^2 \right), \tag{8.55}$$

the extinction efficiency by

$$Q_t = -\frac{2}{\rho^2} \sum_{n=1}^{\infty} (2n+1) \, \Re \left(a_n^r + b_n^r \right) \tag{8.56}$$

and the absorption efficiency by $Q_a = Q_t - Q_s$. Any effect of the surplus electrons on the scattering of light, encoded in a_n^r and b_n^r, is due to the surface conductivity ($\chi < 0$) or the bulk conductivity ($\chi > 0$).

For $\chi < 0$ we use a planar model, which we justify below, to calculate the surface conductivity. For the calculation of the surface conductivity we will employ the $1/z$ image potential. The electron-phonon interaction is given by (8.31). We apply the unitary transformation $H \to UHU^{-1}$ with $U = e^{iS}$, $S = (i/\hbar\omega_s)\sum_{\mathbf{Q}_\parallel} a_{\mathbf{Q}_\parallel}\phi_{\mathbf{Q}_\parallel} - a^\dagger_{\mathbf{Q}_\parallel}\phi^*_{\mathbf{Q}_\parallel}$ to separate static and dynamic coupling [42]. The former leads to the image potential $V = -\Lambda_0 e^2/z$ with $\Lambda_0 = (\varepsilon_0 - 1)/(4(\varepsilon_0 + 1))$ supporting a series of bound Rydberg states whose wave functions read

$$\phi_{n\mathbf{k}}(\mathbf{x}, z) = \frac{1}{\sqrt{A}} e^{i\mathbf{k}\mathbf{x}} \sqrt{\frac{\Lambda_0}{a_B n n!^2}} W_{n,1/2}\left(\frac{2\Lambda_0 z}{n a_B}\right) \tag{8.57}$$

with a_B the Bohr radius, $\mathbf{k} = (k_x, k_y)$, $\mathbf{x} = (x, y)$, A the surface area and $W_{n,m}$ Whittaker's function [55]. Since trapped electrons are thermalised with the surface and the spacing between Rydberg states is large compared to $k_B T$, they occupy almost exclusively the lowest image band $n = 1$. Assuming a planar surface is justified provided the de Broglie wavelength λ_{dB} of the electron on the surface is smaller than the radius a of the sphere. For a surface electron with energy $E_{\text{kin}}/k_B = 300K$ one finds $\lambda_{dB} \approx 8 \times 10^{-7}$ cm. Thus, for particle radii $a > 10$ nm the plane-surface approximation is justified. The residual dynamic interaction enables momentum relaxation parallel to the surface and hence limits the surface conductivity. Introducing annihilation (creation) operators $c_{\mathbf{k}}^{(\dagger)}$ for electrons in the lowest image band, the Hamiltonian describing the dynamic electron-phonon coupling in the lowest image band reads [56]

$$H = \sum_{\mathbf{k}} \varepsilon_{\mathbf{k}} c^\dagger_{\mathbf{k}} c_{\mathbf{k}} + \hbar\omega_s \sum_{\mathbf{Q}} a^\dagger_{\mathbf{Q}} a_{\mathbf{Q}} + H_{int} \tag{8.58}$$

with

$$H_{int} = \frac{1}{\sqrt{A}} \sum_{\mathbf{k},\mathbf{Q}} M_{\mathbf{k},\mathbf{Q}} c^\dagger_{\mathbf{k}+\mathbf{Q}} \left(a_{\mathbf{Q}} - a^\dagger_{-\mathbf{Q}}\right) c_{\mathbf{k}} \tag{8.59}$$

where the matrix element, calculated with the wave function given by (8.57), is

$$M_{\mathbf{kQ}} = \frac{2e\sqrt{\pi\Lambda_0\hbar^3}}{m\sqrt{\omega_s Q}} \left(\frac{2\Lambda_0}{Q a_B + 2\Lambda_0}\right)^3 \left[\mathbf{Q}\cdot\mathbf{k} + \frac{Q^2}{2}\right] \tag{8.60}$$

(m is the electron mass). Within the memory function approach [57] the surface conductivity can be written as

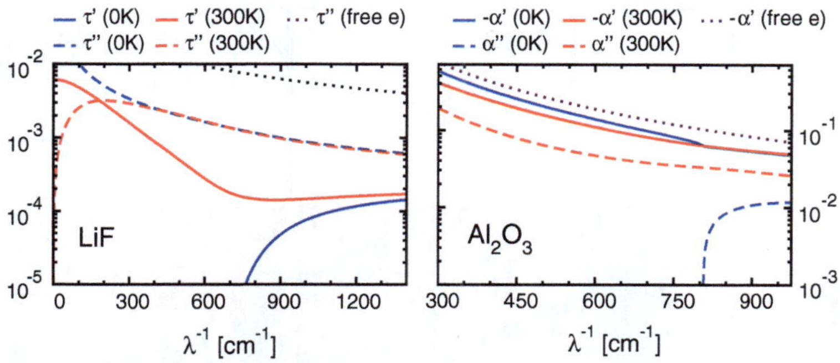

Fig. 8.14 Dimensionless surface conductivity $\tau = \tau' + i\tau''$ for LiF for $n_s = 10^{13}$ cm^{-2} (*left*) and polarisability of excess electrons $\alpha = \alpha' + i\alpha''$ for Al_2O_3 for $n_b = 3 \times 10^{17}$ cm^{-3} (*right*) as a function of the inverse wavelength λ^{-1}

$$\sigma_s(\omega) = \frac{e^2 n_s}{m} \frac{i}{\omega + M(\omega)} \tag{8.61}$$

with n_s the surface electron density. The memory function is then evaluated up to second order in the electron phonon coupling [58]. Since $M(\omega)$ is independent of n_s the surface conductivity is proportional to the surface density of electrons n_s.

For $\chi > 0$ the interaction of the electron with a longitudinal optical bulk phonon limits the bulk conductivity. The coupling of the electron to this mode with frequency ω_{LO} is described by [59]

$$H_{int} = \frac{1}{\sqrt{V}} \sum_{\mathbf{k},\mathbf{q}} \frac{M}{q} c^{\dagger}_{\mathbf{k}+\mathbf{q}} c_{\mathbf{k}} \left(a_{\mathbf{q}} + a^{\dagger}_{-\mathbf{q}} \right), \tag{8.62}$$

where $M = \sqrt{2\pi e^2 \hbar \omega_{LO} \left(\varepsilon_\infty^{-1} - \varepsilon_0^{-1} \right)}$. For the calculation of the bulk conductivity we employ again the memory function approach. In this case the bulk conductivity is proportional to the electron density n_b.

The charge effect on scattering is encoded in the surface conductivity for $\chi < 0$ and the bulk conductivity for $\chi > 0$ which enter through the dimensionless surface conductivity τ or the polarisability α into the scattering coefficients. Figure 8.14 shows $\tau = \tau' + i\tau''$ for LiF and the polarisability for $\alpha = \alpha' + i\alpha''$ for Al_2O_3 as a function of the inverse wavelength λ^{-1}. They turn out to be small even for a highly charged particle with $n_s = 10^{13}$ cm^{-2} (corresponding to $n_b = 3 \times 10^{17}$ cm^{-3} for $\chi > 0$ and $a = 1\mu$m). Compared to a free electron gas where $M(\omega) = 0$ (implying $\tau' = 0$ and $\alpha'' = 0$), the electron-phonon coupling reduces τ'' and α' considerably. For $T = 0K$, $\tau' = 0$ ($\alpha'' = 0$) for $\lambda^{-1} < \lambda_s^{-1} = 675$ cm^{-1}, the inverse wavelength of the surface phonon ($\lambda^{-1} < \lambda_{LO}^{-1} = 807$ cm^{-1}, the inverse wavelength of the bulk LO phonon) since light absorption is only possible above the surface (bulk LO) phonon frequency. At room temperature τ'' and α' still outweigh τ' and α''. The

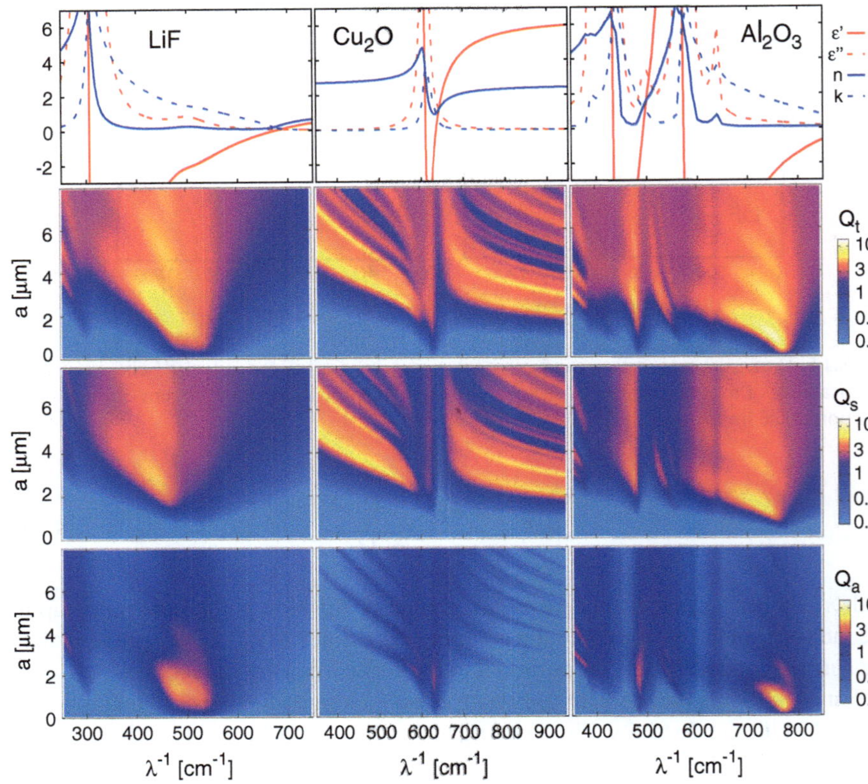

Fig. 8.15 Dielectric constant $\varepsilon = \varepsilon' + i\varepsilon''$, refractive index $N = n + ik$ (*first row*) and extinction efficiency Q_t (*second row*), scattering efficiency Q_s (*third row*) as well as absorption efficiency Q_a (*fourth row*) depending on the particle radius a for LiF, Cu_2O and Al_2O_3 as a function of λ^{-1}

temperature effect on τ'' is less apparent for $\lambda^{-1} > 300$ cm^{-1} than for α' but for $\lambda^{-1} < 300$ cm^{-1} a higher temperature lowers τ'' considerably.

We now turn to the scattering properties of the sphere. For LiF and Al_2O_3 one or two phonon modes dominate the dielectric constant. Far above the highest TO phonon mode ($\lambda^{-1} > 700$ cm^{-1} for LiF and $\lambda^{-1} > 900$ cm^{-1} for Al_2O_3) the refractive index N is real, relatively small and does not vary strongly with frequency. In this regime a micron sized grain would give rise to a typical Mie plot exhibiting interference and ripples which are due to the complicated functional form of a_n^r and b_n^r and not due to the underlying dielectric constant. Surplus electrons would moreover not alter the extinction behaviour in this region because $|\varepsilon| \gg |\tau|$ and $|\varepsilon| \gg |\alpha|$.

However, immediately above the TO phonon resonance in ε, $\varepsilon' < 0$ and $\varepsilon'' \ll 1$. This allows for anomalous optical resonances which are sensitive to small variations of ε and also to τ and α. They are due to resonant excitation of transverse surface modes of the sphere [60] and have been first identified for metallic particles where they lie in the ultraviolet [61, 62]. For a dielectric the TO phonon induces them.

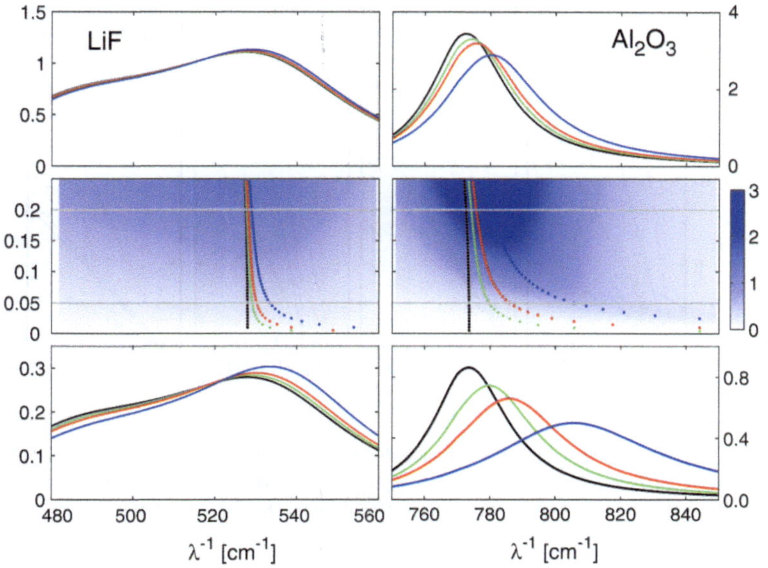

Fig. 8.16 *Middle panel* Extinction efficiency Q_t as a function of the inverse wavelength λ^{-1} and the radius a for a LiF with $n_s = 5 \times 10^{12}\text{cm}^{-2}$ (*left*) and an Al_2O_3 particle with $n_b = 3n_s/a$ (*right*) for $T = 300K$. The dotted lines indicate the extinction maximum for $n_s = 0$ (*black*), 10^{12} (*green*), 2×10^{12} (*red*), and 5×10^{12} cm^{-2} (*blue*). *Top* and *bottom panel* Extinction efficiency Q_t for different surface electron densities as a function of λ^{-1} and two radii $a = 0.2\mu\text{m}$ (*top*) and $a = 0.05\mu\text{m}$ (*bottom*). Adapted from [63]. Copyright (2013) by The American Physical Society

Figure 8.15 shows the dielectric constant ε, the refractive index N as well as the extinction Q_t, scattering Q_s, and absorption efficiency Q_a for LiF, Cu_2O and Al_2O_3 particles as a function of λ^{-1} and the particle radius a. For LiF and Al_2O_3 we find a clearly resolved series of optical resonances. For the smaller particles they are due to absorption while for larger particles they are scattering resonances. The crossover between scattering and absorption occurs in the first resonance. For comparison, we display in Fig. 8.15 also results for Cu_2O particles which do not show clearly resolved resonances. For this material $\varepsilon' < 0$ but ε'' is not sufficiently small to allow for sharp resonances. Nevertheless for submicron-sized particles a small extinction resonance due to absorption can be identified near 630 cm^{-1}.

The extinction resonances are blue-shifted with increasing particle charge [58]. This effect is most significant for small particles. Figure 8.16 shows the small particle tail of the lowest extinction resonance for LiF and Al_2O_3. The main panel shows the extinction efficiency as a function of the particle radius a and the inverse wavelength λ^{-1} for $n_s = 5 \times 10^{12}\text{cm}^{-1}$ (LiF) or $n_b = 3n_s/a$ (Al_2O_3). Superimposed is the extinction maximum for several values of n_s and corresponding n_b. The top and bottom panel show the lineshape of the extinction resonance for $a = 0.2\mu\text{m}$ and $a = 0.05\mu\text{m}$. For both materials the surplus electrons lead to a blue-shift of the resonance.

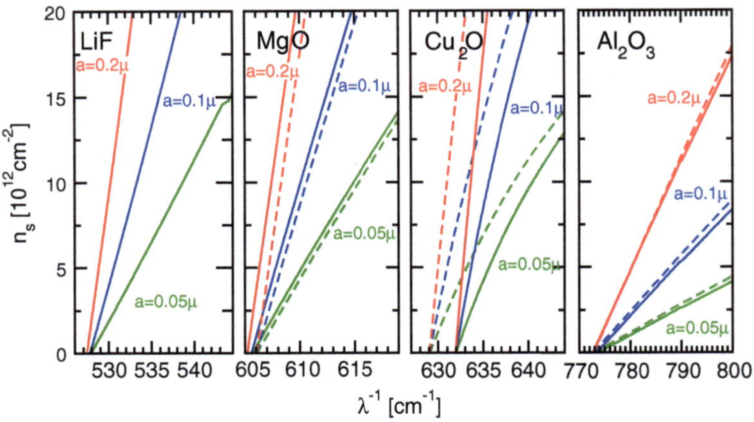

Fig. 8.17 Position of the extinction resonance depending on the surface charge n_s for LiF, MgO, Cu_2O and Al_2O_3 (for equivalent bulk charge $n_b = 3n_S/a$) particles with different radii a. Solid (*dashed*) lines are obtained from the Mie contour [(8.64) and (8.65)]. Adapted from [58]. Copyright (2012) by The American Physical Society

For submicron-sized particles where the resonance shift is most significant $a < \lambda$. In this small particle limit we can expand the scattering coefficients for small ρ. To ensure that in the limit of an uncharged surface, that is, for $\tau \to 0$, a_n^r and b_n^r converge to their known small ρ expansions [54], we substitute $t = \tau/\rho$ prior expanding the scattering coefficients. Up to $\mathcal{O}(\rho^3)$ this yields $a_1^r = a_2^r = b_2^r = 0$ and only $b_1^r \sim \mathcal{O}(\rho^3)$ contributes. Then the extinction efficiency reads

$$Q_t = \frac{12\rho\left(\varepsilon'' + \alpha'' + 2\tau'/\rho\right)}{(\varepsilon' + \alpha' + 2 - 2\tau''/\rho)^2 + (\varepsilon'' + \alpha'' + 2\tau'/\rho)^2}, \tag{8.63}$$

where we have restored τ. Excess charges enter either by τ ($\chi < 0$) or α ($\chi > 0$). For $\tau, \alpha \to 0$ this gives the limit of Rayleigh scattering. The resonance is located at wavenumbers where

$$\varepsilon' + 2 - 2\tau''/\rho = 0 \quad \text{for} \quad \chi < 0, \tag{8.64}$$

$$\varepsilon' + \alpha' + 2 = 0 \quad \text{for} \quad \chi > 0, \tag{8.65}$$

and has a Lorentzian shape provided ε'' and τ' (or α'') vary only negligibly near the resonance wavelength. Figure 8.16 confirms the Lorentzian lineshape for Al_2O_3. For LiF, however, ε'' has a hump (see Fig. 8.15) close to the resonance due to a second TO phonon mode which is much weaker than the dominant TO phonon. This leads to the deviation from the Lorentzian line shape.

For an uncharged surface the resonance is at λ_0^{-1} for which $\varepsilon' = -2$. For $\chi < 0$ the shift of the resonance is proportional to τ'' and thus to n_s, provided ε' is well approximated linearly in λ^{-1} and τ'' does not vary significantly near λ_0^{-1}. In

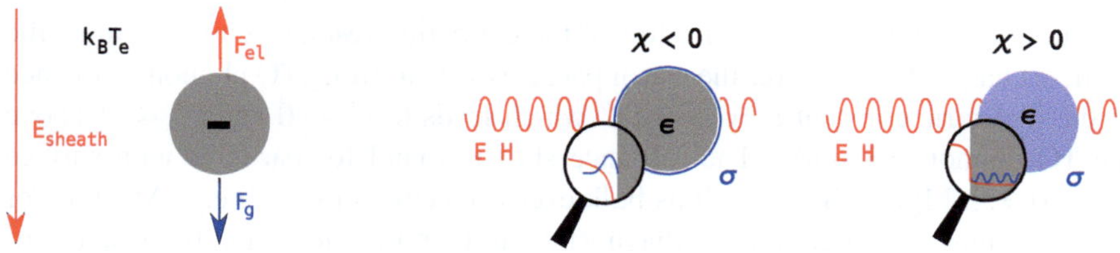

Fig. 8.18 Conventional charge measurement (*left*) and proposed optical charge measurement by Mie scattering (*right*). Conventional charge measurements rely on trapping the particle in the sheath and exploiting a force balance between gravity and the electric force on the particle. The knowledge of the plasma parameters is required to infer the particle charge. The optical measurement does not require the knowledge of plasma parameters

this case, we substitute in (8.64) the expansions $\varepsilon' = -2 + c_\varepsilon(\lambda^{-1} - \lambda_0^{-1})$ and $\tau'' = c_\tau n_s$ where $c_\varepsilon = \partial\varepsilon'/(\partial\lambda^{-1})|_{\lambda_0^{-1}}$ and $c_\tau = \tau''/n_s|_{\lambda_0^{-1}}$. Then the resonance is located at $\lambda^{-1} = \lambda_0^{-1} + c_\tau n_s/(\pi c_\varepsilon a \lambda_0^{-1})$. For $\chi > 0$ the resonance is located at $\lambda^{-1} = \lambda_0^{-1} - c_\alpha n_b/c_\varepsilon$ where $c_\alpha = \alpha'/n_b|_{\lambda_0^{-1}}$.

The proportionality of the resonance shift to n_s for LiF and n_b for Al_2O_3 can also be seen in Fig. 8.17 where we plot on the abscissa the shift of the extinction resonance arising from the surface electron density (or corresponding bulk electron density) given on the ordinate for LiF, MgO ($\chi < 0$) as well as Cu_2O and Al_2O_3 ($\chi > 0$). Both bulk and surface electrons lead qualitatively to the same resonance shift. Note that even for Cu_2O which does not show clearly resolved extinction resonance a shift is discernible. The most promising candidate for an optical charge measurement is Al_2O_3 where the shift is strongest.

To illustrate the similarity of bulk and surface electron effects we consider the resonance condition (8.64) or (8.65) for free electrons, which then becomes

$$\varepsilon' - \frac{2Ne^2}{(ma^3\omega^2)} = -2 \tag{8.66}$$

for $\chi < 0$ and

$$\varepsilon' - \frac{3Ne^2}{(m^* a^3\omega^2)} = -2 \tag{8.67}$$

for $\chi > 0$, where N is the number of electrons on the sphere. The effect of surface electrons is weaker by a factor of $2m^*/3m$. The factor $2/3$ can be understood as a geometric factor as only the parallel component of the electric field acts on the spherically confined electron gas. Most important, however, the fact that τ/ρ and α enter the resonance condition on the same footing shows that the resonance blueshift is in the first place an electron density effect on the polarizability of the grain. We therefore expect the shift to prevail also for a more complex electron distribution on the grain between the two limiting cases of a surface and a homogeneous bulk charge.

Our results suggest to use the shift of the extinction resonance to determine the particle charge. This requires the use of particles with a strong TO phonon resonance in the dielectric constant $\varepsilon = \varepsilon' + i\varepsilon''$ which leads to $\varepsilon' < 0$ and $\varepsilon'' \ll 1$ above the TO phonon frequency. The resoance shift is found for particles with surface ($\chi < 0$, e.g. MgO, LiF) as well as bulk excess electrons ($\chi > 0$ e.g. Al_2O_3). For dusty plasmas an optical charge diagnostic can be rather attractive because established methods for measuring the particle charge [44–46] require plasma parameters which are not precisely known whereas the charge measurement by Mie scattering does not (see Fig. 8.18). Particles showing the resonance shift could be employed as minimally invasive electric probes, which collect electrons depending on the local plasma environment. Determining their charge from Mie scattering and the forces acting on them by conventional means [44–46] would then allow to extract the local plasma parameters. Moreover, the Mie signal would provide a charge diagnostic for nanodust [47], which is too small and light for traditional charge measurements.

8.5 Summary

This chapter introduces a microscopic modelling of the plasma wall which complements traditional sheath models by an interface region—the ESL. In this model the negative charge on the plasma boundary is treated as a wall-thermalised electron distribution minimising the grand canonical potential and satisfying Poisson's equation. Its boundary with the plasma sheath is determined by a force balance between the attractive image potential and the repulsive sheath potential and lies in front of the crystallographic interface. Depending on the electron affinity χ, that is the offset of the conduction band minimum to the potential in front of the surface, two scenarios for the wall-bound electrons are realised. For $\chi < 0$ (e.g. MgO, LiF) electrons do not penetrate into the solid but are trapped in the image states in front of the surface where they form a quasi two-dimensional electron gas. For $\chi > 0$ (e.g. SiO_2, Al_2O_3) electrons penetrate into the conduction band where they form an extended space charge.

These different scenarios are also reflected in the electron physisorption at the wall. For $\chi < 0$ electrons from the plasma cannot penetrate into the solid. They are trapped in the image states in front of the surface. The transitions between unbound and bound states are due to surface vibrations. In this case the sticking coefficient for electrons is relatively small, typically on the order of 10^{-3}. For $\chi > 0$ electron physisorption takes place in the conduction band. For this case sticking coefficients and desorption times have not been calculated yet but in view of the more efficient scattering with bulk phonons, responsible for electron energy relaxation in this case, we expect them to be larger than for the case of $\chi < 0$. This indicates that the electron affinity should be an important parameter for the charging-up of plasma walls. In particular dusty plasmas offer the possibility to study experimentally the charge-up of dust grains depending on this parameter.

Our microscopic model can also be applied to study the optical properties of the wall-bound electrons on a dust particle. Surplus electrons affect the polarisability of the dust particle by their surface ($\chi < 0$) or bulk conductivity ($\chi > 0$). This leads to a blue-shift of an extinction resonance in the infrared for negatively charged dust particles. This effect offers an optical way to measure the particle charge which, unlike tradition charge measurements, does not require the plasma parameters at the position of the particle.

Acknowledgments This work was supported by Deutsche Forschungsgemeinschaft through SFB-TR 24.

References

1. R.N. Franklin, *Plasma Phenomena in Gas Discharges* (Clarendon Press, Oxford, 1976)
2. C.J. Wagner, P.A. Tschertian, J.G. Eden, Appl. Phys. Lett. **97**, 134102 (2010)
3. H. Baumgartner, D. Block, M. Bonitz, Contrib. Plasma Phys. **49**, 281 (2009)
4. A. Piel, O. Arp, D. Block, I. Pilch, T. Trottenberg, S. Kaeding, A. Melzer, H. Baumgartner, C. Henning, M. Bonitz, Plasma Phys. Control. Fusion **49**, 281 (2008)
5. V.E. Fortov, A.V. Ivlev, S.A. Khrapak, A.G. Khrapak, G.E. Morfill, Phys. Rep. **421**, 1 (2005)
6. J. Berndt, E. Kovačević, V. Selenin, I. Stefanović, J. Winter, Plasma Sources Sci. Technol. **15**, 18 (2006)
7. Y.B. Golubovskii, V.A. Maiorov, J. Behnke, J.F. Behnke, J. Phys. D: Appl. Phys **35**, 751 (2002)
8. H.E. Wagner, Y.V. Yurgelenas, R. Brandenburg, Plasma Phys. Control. Fusion **47**, B641 (2005)
9. R. Brandenburg, V.A. Maiorov, Y.B. Golubovskii, H.E. Wagner, J. Behnke, J.F. Behnke, J. Phys. D: Appl. Phys **38**, 2187 (2005)
10. L. Stollenwerk, J.G. Laven, H.G. Purwins, Phys. Rev. Lett. **98**, 255001 (2007)
11. M. Bogaczyk, R. Wild, L. Stollenwerk, H.E. Wagner, J. Phys. D: Appl. Phys **46**, 465202 (2012)
12. M. Bogaczyk, S. Nemschokmichal, R. Wild, L. Stollenwerk, R. Brandenburg, J. Meichsner, H.E. Wagner, Contrib. Plasma Phys. **52**, 847 (2012)
13. K.H. Becker, K.H. Schoenbach, J.G. Eden, J. Phys. D: Appl. Phys **39**, R55 (2006)
14. M.J. Kushner, J. Phys. D: Appl. Phys **38**, 1633 (2005)
15. U. Kogelschatz, Plasma Chem. Plasma Process. **23**, 1 (2003)
16. P.A. Tschertian, C.J. Wagner, T.L. Houlahan, B. Li, D.J. Sievers, J.G. Eden, Contrib. Plasma Phys. **51**, 889 (2011)
17. R. Dussart, L. Overzet, P. Lefaucheux, T. Dufour, M. Kulsreshath, M. Mandra, T. Tillocher, O. Aubry, S. Dozias, P. Ranson, J. Lee, M. Goeckner, Eur. Phys. J. D **60**, 601 (2010)
18. K.G. Emeleus, J.R.M. Coulter, Int. J. Electronics **62**, 225 (1987)
19. J.F. Behnke, T. Bindemann, H. Deutsch, K. Becker, Contrib. Plasma Phys. **37**, 345 (1997)
20. R.L. Heinisch, F.X. Bronold, H. Fehske, Phys. Rev. B **85**, 075323 (2012)
21. F. Stern, Phys. Rev. B **17**, 5009 (1978)
22. E.E. Tkharev, A.L. Danilyuk, Vacuum **35**, 183 (1985)
23. M.A. Lieberman, A.J. Lichtenberg, *Principles of Plasma Discharges and Materials Processing* (Wiley-Interscience, New York, 2005)
24. K.U. Riemann, J. Phys. D: Appl. Phys **24**, 493 (1991)
25. J.D. Jackson, *Classical Electrodynamics* (Wiley, New York, 1998)
26. J.B. Cui, J. Ristein, L. Ley, Phys. Rev. Lett. **81**, 429 (1998)
27. F. Maier, J. Ristein, L. Ley, Phys. Rev. B **64**, 165411 (2001)
28. F. Stern, S.D. Sarma, Phys. Rev. B **30**, 840 (1984)
29. H. Lüth, *Solid Surfaces, Interfaces and Thin Films* (Springer, Berlin, 1992)

30. W. Kohn, L.J. Sham, Phys. Rev. **140A**, 1133 (1965)
31. N.D. Mermin, Phys. Rev. **137A**, 1441 (1965)
32. K.G. Emeleus, J.R.M. Coulter, IEE Proc. **135**, 76 (1988)
33. F.X. Bronold, H. Fehske, R.L. Heinisch, J. Marbach, Contrib. Plasma Phys. **52**, 859 (2012)
34. R.L. Heinisch, F.X. Bronold, H. Fehske, Phys. Rev. B **81**, 155420 (2010)
35. R.L. Heinisch, F.X. Bronold, H. Fehske, Phys. Rev. B **82**, 125408 (2010)
36. R.L. Heinisch, F.X. Bronold, H. Fehske, Phys. Rev. B **83**, 195407 (2011)
37. Z.W. Gortel, H.J. Kreuzer, R. Teshima, Phys. Rev. B **22**, 5655 (1980)
38. H.J. Kreuzer, Z.W. Gortel, *Physisorption Kinetics* (Springer, Berlin, 1986)
39. G. Iche, P. Noziéres, J. Phys. (Paris) **37**, 1313 (1976)
40. W. Brenig, Z. Phys. B: **48**, 127 (1982)
41. E. Evans, D.L. Mills, Phys. Rev. B **8**, 4004 (1973)
42. G. Barton, J. Phys. C **14**, 3975 (1981)
43. B. Bendow, S.C. Ying, Phys. Rev. B **7**, 622 (1973)
44. J. Carstensen, H. Jung, F. Greiner, A. Piel, Phys. Plasmas **18**, 033701 (2011)
45. S.A. Khrapak, S.V. Ratynskaia, A.V. Zobnin, A.D. Usachev, V.V. Yaroshenko, M.H. Thoma, M. Kretschmer, H. Hoefner, G.E. Morfill, O.F. Petrov, V.E. Fortov, Phys. Rev. E **72**, 016406 (2005)
46. E.B. Tomme, D.A. Law, B.M. Annaratone, J.E. Allen, Phys. Rev. Lett. **85**, 2518 (2000)
47. F. Greiner, J. Carstensen, N. Koehler, I. Pilch, H. Ketelsen, S. Knist, A. Piel, Plasma Sources Sci. Technol. **21**, 065005 (2012)
48. C.F. Bohren, D.R. Huffman, *Absorption and Scattering of Light by Small Particles* (Wiley, Newyork, 1983)
49. J. Klačka, M. Kocifaj, Prog. Electromagnet. Res. **109**, 17 (2010)
50. J. Klačka, M. Kocifaj, J. Quant. Spectrosc. Radiat. Transfer **106**, 170 (2007)
51. C.F. Bohren, A.J. Hunt, Can. J. Phys. **55**, 1930 (1977)
52. M. Rohlfing, N.P. Wang, P. Krueger, J. Pollmann, Phys. Rev. Lett. **91**, 256802 (2003)
53. B. Baumeier, P. Krueger, J. Pollmann, Phys. Rev. B **76**, 205404 (2007)
54. J.A. Stratton, *Electromagnetic Theory* (McGraw-Hill, NewYork, 1941)
55. E.T. Whittaker, G.N. Watson, *A Course of Modern Analysis* (Cambridge University Press, Cambridge, 1927)
56. M. Kato, A. Ishii, Appl. Surf. Sci. **85**, 69 (1995)
57. W. Götze, P. Wölfle, Phys. Rev. B **6**, 1226 (1972)
58. R.L. Heinisch, F.X. Bronold, H. Fehske, Phys. Rev. Lett. **109**, 243903 (2012)
59. G.D. Mahan, *Many-Particle Physics* (Plenum, New York, 1990), pp. 703–708
60. R. Fuchs, K.L. Kliewer, J. Opt. Soc. Am. **58**, 319 (1968)
61. M.I. Tribelsky, B.S. Luk'yanchuk, Phys. Rev. Lett. **97**, 263902 (2006)
62. M.I. Tribelsky, Europhys. Lett. **94**, 14004 (2011)
63. R.L. Heinisch, F.X. Bronold, H. Fehske, Phys. Rev. E **88**, 023109 (2013)

Chapter 9
Characterization of Local Structures in Plasma Deposited Semiconductors by X-ray Absorption Spectroscopy

M. Alper Sahiner

Abstract Extended X-ray-Absorption Fine-Structure Spectroscopy (EXAFS) has been used to investigate the subtle local structural variations in plasma deposited semiconductors. Grazing incidence geometry EXAFS is a very effective tool to study the surface layers. Since EXAFS is an element specific sensitive local structural probe, it is advantageous to commonly used structural characterization techniques where there is no long-range crystalline order in material. EXAFS can provide crucial information deposition or post-deposition induced crystallographic structural modifications. The information extracted from EXAFS can be used as an important feedback for the thin film growth mechanisms. In this chapter the fundamental principles of EXAFS will be introduced. The data reduction and analyses with the structural model calculations will be discussed. The application of the EXAFS in plasma deposited silicon wafers and plasma-plume deposited high-k dielectric thin films will be presented.

9.1 Introduction

The continuous down scaling of the semiconductor devices creates challenging materials related problems for the semiconductor researchers. Highly sensitive structural characterization techniques are crucial in searching for materials based solutions to these problems. One of most challenging tasks in semiconductor industry is to keep the dopant levels very high in ever shrinking the p and n-type dopant areas of the complementary metal oxide semiconductor (CMOS) devices. The dopant atom concentrations usually exceed the solid solubility limits of silicon or germanium and post deposition annealing processes are applied to prevent the dopant clustering and increase electrical activation. In addition to conventional beamline ion implantation

M. A. Sahiner (✉)
Department of Physics, Seton Hall University, South Orange, NJ 07079, USA
e-mail: mehmet.sahiner@shu.edu

M. Bonitz et al. (eds.), *Complex Plasmas*, Springer Series on Atomic, Optical, and Plasma Physics 82, DOI: 10.1007/978-3-319-05437-7_9,

methods, alternative plasma based deposition techniques have recently been used to increase the dopant levels and electrically active charge carriers in ultra-shallow junctions. Plasma immersion ion implantation (PIII) technology is shown to be one of the effective techniques in this field [1] and can provide a better conformal doping on 3D structures [2]. An increase of retained dose after annealing has been reported for arsenic implantation using plasma sources [3, 4]. These PIII prepared samples were fabricated using AsH_3/H_2, <2kV bias in a sub 30 m Torr pressure. After the deposition the samples were laser annealed using a pulsed laser by varying the laser power, the total annealing time and the number of laser pulses. EXAFS studies on these samples revealed interesting local structural response to variations in the post deposition annealing conditions.

In another example pulsed laser deposited (PLD) high-k dielectric thin films were studied by EXAFS in order to investigate the present non-equilibrium structural phase present in the deposited thin films. In pulsed laser deposition, the solid target material is evaporated by laser pulses of a high-energy KrF excimer laser, ionized and ejected as a plasma plume. The plasma plume expands outwards and deposits the target material on a substrate. The plasma properties of the plume determine the quality of the thin films deposited on the substrate. These plasma plume properties include ion density, ion flow speed, electron temperature, and plume peaking parameter [5]. Hf based oxide thin films were prepared by PLD and subtle variations in the deposition parameter such as substrate temperature and the thickness of the films were probed by EXAFS giving a detailed picture of the non-equilibrium crystal phases in the thin films.

9.2 EXAFS

Extended X-ray absorption fine structure spectroscopy (EXAFS) is a local structural probe utilizing the measurement of energy dependence of X-ray absorption coefficient $\mu(E)$ of the selected main absorbing atom in the material. EXAFS is element specific, that is, by tuning the incoming X-ray energy, through the use of the beamline's double crystal Si monochromator, to the absorption edge of any specific element in the material local structure around that particular atom can be probed. An incident X-ray is absorbed by a main absorbing atom when the energy of the X-ray is transferred to a core-level electron which is ejected from the atom. Any excess energy from the incident X-ray is given to the ejected photoelectron. The energy dependent X-ray absorption coefficient is modulated due to interference between the outgoing photoelectron waves and the backscattered waves from the near neighbor atoms. Figure 9.1 is a schematic diagram showing to the X-ray absorption process, atomic potentials and the EXAFS data for a main absorber atom A surrounded by near-neighbor B atoms. Therefore, the absorption coefficient intrinsically contains full local structural information around the main absorbing atoms. If the scattering properties (scattering amplitude, phase shift, and mean free path) of the neighbor atom are known or calculated through simulations, then EXAFS can provide detailed

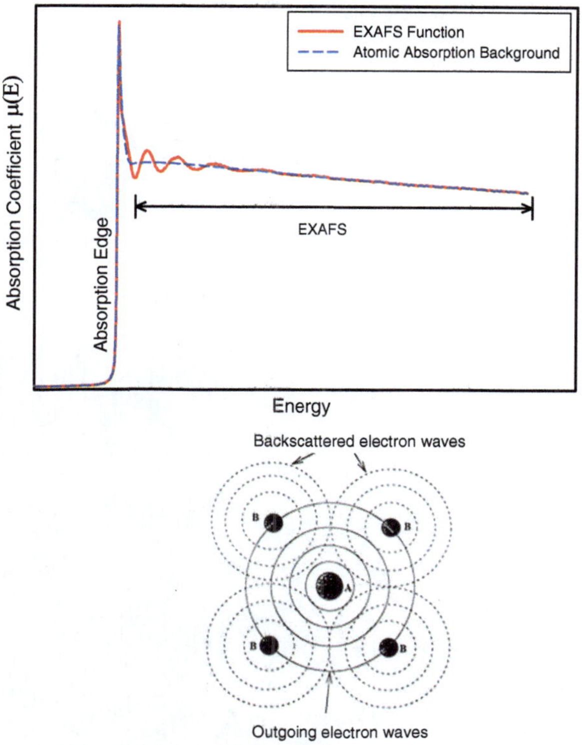

Fig. 9.1 A schematic diagram for EXAFS process and acquired EXAFS data

information on the near neighbor distances, coordination numbers, crystal symmetry, and the structural disorder around the main absorbing atom. Since the scattering is from the first couple of near neighbor atoms EXAFS can provide information on non-periodic structures even on amorphous materials. EXAFS studies on subtle local structural modifications, such as Hf based high-k thin films in this work, is most powerful when EXAFS data is supported by using computer generated models of similar atom clusters and the corresponding scattering simulations of EXAFS functions.

9.2.1 EXAFS Experimental Set-up

In Fig. 9.2, a schematic diagram for a typical EXAFS experiment is shown. Highly collimated, X-ray white beam through the synchrotron source is passed through a monochromator in order to tune the incoming X-ray energies about the absorption edge of the selected atoms. The incoming X-ray intensity (I_0) is measured by a ionization chamber and then the sample is irradiated and either the intensity of the outgoing X-rays (I) or the fluorescent radiation intensity (I_f) is measured either by another ionization chamber or a fluorescence (Lytle) detector.

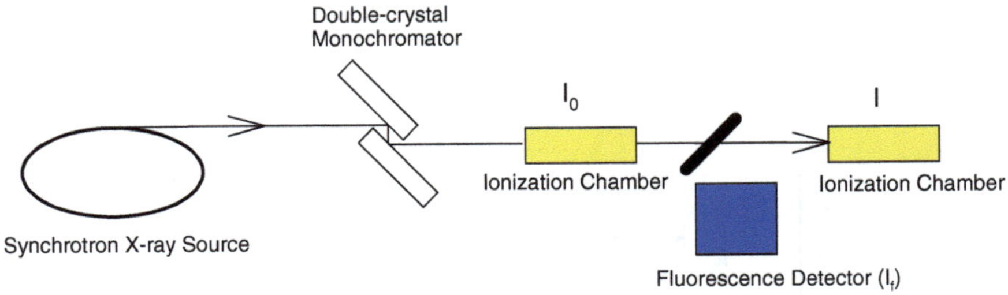

Fig. 9.2 Schematic experimental set-up for an EXAFS experiment

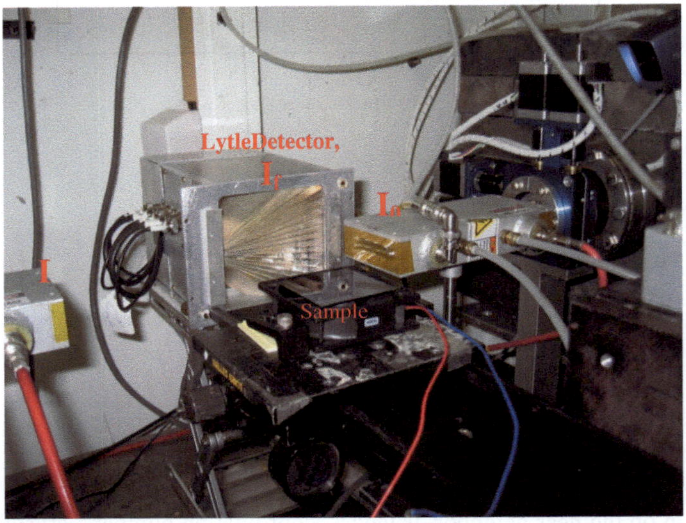

Fig. 9.3 Lytle detector in background measure I_f. A large sample (5 cm × 5 cm) is mounted on a fan motor wired to a variac so it can spin and avoid swamping the detector with Bragg peaks

Some of the EXAFS experiments for this work, were performed at the National Synchrotron Light Source (NSLS) at Brookhaven National Laboratory (BNL) on beamline X23A2, operated by the National Institute of Standards and Technology (NIST). The Hf L_3 absorption edge (9561 eV) or As-K edge (11867 eV) were used in the EXAFS data acquisition in the fluorescence detection mode with a angle of X-ray incidence close to the critical angle of silicon. Figure 9.3 below is a photograph of the EXAFS experimental set-up at X23A2 for these measurements.

9.2.2 EXAFS Data Analysis

The absorption coefficient as a function of energy is then calculated as shown in Fig. 9.4. The spectrum shown in Fig. 9.4 is a typical X-ray absorption spectrum. The spectrum plotted in red is the isolated atomic absorption background and the blue

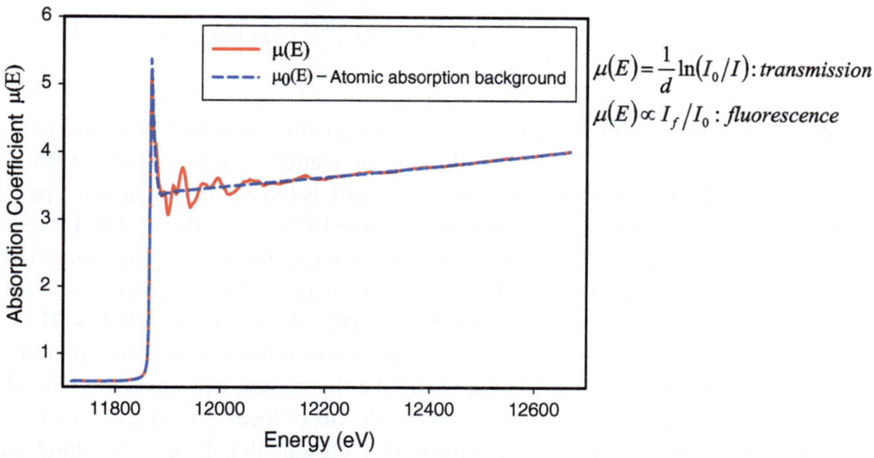

Fig. 9.4 A typical EXAFS spectrum (*blue*) and the atomic absorption background (*red*).

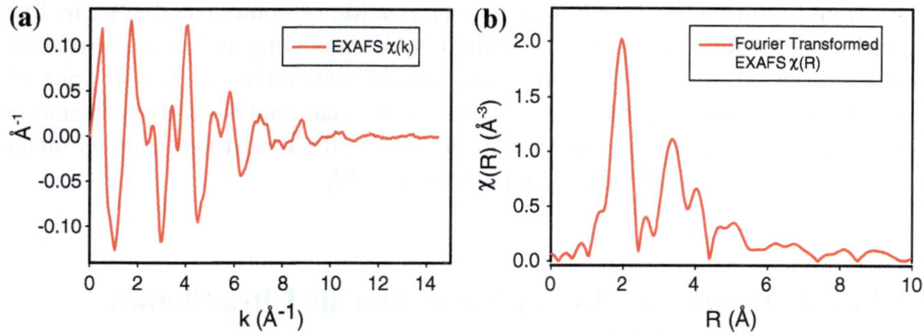

Fig. 9.5 **a** EXAFS function χ(k) **b** Magnitude of the Fourier transform of χ(k)

line plot is the EXAFS spectra exhibiting the oscillatory behavior due to scattering from near-neighbor atoms.

After the atomic absorption background is subtracted, the spectra is plotted in photoelectron wave number, k, using,

$$E - E_0 = \frac{(hk/2\pi)^2}{2m} \Rightarrow k = \sqrt{\frac{2m\,(E - E_0)}{(h/2\pi)^2}}$$

where, E_0 is the absorption threshold energy, m is the electron mass, h is the Planck's constant. This function is called the EXAFS function χ(k). A typical χ(k) and its Fourier Transform is shown in Fig. 9.5.

χ(k) is simply a summation of the scattering contributions (damped sine waves) from all the possible photoelectron scattering paths between the main absorbing atom and its near-neighbors [6, 7].

$$\chi(k) = \sum_j \chi_j(k) = \frac{N_j f_j(k)}{kR_j^2} e^{-2R_j/\lambda(k)} e^{-2k^2\sigma_j^2} \sin\left[2kR_j + \delta_j(k)\right]$$

where, j is the indices for the paths; N_j is coordination number; $R_j$is the near-neighbor distance; f_j scattering amplitude; $\delta_j$is the scattering phase; σ is the Debye-Waller factor indicating the structural disorder; and λ is the electron mean-free path. In Fig. 9.5b the magnitude of the Fourier transformed χ(k) is exhibited. The FT data leads to a pseudo radial distribution function around the main absorbing atom. The FT data does not reflect the actual distances of the near-neighbor atoms because of the FT contains extraneous information such as scattering phases and amplitudes. However, if the scattering amplitudes and phases are known then all the corresponding scattering contributions can be calculated and the theoretical χ(k) can be obtained. To be able to extract reliable information on the details of local structure of a system, detailed EXAFS fitting should be applied based on the calculated theoretical models, χ_j(k). These χ_j(k)'s are then fed into EXAFS fitting routines in order to compute the local structural parameters of the unknown from its experimental EXAFS data.

Previous similar work on EXAFS characterization of complicated systems like layered superconductors [8–10] and similar EXAFS modeling and analysis work on the dopant-related electrically inactive structures in semiconductors [11–16] proved that with careful theoretical modeling and the EXAFS data analysis and interpretation could lead to crucial information about the subtle structural modifications under pressure, ion implantation and post annealing [17–19].

9.3 Local Structural Information in Arsenic Ultra Shallow Junctions by EXAFS

The understanding of the behavior of arsenic in highly doped near surface silicon layers is of crucial importance for the formation of n-type Ultra Shallow Junctions in current and future very-large-scale integration (VLSI) technology. This is of peculiar relevance when studying novel implantation and annealing methods. Past theoretical as well as experimental investigations have suggested that the increase of As concentrations, and therefore the vicinity of the dopant atoms, leads to a drastic increase of electrically inactive defects giving only marginal effects on reducing sheet resistance [11]. Monoclinic SiAs clusters, as well as various arsenic-vacancy aggregates contribute to the deactivation of the arsenic. Giubertoni et al. [11] correlated the results of electrical activation measurements with EXAFS measurements. Specifically, a quantitative interpretation of the EXAFS spectra has been carried out in order to correlate the local atomic order of arsenic to the electrical characteristics as determined by Hall Effect measurement. Moreover, the percentage of substitutional dopant produced by the different annealing processes has been obtained through least squares fits of the EXAFS spectra with simulations of relaxed structures of As_nV defects obtained by density functional theory (DFT) calculations. The results confirm EXAFS as a powerful technique, not only able to correlate atomic structures with macroscopic electrical behaviors, but also to give quantitative information about the defect populations even for ultra-shallow As distributions. For these experiments, EXAFS measurements have been performed at room temperature at the BM08 GILDA beamline of the European Synchrotron Radiation Facility in grazing incidence and fluorescence acquisition with a side-looking 13 element GeHP detector. The sample was positioned horizontally [20]. As K-edge spectra have been acquired in the energy range 11600–12700 eV with variable energy step (0.5 eV in the proximity of the edge, 5 eV at the periphery of the scan) at an incidence angle above the critical for total reflection (about 0.18° measured from the sample surface) for all samples. The critical angle for total reflection for Si varies between 0.154° and 0.140° for energies in the range 11600 and 12700 eV respectively. The chosen angle of incidence allows the sampling of the whole dopant distribution with almost uniform weight across the implant. A 100 keV As implant (fluence 1×10^{15}cm^{-2}) laser (melting) annealed with supposed electrical activation close to 100 % was used as a reference for the EXAFS analyses. Theoretical EXAFS functions were calculated using University of Washington's multiple scattering EXAFS calculation code FEFF8.4 [21]. The structural parameters, which were obtained by the DFT calculations, were used in calculating these EXAFS models. The theoretical EXAFS standards for possible cluster structures and monoclinic SiAs precipitates were used in least-squares EXAFS fits to the Fourier Transformed (FT) data. Experimental EXAFS functions, χ(k)'s, are extracted by subtracting atomic absorption background using the AUTOBK code [22]. The χ(k)'s are then Fourier Transformed (FT) using a Gaussian window for [2.0–10.0 Å^{-1}] k-range with a k^2 weighting in order to fully account the contribution from the larger k region. EXAFS fits were performed assuming the co-presence of all the As_nV(n = 1–4) clusters, SiAs-precipitates and the substitutional-As (arsenic atoms surrounded by silicon atoms in silicon crystal) in the samples.

9.3.1 DFT Calculations

Geometry optimizations were carried out for four systems containing a Si vacancy surrounded by 1–4 As substitutional defects using the CASTEP plane wave density functional code [23]. Initial state structures of As_nV (V = vacancy, n = 1–4) were prepared as follows. A bulk Si crystal ($Fd\bar{3}m$), in which the cell origin was translated by 0.5 a to place a Si atom at its center was first relaxed using the GGA PW-91 exchange-correlation functional [24] to less than 0.01 eV/atom. The planewave basis cutoff energy was 320 eV and a Monkhorst-Pack k-point grid with 0.037 Å^{-1} spacing was used for all energy minimizations. The relaxed Si crystal was modified to contain a Si vacancy defect at the cell center and 1–4 As substitutional defects in the first coordination shell around the vacancy (Fig. 9.6). For each defect state the fractional coordinates were optimized using the same functional and convergence criteria as used for bulk Si. The cubic lattice parameters obtained for relaxed bulk Si were used for initial state structures for each of the As-substitutional defect states.

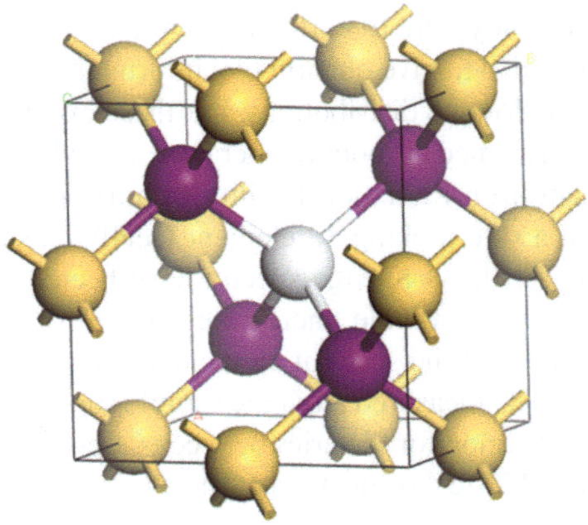

Fig. 9.6 Unit cell schematic showing Si (*yellow*), As substitution sites (*purple*) and Si vacancy (*white*)

Geometry optimizations of the defect states were done using fixed lattice parameters but flexible internal fractional coordinates. Fixing the lattice parameters to relaxed bulk values permits geometry optimization of the defect environment while retaining cubic bulk symmetry in the neighboring cells and bulk crystal lattice.

DFT geometry optimizations yielded a Si bulk lattice constant of 5.382 Å, 0.9 % smaller than the experimental value. DFT geometry optimizations of high atom density crystals, particularly Si, are recognized to yield slightly smaller lattice constants than experimental results although the fractional coordinates are reliable [25]. The lattice parameters and fractional displacements (in x, y and z directions) of the atoms in the first coordination shell around the vacancy for optimized As_nV are shown in Table 9.1. In all cases, the vacancy neighbor shell atoms are displaced toward the vacancy on the order of 0.2–0.35 Å for As and 0.1–0.2 Å for Si. Although other vacancy-substitution defects are possible, we modeled those systems which are expected to exhibit the most stable states. Both the stability and diffusion characteristics of these defect states impact their presence in the lattice, particularly after implantation annealing.

9.3.2 Electrical Data and EXAFS Results

The sheet resistance (R_S) values measured on all thermally treated samples are reported in Table 9.2.

Samples 2 and 3 showed similar values: 679 and 723 Ω/sq, respectively. Due to the similar junction depth (x_j) observed by secondary ion mass spectrometry (SIMS), the R_s difference is expected to be due to a different level of electrical activation. In fact, the measured active carrier dose is 2.9×10^{14} and 2.7×10^{14} cm^{-2} for samples 2 and 3, respectively. When the implanted fluence is increased

Table 9.1 Fractional atomic displacements for geometry optimized As_nV (n = 1–4) states

State	a (Å)	Fractional displacement toward vacancy				
Bulk Si	5.382		–	–	–	–
			(As)	(Si)	(Si)	(Si)
		x	0.0137	0.0051(6)	0.00516	0.00612
AsV	5.382	y	0.0140	0.0059(5)	0.00524	0.00523
		z	0.0140	0.0052(4)	0.00595	0.00523
			(As)	(As)	(Si)	(Si)
		x	0.01377	0.01377	0.01083	0.01083
As_2V	5.382	y	0.01377	0.01377	0.01083	0.01083
		z	0.01451	0.01454	0.00512	0.00512
			(As)	(As)	(As)	(Si)
		x	0.01400	0.01266	0.03578	0.01459
As_3V	5.382	y	0.01433	0.01266	0.01373	0.00793
		z	0.01250	0.01378	0.03607	0.01488
			(As)	(As)	(As)	(As)
		x	0.01768	0.01768	0.01768	0.01768
As_4V	5.382	y	0.01768	0.01768	0.01768	0.01768
		z	0.01768	0.01768	0.01768	0.01768

Table 9.2 Samples description and Hall effect results

Sample ID	Implanted dose (cm^{-2})	Annealing	Etching	Rs (Ω/sq)	Mobility (cm^2/Vs)	Carrier dose (cm^{-2})	Retained dose (cm^{-2})	Active fraction (%)
1	1×10^{15}	As implanted	–	–	–	–	9.6×10^{14}	–
2	1×10^{15}	LA 1150 °C	–	679	31.7	2.9×10^{14}	9.9×10^{14}	29.5
3	1×10^{15}	LA 1300 °C	–	723	32	2.7×10^{14}	1.0×10^{15}	27.3
4	3×10^{15}	LA 1300 °C	–	782	(95)	(9.15×10^{13})	(2.7×10^{15})	(3.4)
5	1×10^{15}	RTP 1050 °C	–	490	64.6	2.0×10^{14}	7.0×10^{14}	28.2
6	1×10^{15}	LA 1150 °C + RTP 1050 °C	–	450.5	60.8	2.2×10^{14}	6.6×10^{14}	33.6
5 etch	1×10^{15}	RTP 1050 °C	Yes	817.9	73.4	1.0×10^{14}	3.1×10^{14}	33.2
6 etch	1×10^{15}	LA 1150 °C + RTP 1050°C	Yes	698.0	68.7	1.30×10^{14}	3.1×10^{14}	41.9

RTP Rapid thermal processing, *LA* Lase annealing

to 3×10^{15} cm^{-2} the R_s value also appears to increase to 782 Ω/sq and the active carrier dose is just 9.1×10^{13} cm^{-2}, indicating poorer electrical activation. Due to the accuracy of the SIMS results, the ratio of the Hall Effect measured active dose to the SIMS determined dose gives a relatively reliable measure of the active fraction of dopant in the junction: this value, reported in Table 9.1, is nearly one third for the 1×10^{15} cm^{-2} samples, whereas it falls to ~3 % for the 3×10^{15} cm^{-2} sample. EXAFS was thus used to investigate the amount of As in substitional position and to understand defects behind the relatively large fractions of inactive dopant. Only As_nV clusters and SiAs precipitates were incorporated in the EXAFS fits together

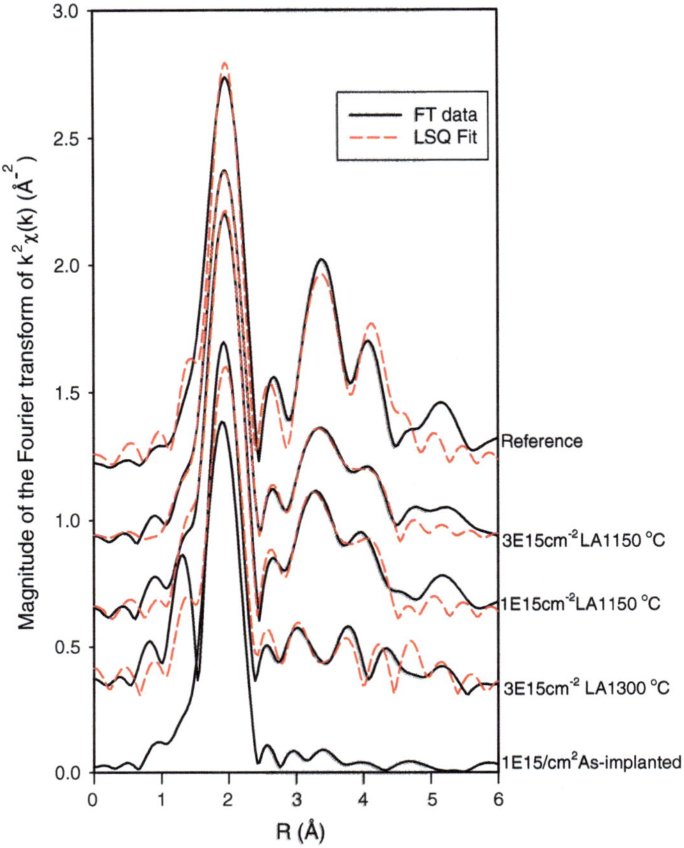

Fig. 9.7 Fourier transformed EXAFS data and the corresponding fits for laser annealed samples. The reference sample and the As-implanted sample (data only) were also plotted for comparison purposes of the second shell structures

with the substitutional arsenic. As discussed before, As_nV complexes are not the only possible As clusters according theoretical studies but most likely ones due to lower formation energies. Not only the formation energies but also the diffusion characteristics of different clusters have important effects on the presence of these structures after post implantation annealing [26, 27]. For As_nV clusters, the local structural parameters obtained from geometry optimization calculations were used in modeling of the EXAFS functions that were used in the least squares fits. For the monoclinic-SiAs, EXAFS modeling the structural parameters from literature was used [28] and for the substitutional-As form, EXAFS model was calculated with arsenic core atom replacing one of the Si atom in a silicon crystal.

Figure 9.7 shows the FT data of the EXAFS function for the laser annealed samples measured above critical angle and the corresponding EXAFS fits except for the

Table 9.3 Active As fraction from Hall effect measurements and fraction of the As complexes obtained from least squares EXAFS fits

Sample	Active fraction (%)	X-ray inc. angle	Substitional As	AsV	As_2V	As_3V	As_4V	SiAs
2	29.5	Above	0.31	0.10	0.00	0.39	0.14	0.07
3	27.3	Above	0.26	0.12	0.02	0.53	0.05	0.00
4	3.4	Above	0.05	0.16	0.05	0.38	0.15	0.21
5	28.2	Above	0.24	0.06	0.08	0.14	0.18	0.14
		Below	0.13	0.08	0.16	0.17	0.29	0.17
6	33.6	Above	0.36	0.09	0.12	0.15	0.14	0.14
		Below	0.14	0.13	0.17	0.18	0.20	0.18
5 etch	33.2	Above	0.57	0.04	0.07	0.04	0.20	0.08
6 etch	41.9	Above	0.58	0.00	0.10	0.08	0.13	0.12
Reference	–	Above	0.59	0.00	0.24	0.00	0.17	0.00

Table 9.4 Coordination numbers, near-neighbor distances and Debye-Waller factors from fits

Sample ID	X-ray inc. angle	First shell			Second shell		
		N	R(Å)	σ(Å^2)	N*	R(Å)	σ(Å^2)
2	Above	3.7(2)	2.38(2)	0.002(1)	10.8(4)	3.85(2)	0.04(1)
3	Above	3.6(2)	2.37(2)	0.002(1)	11.3(3)	3.87(2)	0.03(1)
4	Above	3.3(2)	2.36(2)	0.004(2)	9.9(4)	3.88(2)	0.05(3)
5	Above	3.7(3)	2.38(2)	0.003(2)	10.6(3)	3.86(2)	0.04(2)
	Below	3.4(3)	2.35(1)	0.003(2)	10.3(4)	3.87(1)	0.03(2)
6	Above	3.6(3)	2.38(2)	0.002(1)	10.6(3)	3.85(2)	0.02(2)
	Below	3.5(2)	2.36(1)	0.003(1)	10.2(4)	3.88(2)	0.04(2)
5 etch	Above	3.8(2)	2.39(2)	0.002(1)	11.5(2)	3.85(1)	0.03(2)
6 etch	Above	3.8(2)	2.39(2)	0.002(1)	11.4(2)	3.85(1)	0.03(2)
Reference	Above	3.9(1)	2.40(1)	0.002(1)	11.6(2)	3.84(1)	0.02(1)

as-implanted sample, which exhibits no long range order beyond the first shell as expected for a high dose amorphizing implant.

The final weighting (fraction) of the different structures was determined from the fit results and it is listed in Table 9.3. The arsenic coordination numbers, near-neighbor distances, and Debye-Waller factors are listed in Table 9.4.

Generally the agreement between the active dose obtained from electrical measurement and the fraction of substitutional As obtained with the EXAFS fit results is very good as seen from Table 9.3. The results of the electrical measurements indicate that the application of only laser annealing does not increase activation levels higher than ~30 %: EXAFS results for the substitutional As fraction for samples 2 and 3 are 33 and 26 %, respectively, confirming the findings from combining Hall Effect and SIMS characterization. Increasing the As implant dose (sample 3 compared with sample 4) sharply reduces the activated As fraction. Similarly EXAFS determined substitutional As fraction is significantly lower (5 %) for the 3×10^{15} As implant sample. Furthermore, when electrical activation is lower, the coordination number

Table 9.5 Matrix of analyzed PIII samples; all samples were produced on the same wafer labeled S2

Sample ID	C08: N laser pulses	C09: N × 3 laser pulses	C10: N × 10 laser pulses	C11: N × 30 laser pulse
R1: low laser power	S2R1C08	S2R1C09	S2R1C10	S2R1C11
R6: high laser power	S2R6C08	S2R6C09	S2R6C10	S2R6C11

decreases for both first and second shell. The bond length slightly decreases in the first shell whereas increases in the second one: this behavior is consistent with As_nV structures as already reported by Allain et al. [17] and D'Acapito et al. [29]. The latter obtained bond length values systematically slightly higher than the ones reported in Table 9.4 and from Allain et al. [17], but within the experimental error. Regarding the fractions of the As complexes, the low dose samples do not have relevant presence of SiAs precipitates but most As is in AsV and As_3V defects. The latter prevails as main deactivating defect even on the more thermodynamically stable As_4V, probably because of kinetics constraints and because entropy does not favor the formation of large complexes in line with what suggested by Mueller et al. [30]. When the As dose is increased (sample 4), the fractions of dopant in As_4V and SiAs complexes increase as expected being the concentration higher.

9.3.3 *Arsenic PIII Structures by EXAFS*

Arsenic implants were fabricated by plasma ion immersion implantation (PIII) using AsH_3/H_2, <2kV bias in a sub 30 m Torr pressure [2]. The samples were subsequently laser annealed using a pulsed laser to a range of laser powers and varying the total annealing time by varying the number of pulses. A single 300 mm (100) Si wafer (labeled 'S2') was processed adjusting the doping parameters to implant a nominal Arsenic dose of $\sim 1 \times 10^{15}$ at/cm^2. On the wafer a pattern of differently annealed areas of $\sim 1 \times 3$ cm^2 was created by the laser thermal treatment using different laser energy per area [J/cm^2]. Varying two main parameters of the laser annealing, namely laser power and the number of laser pulses, two series of samples were generated, one annealed using low laser energy (R1) and the other using high laser energy (R6). Within each series the number of pulses was varied to investigate whether this affects the annealing. Table 9.5 shows the matrix of analyzed samples.

Fourier transformed EXAFS data collected for selected PIII samples are displayed in Fig. 9.8. They agree very well with the findings obtained by SIMS and XPS in terms of As concentrations in the oxide and silicon: clear amplitudes corresponding to As having Oxygen as nearest neighbor (labeled as 'As-O') for all samples of series R1 (except for sample R1C11) indicate a significant amount of As in SiO_2.

The standard XA4 (80 % As in SiO_2) is plotted for comparison. The FT data show that the As-Si peaks are slightly shifted towards larger near-neighbor distances in the R1 samples indicating a change in the As local environment compared with the

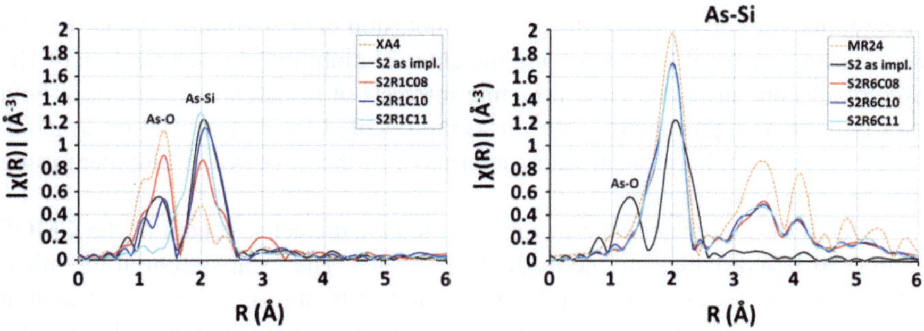

Fig. 9.8 Fourier transform of the EXAFS collected for selected PIII samples and standards XA4 (∼80 % of As in SiO_2) and MR24 (substitutional As in Si). The labeled peaks indicate the amplitudes corresponding to the first coordination shells of As-O and As-Si structures

reference samples. It is interesting to note that the As-O peaks on the other hand agree very well with the corresponding amplitudes of the standard suggesting that it is the part of As in Si (at the Si interface) which is (chemically) different. However, this is not surprising as the As concentrations in the PIII samples are more than one order of magnitude larger than the ones in the standard making it reasonable that the As in that case is part of a different chemical complex. As before sample R1C11 shows different EXAFS than all other samples: in agreement with XPS (oxide thickness) and SIMS (oxide thickness and profile) the EXAFS shows no detectable As-O signal, suggesting that all (most) As is in Si. The remaining part of As found in the oxide layer by SIMS has to be considered below the detection limit of EXAFS analysis for this sample, because the weak As-O signal is covered by the much stronger signal corresponding to As-Si. (The same is obviously true for the EXAFS collected for the samples of series R6 where the As-O to As-Si signal ratio is even smaller.)

Finally, none of the samples of series R1 show any detectable higher local order, i.e. crystallinity, like in the standard MR24 suggesting that the annealing was able to redistribute part of the As (diffusion into Si, redistribution within As rich oxide layer) but was not sufficient to significantly heal the crystal damage (within the EXAFS detection limit).

From a technological point of view, when aiming for ultra-shallow As distributions with very high retained doses, we can conclude that a combination of PIII and laser annealing with carefully optimized laser parameters seems a very promising approach as revealed from structural EXAFS analyses [2].

9.4 Local Structural Information in High-k Dielectrics by EXAFS

Although, the use of high-k dielectrics in the SiO_2 in the gate dielectric region of the complementary metal oxide (CMOS) devices have already started in industry in 2007, some of the materials related issues described by Wilk [31] such as the determination of the stable crystal phases of the formed oxides under high processing temperatures still remain to be solved. Since the possible inclusion of Ge in future CMOS devices due to promising results in increasing the channel mobility, characterization of Hf-based oxides on germanium are attracting interests of researchers [32, 33]. In both cases (silicon and germanium substrates) the detailed local structural information around the Hf atom has crucial value in terms of further process development using these exciting new structures.

In search of the replacement of SiO_2 at the gate dielectric region of the CMOS devices, Hf based oxides are the leading candidates. The local structural modifications around the Hf atom and a reliable method of monitoring the local structural modifications as response to synthesis conditions are crucial in studying these materials. One of the key questions to be addressed in studying the Hf based oxides is determining the stabilized crystal structure formed under various synthesis conditions. Previously, crystal phase transformations due to post deposition treatments in Hf based oxide thin films [34] or Zr content $Hf_{1-x}Zr_xO_2$ nanocrystals [35] have been reported.

EXAFS analyses were used in order to study the thin films of Hf based oxides deposited on silicon and germanium by pulsed laser deposition technique [36]. It has been observed that the HfO_2 crystal structure is highly dependent on parameters of the synthesis such as substrate temperature during the deposition and Zr concentration in $Hf_{1-x}Zr_xO_2(x = 0.0–1.0)$. The local structural modifications due to substrate temperature variations during the deposition and the Zr inclusion and the local structure of HfO_2 thin films on germanium were probed by X-ray absorption fine-structure spectroscopy (EXAFS) [36]. Specifically, pulsed laser deposition (PLD) technique has been used in the deposition of the thin films with systematic variations of substrate temperature, Zr content of the targets and substrate selection. Non-equilibrium processes between plasma plume produced by the high energy KrF laser pulses on the solid HfZrO targets and the single crystal silicon or germanium substrate is shown to lead to non-equiblirum crystal phases in these high-k dielectric thin films [37].

The local structural information acquired from extended X-ray absorption spectroscopy (EXAFS) is correlated with the thin film growth conditions. The response of the local structure around Hf and Zr atoms to growth parameters was investigated by EXAFS experiments performed at the National Synchrotron Light Source of Brookhaven National Laboratory. The competing crystal phases of oxides of Hf were identified and the intricate relation between the stabilized phase and the parameters as: the substrate temperature; Hf to Zr ratio; have been revealed. Specifically, HfO_2 thin films on Si(100) exhibit a tetragonal to monoclinic phase transformation upon increase in the substrate temperature during deposition whereas, HfO_2 PLD films on Ge(100) substrates remain in tetragonal symmetry regardless of the substrate temperature [36].

Previous studies in determining the crystal symmetry of the HfO_2 with respect to thin film layer thickness and annealing conditions revealed existence of non-equilibrium phases (tetragonal) of HfO_2 [34] under certain annealing conditions. Specifically, 1.4, 1.8, and 4.0 nm thick HfO_2 films on 1.0 nm SiO_2 interfacial layers on Si(100) substrates were subjected to different annealing treatments. Figure 9.9

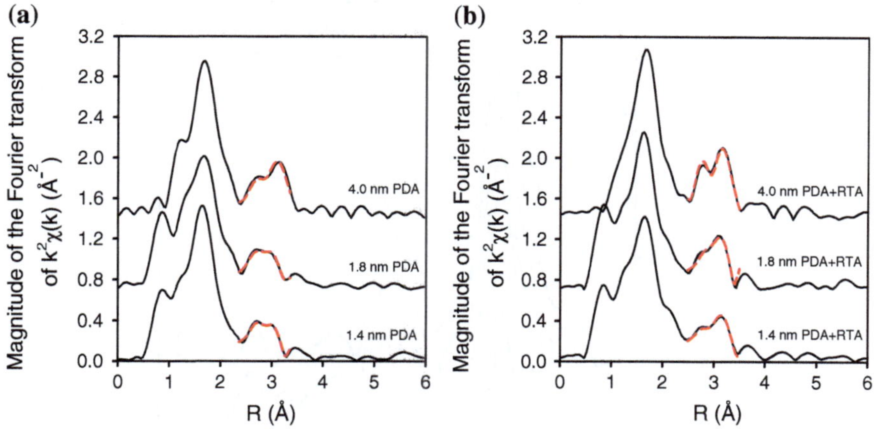

Fig. 9.9 HfO_2 thin films on Si(100) and the second shell EXAFS fits. *PDA* Post deposition annealing, *RTA* Rapid thermal annealing

Table 9.6 Fraction of different crystal phases in HfO_2 thin films

	Fraction		
	Monoclinic	Tetragonal	Orthorhombic
1.4 nm PDA	0.00	0.96	0.04
1.8 nm PDA	0.00	0.99	0.01
4.0 nm PDA	0.67	0.33	0.00
1.4 nm PDA + RTA	0.57	0.43	0.00
1.8 nm PDA + RTA	0.60	0.40	0.00
4.0 nm PDA + RTA	0.96	0.04	0.00

shows the FT data and corresponding second shell fits and Table 9.6 lists the fraction of the different phases present in these films [34].

During the pulsed laser deposition process the thin films of $Hf_{1-x}Zr_xO_2$(x = 0, 0.10, 0.25, 0.50, 0.75), were deposited on 2 inch p-type Si(100) and wafers using a KrF excimer laser with a wavelength of 248 nm. The surface oxide on the substrate wafers were removed by a 1:10 of solution of HF + H_2O and rinsed using deionized water solution before the deposition. The partial O_2 pressure was kept at 100 mTorr during the deposition to prevent oxygen deficiency in the thin films. The laser energy density was set to 1.1 J/cm^2 and laser pulse frequency was at 10 Hz. The substrate temperature was varied between 100 and 800 °C. The target to substrate distance was set to 5 cm in all the depositions. The orientation/quality of the films was checked by X-ray diffraction. The thickness of the films varied between 15–20 nm as verified with a thin film reflectometry system with a resolution of 0.1 nm.

Hf L_3-edge X-ray absorption fine structure spectroscopy (EXAFS) experiments were performed at the National Institute of Standards and Technology's (NIST) beamline (X23A2) at National Synchrotron Light Source (NSLS) at Brookhaven

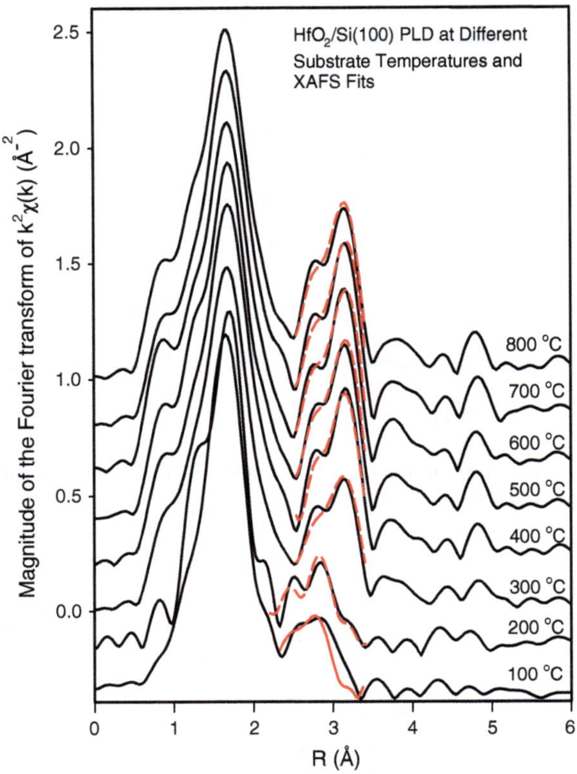

Fig. 9.10 The Fourier-transformed EXAFS data and the least-square EXAFS fits to the data for the PLD deposited HfO2/Si(100) films at substrate temperatures of 100–800 °C

National Laboratory (BNL). EXAFS data were acquired in the fluorescence detection mode and the X-ray angle of incidence was set to 0.5° during the measurements.

All the EXAFS measurements were performed at the Hf L_3 absorption edge and the atomic absorption background were subtracted by AUTOBK code [22]. EXAFS functions $\chi(k)$'s are Fourier Transformed (FT) using a Gaussian window for (2.0–12.5 Å^{-1}) k-range with a k^2 weighting in order to fully account the larger k region. University of Washington's multiple scattering code FEFF8.4 [21], has been used for the EXAFS calculations of the theoretical standards to be used in least-square EXAFS fits to the data. The theoretical standards for all the possible crystal phases of HfO_2 were created using lattice parameters and fractional coordinates from literature [38]. These were monoclinic, tetragonal, cubic and orthorhombic structures. The stable structure for HfO_2 at the room temperature is the monoclinic phase [35]. For the least square fits of the Fourier Transformed (FT) HfO_2 EXAFS data all of these phases are used for the identification of the final crystal phases present in the thin films. The first shell is dominated by the Hf-O near-neighbors and the second shell mostly involves Hf-Hf scattering in the EXAFS functions for all the different crystal symmetries. The second shell is more sensitive to modifications on the local structure around Hf atom.

Figure 9.10 shows the FT data for HfO_2/Si(100) PLD films deposited at different substrate temperatures and the EXAFS fits to the data.

The FT data in the second shell (2.2–3.5 Å) indicate a difference for the two lowest substrate temperature deposition samples (100 and 200 °C) and the samples, which were deposited at 300 °C or higher substrate temperatures. The differences are attributed to the slight modifications of the crystal symmetry of Hf atoms. In order to probe these subtle changes all the calculated possible phases of the HfO_2 are used. The double peak structure in the second shell (smaller peak around 2.8 Å and a slightly larger peak around 3.1 Å) is a signature of the HfO_2 monoclinic phase [34]. The EXAFS fits results for the near-neighbor distances (R) and the coordination numbers (N) are tabulated in Table 9.7. The fit region is set to the second shell and the multiple scattering paths were negligible. The uncertainties in the near neighbor distances, and the coordination numbers are ±0.02 Å and ±0.1, respectively. The fraction of the two found phases (tetragonal and monoclinic) and the near neighbor distances and coordination numbers around the Hf atom are also listed in Table 9.7.

For the films deposited at 100 and 200 °C substrate temperature the crystal symmetry around the Hf atom is tetragonal when the substrate temperature is raised to 300 °C the monoclinic structure starts to appear and when higher substrate temperatures are used the HfO_2 structure settles in pure monoclinic phase as indicated in Table 9.7. This indicates that the HfO_2 tetragonal phase can be stabilized on Si(100) by setting the substrate temperature at the deposition below 300 °C.

Figure 9.11 shows the FT EXAFS data for the $Hf_{0.9}Zr_{0.1}O_2$/Si(100) PLD thin films. For this series PLD films were prepared using a mixture of HfO_2 and ZrO_2 powder and in a fine mesh and using the resulting mixture as the target material for the deposition. As can be observed from the evolution of the second shell peaks the settling of the monoclinic phase is delayed up to 500 °C when 10 % Zr is incorporated into the target material.

In Fig. 9.12, overlays of FT data for $Hf_{1-x}Zr_xO_2$/Si(100){x = 0, 0.10, 0.25, 0.50, 0.75} deposited at 200 and 500 °C are plotted side by side. For the 200 °C substrate temperature deposition, the increase in the Zr content does not change the structures of the second shell and the symmetry is tetragonal for all x ranges.

However, at 500 °C the Zr effects are seen. HfO_2 monoclinic structure is observable in the second shell up to $x = 0.10$ but the structure gradually shifts away monoclinic for $x \geq 0.25$.

Figure 9.13 is an overlay of the FT data for HfO_2/Ge(100) films deposited at substrate temperature between 100 and 400 °C and HfO_2/Si(100) films deposited at 100, 200, and 400 °C for comparison purposes.

All the films on Ge(100) exhibits a very similar structure to those on Si(100) for the two lowest substrate temperatures of deposition i.e., 100 and 200 °C. As discussed previously, the HfO_2/Si(100) structure is exhibits tetragonal symmetry for 100 and 200 °C and settles down in monoclinic structure for temperatures higher than 200 °C. In contrast HfO_2/Ge(100), remains in tetragonal symmetry for all temperature ranges up to 500 °C. An overlay of the FT data of HfO_2/Si(100) at 400 °C is plotted just to show the difference between the settled monoclinic phase for films on silicon and the sustained tetragonal phases on germanium. Debernardi et al. [39]

Table 9.7 2nd shell EXAFS fit results to Fourier transformed data of HfO_2/Si(100)

	2nd shell fits for HfO_2 on silicon					
	Monoclinic ($P2_1/c$)			Tetragonal ($P4_2/nmc$)		
Sample ID	R(Å)	N_{Hf-Hf}	Fraction	R(Å)	N_{Hf-Hf}	Fraction
HfO_2/Si(100) @100 °C	3.36	6.3	0.03	3.61	10.9	0.97
HfO_2/Si(100) @200 °C	3.38	6.2	0.05	3.59	10.3	0.95
HfO_2/Si(100) @300 °C	3.37	6.5	0.65	3.56	9.8	0.35
HfO_2/Si(100) @400 °C	3.35	6.5	0.88	3.53	9.5	0.12
HfO_2/Si(100) @500 °C	3.40	6.2	0.94	3.52	9.6	0.06
HfO_2/Si(100) @600 °C	3.41	6.7	1.00			0.00
HfO_2/Si(100) @700 °C	3.41	6.6	1.00			0.00
HfO_2/Si(100) @800 °C	3.43	6.8	1.00			0.00

R is the near-neighbor distance and the N is the coordination number around Hf atom

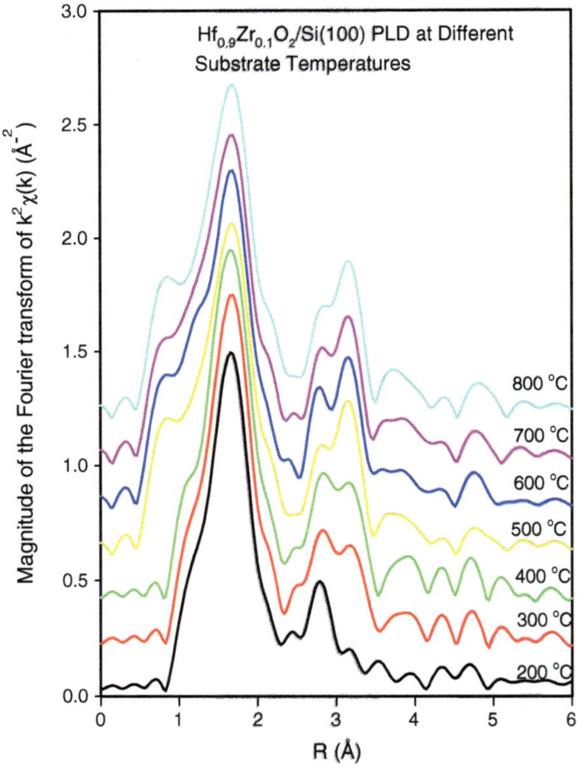

Fig. 9.11 The Fourier transformed EXAFS data for the PLD deposited $Hf_{0.9}Zr_{0.1}O_2$/Si(100) films at substrate temperatures of 200–800 °C

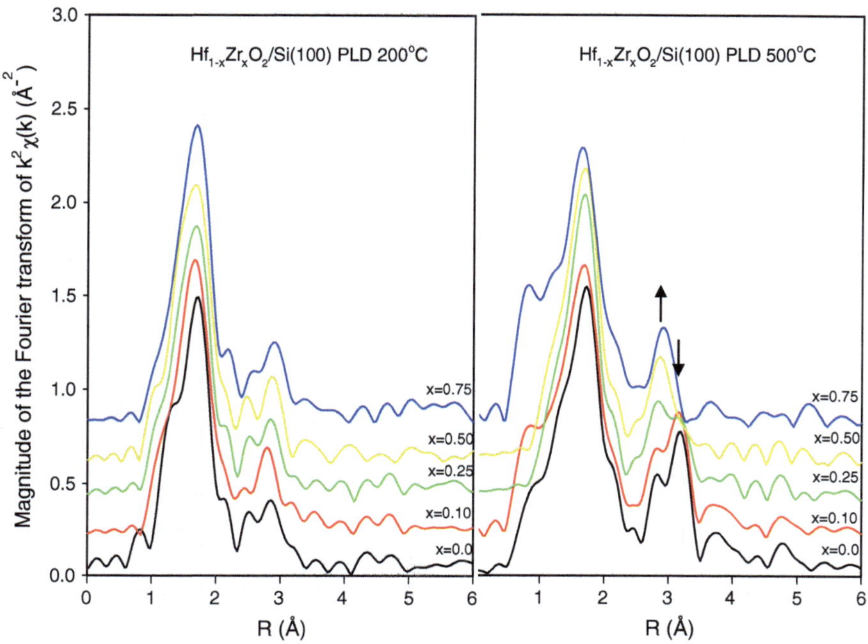

Fig. 9.12 The Fourier transformed EXAFS data for the PLD deposited $Hf_{1-x}Zr_xO_2$/Si(100) {x = 0.0–0.75} at T = 200 °C and T = 500 °C

have suggested epitaxial tetragonal structure of HfO_2/Ge(100) by depending on their theoretical calculations. They based their arguments on the better lattice match between the tetragonal phase of HfO_2 and the Ge but failed to observe it experimentally by X-ray diffraction. To our knowledge, our work is the first experimental evidence of tetragonal growth of HfO_2 on Ge(100). The EXAFS fit results for Ge substrate films are tabulated in Table 9.8 confirm the tetragonal symmetry in these films.

9.5 Summary

Extended X-ray absorption fine structure spectroscopy has been used to investigate the subtle local structural modifications and thin-film formation process caused by the variations in the plasma process parameters in semiconducting materials.

The effectiveness of the EXAFS in non-destructively probing microstructural variations were shown in two materials related challenges in semiconductor materials. The first example was on the dopant clustering and deactivation problem in high dose plasma immersion ion implants (arsenic). The second example was on the non-equilibrium crystal phase identificationson the high-k dielectric thin films prepared by

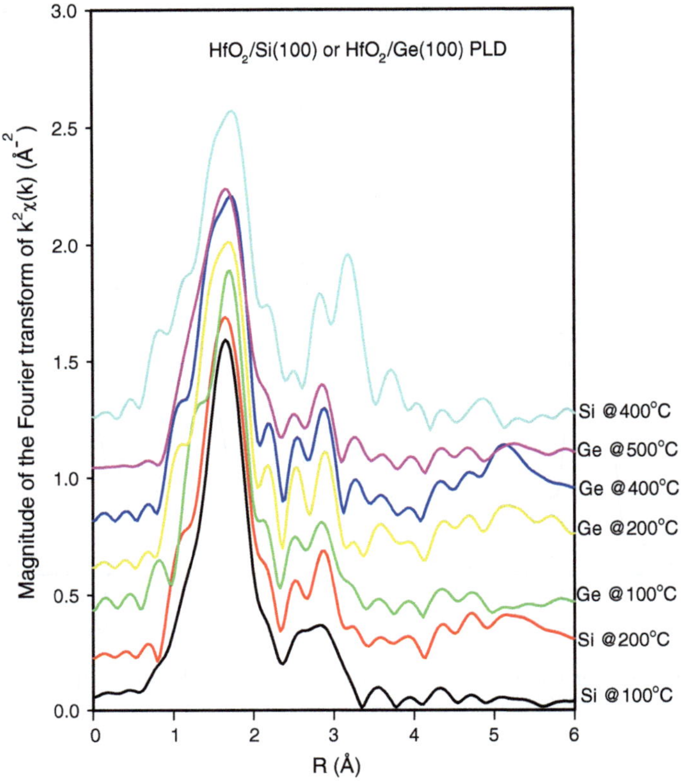

Fig. 9.13 The Fourier transformed EXAFS data for the PLD deposited $HfO_2/Ge(100)$ at T = 100, 200, 400, and 500 °C. FT Data for $HfO_2/Si(100)$ films at T = 100, 200, and 400 °C are overlayed for comparison

Table 9.8 2nd shell EXAFS fit results to Fourier transformed data of $HfO_2/Ge(100)$

	2nd shell fits for HfO_2 on germanium-tetragonal ($P4_2/nmc$)	
Sample ID	R (Å)	N_{Hf-Hf}
$HfO_2/Ge(100)$ @100 °C	3.58	10.2
$HfO_2/Ge(100)$ @200 °C	3.57	10.5
$HfO_2/Ge(100)$ @400 °C	3.63	10.8
$HfO_2/Ge(100)$ @500 °C	3.62	10.4

R is the near-neighbor distance and the N is the coordination number around Hf atom

the pulsed laser deposition (through plasma plume of the target materials). Through these examples EXAFS has been shown to be a very effective non-destructive local X-ray technique in studying the physics of the complex plasma processes.

Acknowledgments This work was supported by Research Corporation Award # CC6405 and NSF DMI 0420952, and SEMATECH.

References

1. D. Gupta, Int. J. Adv. Technol. **2**(4), 507–529 (2011)
2. F. Meirer et al., AIP Conf. Proc. **1496**, 183 (2012)
3. S. Qin et al., IEEE Trans. Plasma Sci. **37**(10), 2082 (2009)
4. S.B. Felch et al., Surf. Coat. Tech. **156**, 229 (2002)
5. J. Haverkamp, R.M. Mayo, M.A. Bourham et al., Plasma plume characteristics and properties of pulsed laser deposited diamond-like carbon films. J. Appl. Phys. **93**, 3627 (2003)
6. D.C. Koningsberger, R. Prins (eds.), *X-ray absorption: principles, applications, techniques of EXAFS, SEXAFS, and XANES, in chemical analysis 92* (John Wiley, NJ, 1988)
7. G. Bunker (ed.), *Introduction to XAFS: a practical guide to X-ray absorption fine structure spectroscopy* (Cambridge University Press, Cambridge, 2010)
8. A. Sahiner, D. Crozier, D.T. Jiang, R. Ingalls, Phys. Rev. B **59**, 3902 (1999).
9. A. Sahiner, M. Croft, S. Guha, I. Perez, Z. Zhang, M. Greenblatt, P.A. Metcalf, H. Jahns, G. Liang, Phys. Rev. B **51**, 5879 (1995)
10. A. Sahiner, M. Croft, S. Guha, Z. Zhang, M. Greenblatt, I. Perez, P.A. Metcalf, H. Jahns, G. Liang, Phys. Rev. B **53**, 9745 (1996)
11. D. Giubertoni, G. Pepponi, S. Gennaro, M. Bersani, M.A. Sahiner, S.P. Kelty, R. Doherty, M.A. Foad, M. Kah, K.J. Kirkby, J.C. Woicik, P. Pianetta, J. Appl. Phys. **104**, 103716 (2008)
12. M.A. Sahiner, S.W. Novak, J.C. Woicik, Y. Takamura, P.B. Griffin, J.D. Plummer, MRS Proc. **717**, C3.6.1 (2003)
13. M.A. Sahiner, D.F. Downey, S.W. Novak, J. Liu, V. Krishnamoorthy, IEEE Trans. Ion Implant. Technol. **432**, 600 (2000)
14. M.A. Sahiner, D.F. Downey, S.W. Novak, J.C. Woicik, D.A. Arena, Microelectron. J. (Elsevier) **36**, 3 (2005)
15. M.A. Sahiner, P. Ansari, M.S. Carroll, C.A. King, Y.S. Suh, R.A. Levy, T. Buyuklimanli, M. Croft, MRS Proc. **864**, E7.8.1 (2005)
16. M.A. Sahiner, S.W. Novak, J.C. Woicik, J. Liu, V. Krishnamoorty, MRS Proc. **669**, J5.8.1 (2001)
17. J.L. Allain, A. Bourret, J.R. Regnard, A. Armigliato, Appl. Phys. Lett. **61**, 264 (1992)
18. A. Armigliato, F. Romanato, A. Drigo, A. Carnera, C. Brizard, J.R. Regnard, J.L. Allain, Phys. Rev. B **52**, 1859 (1995)
19. C. Revenant-Brizard, J.R. Regnard, S. Solmi, A. Armigliato, S. Valmorri, C. Cellini, F. Romanato, J. Appl. Phys. **79**, 9037 (1996)
20. G. Pepponi, D. Giubertoni, S. Gennaro, M. Bersani, M. Anderle, R. Grisenti, M. Werner, J.A. van den Berg, *Ion Implantation Technology*, Marseille, France, p. 117, 2006, ed. by K.J. Kirkby et al., American Institute of Physics
21. A.L. Ankudinov, B. Ravel, J.J. Rehr, S.D. Conradson, Phys. Rev. B **58**, 7565 (1998)
22. M. Newville, P. Livins, Y. Yacoby, J.J. Rehr, E.A. Stern, Phys. Rev. B **47**, 14126 (1993)
23. M.D. Segall, P.J.D. Lindan, M.J. Probert, C.J. Pickard, P.J. Hasnip, S.J. Clark, M.C. Payne, J. Phys. Cond. Matt. **14**, 2717 (2002)
24. J.P. Perdew, Y. Wang, Phys. Rev. B **45**, 13244 (1992)
25. C. Skylaris, P.D. Haynes, J. Chem. Phys. **127**, 164712 (2007)
26. M. Ramamoorthy, S.T. Pantelides, Phys. Rev. Lett. **76**(25), 4753 (1996)
27. D.C. Mueller, E. Alonso, W. Fichtner, Phys. Rev. B **68**, 045208 (2003)
28. T. Wadsten, Acta Chem. Scand. **19**, 1232 (1965)
29. F. d'Acapito, S. Milita, A. Satta, L. Colombo, J. Appl. Phys. **102**, 043524 (2007)
30. D.C. Mueller, E. Alonso, W. Fichtner, Phys. Rev. B **68**, 045208 (2003)
31. W.R.M. Wilk, J.M. Anthony, J. Appl. Phys. **89**, 5243 (2001)
32. C. Beer et al., Appl. Phys. Lett. **91**, 263512 (2008)
33. Y.J. Yang, W.S. Ho, C.F. Huang, S.T. Chang, C.W. Liu, Appl. Phys. Lett. **91**, 102103 (2007)
34. P. Lysaght, J.C. Woicik, M.A. Sahiner, B.-H. Lee, R. Jammy, Appl. Phys. Lett. **91**, 122910 (2007)

35. J. Tang, J. Fabbri, R.D. Robinson, Y. Zhu, I.P. Herman, M.L. Steigerwald, L.E. Brus, Chem. Mater. **16**, 1336 (2004)
36. M. Alper Sahiner, J.C. Woicik, T. Kurp, J. Serfass, M. Aranguren, Mater. Sci. Semicond. Process. **11**, 245 (2008)
37. M.A. Sahiner, J.C. Woicik, P. Gao, P. McKeown, M.C. Croft, M. Gartman, B. Benapfla, Thin Solid Films **515**, 6548 (2007)
38. X. Zhao, D. Vanderbilt, Phys. Rev. B **65**, 233106 (2002)
39. A. Debernardi, C. Wiemer, M. Fanciulli, Phys. Rev. B **76**, 155405 (2007)

Chapter 10
Kinetic Monte Carlo Simulations of Cluster Growth and Diffusion in Metal-Polymer Nanocomposites

Lasse Rosenthal, Thomas Strunskus, Franz Faupel, Jan Willem Abraham and Michael Bonitz

This chapter is dedicated to the memory of Vladimir Zaporojtchenko.

Abstract This chapter presents computer simulations of cluster growth and diffusion in metal-polymer nanocomposites based on the kinetic Monte Carlo method. The goal is to condense the crucial processes taking place during polymer metallization into a self-consistent simulation scheme which is able to reproduce and predict qualitative changes caused by variations of deposition conditions. The focus of this chapter lies on the discussion of the simulation model and the mathematical framework of kinetic Monte Carlo methods. The first section gives a brief overview over important physical aspects during metal-polymer nanocomposite formation. It is followed by a compact introduction into the theory of continuous-time Markov chains which is the mathematical foundation of the kinetic Monte Carlo method. In another section, the basic assumptions and simplifications of the simulation model are extensively discussed and physically motivated. The chapter closes with a discussion of simulation results of metal-polymer interface formation and co-deposition of metal

L. Rosenthal (✉) · J. W. Abraham · M. Bonitz
Institut für Theoretische Physik und Astrophysik, Christian-Albrechts-Universität zu Kiel, Leibnizstraße 15, 24098 Kiel, Germany
e-mail: rosenthal@theo-physik.uni-kiel.de

J. W. Abraham
e-mail: abraham@theophysik.uni-kiel.de

M. Bonitz
e-mail: bonitz@theo-physik.uni-kiel.de

T. Strunskus · F. Faupel
Institut für Materialwissenschaft, Lehrstuhl für Materialverbunde, Christian-Albrechts-Universität zu Kiel, Kaiserstraße 2, 24143 Kiel, Germany

F. Faupel
e-mail: ff@tf.uni-kiel.de

M. Bonitz et al. (eds.), *Complex Plasmas*, Springer Series on Atomic, Optical, and Plasma Physics 82, DOI: 10.1007/978-3-319-05437-7_10,

and polymer. Yet, the intention of these simulations is not to provide a microscopic understanding of metal-polymer nanocomposites formation.

10.1 Introduction

Due to the ongoing success of nanochemistry and nanotechnology, the research field of nanocomposite materials has received increased attention since the end of the twentieth century. This results from the fact that the combination of the complementary properties of the nm-sized filler particles and the host material allows for the production of multifunctional materials that can serve various functions. Examples include numerous applications in optics based on surface plasmon resonance [1, 2], electronics [3–6], food packaging [7], and medicine [8, 9]. However, nanocomposites are also widespread in biological systems. One striking example of a naturally occurring composite material is the inner shell layer produced by some molluscs, known as nacre (mother of pearl). Despite the brittle structure of its constituents, nacre exhibits a very high resilience and toughness which exceeds the toughness of its constituent materials by a factor of 1000 [10]. Due to its remarkable mechanical properties, nacre not only serves for decorative use, but is also an inspiration for the development of composite materials that mimic its microscopic structure [11, 12].

An important subcategory of nanocomposites are metal-polymer nanocomposites [13] which are based on a synthesis of nm-sized metallic filler particles dispersed in a polymer matrix. They are of particular interest due to their novel properties offering hosts of new applications [9, 14, 15]. Functional properties obtained by embedding metallic nanoparticles in a dielectric matrix include electronic conductivity ranging from single-electron hopping and tunneling to percolation [4], particle surface plasmon resonances giving rise to characteristic optical features [16] explored in the new field of plasmonics [17], magnetic properties governed by ferromagnetic single domain behavior or superparamagnetism [18], granular giant magnetoresistance [19], and enhancement of catalytic activity owing to the large effective surface area of the nanoparticles and the marked contribution of the surface energy to the chemical potential [20].

A striking example of electronic properties of a metal-polymer nanocomposite based on Au and Teflon AF is reported in Takele et al. [21], see Fig. 10.1. Here, Takele et al. report about a nanocomposite prepared by vapor phase co-deposition which spans ten magnitudes of electrical resistance ranging from 10^{12} to $10^{2}\,\Omega$. The investigated nanocomposite exhibited a percolation threshold at a filling factor around 0.43. Once the filling factor exceeded this value, the electrical resistance quickly dropped by an amount of $10^{10}\,\Omega$.

Several approaches such as reduction of a metal salt [22], ball milling [23], plasma polymerization in conjunction with metal deposition [24–26], co-evaporation of a metal and an organic component [27, 28], and co-sputtering [29, 30] have been reported to produce functional polymer composites containing metal nanoparticles. Many applications require good control of the metal filling factor, filling factor

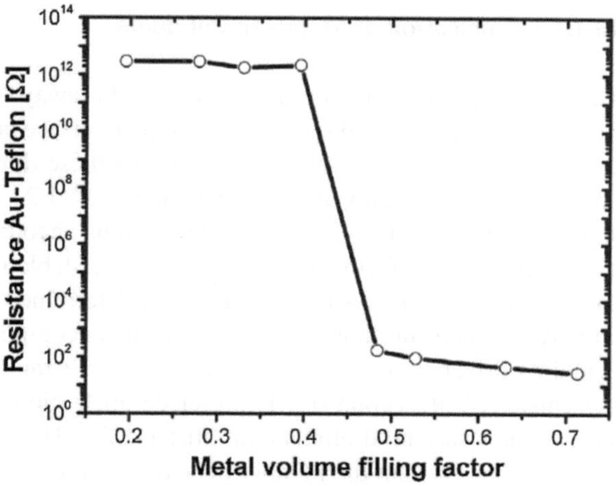

Fig. 10.1 Change of the electrical resistance in an Au-Teflon AF nanocomposite due to variation of the metallic filling factor. With kind permission reprinted from Ref. [21]

profile, and composition of the particles. Particularly with respect to the two latter points, wet chemical approaches are usually less or not suited. Here, physical vapor deposition has proven to be very successful [29, 31].

10.1.1 Physical Aspects of Polymer Metallization

Metals and polymers are materials with very contrasting properties. Polymers are covalently bonded chain molecules built from repeating structural units, the so-called monomers, which can be very simple chemical structures like in polyethylene, $-CH_2-$, or very complex compounds itself [32]. Many polymers exist in the glass state with no long-range order. There are polymers such as polyethylene oxide which exhibit crystalline structures; however, interconnection of polymer chains prevents crystalline regions from extending over large areas [32]. Polymers usually have small cohesive energies due to the weak interaction between polymer chains which is mediated by the weak van der Waals force. Due to the large number of monomers a polymer is made of (10^2–10^5) and the degrees of rotational freedom of covalent bonds at nearly constant bond angle and energy, polymer chains may attain an almost uncountable number of spatial arrangements allowing for a description in terms of the random coil model [33].

In contrast to that, metals tend to form densely packed crystalline structures whose physical characteristics arise from the strong metallic bonds between the atoms. Therefore, the cohesive energy of metals exceeds the cohesive energy of polymers by typically two orders of magnitude [13, 33, 34] and they exhibit a strong tendency to aggregate and form clusters (Volmer-Weber growth) during the metallization process.

10.1.1.1 Nucleation of Metals in Polymers

The initial stage of polymer metallization is far away from thermodynamic equilibrium as metal atoms impinge on the polymer surface and undergo a rather complex dynamics afterwards. Essential processes of these dynamics are random diffusion on the surface and into the bulk (see Sect. 10.1.1.2) and re-evaporation from the surface (see Sect. 10.1.1.3). If two metal atoms encounter each other on their path, they aggregate and form a cluster. Clusters are stable if their size exceeds the critical nucleus size which was experimentally determined to be one [36], implying that already dimers form stable clusters and do not decompose. The nucleation of clusters is mainly driven by two processes: the first one is random nucleation by encountering clusters. The second one is preferred nucleation which takes place at special surface sites that bind atoms and clusters [37]. The exact nature of these surface site is not known, but the number density of these sites can be influenced with ion-beam treatment [36, 38]. Preferred nucleation is expected to be the dominant growth mechanism in the early stages of polymer metallization and for low condensation coefficients. Under different conditions, random nucleation is expected to dominate [34, 37]. Both growth mechanisms have been observed in experiments [38–40].

Cluster growth during polymer metallization can be categorized into different morphological sequences. The first stage can be identified by the growth of compact isolated islands which develop towards elongated islands in the second stage. These elongated islands start to overlap and build a percolating network in the third stage and, finally, form a continuous film spanning the entire substrate in the fourth stage [41, 42]. Figure 10.2 shows TEM micrographs illustrating these sequences. Recent experiments have shown that agglomeration of metal atoms is not restricted to growth of spherical clusters: under specific conditions, co-deposition of the metallic and the polymer component can be applied to produce parallel metallic nanocolumns as it was reported for a Fe-Ni-Co alloy in a Teflon AF matrix [31]. TEM images of these nanocolumns, see Fig. 10.3, have shown that they are parallelly arranged and exhibit a homogeneously distributed aspect ratio.

10.1.1.2 Diffusion of Metals in Polymers

From the technological point of view, diffusion of metals in polymers is of great interest. Especially in microelectronics, where polymers are used as low-k dielectrics (a material with a small dielectric constant) to build electronic devices, the need for further miniaturization has triggered intensive research on diffusion of metals in polymers [43–45]. Since even small amounts of metal diffusing from the conductance lines act as deep impurities and increase the parasitic capacitance between the conductance lines, one central task of this research is to develop efficient methods to block metal diffusion in polymers by some barrier [37, 45]. While metals are known to form thin films and clusters obeying Volmer-Weber growth [47], when they are deposited on a polymer surface, the intermixing of metals and polymers depends on their chemical interaction. For low reactivity metals like Cu, Ag, Au, and Pd,

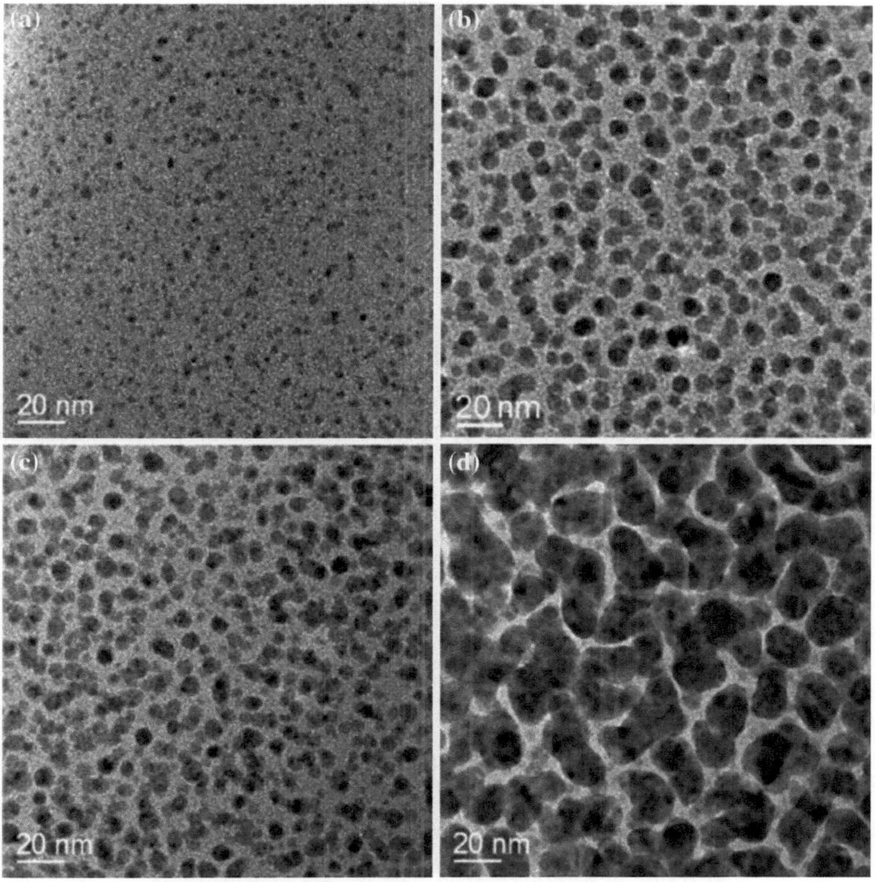

Fig. 10.2 TEM micrographs of Nylon-Ag nanocomposites with a thickness of about 60 nm and different metallic filling factors: **a** 4.4, **b** 14, **c** 21, and **d** 41.5 %. The transition from isolated clusters growth (**a**) to elongated islands emerged from overlapping clusters (**b**, **c**) and finally the percolation network (**d**) is clearly illustrated. With kind permission reprinted from Ref. [35]

which exhibit weak chemical interactions with polymers [37], almost no intermixing with polymers should be expected due to their high aggregation tendency [33, 48] (Fig. 10.4).

In contradiction to these predictions, LeGoues et al. performed TEM studies of the Cu–PMDA-ODA interface in the 1980s, which revealed a remarkable migration of Cu into the PMDA-ODA and subsequent formation of spherical agglomerates below the surface at temperatures between 293 and 573 K [43, 44]. These investigations have also shown that diffusion of Cu into polyimide is restricted to low deposition rates and elevated temperatures; a high deposition rate at elevated temperature effectively impeded diffusion and clustering of Cu under the surface as well as room temperature deposition and subsequent annealing. Tromp et al. drew the conclusion

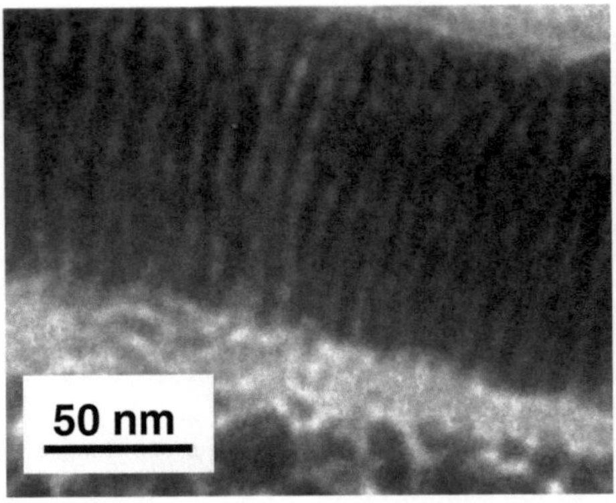

Fig. 10.3 Cross-sectional TEM image showing the formation of Fe-Ni-Co nanocolumns in Teflon AF on top of a layer of Ag clusters isolated by 20 nm of the same matrix. With kind permission reprinted from Ref. [95]

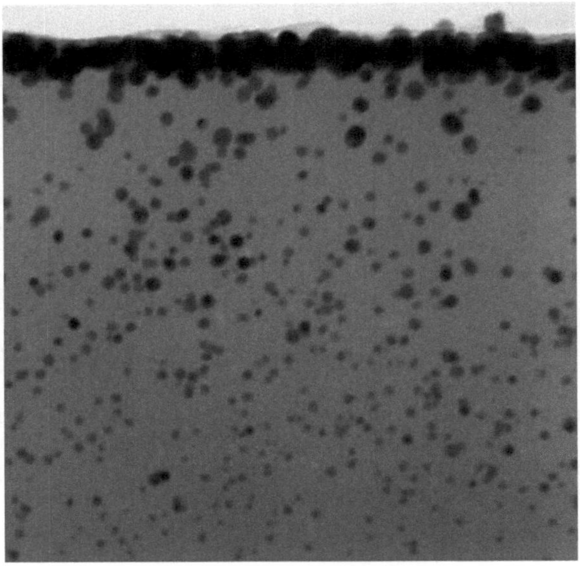

Fig. 10.4 Cross-sectional TEM micrograph of an Au-TMC-PC interface. Au was deposited with a rate of about 0.03 nm/min at 235 °C. The micrograph shows the interface after annealing for 12 h at 290 °C. The glass transition temperature of TMC-PC lies around 235 °C. The depicted sample has a width of about 320 nm. With kind permission reprinted from Ref. [46]

that an accelerated cluster growth on the surface due to enhanced deposition rates impedes diffusion of Cu under the surface, implying that diffusion into polymers is dominated by atoms and small clusters [43]. These conclusions are supported by experiments verifying the aggregation-induced immobility of Cu on polyimide [49].

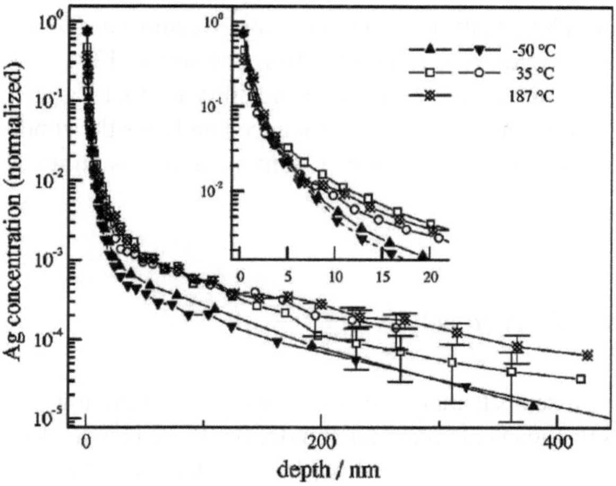

Fig. 10.5 Experimental concentration profile of Ag in trimethylcyclohexane polycarbonate obtained by radiotracer measurements in conjunction with ion-beam sputtering at different temperatures. With kind permission reprinted from Ref. [45]

Similar behavior for the Ag-polyimide interface [50, 51] and other metal/polymer combinations was observed in further experiments [37]. For metals forming strong chemical bindings with polymers, diffusion inside the polymer was found to be significantly reduced [44].

Atomic diffusion coefficients of metals in polymers have been studied by employing radiotracer measurements in conjunction with ion-beam sputtering [45, 51, 52]. This technique allows to determine depth-resolved concentration profiles from which the diffusion coefficients can be deduced by fitting the linear tail of the profile to the thin-film solution of Fick's second law [45, 52]. An experimental concentration profile is depicted in Fig. 10.5. Thran et al. have reported radiotracer measurements of the diffusion of Ag in TMC-PC, which clearly proved a migration of Ag into TMC-PC over large distances even at room temperature [45].

10.1.1.3 Condensation of Clusters

When metal atoms are deposited on a metal surface, they usually exhibit a condensation coefficient C close to unity, even at elevated temperatures [34, 37]. The condensation coefficient is defined as the ratio of the number of adsorbed atoms and the total number of atoms impinged on the surface. The behavior of metal atoms which are deposited on a polymer surface is usually different from those deposited on metal surfaces. The condensation coefficient strongly depends on the metal/polymer combination as well as on the temperature and shows extreme variations by three orders of magnitude ranging from 0.002 for Ag on Teflon AF to 0.95 for Ag on polyimide at room temperature [36, 39, 40]. Radiotracer measurements performed by Thran et al. revealed that incomplete condensation of metal atoms on polymers is not due to direct backscattering of atoms from the polymer surface. Instead, metal atoms perform a random walk on the surface prior to reemission and have to overcome an energy barrier in order to be re-emitted from the surface [39].

If the metal coverage is very high, the probability for metal atoms to impinge on pre-existing metal clusters is drastically enhanced and thus the condensation coefficient reaches almost unity as it is known from metal atoms on metal surfaces [34].

10.1.2 Chapter Overview

The aim of the present chapter is to investigate the formation of metal-polymer nanocomposites by means of computer simulations [53]. This work essentially gains its motivation by the research carried out by the Chair for Multicomponent Materials at the University of Kiel. Its aim is to contribute to the understanding of the coupled processes determining the microstructure of metal-polymer nanocomposites from a macroscopic point of view. Since the complexity of the coupled physical and chemical processes occurring during metal-polymer nanocomposites formation does not allow for a treatment based on microscopic approaches, a simpler simulation approach neglecting the microscopic structure of the constituents has to be applied. Here, the method of choice is the kinetic Monte Carlo (KMC) method [54] which has seen a remarkable progress over the last decades and which has been successfully applied to study the dynamics of coupled processes leading to structure formation of matter which are not accessible with *ab-initio* methods or continuum descriptions. Within in this framework, a simulation scheme including and combing several approaches of previous simulations [41, 55–57] was developed, implemented and applied to different experimental scenarios such as the formation of a metal-polymer interface, co-deposition of metal and polymer and the self-organized growth of metallic nanocolumns. The implemented model allows for simulations covering all morphological sequences of cluster growth reaching from growth of isolated clusters to percolation.

Substantial simplifications of the simulation model are the neglecting of the inner atomic structure of the clusters and the polymer matrix. Instead, the clusters are modeled within the framework of the liquid drop model [58] and basic effects of the polymer matrix on the clusters are condensed into the diffusion and evaporation rates of clusters. Due to the complexity of the studied system on the one hand and the simplicity of the simulation on the other hand, the simulations are not intended to reproduce experimental outcomes quantitatively. Rather, the intention is to understand and predict qualitative changes due to variations of the deposition conditions. To this end, the simulation model is designed in such a way that the simulations allow for a detailed investigation of systems containing up to several million metal clusters.

10.2 Theoretical Foundations of Kinetic Monte Carlo Methods

The Markov chain method is one of the basic mathematical tools used in numerical simulations in statistical physics. Markov chain methods can be divided into two main branches: discrete-time Markov chain (DTMC) methods and continuous-time Markov chain (CTMC) methods. While DTMCs methods mainly find application in simulations of systems in thermodynamic equilibrium and provide the mathematical framework of the Metropolis algorithm [59], CTMCs methods can be applied to model the time development of physical systems out of equilibrium. The CTMC method therefore provides the mathematical and theoretical foundation of the kinetic Monte Carlo (KMC) algorithm (also known as Bortz-Kalos-Lebowitz (BKL) algorithm) as it was invented by Bortz and co-workers in the 1970s [54], as well as several other algorithms built upon it.

The aim of this section is to provide a brief introduction into the theory of continuous-time Markov chains and to discuss two fundamental algorithms to simulate CTMCs.

10.2.1 Fundamental Terms of Probability Theory

10.2.1.1 Basic Concepts of Probability

Let $\{A_i\}_{i\in I}$ denote a countable set of pairwise mutually exclusive events and P denote a probability measure satisfying the Kolmogorov axioms. Then the probability of *either occurring* satisfies

$$P(\cup_{i\in I}A_i) = \sum_{i\in I} P(A_i). \tag{10.1}$$

Consider two events A and B of an arbitrary probability space. A and B are defined to be stochastically independent if the occurrence of A does not affect the probability of the occurrence of B and vice versa. In this case, the joint probability of the occurrence of A and B is given by the product of their probabilities:

$$P(A \cap B) = P(A)P(B). \tag{10.2}$$

The *conditional probability* $P(A \mid B)$ represents the probability of event A to occur on the condition that event B already occurred:

$$P(A \mid B) := \frac{P(A \cap B)}{P(B)}. \tag{10.3}$$

10.2.1.2 The Law of Total Probability and Bayes Sequential Formula

From the definition of the conditional probability (10.3), two important conclusions can be deduced. Consider the events $B_1, \ldots, B_n$ to be a partition[1] of a given event set Σ. Then it follows from $A = \cup_{i=1}^{n}(A \cap B_i)$, Def. (10.3) and (10.1), the *law of total probability*

$$P(A) = \sum_{i=1}^{n} P(A \mid B_i)P(B_i). \tag{10.4}$$

The law of total probability expresses the probability of the occurrence of event A in terms of the realization of several distinct events.

The second conclusion is known under the name *Bayes sequential formula.* For any collection of events $A_1, \ldots A_n$ it holds

$$\begin{aligned} P(\cap_{i=1}^{n} A_i) &= P(A_1) \cdot \frac{P(A_1 \cap A_2)}{P(A_1)} \cdot \frac{P(A_1 \cap A_2 \cap A_3)}{P(A_1 \cap A_2)} \cdot \ldots \cdot \frac{P(\cap_{i=1}^{n} A_i)}{P(\cap_{i=1}^{n-1} A_i)} \\ &= P(A_1) \prod_{i=2}^{n} P(A_i \mid \cap_{k=1}^{i-1} A_k). \end{aligned} \tag{10.5}$$

10.2.1.3 The Exponential Distribution

The *exponential distribution* is a class of continuous distribution functions and of great importance for the modeling of processes in which events occur continuously and independently at a constant averaged rate. Commonly, the exponential distribution is defined via its probability density function $f_X(x; \lambda)$:

$$f_X(x; \lambda) := \begin{cases} \lambda \mathrm{e}^{-\lambda x} & x \geq 0 \\ 0 & x < 0, \end{cases} \tag{10.6}$$

where λ is the parameter of the distribution and usually called the rate parameter. The exponential distribution has the following properties:

(i) The expectation value of the exponential distribution is given by

$$\mathrm{E}[X] = \int_0^\infty x\lambda \mathrm{e}^{-\lambda x}\, dx = \frac{1}{\lambda}. \tag{10.7}$$

(ii) The cumulative distribution function $F_X()$ of the exponential distribution reads

$$F_X(x; \lambda) := P(X \leq x) = \int_0^x \lambda \mathrm{e}^{-\lambda x'}\, dx' = 1 - \mathrm{e}^{-\lambda x}. \tag{10.8}$$

[1] A collection of pairwise disjoint sets overlapping the whole event set Σ.

(iii) $P(X > x) = 1 - F_X(x; \lambda) = \mathrm{e}^{-\lambda x}$.

The key feature which makes the exponential distribution important for many applications is its memorylessness: the distribution of an exponentially distributed random variable X conditioned on $\{X > u\}$ is again exponential with the same rate parameter:

$$\begin{aligned} P(X > t + u \,|\, X > u) &= \frac{P(X > t + u)}{P(X > u)} = \frac{\mathrm{e}^{-\lambda(t+u)}}{\mathrm{e}^{-\lambda u}} \\ &= \mathrm{e}^{-\lambda t} = P(X > t). \end{aligned} \tag{10.9}$$

It can easily be shown that the exponential distribution is the only distribution fulfilling the memoryless property.

Another important property of the exponential distribution, which is crucial for the applicability of the BKL algorithm, is the fact that the *distribution of the minimum* of independent exponentially distributed (i.e.d.) random variables is again exponential. Consider $\{X_i\}_{i\in I}, I = 1, \ldots, n$ to be i.e.d. random variables with corresponding rate parameters $\{\lambda_i\}_{i\in I}$. Then, One obtains

$$\begin{aligned} P(\min\{X\}_{i\in I} > x) &= P(\cap_{i=1}^{n}\{X_i > x\}) = \prod_{i=1}^{n} P(X_i > x) \\ &= \mathrm{e}^{-x\sum\limits_{i=1}^{n}\lambda_i}, \end{aligned} \tag{10.10}$$

which proves the statement.

The so-called *competition theorem* is complementary to the distribution of the minimum of i.e.d. random variables and provides the distribution of the index of the random variable which achieves the minimum in (10.10). Again, $\{X_i\}_{i\in I}$ are i.e.d. random variables with corresponding rate parameters $\{\lambda_i\}_{i\in I}$. Define $K := j$ if $X_j = \min\{X_i\}_{i\in I}$ and $U := \min\{X_i\}_{i\in I\setminus\{K\}}$. Further, put $f_u(u) = \lambda_u \mathrm{e}^{-\lambda_u u}$ with $\lambda_u = \sum_{i\in I\setminus\{K\}} \lambda_i$. Then the distribution $P(K = j)$ reads

$$\begin{aligned} P(K = j) &= \lim_{x\to\infty} P(X_j < U, X_j < x) = \lim_{x\to\infty} \int_0^x \left(\int_y^\infty f_u(u)\,\mathrm{d}u \right) \lambda_j \mathrm{e}^{-\lambda_j y}\,\mathrm{d}y \\ &= \lim_{x\to\infty} \lambda_j \int_0^x \mathrm{e}^{-y\sum\limits_{i=1}^{n}\lambda_i}\,\mathrm{d}y = \frac{\lambda_j}{\sum_{i=1}^{n}\lambda_i}. \end{aligned} \tag{10.11}$$

Equations (10.10) and (10.11) provide the basic components of algorithms for the numerical treatment of CTMCs.

10.2.2 Continuous-Time Markov Chains

10.2.2.1 Definition

A Markov chain is the mathematical model of a system undergoing transitions from one state to another in a countable set of states. In contrast to DTMCs where no transition times are defined, CTMCs are processes including a distribution function for the time interval between two transitions. Formally, a CTMC is defined as a stochastic process $\{X_t\}_{t\in\mathbb{R}^+}$ on a countable[2] sample space $\Omega = \{\omega_1, \ldots, \omega_k\}$ such that:

(i) The sample paths $t \to X_t$ are right-continuous step functions.
(ii) The process $\{X_t\}_{t\in\mathbb{R}^+}$ fulfills the *Markov property*, that is

$$P(X_{t+u} = \omega_j \,|\, X_t = \omega_i, \cap_{s=1}^{n}\{X_{t_s} = \omega_{i_s}\}) = P(X_{t+u} = \omega_j \,|\, X_t = \omega_i) \quad (10.12)$$

for all $\omega_j, \omega_i, \omega_{i_1}, \ldots, \omega_{i_n} \in \Omega$ and any nondecreasing sequence of times $0 \leq t_1 \leq t_2 \leq \cdots \leq t_n \leq t \leq t + u$.

Condition (1) ensures that no explosions occur, i.e. in every time interval Δt, only a finite number of transitions can occur. Condition (2) states that at any time the future of the Markov process is only affected by its current state and not by the sequence of states that preceded it.

A special class of Markov chains are *time homogeneous* Markov chains. Time homogeneous Markov chains are characterized by an isotropic distribution of transition times, i.e. for all $\omega_i, \omega_j \in \Omega$ and any $0 \leq u \leq t$ it is

$$P(X_{t+u} = \omega_j \,|\, X_t = \omega_i) = P(X_t = \omega_j \,|\, X_{t-u} = \omega_i). \quad (10.13)$$

10.2.2.2 The Transition Matrix

The time evolution of a Markov chain is described by the transition matrix $\mathbf{P}(t) = \{p_{ij}(t)\}_{\omega_i,\omega_j\in\Omega}$,

$$p_{ij}(t) := P(X_{u+t} = \omega_j \,|\, X_u = \omega_i), \quad \lim_{\varepsilon\to 0^+} \mathbf{P}(\varepsilon) = \mathbb{1}, \quad (10.14)$$

which is a continuous function of t. Here it is important to realize that the entry $p_{ij}(t)$ of the transition matrix just renders the probability to find the system in state ω_i at any given time u and in state ω_j a time interval t later. Note that the process does not necessarily evolve directly from state ω_i to state ω_j; instead, it can evolve from state ω_i to state ω_j by occupying an arbitrary number of states in between the time interval t.

[2] Here, only finite sample spaces are considered.

The distribution of states is governed by the time-dependent probability vector

$$\mathbf{w}(t) = \big(w_i(t), \ldots, w_k(t)\big), \quad w_i(t) := P(X_t = \omega_i), \tag{10.15}$$

which can be computed, once the transition matrix and the initial distribution $w(t=0)$ are known, by applying the law of total probability (10.4),

$$\begin{aligned} w_i(t) &= P(X_t = \omega_i) = \sum_j P(X_t = \omega_i, X_0 = \omega_j) \\ &= \sum_j P(X_0 = \omega_j) P(X_t = \omega_i \,|\, X_0 = \omega_j) = \sum_j w_j(0) p_{ji}(t) \\ &\Longrightarrow \mathbf{w}(t) = \mathbf{w}(0)\mathbf{P}. \end{aligned} \tag{10.16}$$

Here, the notation $\sum_i \equiv \sum_{i,\omega_i \in \Omega}$ was introduced for better readability. In line with this notation, the abbreviation $\sum_{i \neq j} \equiv \sum_{i,\omega_i \neq \omega_k}$ will also be used in the following.

Under the same conditions, the probability of an evolution path, i.e. a sequence of occupied states at fixed times, can be computed by applying Bayes sequential formula (10.5), in combination with the law of total probability (10.4):

$$\begin{aligned} P(X_{t_1} = \omega_{i_1}, \ldots, X_{t_n} = \omega_{i_n}) &= \sum_j P(X_{t_0=0} = \omega_j, X_{t_1} = \omega_{i_1}, \ldots, X_{t_n} = \omega_{i_n}) \\ &= \sum_j P(X_0 = \omega_j) \prod_{k=1}^{n} P(X_{t_k} = \omega_{i_k} \mid \cap_{l=0}^{k-1} \{X_{t_l} = \omega_{i_l}\}) \\ &= \sum_j w_j(0) \prod_{k=1}^{n} P(X_{t_k} = \omega_{i_k} \,|\, X_{t_{k-1}} = \omega_{i_{k-1}}) \\ &= \sum_j w_j(0) \prod_{k=1}^{n} p_{i_{k-1},i_k}(t_k - t_{k-1}), \quad \omega_0 \equiv \omega_j. \end{aligned} \tag{10.17}$$

One central equation containing the key property of the transition matrices $\mathbf{P}(t)$ is the *Chapman-Kolmogorow equation* which renders the transition probability from state ω_i to state ω_j as the sum of all possible ways with intermediate states ω_k in between:

$$\begin{aligned} p_{ij}(t+u) &= P(X_{t+u} = \omega_j \,|\, X_0 = \omega_i) \\ &= \sum_k P(X_{t+u} = \omega_j, X_t = \omega_k \,|\, X_0 = \omega_i) \\ &= \sum_k P(X_{t+u} = \omega_j \,|\, X_t = \omega_k, X_0 = \omega_i) P(X_t = \omega_k \,|\, X_0 = \omega_i) \\ &= \sum_k p_{kj}(u) p_{ik}(t) \quad \Longleftrightarrow \quad \mathbf{P}(t+u) = \mathbf{P}(t)\mathbf{P}(u). \end{aligned} \tag{10.18}$$

Thus, the transition matrices obey the exponentiation identity

$$\mathbf{P}\Big(\sum_{i=1}^{n} t_i\Big) = \prod_{i=1}^{n} \mathbf{P}(t_i), \tag{10.19}$$

which follows from (10.18) by induction.

10.2.2.3 Two Properties of Transition Times

As stated at the beginning of Sect. 10.2.2.1, the main extension of CTMCs compared to DTMCs, is the incorporation of transition times between two consecutive states. Consequently, one of the main questions is how the distribution of transition times can be computed. By making use of the Markov property (10.12), it can be shown that the distribution of transition times fulfills the memoryless property. To prove this statement, consider the process to reach state ω_i at time t_i and let $\tau_i := \sup\{t \geq t_i \mid X_t = \omega_i\} - t_i$ denote the time interval after which the process transitions away from state ω_i. Then the event $\{\tau_i > u\}$ is equivalent to the event $\{X_s = \omega_i, s \in [t_i, t_i + u]\}$. By applying the Markov property (10.12), and the definition of the time homogeneity (10.13), one obtains[3]

$$\begin{aligned} P(\tau_i > u + t \mid \tau_i > u) &= P(X_s = \omega_i, s \in [t_i, t_i + u + t] \mid X_s = \omega_i, s \in [t_i, t_i + u]) \\ &= P(X_s = \omega_i, s \in [t_i + u, t_i + u + t] \mid X_{t_i + u} = \omega_i) \\ &= P(X_s = \omega_i, s \in [t_i, t_i + t] \mid X_{t_i} = \omega_i) \\ &= P(\tau_i > t). \end{aligned} \tag{10.20}$$

Thus, the distribution of transition times has the memoryless property implying that it is exponentially distributed with rate λ_i.

Another important property which requires a more formal proof (see e.g. [60]) is the independence of the transition time τ_i and the state ω_j the process reaches after the next transition:

$$P(X_{\tau_i} = \omega_j \mid \tau_i \leq t, X_0 = \omega_i) = P(X_{\tau_i} = \omega_j \mid X_0 = \omega_i) =: a_{ij}. \tag{10.21}$$

To distinguish this definition from the definition of the transition matrix (10.14), one has to recall that τ_i denotes the time interval the process needs to leave state ω_i and not an arbitrary time interval as in (10.14).

10.2.2.4 The Q-Matrix

Having stated that the transition times of CTMCs are exponentially distributed, the remaining question is how the transition from one state to another can be computed. It can be shown that for every transition $\omega_i \to \omega_j$ a rate λ_{ij} can be constructed, giving rise to the notion of a CTMC as a sequence of competing transitions with exponentially distributed transition times. The corresponding rates are given by the entries of the *Q-matrix* (also called *infinitesimal generator*), $\mathbf{Q} = \{q_{ij}\}_{\omega_i, \omega_j \in \Omega}$, which is defined as

$$\mathbf{Q} := \lim_{\varepsilon \to 0^+} \frac{\mathbf{P}(\varepsilon) - \mathbf{P}(0)}{\varepsilon} = \lim_{\varepsilon \to 0^+} \frac{\mathbf{P}(\varepsilon) - \mathbb{1}}{\varepsilon}, \tag{10.22}$$

---

[3] Strictly speaking, the argumentation is not completely correct because the Markov property (10.12), only holds for finite sequences of states. For a rigorous proof, the reader is referred to [60].

where $\mathbb{1}$ denotes the identity matrix. From the definition of the Q-matrix, it follows that the entries of the transition matrix can be written as a Taylor series,

$$p_{ij}(\varepsilon) = q_{ij}\varepsilon + \delta_{ij} + \mathcal{O}(\varepsilon^2), \quad \varepsilon > 0. \tag{10.23}$$

The diagonal elements of **Q** are given by

$$\begin{aligned} q_{ii} &= \lim_{\varepsilon\to 0^+} \tfrac{p_{ii}(\varepsilon)-1}{\varepsilon} = \lim_{\varepsilon\to 0^+} \frac{\left(1-\sum_{j\neq i} p_{ij}(\varepsilon)\right)-1}{\varepsilon} = -\sum_{j\neq i} q_{ij} \\ &=: -q_i. \end{aligned} \tag{10.24}$$

By expanding the distribution of the transition time τ_i in a Taylor series, it can easily be shown that q_i is equal to the rate λ_i of the transition time τ_i. From (10.20), it follows that

$$P(\tau_i \leq \varepsilon) = 1 - \mathrm{e}^{-\lambda_i\varepsilon} = \lambda_i\varepsilon + \mathcal{O}(\varepsilon^2). \tag{10.25}$$

Alternatively, $P(\tau_i \leq \varepsilon)$ can be written in terms of the probability matrix **P**,

$$\begin{aligned} P(\tau_i \leq \varepsilon) &= \sum_{j\neq i} P(X_\varepsilon = \omega_j \,|\, X_0 = \omega_i) \sum_{j\neq i} p_{ij}(\varepsilon) \\ &\overset{(23)}{=} \sum_{j\neq i} q_{ij}\varepsilon + \mathcal{O}(\varepsilon^2) \overset{(24)}{=} q_i\varepsilon + \mathcal{O}(\varepsilon^2), \end{aligned} \tag{10.26}$$

where ε has to be sufficiently small in order to guarantee that only one transition (from ω_i to the subsequent state) occurs during the time interval ε. Equating the linear coefficients of (10.25) and (10.26) yields

$$q_i = \lambda_i, \tag{10.27}$$

and thus

$$P(\tau_i \leq \varepsilon) = 1 - \mathrm{e}^{-q_i\varepsilon}. \tag{10.28}$$

In order to work out the meaning of the off-diagonal elements of **Q**, one has to use (10.23) and the definition of the transition matrix (10.14). Again, ε is assumed to be sufficiently small, such that only one transition may occur during the time interval ε,

$$\begin{aligned} p_{ij}(\varepsilon) &= P(X_\varepsilon = \omega_j \,|\, X_0 = \omega_i) = P(X_{\tau_i} = \omega_j,\, \tau_i \leq \varepsilon \,|\, X_0 = \omega_i) \\ &= P(\tau_i \leq \varepsilon \,|\, X_0 = \omega_i) P(X_{\tau_i} = \omega_j \,|\, \tau_i \leq \varepsilon, X_0 = \omega_i) \\ &= P(\tau_i \leq \varepsilon) P(X_{\tau_i} = \omega_j \,|\, X_0 = \omega_i) \\ &= \left(q_i\varepsilon + \mathcal{O}(\varepsilon^2)\right) a_{ij} = q_i a_{ij}\varepsilon + \mathcal{O}(\varepsilon^2), \quad i \neq j. \end{aligned} \tag{10.29}$$

Here it is important to note that $P(\tau_i \leq \varepsilon \,|\, X_0 = \omega_i) \equiv P(\tau_i \leq \varepsilon)$ since the definition of the transition time already implies the process to be in state ω_i at $t = 0$, see Sect. 10.2.2.3. By equating the linear coefficients of (10.23) and (10.29), one obtains

$$q_{ij} = q_i a_{ij} \quad \Longrightarrow \quad a_{ij} = \frac{q_{ij}}{q_i}, \quad i \neq j. \tag{10.30}$$

Equations (10.21, 10.28 and 10.30) are the key results of this section. They reveal that the entries of the Q-matrix determine both the transition time τ_i and the next transition $\omega_i \to \omega_j$ given a state ω_i:

$$\begin{aligned} &P(X_{\tau_i} = \omega_j, \tau_i \leq t \,|\, X_0 = \omega_i) \\ &\quad = P(\tau_i \leq t) P(X_{\tau_i} = \omega_j \,|\, X_0 = \omega_i) = \left(1 - \mathrm{e}^{-q_i t}\right) \frac{q_{ij}}{q_i}. \end{aligned} \tag{10.31}$$

In addition, comparison of (10.30) and the competition theorem (10.11), suggests to interpret the off-diagonal elements q_{ij} of the Q-matrix as transition rates determining the time the process needs to reach state ω_j starting from ω_i.

10.2.3 Two Algorithmic Constructions of Continuous-Time Markov Chains

Having introduced the fundamentals of probability theory and the theory of CTMCs in the previous sections, this section is devoted to the discussion of two algorithms for the construction of CTMCs. Both algorithms have in common that in each simulation step the knowledge of all possible transitions $\omega_i \to \omega_j, j \neq i$ and their corresponding rates $\{q_{ij}\}$ is required.

10.2.3.1 The BKL Algorithm

The first algorithm to be discussed is the BKL algorithm [54], also known as residence-time algorithm or the n-fold way. It mainly consists of two separate units. The first one is the building of the *embedded Markov chain* $\{X_n\}_{n \in \mathbb{N}}$ of the process:

$$X_n := X_{t_n}, \tag{10.32}$$

where $\{t_n\}_{n \in \mathbb{N}}$ denotes the sequence of transition times. The second one is the generation of the sequence of transition times. Due to the independence of the transition time and the subsequent transition path (10.21), these two units can be treated separately.

Algorithm I (BKL algorithm)

Initialization:

(i) Set the initial time $t_0 = 0$.
(ii) Select the initial state X_0 according to an arbitrary initial distribution.

Simulation step:

(i) Set up a list of all possible transitions $\omega_i \rightarrow \omega_j$ and their corresponding rates $\{q_{ij}\}$.
(ii) Compute the diagonal element $q_i = -\sum_j q_{ij} = -q_{ii}$ of the Q-matrix, where i labels the current state of the process.
(iii) Sample the residence time τ_n of the process in state ω_i from an exponential distribution with rate q_i and set $t_{n+1} = t_n + \tau_n$.
(iv) Sample the next transition $\omega_i \rightarrow \omega_j$ according to the probability distribution

$$P(X_{t_{n+1}} = \omega_j \,|\, X_{t_n} = \omega_i) = a_{ij} = \frac{q_{ij}}{q_i}. \tag{10.33}$$

(v) Perform the transition $\omega_i \rightarrow \omega_j$ and continue with step 1.

With regard to random number sampling, the BKL algorithm is very efficient since it requires only two random numbers in each simulation step (step 3 and 4). However, In many cases, the most time-consuming part of this algorithm is the computation of the diagonal element q_i of the Q-matrix (step 2).

10.2.3.2 The First Reaction Method

The second algorithm to be covered is the First Reaction (FR) method. It was invented by D. Gillespie in the 1970s in order to simulate the time evolution of coupled chemical reactions [61–63]. Mathematically, it is based on the formal equivalence of (10.30) and the competition theorem, (10.11): in Sect. 10.2.2.4 it was shown that the elements of the Q-matrix determine the local behavior of the CTMC:

$$P(X_{\tau_i} = \omega_j, \tau_i \leq t \,|\, X_0 = \omega_i) = \left(1 - \mathrm{e}^{-q_i t}\right) \frac{q_{ij}}{q_i}. \tag{10.34}$$

Combination of (10.10) and (10.11) shows that (10.34) is formally equivalent to

$$\begin{aligned} P(\min\{X\}_{i\in I} \leq x, K = j) &= P(\min\{X\}_{i\in I} \leq x) P(K = j) \\ &= \left(1 - \mathrm{e}^{-x\sum\limits_{i=1}^{n} \lambda_i}\right) \frac{\lambda_j}{\sum_{i=1}^{n} \lambda_i}, \end{aligned} \tag{10.35}$$

with K being defined as in Sect. 10.2.1.3.

As already suggested in Sect. 10.2.2.4, this equivalence allows for an interpretation of the time development of a CTMC as a competition of all possible transitions $\omega_i \rightarrow \omega_j, j \neq i$ in terms of competing i.e.d. random variables with rates q_{ij}. Instead of creating the embedded Markov chain and sampling the residence time independently from each other, as it is done within the BKL algorithm, it is possible to sample a transition time for every possible transition and choose the transition with the smallest residence time as the next transition and the corresponding time as the next transition time.

Algorithm II (FR method)
As in the BKL algorithm, $\{t_n\}_{n \in \mathbb{N}}$ denotes the sequence of transition times.

Initialization:

(i) Set the initial time $t_0 = 0$.
(ii) Select the initial state X_0 according to an arbitrary initial distribution.

Simulation step:

(i) Set up a list of all possible transitions $\omega_i \rightarrow \omega_j$ and their corresponding rates $\{q_{ij}\}$, where ω_i denotes the current state of the process.
(ii) Sample a transition time τ_{ij} for every possible transition $\omega_i \rightarrow \omega_j, j \neq i$ and store them in ascending order.
(iii) Choose the transition $\omega_i \rightarrow \omega_k$ for which $\tau_{ik} < \tau_{ij}$, $\forall j \neq k$ and set $t_{n+1} = t_n + \tau_{ik}$
(iv) Perform the transition $\omega_i \rightarrow \omega_k$ and continue with step 1.

In each simulation step, this algorithm requires to sample one random number for every possible transition $\omega_i \rightarrow \omega_j$. In comparison to the BKL algorithm which only requires *two* random numbers in each simulation step, this algorithm seems not to be very efficient at first glance. However, in many cases most of the transitions and corresponding rates stay unaffected during a transition step.[4] In this case, the FR method can be modified in a very effective way by utilizing the memoryless property of the exponential distribution (10.9). Then, step 2 of the algorithm requires to sample transition times only for new transitions. These transition times have to be sorted into the preexisting list of transition times which correspond to transitions unaffected by the last transition. For these unchanged transitions no new residence times need to be sampled due to the memoryless property of the exponential distribution. For many systems, this modified version of the FR method turns out to be superior to the BKL algorithm.

10.3 Description of the Simulations

The section is devoted to the description of a simulation model that covers the main aspects of cluster growth and diffusion during nanocomposite formation. Due to the complexity of the processes which are crucial for the formation of metal polymer

[4] In this relation, it is useful to think of the simulation of surface diffusion processes, where a diffusion jump of one single particle is unlikely to affect *all* other particles, but rather some particles in its local vicinity.

nanocomposites, an *ab-initio* approach is not applicable. Alternatively, the system is modeled by using an algorithm based on the KMC method [54, 64] whose mathematical foundations were discussed in Sect. 10.2. Since many of the required input parameters are not known exactly, it is not the aim of the simulations to reproduce experiments quantitatively. Instead, the goal was to develop a feasible simulation scheme that covers the main aspects of nanocomposite formation and which is able to reproduce and predict qualitative changes caused by variations of experimental conditions. Therefore, in terms of KMC simulations, the complex processes taking place during polymer metalization have to be condensed into a self-consistent simulation scheme containing a catalog of the main cluster processes and their corresponding rates.

10.3.1 Modeling the Metal-Polymer Nanocomposite

10.3.1.1 The Polymer Matrix

The complex microscopic properties of the polymer substrate, its atomic and chemical structure are not directly included in the simulation model. Instead, its influence is reflected in averaged cluster mobilities (expressed in rate constants), atomic evaporation rates and diffusion jump lengths of clusters. In general, the polymer is considered as a continuum.

Two different experimental scenarios are covered by the simulations: the first one is the formation of a metal-polymer interface. In this case, the polymer is located in the half-space $\{(x, y, z) \in \mathbb{R}^3 \,|\, z \leq 0\}$ with its surface located at $z = 0$. The second scenario is co-deposition of metal and polymer [65]. To model this process, the polymer surface is shifted in positive z-direction with a constant velocity v_p corresponding to the incoming flux of polymer. This shift is superimposed by a second shift which is due to the arrival of new metal atoms on the surface. This is necessary to avoid metallic filling factors larger than 1, see also Sect. 10.5. To this end, monomers which are deposited on the surface are assigned an effective volume V_{eff} which is defined as the volume one atom contributes to one ML of atoms (one ML is defined as certain density of atoms covering the surface. For details see Sect. 10.3.1.3). The polymer surface is then shifted by $\delta z = V_{eff}/A$ (A is the area of the surface) every time a monomer is deposited on the surface and shifted by -δz once a monomer evaporates from the surface.

In experiments, the polymer substrate is often pretreated with ion bombardment [38] or reactive metals [45] in order to produce point-like surface defects with a desired surface density. To incorporate these effect in the simulations [66], distinguished points with a predefined trapping radius $r_t = 0.3$ nm are randomly distributed on the surface. The effect of defects is to fix clusters as soon as they diffuse into their trapping region. These clusters serve as preferred nucleation sites for cluster growth.

10.3.1.2 Clusters

Cluster Geometry and Growth

Clusters are considered within the framework of the liquid drop model [58]. The liquid drop model assumes clusters to have a constant density regardless of their size and a spherical shape, which is a good assumption supported by TEM measurements [67, 68] and MD simulations of small Ag_n clusters [69]. Then the radius r_n of a cluster C_n containing n atoms is equal to

$$r_n = r_a n^{1/3}, \tag{10.36}$$

where r_a is the atomic radius of the cluster species. According to experimental results, clusters are assumed to be stable in the sense that they do not decompose into smaller clusters [36]. In general, clusters may agglomerate and form larger clusters if the distance of their surfaces falls below a critical distance r_{crit}. The nucleation of clusters starts from a two-body process and obeys the following reaction scheme:

$$C_1^{(t)} + C_1 \rightarrow C_2^{(t)}, \qquad C_n^{(t)} + C_m \rightarrow C_{n+m}^{(t)}, \tag{10.37}$$

where the subscripts denote the number of atoms the cluster contains and the optional superscript t indicates that the cluster is trapped by a defect site. As one can see from the reaction scheme, formation of clusters is mainly due to two processes. The first one is the formation of large clusters by merging of two free smaller clusters. The second process that has to be taken into account is the growth of clusters at defect sites. Within this process, one of the two participating clusters (and the resulting cluster) is trapped by a surface defect. After agglomeration, the radius r_{m+n} of the new cluster is given by:

$$r_{m+n} = \left(r_m^3 + r_n^3\right)^{1/3}, \tag{10.38}$$

which is in accordance to (10.36). This model of cluster growth goes back to P. Meakin and F. Family and is known as the Family and Meakin model (FMM). It was invented in order to study liquid droplet deposition, growth, and coalescence on two-dimensional substrates [70, 71]. Obviously, this model can only describe the first stage of cluster growth which is characterized by growth of separated spherical islands. In order to describe the transition from spherical growth to the formation of elongated structures, consisting of partially coalesced clusters, and finally the full percolation stage, a cutoff radius r_c is introduced above which two overlapping clusters do not fully coalesce but stay side by side. This growth model is an extension of the FMM and usually referred to as the Interrupted Coalescence model (ICM) [41]. The ICM involves the following rules for the growth of two clusters with radiuses r_m and r_n:

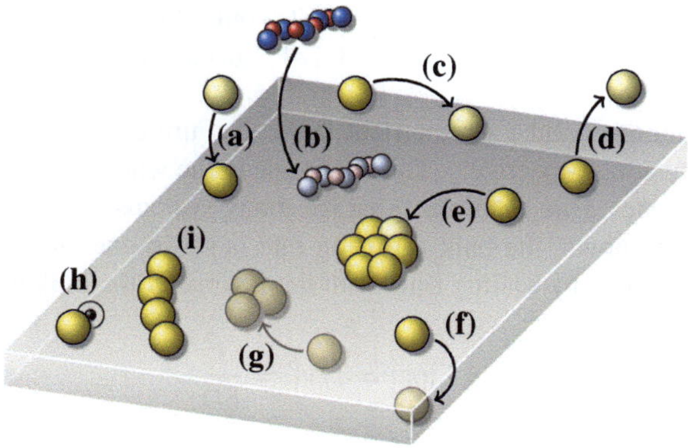

Fig. 10.6 Illustration of the elementary cluster processes considered in the KMC simulations of nanocomposite formation including deposition of atoms and polymer (*a*, *b*), surface and bulk diffusion (*c*, *f*, *g*), reemission of monomers (*d*), cluster agglomeration (*e*, *g*), trapping at surface defects (*h*), and formation of percolating networks induced by interrupted coalescence of clusters (*i*)

(i) If $r_m < r_c$ *or* $r_n < r_c$, the clusters agglomerate according to (10.38) and the new cluster is located at the center of mass of the two clusters.
(ii) If $r_m \geq r_c$ *and* $r_n \geq r_c$, the clusters do not coalesce but stay juxtaposed.

Cluster Processes

After being deposited on the surface, atoms (clusters) may perform several competing processes, see Fig. 10.6 for an illustration. Clusters may perform two different diffusion processes: surface and bulk diffusion. Surface diffusion leads from one point on the surface to another one. Bulk diffusion is diffusion from the surface into the bulk or diffusion within the bulk. According to MD simulations indicating that diffusion in polymers is a hopping process [72–76], clusters perform isotropically distributed diffusion jumps of constant length $l = 0.6\,\text{nm}$, which is approximately the diameter of a polycarbonate chain. Diffusion jumps occur at discrete points in time and a jump frequency depending on the material as well as the cluster size, for details see Fig. 10.6.

Radiotracer measurements revealed that, depending on the temperature and the metal/polymer combination, metal atoms do not fully condensate on the polymer surface but partially re-evaporate. According to these measurements, metal atoms perform a random walk on the surface prior to re-emission, rather than being backscattered directly [39]. To capture this effect, monomers are allowed to evaporate from the surface with a certain rate ν_e that has to be adjusted to the desired values of condensation coefficients C known from experiments.

Time Evolution and Diffusion Rates

The time evolution of the system is governed by the frequencies $\nu_n^{b/s}$ of diffusion events which are related to the diffusion coefficients $D_n^{b/s}$ by

$$D_n^{b/s} = \beta l^2 \nu_n^{b/s}, \qquad \beta = \begin{cases} 1/6 : \text{bulk diffusion} \\ 1/4 : \text{surface diffusion}, \end{cases} \tag{10.39}$$

with the jump length l and a geometrical factor β. Further, b and s stand for bulk and surface diffusion, respectively. The time scale in the whole simulation is then set by one of these rates. Due to the fact that for many systems exact atomic diffusion constants are not known, the bulk diffusion rate is set to one and a dimensionless time τ scale is defined by scaling time with the inverse of the bulk diffusion rate of monomers:

$$\tau = 1\,\text{jpa} := 1/\nu_1^b = \frac{l^2}{6D_1^b}. \tag{10.40}$$

By definition, 1 jpa is the average time between two bulk diffusion events of a monomer, see Sect. 10.2.1.3 (10.7). Assuming a bulk diffusion coefficient D_1^b of $10^{-15}\,\text{cm}^2/\text{s}$, which is a typical magnitude for metal/polymer systems [34, 46, 77], and the jump length to be $l = 0.6\,\text{nm}$, 1 jpa translates into 6×10^{-1} s.

Furthermore, all rates of processes incorporated in the simulations are given in units of the bulk diffusion rate.

The diffusion coefficients obviously depend on the cluster size which is also indicated by the subscript n denoting the number of atoms a cluster consists of. To model the bulk diffusion of clusters in polymer systems below the glass transition temperature T_g, different approaches were developed. A central role has been played by the so-called Free Volume theory (FVT) introduced by Cohen and Turnbull [78]. It was developed in the 1950s to describe self-diffusion in liquids of hard spheres. The main idea of the FVT is to explain diffusion in terms of the redistribution of vacancies. The assumptions of the FVT lead to the following prediction for the diffusion coefficient D of the diffusing species:

$$D \sim e^{-\gamma \frac{V^*}{V_f}}, \tag{10.41}$$

where γ is a geometrical factor, V^* is the minimum volume into which a molecule of the diffusing species can jump, and V_f is the average free volume per spherical molecule in the liquid. Cohen and Turnbull stated that V^* is approximately the specific volume of the diffusing molecule which means that the FVT, in its original formulation, predicts an exponential decay of the diffusion coefficient with the size of the diffusing particle. In the 1970s, the concepts of the FVT were extended to describe binary diffusion processes in systems consisting of a polymer species and a solvent [79–82]. These extensions also predict an exponential dependence of the diffusion coefficient on the particle size.

Based on the outcomes of the FVT, the following scaling of the bulk diffusion rate with the cluster size n is chosen:

$$\nu_n^b = 2^{-n} \cdot \nu_1^b, \qquad \nu_1^b \equiv 1. \tag{10.42}$$

In order to describe surface diffusion adequately, the FVT is not applicable since the concept of vacancies is not appropriate. As a first approximation a simple power law to connect the surface diffusion rate of a cluster with its size is assumed:

$$\nu_n^s = n^{-\alpha} \cdot \nu_1^s, \qquad \nu_1^s = m \cdot \nu_1^b, \tag{10.43}$$

where the coefficient m denotes the ratio between the surface and bulk diffusion rate of monomers. Typically, surface diffusion is many orders of magnitude faster than bulk diffusion, which is also the case for metal diffusion in polymer systems [37], where the ratio of surface and bulk diffusion coefficients might be substantially smaller than for typical solids due to the large surface roughness of polymers [33]. However, a value around $m = 40$ appeared to be a reasonable choice in order to reproduce the typical bimodal shape of cluster size distributions observed in experiments [33]. For the simulation data presented in this work, a value $\alpha = 1$ was used. MD simulations of cluster diffusion on crystalline surfaces have also predicted a diffusivity following a power law of this form, where α is close to 1 [56].

The glass transition temperature T_g is related to the long range mobility of polymer chains [33, 37]. MD simulations [83] and experiments [84, 85] have shown that diffusion of solvents in polymers divides into two regimes, depending on the ratio σ of the particle size and the gyration radius of the polymer chains. For large particles, ($\sigma > 1$), theoretical investigations [86] and MD simulations [83] predict the polymer melt to behave like a continuum on the length scale of the particles and the diffusion coefficient D_n^b to be in good agreement with the *Stokes-Einstein* relation:

$$D_n^b = \frac{k_B T}{6\pi \eta r_n} \quad \Longrightarrow \quad D_n^b \sim n^{-1/3}. \tag{10.44}$$

If the particles are smaller than the polymer chains, ($\sigma < 1$), the prediction of the Stokes-Einstein relation usually fails and underestimates the diffusivity since the clusters start to see the microscopic viscosity [83] which is expected to be reduced in comparison to the global viscosity due to depletion of polymer around those particles' surfaces [87, 88]. In this regime, MD simulations suggest the diffusion coefficient to be inversely proportional to the cube of the radius [83], contradicting the predictions of the Stokes-Einstein relation,

$$D_n^b \sim R_n^{-3} \quad \Longrightarrow \quad D_n^b \sim n^{-1}, \tag{10.45}$$

with R_n being the particles hydrodynamical radius which is defined as the radius of a hard sphere that diffuses at the same rate as the particle. In many cases R_n is orders of magnitude larger than the actual particle radius [89]. However, since clusters are considered as hard spheres in the simulations, R_n is identical to the cluster radius.

10.3.1.3 Deposition

As in real experiments, metal atoms are deposited on the surface during a finite deposition time t_d with a certain flux R_m of atoms which is assumed to be stationary. The deposition rate is given units of ML/jpa, where one ML corresponds to a surface density of 10 atoms per nm^2 [36]. Assuming that one ML has the thickness of the atomic diameter d_a, 1 ML/jpa can be translated into SI units using the definition of jpa, Def. (10.40),

$$1\,\mathrm{ML/jpa} = \frac{6 d_a D_1^b}{l^2}. \tag{10.46}$$

Assuming the same bulk diffusion coefficient D_1^b and jump length l as above, 1 ML/jpa translates into 3×10^1 nm/min.

10.3.2 The Simulation Algorithm

As already mentioned in Sect. 10.2, the algorithm used in the simulations is based on the FR method introduced by Gillespie in the 1970s [61]. The main idea of this algorithm is to sample time points for every possible elementary process i of the system with its associated rate ν_i. In contrast to the notation in Sect. 10.2.3, here the transition rates are labeled by just one number and not by two as in Sect. 10.2.3. This is possible since the transitions the system undergoes do not depend on the current state of system: the diffusion rate of a cluster of a certain size stays unchanged during the evolution of the system. These time points are stored in ascending order and in every simulation step, the process associated with the smallest process time t_{min} is carried out and the system clock is advanced to the time point t_{min}. Next, the time points of processes belonging to those clusters that were involved in that transition (by cluster agglomeration) are removed from the list, whereas for the remaining clusters, whose state changed due to the last process, a new process time t_i is sampled from an exponential distribution and sorted into the list. At this point, the FR method reveals its advantage over the BKL algorithm; due to the memoryless property of the exponential distribution, the process times of those clusters that were not affected by the last process do not have to be re-sampled. In contrast to that, within the BKL algorithm, the computation of the diagonal element q_i of the Q-matrix and the subsequent sampling of the next process is usually computationally very expensive. In case of the present simulations, the most time-consuming part is the organization of the time-ordered list of processes.

Figure 10.7 shows an algorithm flowchart illustrating the numerical procedure for the case of metal-polymer interface formation with unrestricted cluster growth. As one can see from Fig. 10.7, the iteration cycle (which is framed by the dotted line) always begins with the execution of the process associated with the smallest process time. In case of atomic evaporation, the dataset of the corresponding cluster is simply

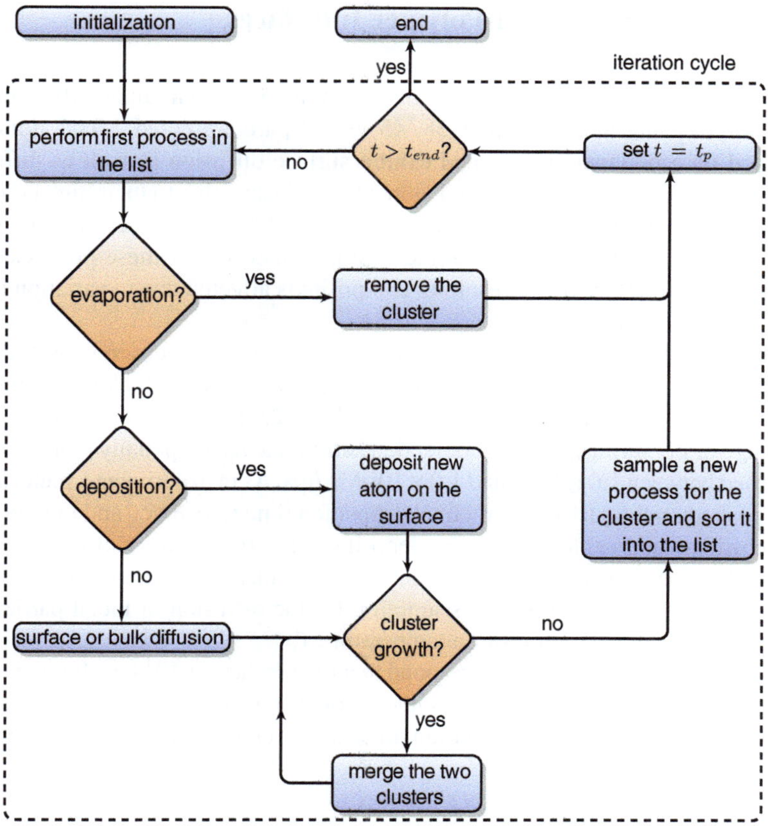

Fig. 10.7 A flowchart of the simulations illustrating the numerical procedure within one iteration cycle

destroyed and the clock is advanced to the point in time at which the evaporation takes place. If the process is a diffusion process or deposition of a new cluster, the diffusion process is executed or a new cluster is deposited on the surface of the polymer matrix, respectively. Next it is checked if the distance of the processing cluster to any other cluster lies below the critical distance r_{crit}. If yes, the two clusters merge according to (10.37) and (10.38) and the data set of one of the clusters is destroyed.[5] This step has to be performed in a loop since the distance to any other cluster may fall below the critical distance due to agglomeration. This loop is left if the distance to all other clusters lies above the critical distance. Afterwards, a new process and an associated process time are sampled and sorted into the time-ordered list of processes. The algorithm terminates once the time exceeds the total simulation time t_{end} which is given as an input parameter to the simulations.

[5] In effect that means the dataset of the non-processing cluster is destroyed and its process time and the corresponding process are removed from the time-ordered list of processes.

10.4 Formation of Metal-Polymer Interfaces

In this chapter, numerical results of kinetic Monte Carlo simulations [65, 66] of metal-polymer interface formation [38, 45, 67, 90] are presented. Metal atoms are deposited on a polymer surface and exhibit surface diffusion as well as diffusion into the bulk. At the same time metal atoms form clusters due to their high cohesive energy, re-evaporate from the surface, or get trapped by surface defects. The focus of the present chapter is to discuss the complex interplay of these processes and how the microstructure of the resulting composite is affected due to variation of the deposition conditions.

For all results discussed below, the size of the surface lies between 2500×2500 and 4500×4500 nm. If not explicitly mentioned, the deposited thickness δ amounts 2 ML, corresponding to a number of 6.25×10^7 to 2.025×10^8 atoms for the given size range of the surface. The monomer deposition rate R_m is spatially homogeneous and varied between 1.0×10^{-5} and 1.0×10^3 ML/jpa. As in experiments, metal atoms are deposited on the surface over a final deposition time $t_d = \delta / R_m$ and immediately start to diffuse and agglomerate after deposition. For the majority of the presented results, the size dependence of the bulk diffusion rate ν_n^b is chosen according to (10.42), which is an appropriate assumption for the diffusion of metal particles in polymers below the glass transition temperature T_g.

All results presented below correspond to a time instant of 10^4 jpa after the termination of the deposition. The time dependence of the results was carefully studied and it has been found that longer simulation times do not change the results significantly.

10.4.1 Complete Condensation without Surface Defects

The interplay of cluster agglomeration and diffusion depending on the deposition conditions is investigated for a polymer surface free of defects. Atomic evaporation is neglected.

10.4.1.1 Concentration Profiles

The first quantity which will be discussed is the depth-resolved concentration of metal inside the polymer bulk. In order to compute the metal concentration, the polymer was divided in thin layers of thickness 0.015 nm parallel aligned to the surface. The concentration in a given layer at depth z is then defined as the layer volume occupied by clusters divided by the total volume of that layer.

Concentration profiles for two different deposition rates differing by a factor of 20 are depicted in Fig. 10.8. The solid line shows the total concentration and the other lines the fraction of the concentration attributed to clusters in a certain size range as it is indicated in the figure. The total concentration in Fig. 10.8 can, in general,

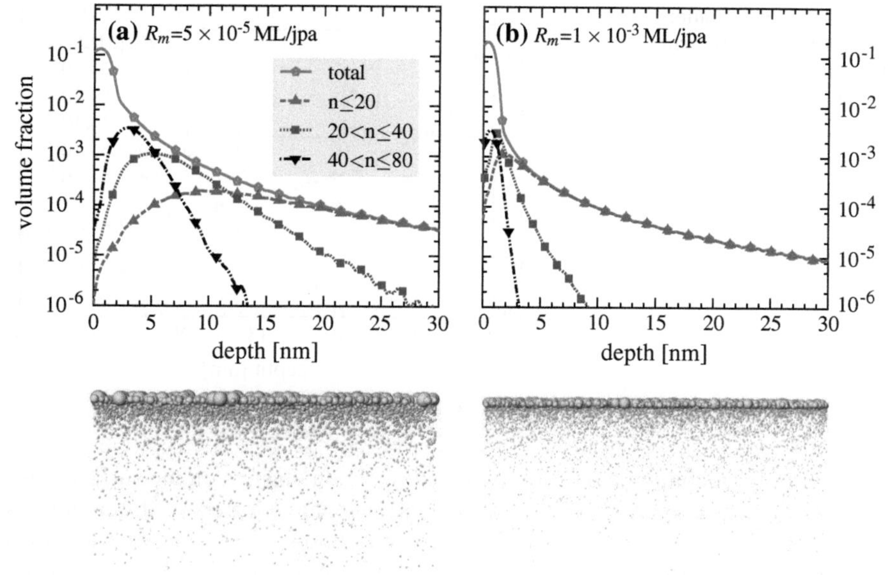

Fig. 10.8 *Upper panel* Depth-resolved Concentration profiles for two different values of the atomic deposition rate R_m as indicated in the figure. *Lower panel* Illustration of the corresponding cluster density in a 120 × 60 nm extract of the system. The line of vision is directed onto the x-z plane

be divided into three parts whose shape is strongly influenced by the deposition rate R_m. The first region is a narrow area just below the surface with a width of about 1 to 2 nm. Here, the cluster concentration is very high with a maximum located about one diffusion length underneath the surface. The second region is characterized by an abrupt decay of the concentration by one to three orders of magnitude and extends over about 1 to 2 nm. The last region which begins in a depth of about 2 to 4 nm, is very broad, and shows a much slower decay of the metal concentration in the bulk. Here the density decay is close to a Gaussian profile which would be the profile in the case of free diffusion of non-interacting particles according to the thin-film solution of Fick's second law. Consequently, all deviations from this behavior result from agglomeration and cluster growth which is most distinctive in the vicinity of the surface.

Figure 10.9 shows how the deposition rate affects the concentration profiles. With an increase of the rate, also the concentration in the first region near the surface grows monotonically. From the second region on, the picture turns into its opposite: an increase of the deposition rate leads to a monotonic reduction of the metal fraction in the bulk. The explanation is that an increased deposition rate leads to a more homogeneous growth of smaller, more or less immobile, clusters on the surface (see Sect. 10.4.1.2), which results in a higher coverage of the surface. Consequently, the probability for small mobile clusters to encounter larger immobile clusters on the surface before diffusing into the bulk is enhanced. In addition to that, there is a higher

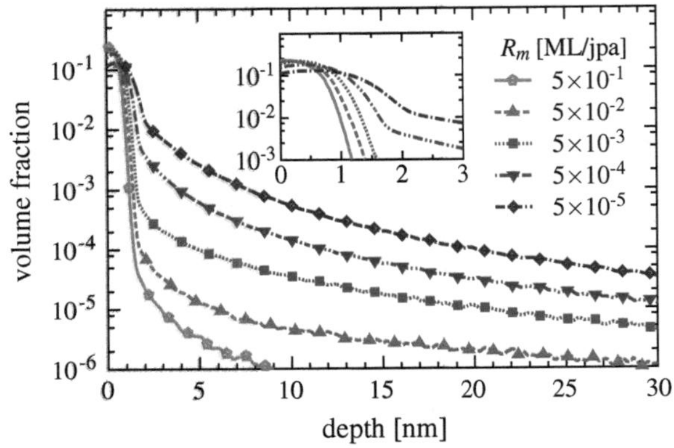

Fig. 10.9 Concentration profiles for five different deposition rates R_m spanning five orders of magnitude

probability for incoming atoms to impinge directly on clusters on the surface. This explanation is readily verified by considering how many clusters of different sizes are found in a given depth. In Fig. 10.8 the contributions of clusters with less than 20 atoms, with 20 to 40, 40 to 80 atoms and of larger clusters, to the total density are depicted separately. One clearly sees that, near the surface, the concentration is dominated by the largest clusters with a size of more than 80 atoms. This has been observed for all simulated deposition rates. For the smaller cluster groups one observes a maximum concentration in a depth of few nanometers, cf. Fig. 10.8: first, clusters with a size between 40 and 80 atoms reach a maximum concentration in a depth of about 3 nm, followed by clusters with 20 to 40 atoms the concentration of which peaks in about 5 nm depth. Finally, clusters smaller than 20 atoms have their highest concentration in a depth of about 10 nm. Deeper in the bulk practically only the smallest clusters are found.

10.4.1.2 Cluster Size Distributions

In addition to the concentration profiles discussed above, the simulations also yield the complete information about the size spectrum of the grown clusters. To this end, size distributions were calculated from the resulting cluster configuration as the fraction of clusters with a particular size n (measured in number of atoms) related to the total number of clusters. Despite the fact that the size distributions are defined only on integers, they are depicted as continuous functions resulting from spline approximations of the given data sets. Since the simulations usually contain several million clusters, the size distributions are a priori very smooth such that the depicted splines do not distort the simulated size distributions. In Fig. 10.10 size distributions for three different deposition rates ranging from 2.0×10^{-5} to 1.0×10^{-3}ML/jpa are depicted. Figure 10.10a shows the global distribution of cluster sizes which is typically characterized by three maximums whose shape and position vary with the deposition rate. A bimodal shape of the cluster size distribution with one maximum

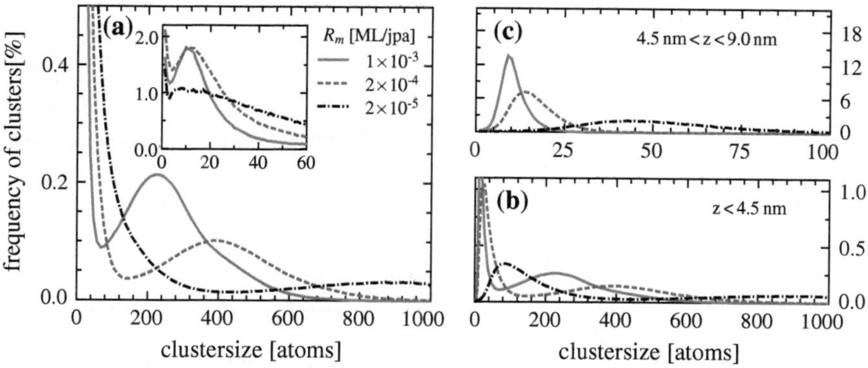

Fig. 10.10 Cluster size distributions for three different deposition rates R_m spanning three orders of magnitude as indicated in the figure. No surface defects. *Left panel* global cluster size distribution. *Right panel* depth-resolved size distributions in two different thin layers of thickness 4.5 nm below the surface. Note the different vertical scales

at small sizes and one maximum at larger sizes is very well known from experiments [33, 91, 92]. The finite resolution of experimental measurements of cluster sizes might be the reason for the indistinguishability of the first two peaks which are very close to each other.

At all deposition rates, the first peak is contributed by monomers and the second peak by clusters of a size around $n = 10$, whereby the second peak is very broadened for the lowest deposition rate. Both peaks exhibit about the same intensity and dominate the third peak by a factor of around 10. Thus, there is a large fraction of dimers and trimers and a rapid decay of the size distribution towards larger clusters. An increase of the deposition rate leads to a sharpening of the first peak, see inset of Fig. 10.10a. This maximum is mainly attributed to monomers deep in the bulk which is verified by Fig. 10.10b, c which show the size distribution in two layers of thickness 4.5 nm under the surface. here, one can see that monomers practically do not make a contribution to the size distributions up to a depth of 9 nm, and thus are located deeper in the bulk.

The size distributions at the surface layer(Fig. 10.10b) exhibit a bimodal shape with two maximums. The position of the second maximum corresponds to the position of the third maximum in the global size distributions. Since this second maximum already vanishes in the second layer under the surface, the third maximum of the global distribution relates to clusters located in the direct vicinity of the surface. Its position and shape exhibit a strong dependence on the atomic deposition rate; an increase of the deposition rate causes a shift of the peak to smaller cluster sizes and a simultaneous narrowing. This reflects the tendency towards homogeneous nucleation at the surface and a narrowing of the growth region of larger clusters in case of high deposition rates(see also the lower panel of Fig. 10.8).

Deeper in the bulk, the size distributions only consist of one maximum whose position shifts to smaller sizes in larger depths. The position of this peak coincides

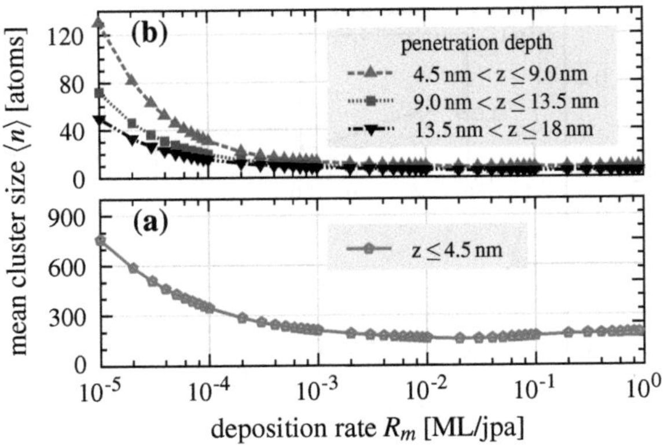

Fig. 10.11 Depth resolved mean cluster size $\langle n \rangle$ in thin layers of thickness 4.5 nm as a function of the deposition rate R_m for a surface without defects. The *bottom figure* shows the top layer just below the surface, the *upper figure* shows three layers deeper in the bulk

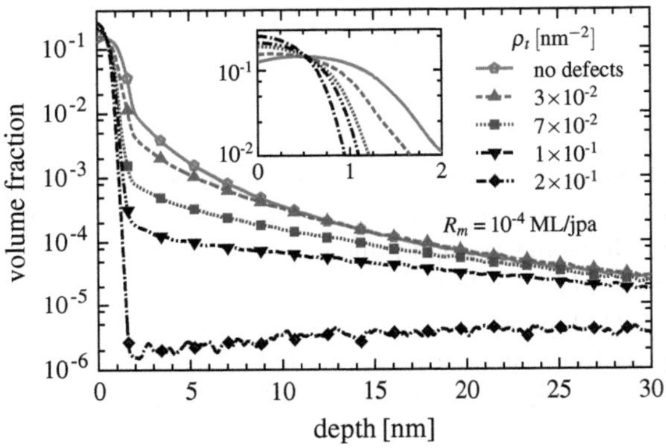

Fig. 10.12 Concentration profiles for five different densities ρ_t of surface defects ranging from 0 to $2 \times 10^{-1}\,\text{nm}^{-2}$ and a fixed deposition rate $R_m = 10^{-4}$ ML/jpa

more or less with the mean cluster size $\langle n \rangle$ in the corresponding layer. This is clearly confirmed by Fig. 10.11 which depicts the mean cluster size in four layers of thickness 4.5 nm under the surface as a function of the deposition rate. The general trend one observes is a clear reduction of the mean cluster size with growing distance from the surface. The decay is particularly strong immediately below the surface: from the first to the second layer the mean cluster size drops approximately by a factor of six.

As already mentioned before, the values of the mean cluster size are close to the peak positions in Fig. 10.10b, c. Again, there is a clear trend of a monotonic decrease of the mean cluster size with the deposition rate in any depth. At a value around 1.0×10^{-3} ML/jpa, the mean cluster size starts to saturate in all layers. The decrease of the mean cluster size may seem surprising, but it is consistent with the observations made in Sects. 10.4.1.1 and 10.4.1.2. As one can see from Fig. 10.9, high deposition rates lead to enhanced metal concentrations in the vicinity of the surface and narrow cluster size distributions, cf. Fig. 10.10. The enhanced concentration of clusters near the surface inhibits the penetration of metal into the bulk such that an increase of the deposition rate is accompanied by an overall decrease of the mean cluster size.

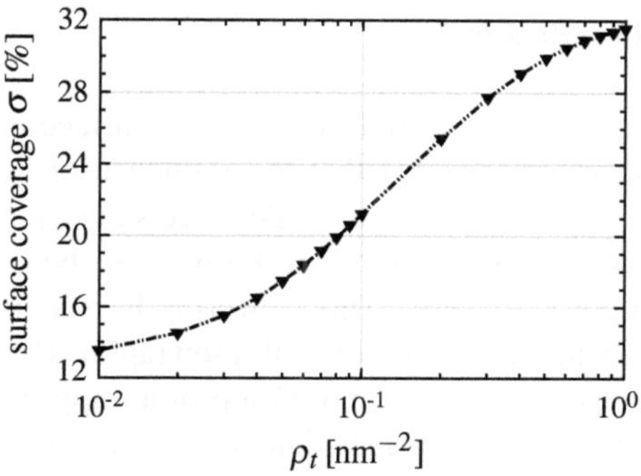

Fig. 10.13 Surface coverage as a function of the density ρ_t of surface defects at constant deposition rate $R_m = 1.0 \times 10^{-4}$ ML/jpa

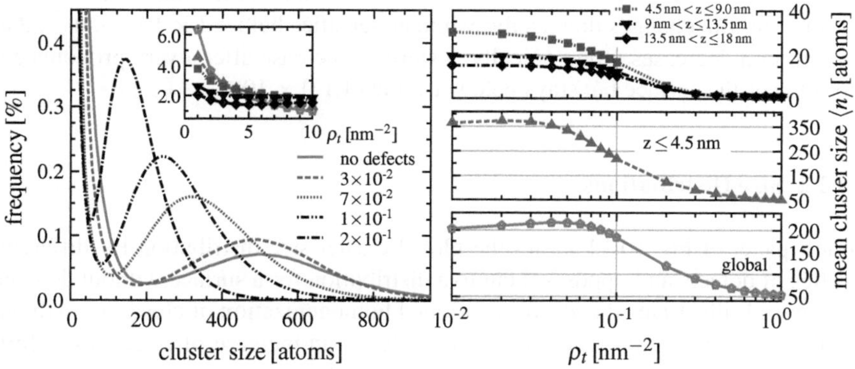

Fig. 10.14 *Left* cluster size distributions for five different densities ρ_t of surface defects as indicated in the figure. *Right* mean cluster size as a function of the density of surface defects. the *bottom figure* shows the global average. The *upper figures* show the averaged cluster size in thin layers of thickness 4.5 nm as indicated in the figure

10.4.2 Complete Condensation with Surface Defects

In experiments, the polymer surface is often exposed to ion bombardment [38] or pretreated with reactive metals [45] in order to gain a certain density ρ_t of surface defects. The main effects of these defects are trapping of clusters and a significant reduction of metal diffusion into the polymer bulk. In [45], Thran et al. report about experimental investigations of silver diffusion into TMC-PC where only 0.08 ML (which corresponds to a density of $8.0 \times 10^{-1}\text{nm}^{-2}$) of Cr pre-deposited on the surface act as an almost perfect barrier against diffusion of silver into the polymer matrix.

10.4.2.1 Concentration Profiles

Concentration profiles for five different densities of surface defects are illustrated in Fig. 10.12. One clearly recognizes that the influence of defects is different in the bulk and in the vicinity of the surface: while defects raise the concentration in the direct vicinity of the surface up to a depth of 0.5 nm, see the inset of Fig. 10.12, the concentration in the bulk is lowered. This Lowering is especially pronounced in a depth between 0.5 and 3 nm where the concentration rapidly decreases. Deeper in the bulk, this effect weakens and the concentration profiles approach each other. Obviously, the reason for this effect of surface defects is the enhanced growth of trapped clusters at the surface which act as nucleation centers and cause a more homogeneous nucleation of clusters, see Fig. 10.14. This homogenization is accompanied by an enhanced surface coverage[6] σ, see Fig. 10.13, leading to a reduced diffusion of metal into the bulk. For defect densities $\rho_t > 2 \times 10^{-1}\,\mathrm{nm}^{-2}$, the behavior of the concentration profile changes drastically; while at densities below $2 \times 10^{-1}\,\mathrm{nm}^{-2}$, the concentration decreases monotonically, it starts to increase after a very pronounced drop off near the surface but stays below a value of 1.0×10^{-5}.

10.4.2.2 Size Distributions

The left panel of Fig. 10.14 shows the global cluster size distribution for different densities of defects and opposes them to a distribution for a surface without defects (solid line). It illustrates the above-mentioned homogenization of cluster growth on the surface due to the presence of surface defects: an increase of ρ_t causes a shift of the secondary maximum (which is attributed to clusters on the surface and in the direct vicinity of the surface) to smaller sizes. At the same time, the maximum is considerably slandered.

The upper panel of the right-hand side of Fig. 10.14 shows that the mean cluster size is not significantly affected up to a value of $\rho_t \approx 6 \times 10^{-2}\,\mathrm{nm}^{-2}$; at small densities, defects cause an increase of the mean cluster size which becomes more evident by comparing the solid and the dashed line ($\rho_t = 3 \times 10^{-2}\,\mathrm{nm}^{-2}$) in the left panel of Fig. 10.14. At values above $6 \times 10^{-2}\,\mathrm{nm}^{-2}$ the mean cluster size starts to decrease which is associated with a clear reduction of metal diffusion into the bulk, see Fig. 10.12.

10.4.3 Incomplete Condensation

As already mentioned in Sect. 10.1.1.3, the condensation coefficient C of metals on polymers strongly depends on the metal/polymer combination as well as on the temperature of the substrate. In the simulations, the incomplete condensation of metal

[6] Defined as the fraction of the surface being covered by clusters.

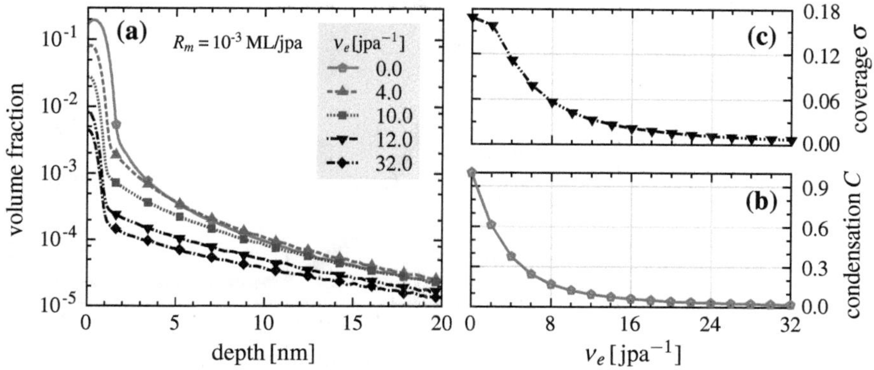

Fig. 10.15 *Left* concentration profiles for five different values of the monomer evaporation rate ν_e as indicated in the figure. *Right* condensation coefficient C (*bottom*) and surface coverage σ (*top*) as a function of the monomer evaporation rate ν_e

on polymer is achieved by allowing monomers to re-evaporate from the surface with a certain rate ν_e as described in Sect. 10.3.1.2.

In Fig. 10.15a the influence of monomer evaporation on the concentration profile is shown for five different evaporation rates ν_e lying in the parameter range $\nu_e = 0.0 \ldots 32.0\,\text{jpa}^{-1}$. Obviously, the main effect of monomer evaporation is an overall reduction of the metal concentration which is caused by desorption of monomers from the surface and the resulting decrease of the total amount of metal in the polymer. Most affected is the region in the vicinity of the surface. Here, the reduction amounts up to a factor of about 60, which roughly corresponds the reduction of the condensation coefficient, as shown in Fig. 10.15b. Deeper in the bulk, the reduction is clearly reduced. For very low evaporation rates $\nu_e < 10\,\text{jpa}^{-1}$, this effect is not noticeable or even reversed, see the dashed curve ($\nu_e = 4.0$) in Fig. 10.15a which exceeds the solid curve at a depth of 5 nm. The reason for this increase of the concentration in the bulk is the reduction of the concentration near the surface which increases the probability of clusters on the surface to diffuse into the bulk.

10.4.4 Interrupted Coalescence and Percolation

For high deposited thicknesses δ, metal clusters are expected to form a percolating network on the surface. By definition, percolation in thin films sets in if there exists a chain of overlapping clusters that connects one side of the surface with its opposite side [93]. Usually, systems that are studied with regard to percolation are characterized by the percolation coverage (sometimes also referred to as percolation threshold) ρ_c, which is the coverage of the surface at the moment when percolation sets in. Since diffusion of clusters into the bulk is possible in the present simulations, it is appropriate to additionally characterize the percolation transition by the amount

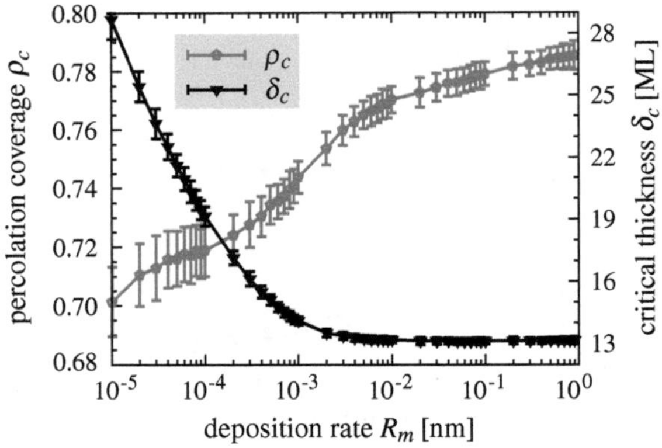

Fig. 10.16 Percolation coverage ρ_c and critical thickness δ_c as a function of the monomer deposition rate R_m

of metal δ_c (referred to as critical thickness) that was deposited until percolation sets in.

To capture the effect of percolation in the simulations, it is crucial to impose further restrictions on the growth process leading to the formation of elongated structures of clusters (in the following referred to as chains). To this end, the assumption of unrestricted cluster growth is dropped and a cut-off radius r_c is introduced above which two overlapping clusters do not agglomerate. In the literature, this model is referred to as the ICM. It was introduced in the early 1990s by Yu et al. in order to study the coalescence of metal clusters in thin films [41]. In their simulations, Yu et al. neglected diffusion of clusters and investigated the dependence of the percolation coverage on the cut-off radius r_c. The ICM involves the following rules for two overlapping clusters with radii r_m and r_n:

(i) If $r_m < r_c$ *or* $r_n < r_c$, the clusters agglomerate according to (10.38) and the new cluster is located at the center of mass of the two clusters.
(ii) If $r_m \geq r_c$ *and* $r_n \geq r_c$, the clusters do not coalesce but stay juxtaposed.

10.4.4.1 Percolation Transition

The percolation transition is investigated for different deposition rates R_m ranging from 1.0×10^{-5} to 9.0×10^{-1} ML/jpa. The results are averaged over 200 runs with a constant parameter set but different initial conditions. Figure 10.16 presents the percolation threshold ρ_c and the critical thickness δ_c as a function of the deposition rate R_m. The percolation threshold ρ_c increases monotonically with the deposition rate R_m, which results from the trend towards more homogeneous cluster growth at high deposition rates. This naturally causes higher surface coverages, cf. Fig. 10.17. In contrast to that, the critical thickness δ_c exhibits the opposite behavior: δ_c starts at value of around 28 ML at $R_m = 1.0 \times 10^{-5}$ ML/jpa and rapidly decreases towards 13 ML at $R_m = 2.0 \times 10^{-3}$ ML/jpa where it starts to saturate. The reason for the high values of δ_c at low deposition rates is an enhanced diffusion of metal into the bulk,

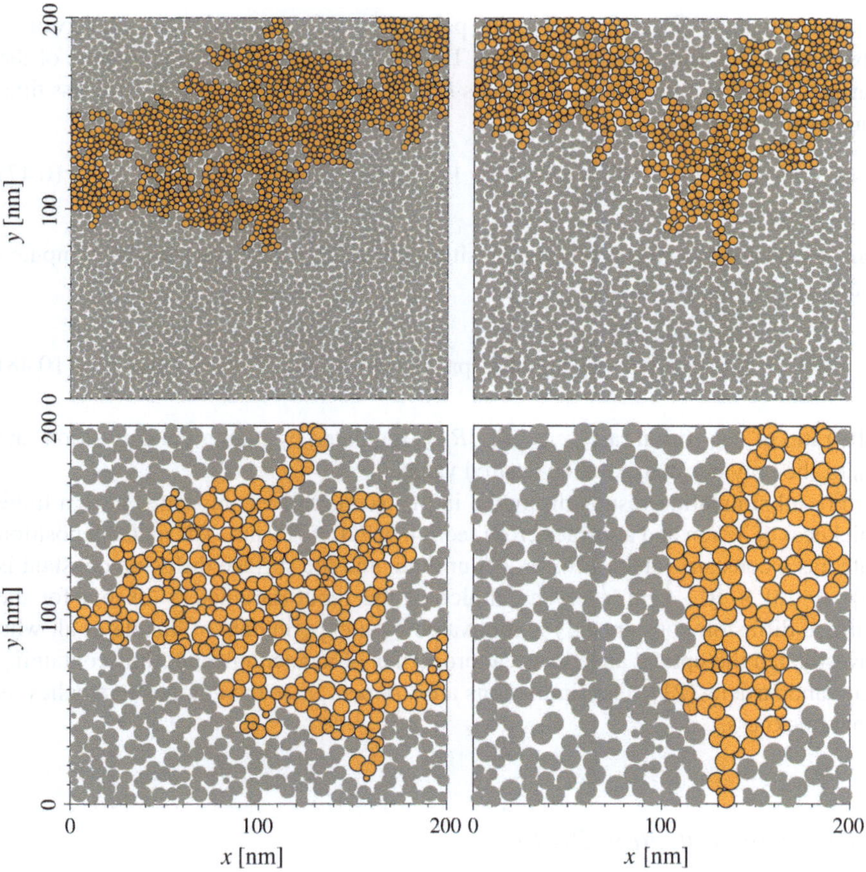

Fig. 10.17 *Top view* of the surface at the percolation threshold for four different deposition rates R_m and a critical radius $r_c = 8r_a = 1.2$ *nm*. *Top left* $R_m = 9.0 \times 10^{-1}$ ML/jpa. *Top right* $R_m = 5.0 \times 10^{-4}$ ML/jpa. *Bottom left* $R_m = 1.0 \times 10^{-5}$ ML/jpa. *Bottom right* $R_m = 2.0 \times 10^{-6}$ ML/jpa. Cluster islands that connect opposite sides of the surface are colored in *orange*

see Sect. 10.4.1.1, and a comparably slow cluster growth. Hence, a large fraction of clusters does not participate in the formation of the percolating network on the surface.

10.5 Co-Deposition of Metal and Polymer

This section is devoted to simulations of co-deposition of metal and polymer. The main focus lies on the influence of two important parameters: the ratio of deposition rates $\kappa := R_m/R_p$ and the monomer evaporation rate ν_e.

In contrast to Sect. 10.4, all input parameters of the simulations are given in absolute units throughout this chapter. This is achieved by assuming a value of the surface diffusion constant of monomers D_1^s and determining the dimensionless time unit by means of D_1^s,

$$1\,\mathrm{ML} = 1/\nu_1^s = \frac{l^2}{4D_1^s}. \tag{10.47}$$

The dimensionless values of the deposition rates R_m and R_p can then be computed via the definition of ML/jpa,

$$1\,\mathrm{ML/jpa} = \frac{4d_a D_1^s}{l^2}. \tag{10.48}$$

The ratio of deposition rates $\kappa := R_m/R_p$ is regulated by tuning the deposition rate R_p of the polymer matrix to the desired value.

For all simulation results discussed in this section, the composites have a finite thickness of 32 nm and a surface cross section of 700 × 700 nm. The metal deposition rate has a constant value of $R_m = 1.6$ nm/min and the surface diffusion constant is $D_1^s = 2.0 \times 10^{-12}\,\mathrm{cm}^{-2}/\mathrm{s}$. The choice of the input parameters (except for the surface diffusion constant D_s^1) is motivated by a experiments of Takele et al. who investigated the optical and electrical properties of nanocomposite films containing Au nanoparticles [21]. The simulations are stopped, once the composite reaches its final thickness.

10.5.1 Concentration Profiles

Figure 10.18 shows concentration profiles for eight different values of the deposition rate ratio κ between 0.5 and 4.0. Contrary to the concentration profiles for pure monomer deposition discussed in Sect. 10.4, where the concentration spans several orders of magnitude, the metal is almost constantly distributed over the whole composite.

For all investigated values of κ the concentration profile can be divided into three parts: first a very small area in the direct vicinity of the surface with a strong increase of the concentration from a low value directly at the surface to a higher value a few nanometers below the surface. The second part is a broad region of constant concentration which is followed by a thin concentration peak near the bottom of the composite. This peak results from a high instantaneous condensation coefficient $\beta(t)$ at the beginning of the deposition, cf. Fig. 10.19. The instantaneous condensation coefficient is defined as the ratio of the number of adsorbed atoms per unit of time and the deposition rate [36, 94]. The reason for this high instantaneous condensation coefficient is the missing opportunity of clusters to diffuse deep into the bulk when the composite has a thickness of only a few nanometers.

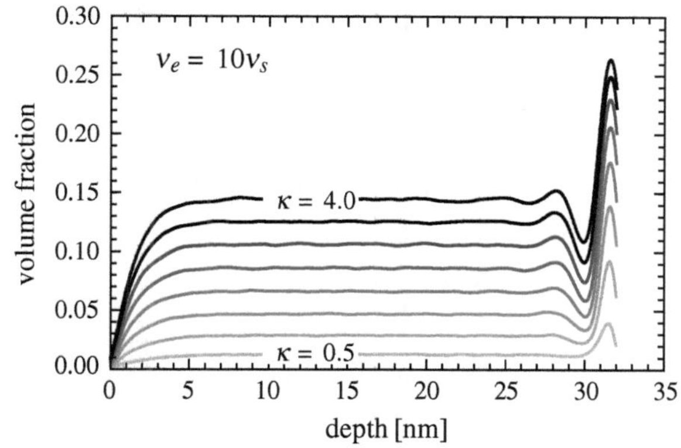

Fig. 10.18 Concentration profiles for eight different values of the deposition rate ratio κ in the parameter range $\kappa = 0.5, 1.0, \ldots 4.0$. The monomer evaporation rate is $\nu_e = 10\nu_s = 10\,\text{ML/jpa}$

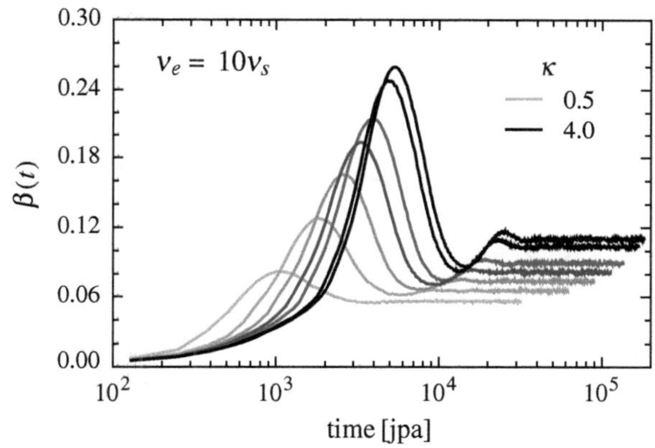

Fig. 10.19 Instantaneous condensation coefficient $\beta(t)$ as a function of time. The monomer evaporation rate ν_e is the same as in Fig. 10.18

10.5.2 Size Distributions

Figure 10.20 shows cluster size distributions for four different values of κ. The distributions exhibit a shape with two maximums whose positions show no variation with κ. The main peak of the distributions is contributed by monomers near the surface, which is not surprising since the simulations are stopped once the composite reaches a thickness of 32 nm. Consequently, monomers located at the surface do not have enough time to grow larger. The secondary maximum of all distributions shown in Fig. 10.20 is located at $n \approx 50$, whereby merely the intensity of the maximum varies with κ: the intensity decreases when κ gets larger. This is due to the fact that higher deposition rate ratios cause an accelerated cluster growth since direct impingement of monomers on clusters is more likely to occur. Thus, the fraction of clusters of sizes around the secondary maximum becomes smaller in support of larger clusters, which can be seen from the long-drawn-out tail of the distribution at higher values of κ.

The effect of different monomer evaporation rates on different quantities is depicted in Fig. 10.21. Figure 10.21c shows the mean cluster size $\langle n \rangle$ as a function of

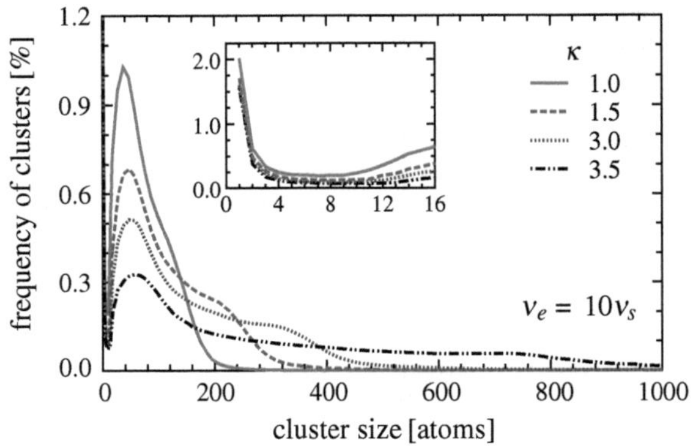

Fig. 10.20 Cluster size distributions for four different values of κ as indicated in the figure. The monomer evaporation rate is the same as in Fig. 10.18

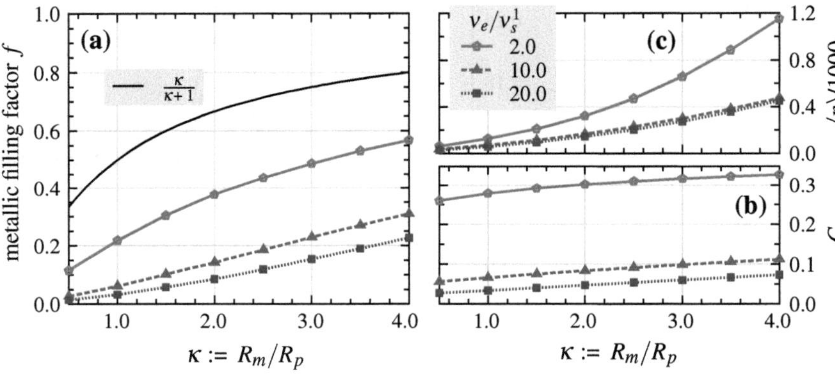

Fig. 10.21 *Left* metallic filling factor as a function of the ratio of deposition rates κ for different evaporation rates as indicated in the *right panel*. *Right* condensation coefficient C (*bottom*) and mean cluster size $\langle n \rangle$ divided by 1000 (*top*) as a function of κ

κ. The monomer evaporation rate is thereby given is units of the surface diffusion rate of monomers ν_1^s. A monomer evaporation rate $\nu_e = 2\nu_s^1$ means that the average time a monomer needs to evaporate from the surface is half of the average time between two surface diffusion jumps, see also (2.2). Expectably, the mean size $\langle n \rangle$ decreases at higher evaporation rates and increases monotonically with κ (for the same reason mentioned in the discussion of the cluster size distributions). For $\nu_e = 2\nu_s^1$ the mean size reaches a value near $\langle n \rangle \approx 1200$ at $\kappa = 4.0$, whereas for $\nu_e = 10(20)\nu_s^1$ $\langle n \rangle$ is lowered by a factor of about 3. The same effect is observable for the condensation coefficient C, cf. Fig. 10.21b. Here, the condensation coefficient has a value of about 0.3 for $\nu_e = 2.0\nu_s^1$ and $\kappa = 4.0$ and decreases towards a value around 0.1 (which is around the value Takele et al. have measured for Au co-deposited with Teflon AF[21]) if ν_e is increased to $\nu_e = 10.0\nu_s^1$.

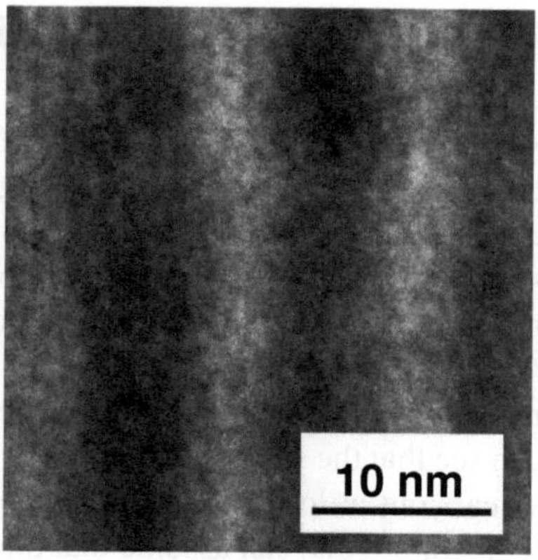

Fig. 10.22 Cross-sectional higher magnified TEM image of self-organized nanocolumns of Fe-Ni-Co in Teflon AF. With kind permission reprinted from Ref. [95]

The left panel of Fig. 10.21 shows the metallic factor as a function of κ. The metallic filling factor f is calculated from the ratio of the volume of the deposited metal V_m and the total volume V_{tot} of the composite:

$$f = \frac{V_m}{V_{tot}} = \frac{V_m}{V_m + V_p}, \tag{10.49}$$

where the total volume V_{tot} is the sum of the metal volume and the polymer volume. V_m and V_p can be expressed via the corresponding deposition rates of the metallic and the polymer component, R_m and R_p, respectively:

$$V_p = R_p L^2 t_d\,, \tag{10.50}$$

$$V_m = C R_m L^2 t_d\,, \tag{10.51}$$

where L is the side length of the quadratic surface and t_d is the deposition time. For the calculation of the metal volume, the deposition rate R_m has to be multiplied by the condensation coefficient C. Inserting (10.50) and (10.51) into the definition of the filling factor (10.49), and taking into account that the condensation coefficient C is a function of κ yields the following expression:

$$f(\kappa) = \frac{C(\kappa)R_m}{C(\kappa)R_m + R_p} = \frac{C(\kappa)\kappa}{C(\kappa)\kappa + 1} = \frac{1}{1 + \frac{1}{C(\kappa)\kappa}}\,. \tag{10.52}$$

The solid black curve in Fig. 10.21 illustrates the course of the filling factor in case of full condensation, i.e. $C = 1$. From (10.52) and Fig. 10.21, one can immediately see, how the condensation coefficient C and κ affect the filling factor: an increase of C as well as κ causes an increase of the metallic filing factor, which is clear because an enhanced condensation increases the fraction of metal by definition, and a high value of κ involves a lower polymer deposition rate R_p (Fig. 10.22).

10.6 Self-Organized Formation of Metallic Nanocolumns

As already stated in Sect. 10.1.1.1, co-deposition of a Fe-Ni-Co alloy and Teflon AF has been found to lead to the self-organized formation of thin metallic nanocolumns under specific deposition conditions [31, 95]. It was found that the transition from spherical to columnar growth takes place at a critical value of the ratio of deposition rates $\kappa := R_m/R_p$ and is accompanied by a steep increase of the metallic filling factor f. Figure 10.23 shows the metal volume filling factor f as a function of the deposition rate ratio $\kappa = R_m/R_p$ at different temperatures ranging from −70 to 300 °C. One can see that the steep increase of the filling factor is most pronounced at 300 °C and smears out at lower temperatures until it vanishes completely at −70 °C. For this temperature one can assume that the condensation coefficient is close to one for all values of κ. Therefore, the data were fitted to the function

$$f(\kappa) = \frac{1}{1 + \frac{1}{C\kappa}}, \tag{10.53}$$

which is (10.52) for the case of condensation coefficients independent of κ. The fitting parameter is the metal condensation coefficient C. The fit yields $C = 0.94 \pm 0.08$ which is in good agreement with the expectation that the condensation coefficient approaches unity at low temperatures [67]. It has to be pointed out, however, that the condensation coefficient depends on the metal coverage at the surface of the growing composite and, hence, on κ because metal atoms stick with a probability of unity if they directly impinge on a metal nanoparticle or if they reach a metal nanoparticle via surface diffusion. Therefore (10.53) is not applicable at higher temperatures, where the condensation coefficient on Teflon AF is expected to be very low due to thermal activation [39, 67]. Even the value of $C = 0.94$ obtained at −70 °C probably overestimates the condensation coefficient for the pure polymer [95].

10.6.1 Simulation Details

The simulations are based on an extended version of the model described in Chap. 3 and involve the melting point depression of metal clusters. The main assumption is that clusters exist in the liquid state if their radius lies below the melting radius r_m and start to solidify by exceeding the melting radius. If clusters are liquid, they are assumed to merge according to (10.37). For clusters that grew beyond the melting radius, a second growth mechanism is introduced which leads to the formation of elongated columnar structures growing into the direction of the surface. It is initialized when the radius of one of two merging clusters is above the melting radius r_m, i.e. when the number of atoms N exceeds a certain value N_m. This growth occurs, for instance, when a monomer impinges directly into the interaction region of a pre-existing cluster with $N > N_m$ atoms which is partially buried by the surface. This

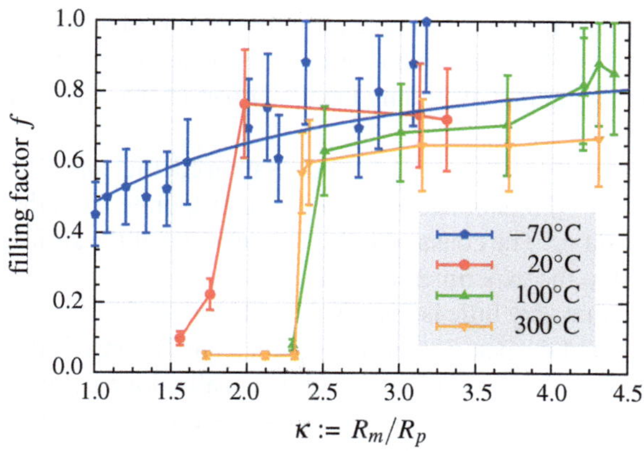

Fig. 10.23 Experimental results of the metallic filling factor as a function of the ratio of deposition rates κ of Fe-Ni-Co and Teflon AF for different temperatures. Reprinted from Ref. [95]

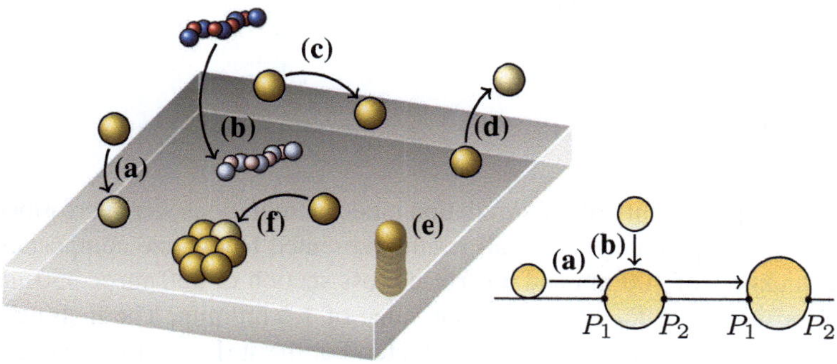

Fig. 10.24 *Left*: Sketch of the cluster processes included in the simulations: deposition of monomers (*a*) and polymer (*b*), surface diffusion of clusters (*c*), evaporation of monomers (*d*), formation of metallic nanocolumns (*e*), cluster growth induced by surface diffusion (*f*). *Right*: Illustration of the two basic mechanisms of nanocolumn growth: (*a*) growth induced by surface diffusion and (*b*) by direct impingement of a monomer on a preexisting cluster. The polymer surface moves upwards during the deposition process due to arrival of new organic molecules from the gas phase

process also occurs when a cluster merges with a partially buried cluster as a result of surface diffusion. It is assumed that the new cluster does not reach a spherical shape after equilibration. Instead, the incoming cluster coalesces with the part of the buried cluster which extends above the surface (see Fig. 10.22).

These growth mechanisms are subject to two boundary conditions: the first one is volume conservation (due to the assumption of constant density). Furthermore the points P_1 and P_2 in Fig. 10.24 (denoting the circular intersection line of the cluster with the surface) are assumed to remain immobile. Hence, the resulting initial column consists of two spherical caps separated by the surface dividing the column into a buried part and a part located above the surface. Note that the intersection line of the clusters with the surface moves upwards during the deposition process due to the arrival of new organic molecules. These two mechanisms of nanocolumn growth are depicted in Fig. 10.24. These mechanisms can repeat the same way with a free cluster and a pre-existing column, where only the upper part of the column is involved into the growth process.

To incorporate the effect of very low condensation coefficients C known for metals on Teflon AF [39, 67], monomers are allowed to evaporate from the surface with a certain rate constant ν_e. The constant ν_e is given in units of the surface diffusion rate ν_s and is adjusted to the experimental values of C.

Due to the fact that very little is known about the rates of surface diffusion of clusters on Teflon AF and evaporation of monomers from Teflon AF, the temperature is not incorporated into the simulations which is usually done by applying an Arrhenius law for the description of these processes. Instead, the main input parameters (e.g. surface diffusion and evaporation rate of monomers and the melting radius r_m) are treated as free parameters to achieve the best accordance with the experiments. To incorporate the size-dependence of the surface diffusion coefficients of metal clusters, the power law given by (10.43) is applied.

10.6.2 Results

To check the applicability of the simulation model, simulations were performed over a broad range of parameters. For all results presented below, the composites have a finite thickness of 100 nm and a surface cross section of 350 nm × 350 nm. The metal deposition rate R_m has a constant value of 1.5 nm/min. The melting radius r_m of the clusters is treated as a free parameter. For the surface diffusion coefficient of monomers D_1 and the evaporation rate ν_e of monomers, the following parameter range is used: $D_1 = 1 \times 10^{-11} \ldots 2 \times 10^{-10}\,\mathrm{cm}^2/\mathrm{s}$ and $\nu_e = 1 \times 10^3 \ldots 2 \times 10^4\,\mathrm{s}^{-1}$. In order to obtain good statistics, the results presented below are averaged over 20 runs with a constant parameter set but different initial conditions. The deviations are usually less than one percent, so no error bars are included in the figures. The focus of the investigations lies on the influence of the monomer evaporation rate (desorption from the surface) and the surface diffusion rate, which depend on the metal-polymer interaction and hence on the condensation (or sticking) coefficient C.

The main effect which was observed during experiments is a dramatic increase of the volume filling factor f with the deposition rate ratio κ. In contrast to the experiments, the simulations provide additional data such as the size distribution of clusters, their spatial distribution as well as the exact number of the nanocolumns. This allows for a more complete understanding of the self-organized process of nanocolumn growth. In Fig. 10.25 simulation results of different quantities as a function of the deposition rate ratio κ are shown for a system with a surface area of 350 × 350 nm and a final thickness (after the termination of both deposition processes) of 100 nm.

The simulations clearly show a strong increase of the metal filling factor f for values of $\kappa \geq 1.5$ which is related to the formation of nanocolumns cf. Fig. 10.25a, c. The upper panel of Fig. 10.25 which shows the number of nanocolumns indicates

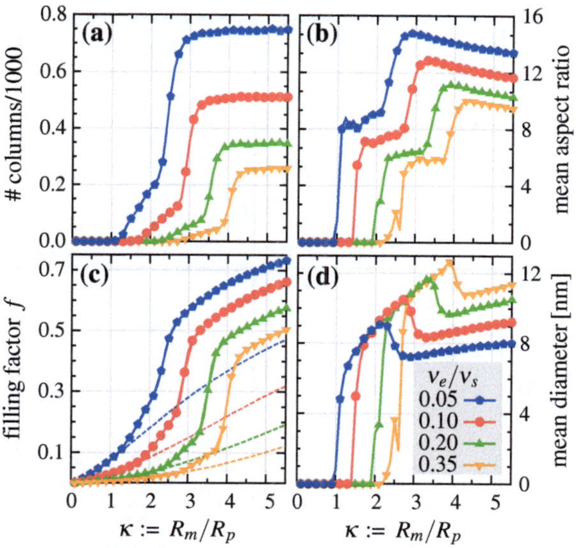

Fig. 10.25 Nanocolumn properties as a function of the deposition rate ratio κ, for four different values of the evaporation rate ν_e. The melting radius r_m is set to 2.23 nm and the monomer surface diffusion coefficient D_1 to $1.845 \times 10^{-11}\,\text{cm}^2/\text{s}$ (corresponding to the rate $\nu_s = 2.05 \times 10^4\text{s}^{-1}$). The figure parts show **a** the total number of nanocolumns, **b** the mean aspect ratio (defined as the ratio of the mean length to the mean diameter of the columns). **c** the metal filling factor with (*solid lines*) and without (*dashed lines*) column growth **d** the mean column diameter

that there is a sharp transition from the pure spherical growth regime to a regime of column growth that coincides with the strong increase of the filling factor. Within our model the explanation of this phenomenon is as follows: When atoms impinge on the surface, they may undergo various competing processes like surface diffusion, re-emission and nucleation after encountering each other. One crucial point for the observed transition is the low condensation coefficient of metal on Teflon AF, which is caused by the weak chemical interaction of the two components. Metal atoms (clusters) have to encounter each other and form nuclei that can be stabilized in the polymer matrix and initiate the column growth. The simultaneous deposition of the polymer matrix works against the growth of nuclei and isolates the clusters from each other. For low values of κ the re-emission of atoms and the growth of the polymer matrix are the dominant processes and prevent the growth of clusters that are big enough to initiate column growth. When κ exceeds a critical value, the deposition of metal atoms plays the dominant role and the re-emission and isolation of clusters by the growing matrix is compensated by agglomeration of atoms with pre-existing clusters. As a consequence, the growth of clusters is strongly accelerated and some clusters reach the critical cluster size to initiate column growth.

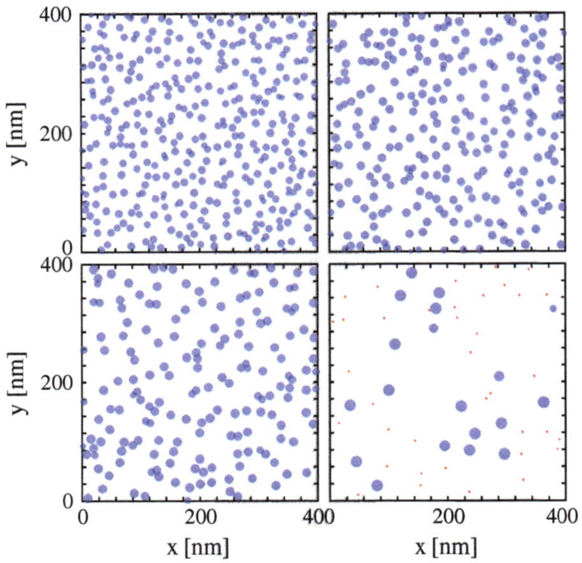

Fig. 10.26 *Top view* of the surface microstructure after termination of both deposition processes for $\kappa = 0.4$ and four different evaporation rates ν_e. *Top left* $\nu_e = 2.05 \times 10^3\,\mathrm{s}^{-1}$, *top right* $\nu_e = 4.1 \times 10^{3-1}$, *bottom left* $\nu_e = 8.2 \times 10^3\,\mathrm{s}^{-1}$, *bottom right* $\nu_e = 1.64 \times 10^4\mathrm{s}^{-1}$. Other parameters are the same as in Fig. 10.25. Columns (clusters) are depicted as *blue* (*red*) circles. The size of the surface is the same in all figures

10.6.2.1 Effect of Atomic Desorption

Figure 10.25 shows how the filling factor and geometrical properties of the nanocolumns are affected by the evaporation rate ν_e and the deposition rate ratio. The whole range of investigated values of κ can be divided into four regions: The first region is characterized by pure spherical growth and an almost linear increase of the metal filling factor, see Fig. 10.25c. In the second region column growth sets in, see Fig. 10.25a. With an increasing desorption rate and the concomitant increase in the surface diffusivity, the filling factor starts to increase nonlinearly and the transition to the columnar growth regime is shifted to larger values of κ, cf. Fig. 10.25a. This effect can be easily understood in terms of the underlying column growth model: As explained in Sect. 10.6.1, clusters have to grow beyond the melting size r_m to act as initial nuclei for column growth. When the desorption rate of atoms ν_e from the surface increases the growth of clusters is slowed down. Consequently, the probability of cluster growth can only be enhanced by slowing down the embedding of clusters into the matrix via increasing the mobility of clusters which occurs when κ is increased (recall that κ is increased via reduction of the polymer deposition rate R_p).

A further effect of increasing the atomic desorption is a reduction of the total number of columns. This is a direct consequence of the enhanced probability of

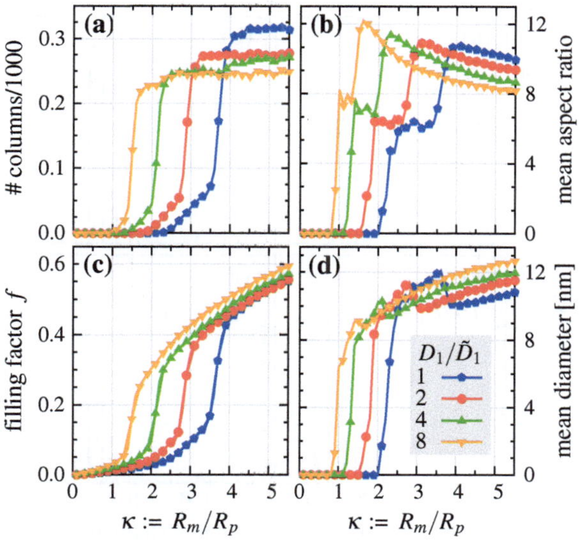

Fig. 10.27 Same as Fig. 10.25, but for the case of different surface diffusion coefficients D_1 of monomers. The constant $\tilde{D}_1$ is set to $1.845 \times 10^{-11}\,\text{cm}^2/\text{s}$. The evaporation rate equals $\nu_e = 4.92 \times 10^3\,\text{s}^{-1}$

atom re-emission. This effect can also be seen in Fig. 10.26 which shows a top view of the composite after termination of deposition: clearly the columns grow thicker with higher desorption rates.

The second stage is characterized by a relatively small rise of the number of columns with increasing κ. During this stage the columns are steadily growing thicker and whereas their aspect ratio remains within a relatively narrow range see Fig. 10.25b. This, obviously, implies that the mean column length is also increasing at a comparable rate. Also, the diameter of the columns increases when the desorption rate is increased.

The third stage can be identified by an abrupt increase of the total number of columns. From Fig. 10.25c one can see that the accelerated growth of columns is accompanied by a steep rise of the filling factor, which is more pronounced for higher desorption rates. Our simulations show that this stage comes along with a growing number of columns extending over the whole height of the final composite, resulting in a higher metal coverage of the surface. The mean column diameter is decreasing during this stage due to the increased number of nuclei available for column growth. This gives rise to a significantly higher aspect ratio. Our simulations show that the average height of the nanocolumns is even some nanometers larger than the thickness of the substrate, see Fig. 10.28, i.e. the majority of the columns extends above the surface. Consequently the incoming atoms are distributed among a larger number of columns compared to the previous stage.

During the fourth growth stage that starts between $\kappa \approx 3.0$ (blue curve in Fig. 10.25c) and $\kappa \approx 4.2$ (brown curve), the number and average length of columns

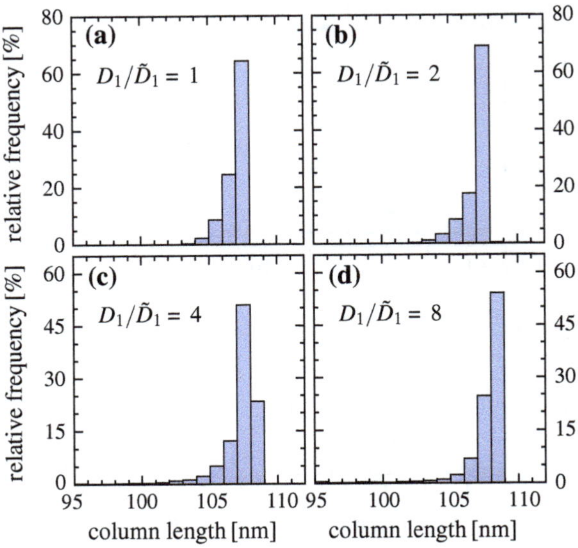

Fig. 10.28 Distribution of column lengths for $\kappa = 4.0$ depicted in a histogram with a resolution of 1 nm. The parameter set is the same as in Fig. 10.27

remain constant. Only the filling factor and the column diameter increase (and, therefore, the aspect ratio decreases), due to the higher amount of metal atoms impinging on the surface.

10.6.2.2 Effect of Surface Diffusion

In Fig. 10.27, the same quantities are depicted as in Fig. 10.25, but for a fixed desorption rate $\nu_e = 4.92 \times 10^3\ \mathrm{s}^{-1}$ and different surface diffusion coefficients D_1. The main effect of increasing D_1 is a shift of the transition from spherical growth to columnar growth to smaller values of κ, see Fig. 10.27a, c. The reason is obviously the accelerated cluster growth caused by a higher mobility. While for slow surface diffusion the cluster growth is dominated by direct impingement of atoms on pre-existing clusters, for fast surface diffusion the probability of two clusters to encounter each other during surface diffusion is considerably enhanced. As a consequence, clusters reach the critical nucleus size to induce the columnar growth already for smaller values of κ. Interestingly, the filling factors for higher values of κ ($\geq$4.4) lie very close together. The simulations do not show a tendency of the surface diffusion coefficient to influence the filling factor. Other quantities are also weakly affected by the diffusion constant, when columnar growth is observed. Figure 10.27a shows that for all simulations the number of columns lies between 200 and 300 what corresponds to a column surface density between 1.63×10^{-3} and $2.12 \times 10^{-3}\ \mathrm{nm}^{-2}$, where the density is higher the lower is D_1. As one can see in Fig. 10.27d, the thickness of the columns follows the opposite trend: A higher surface diffusivity leads to thicker columns, which is not surprising since clusters grow faster. The aspect ratio shows the opposite trend, cf. Fig. 10.27b. This implies that, in the simulated range of D_1, the average length of the nanocolumns stays nearly constant.

Figure 10.28 shows the distribution of column lengths after termination of both deposition processes, for four different surface diffusion coefficients. All columns extend some nanometers above the surface (which is located at 100 nm). Furthermore, for all values of D_1 the length distribution is very narrow with a maximum at about 108 nm.

10.7 Summary and Outlook

In this chapter, we gave a detailed overview on recent simulations of cluster growth and diffusion in metal-polymer nanocomposites including deposition of metal atoms on a polymer surface and co-deposition of metal and polymer. Emphasis was placed on the discussion of the simulation model and its physical justification. Therefore, the first section of this chapter briefly outlines the main physical aspects of polymer metallization. This section is followed by a brief introduction into the theory of continuous-time Markov chains which form the theoretical foundation of kinetic Monte Carlo methods. In the center of this chapter is the detailed discussion of the simulation model which is provided in the third section. The simulation scheme includes a large catalog of nine different processes that are listed in Fig. 10.6. The simplicity of these processes allows for a study of rather large systems containing up to several 10^8 atoms thus yielding results with excellent statistics. The price one has to pay for these kind of simulations is the phenomenological character of the processes and their corresponding rates which have to be given as input parameters to the simulations provided by experiments or simulations based on microscopic approaches.

Furthermore, future simulations have to take into account the size spectrum as well as the energy distribution of the deposited clusters. Therefore, it is subject of ongoing work to develop a similar kinetic Monte Carlo scheme for the simulation of cluster growth in the plasma phase [58, 96] prior to deposition.

In the last section, simulation results of cluster-polymer interface formation and co-deposition of metal atoms and polymer were presented and discussed. Here, the focus lies on the analysis of how variations of the deposition conditions affect the distribution of metal inside the polymer matrix and the size spectrum of clusters. Future simulations will take these quantities as input parameters and compute the optical as well as the electrical properties of the nanocomposites.

A further application of the presented model is the simulation of metallic nanocolumn formation which was discovered during co-deposition of a Fe-Ni-Co alloy and Teflon AF [31]. A detailed report of these simulations can be found in [53, 95].

References

1. A. Biswas, O. Aktas, J. Kanzow, U. Saeed, T. Strunskus, V. Zaporojtchenko, F. Faupel, Mater. Lett. **58**(9), 1530 (2004)
2. A. Biswas, O.C. Aktas, U. Schuürmann, U. Saeed, V. Zaporojtchenko, F. Faupel, T. Strunskus, Appl. Phys. Lett. **84**(14), 2655 (2004)
3. T.A. Fulton, G.J. Dolan, Phys. Rev. Lett. **59**, 109 (1987)
4. H. Graf, J. Vancea, H. Hoffmann, Appl. Phys. Lett. **80**(7), 1264 (2002)
5. H. Takele, S. Jebril, T. Strunskus, V. Zaporojchenko, R. Adelung, F. Faupel, Appl. Phys. A **92**(2), 345 (2008)
6. M. Karttunen, P. Ruuskanen, V. Pitkänen, W. Albers, J. Electron. Mater. **37**(7), 951 (2008)
7. A. Emamifar, *Applications of Antimicrobial Polymer Nanocomposites in Food Packaging* (InTech, Vienna, 2011), pp. 299–318
8. F.R.F. Fan, A.J. Bard, J. Phys. Chem. B **106**(2), 279 (2002)
9. F. Faupel, V. Zaporojtchenko, H. Greve, U. Schürmann, V.S.K. Chakravadhanula, C. Hanisch, A. Kulkarni, A. Gerber, E. Quandt, R. Podschun, Contrib. Plasma Phys. **47**(7), 537 (2007)
10. A.P. Jackson, J.F.V. Vincent, R.M. Turner, Proc. R. Soc. Lond. B **234**(1277), 415 (1988)
11. A. Finnemore, P. Cunha, T. Shean, S. Vignolini, S. Guldin, M. Oyen, U. Steiner, Nat. Commun. **3**, 966 (2012)
12. A. Sellinger, P.M. Weiss, A. Nguyen, Y. Lu, R.A. Assink, W. Gong, C.J. Brinker, Nature **394**(6690), 256 (1998)
13. A.D. Pomogailo, V.N. Kestelman, *Springer Series in Materials Science*, vol. 81, 1st edn. (Springer, Berlin, 2005)
14. G.V. Ramesh, S. Porel, T.P. Radhakrishnan, Chem. Soc. Rev. **38**, 2646 (2009)
15. Q. Wang, L. Zhu, *ACS Symposium Series 1034, Functional Polymer Nanocomposites for Energy Storage and Conversion* (American Chemical Society, Washington, 2010)
16. U. Kreibig, M. Vollmer, *Optical Properties of Metal Clusters. No. Bd. 25 in Springer series in Materials Science* (Springer, Berlin, 1995)
17. S.A. Maier, *Plasmonics: Fundamentals and Applications* (Springer, New York, 2007)
18. C.B. Murray, S. Sun, H. Doyle, T. Betley, MRS Bull. **26**, 985 (2001)
19. A.E. Berkowitz, J.R. Mitchell, M.J. Carey, A.P. Young, S. Zhang, F.E. Spada, F.T. Parker, A. Hutten, G. Thomas, Phys. Rev. Lett. **68**, 3745 (1992)
20. M. Okumura, Y. Kitagawa, M. Haruta, K. Yamaguchi, Chem. Phys. Lett. **346**(12), 163 (2001)
21. H. Takele, H. Greve, D. Paretkar, V. Zaporojtchenko, F. Faupel, Eur. Phys. J. Appl. Phys. **33**, 83–89 (2006)
22. L. Tröger, H. Hünnefeld, S. Nunes, M. Oehring, D. Fritsch, J. Phys. Chem. B **101**(8), 1279 (1997)
23. K. Günther-Schade, H.L. Castricum, H.J. Ziegler, H. Bakker, F. Faupel, Polym. Eng. Sci. **44**(7), 1351 (2004)
24. J. Hong, E. Kay, S.X. Wang, IEEE Trans. Magn. **32**(5), 4475 (1996)
25. L. Martin, H. Biederman, J. Zemek, Vacuum **35**(45), 171 (1985)
26. D. Salz, R. Lamber, M. Wark, A. Baalmann, N. Jaeger, Phys. Chem. Chem. Phys. **1**, 4447 (1999)
27. K. Behnke, T. Strunskus, V. Zaporojtchenko, F. Faupel, *Metal/Polymer Interfaces and Composite Materials Prepared by Vapor Phase Deposition* (Ddp Goldenbogen, Dresden, 2000), p. 1052
28. A. Biswas, Z. Marton, J. Kanzow, J. Kruse, V. Zaporojtchenko, F. Faupel, T. Strunskus, Nano. Lett. **3**(1), 69 (2003)
29. H. Biederman, Vacuum **59**(23), 594 (2000)
30. U. Schürmann, W. Hartung, H. Takele, V. Zaporojtchenko, F. Faupel, Nanotechnology **16**(8), 1078 (2005)
31. H. Greve, A. Biswas, U. Schürmann, V. Zaporojtchenko, F. Faupel, Appl. Phys. Lett. **88**(12), 123103 (2006)

32. H.J. Butt, M. Kappl, *Surface and Interfacial Forces*, 1st edn. (Wiley-VCH, Weinheim, 2010)
33. F. Faupel, R. Willecke, A. Thran, Mat. Sci. Eng. R **22**(1), 1 (1998)
34. F. Faupel, V. Zaporojtchenko, A. Thran, T. Stunskus, M. Kiene, *Metal Diffusion in Polymers and on Polymer Surfaces* (Springer, Berlin, 2005), pp. 333–363
35. H. Takele, H. Greve, C. Pochstein, V. Zaporojtchenko, F. Faupel, Nanotechnology **17**(14), 3499 (2006)
36. V. Zaporojtchenko, K. Behnke, A. Thran, T. Strunskus, F. Faupel, Appl. Surf. Sci. **144**, 355 (1999)
37. F. Faupel, V. Zaporojtchenko, T. Stunskus, J. Erichsen, K. Dolgner, A. Thran, M. Kiene, *Fundamental Aspects of Polymer Metallization* (Kluwer Academic/Plenum Publishers, New York, 2002), pp. 73–96
38. V. Zaporojtchenko, T. Strunskus, K. Behnke, C. Bechtolsheim, A. Thran, F. Faupel, Microelectron. Eng. **50**(14), 465 (2000)
39. A. Thran, M. Kiene, V. Zaporojtchenko, F. Faupel, Phys. Rev. Lett. **82**(9), 1903 (1999)
40. V. Zaporojtchenko, K. Behnke, T. Strunskus, F. Faupel, Surf. Interface Anal. **30**(1), 439 (2000)
41. X. Yu, P.M. Duxbury, G. Jeffers, M.A. Dubson, Phys. Rev. B **44**, 13163 (1991)
42. G. Jeffers, M.A. Dubson, P.M. Duxbury, J. Appl. Phys. **75**(10), 5016 (1994)
43. R.M. Tromp, J. Vac. Sci. Technol. A **3**(3), 782 (1985)
44. F.K. LeGoues, J. Vac. Sci. Technol. A **6**(4), 2200 (1988)
45. A. Thran, T. Strunskus, V. Zaporojtchenko, F. Faupel, Appl. Phys. Lett. **81**, 244 (2002)
46. C. Bechtolsheim, V. Zaporojtchenko, F. Faupel, Applied. Surf. Sci. **151**, 119 (1999)
47. J. Wetzel, D. Smith, G. Appleby-Mougham, Res. Soc. Symp. Proc. **40**, 271 (1984)
48. F. Faupel, R. Willecke, A. Thran, M. Kiene, C. Bechtolsheim, T. Strunskus, Defect Diffus. Forum **143–147**, 887 (1997)
49. M. Kiene, T. Strunskus, R. Peter, F. Faupel, Adv. Mater. **10**(16), 1357 (1998)
50. A. Foitzik, F. Faupel, Mater. Res. Soc. Symp. Proc. **203**, 59 (1991)
51. F. Faupel, Phys. Stat. Sol. (a) **134**(1), 9 (1992)
52. R. Willecke, F. Faupel, J. Polym. Sci. Poly. Phys. **35**(7), 1043 (1997)
53. L. Rosenthal, Kinetic Monte Carlo Simulations of Metal-Polymer Nanocomposite Formation. PhD thesis, Christian-Albrechts-Universität zu Kiel, 2013
54. A.B. Bortz, M.H. Kalos, J.L. Lebowitz, J. Comput. Phys. **17**(1), 10 (1975)
55. A. Thran, F. Faupel, Defect Diffus. Forum **143–147**, 903 (1997)
56. P. Jensen, Rev. Mod. Phys. **71**(5), 1695 (1999)
57. A. Thran, Diffusion von Gasen und Metallen in Polymeren: Datenanalysen, Experimente und Simulationen. PhD thesis, Christian-Albrechts-Universität zu Kiel, 2000
58. B.M. Smirnov, *Cluster Processes in Gases and Plasmas* (Wiley-VCH, Weinheim, 2010)
59. N. Metropolis, A.W. Rosenbluth, M.N. Rosenbluth, A.H. Teller, E. Teller, J. Chem. Phys. **21**, 1087 (1953)
60. S.P. Lalley, *Continuous Time Markov Chains*, Lecture Notes (University of Chicago, Chicago, 2012)
61. D.T. Gillespie, J. Comput. Phys. **22**(4), 403 (1976)
62. D.T. Gillespie, J. Phys. Chem. **81**(25), 2340 (1977)
63. D.T. Gillespie, J. Comput. Phys. **28**(3), 395 (1978)
64. K.A. Fichthorn, W.H. Weinberg, J. Chem. Phys. **95**(2), 1090 (1991)
65. M. Bonitz, L. Rosenthal, K. Fujioka, V. Zaporojtchenko, F. Faupel, H. Kersten, Contrib. Plasma Phys. **52**(10), 890 (2012)
66. L. Rosenthal, A. Filinov, M. Bonitz, V. Zaporojtchenko, F. Faupel, Contrib. Plasma Phys. **51**(10), 971 (2011)
67. V. Zaporojtchenko, K. Behnke, T. Strunskus, F. Faupel, Surf. Sci. **454**, 412 (2000)
68. V. Zaporojtchenko, T. Strunskus, J. Erichsen, F. Faupel, Macromolecules **34**(5), 1125 (2001)
69. J. Koutecký, V. Bonačić-Koutecký, L. Češpiva, P. Fantucci, J. Chem. Phys. **98**(10), 7981 (1993)
70. F. Family, P. Meakin, Phys. Rev. Lett. **61**(4), 428 (1988)
71. F. Family, P. Meakin, Phys. Rev. A **40**(7), 3836 (1989)
72. J. Sonnenburg, J. Gao, J.H. Weiner, Macromolecules **23**(21), 4653 (1990)

73. F. Müller-Plathe, J. Chem. Phys. **94**(4), 3192 (1991)
74. R.H. Boyd, P.V.K. Pant, Macromolecules **24**(23), 6325 (1991)
75. R.M. Sok, H.J.C. Berendsen, W.F. van Gunsteren, J. Chem. Phys. **96**(6), 4699 (1992)
76. A.A. Gusev, U.W. Suter, J. Chem. Phys. **99**(3), 2228 (1993)
77. F. Faupel, Adv. Mater. **2**(5), 266 (1990)
78. M.H. Cohen, D. Turnbull, J. Chem. Phys. **31**(5), 1164 (1959)
79. J.S. Vrentas, J.L. Duda, J. Polym. Sci. A1 **15**(3), 403 (1977)
80. J.S. Vrentas, J.L. Duda, J. Polym. Sci. A1 **15**(3), 417 (1977)
81. J.S. Vrentas, J.L. Duda, J. Polym. Sci. A1 **15**(3), 441 (1977)
82. J. Vrentas, C. Vrentas, Eur. Polym. J. **34**(56), 797 (1998)
83. J. Liu, D. Cao, L. Zhang, J. Phys. Chem. C **112**(17), 6653 (2008)
84. A. Tuteja, M.E. Mackay, S. Narayanan, S. Asokan, M.S. Wong, Nano Lett. **7**(5), 1276 (2007)
85. J. Won, C. Onyenemezu, W.G. Miller, T.P. Lodge, Macromolecules **27**(25), 7389 (1994)
86. M. Krüger, M. Rauscher, J. Chem. Phys. **131**(9), 094902 (2009)
87. R. Tuinier, J.K.G. Dhont, T.H. Fan, Europhys. Lett. **75**(6), 929 (2006)
88. R. Tuinier, T.H. Fan, Soft Matter **4**(2), 254 (2008)
89. D.H. Cole, K.R. Shull, L.E. Rehn, P. Baldo, Phys. Rev. Lett. **78**, 5006 (1997)
90. F. Faupel, A. Thran, V. Zaporojtchenko, M. Kiene, T. Strunskus, K. Behnke, AIP Conf. Proc. **491**, 201 (1999)
91. Y.N.C. Chan, G.S.W. Craig, R.R. Schrock, R.E. Cohen, Chem. Mater. **4**(4), 885 (1992)
92. J. Röder, H.U. Krebs, Appl. Phys. A **85**(1), 15 (2006)
93. D. Stauffer, A. Aharony, *Introduction to Percolation Theory*, 2nd edn. (Taylor & Francis, London, 2003)
94. J.A. Venables, G.D.T. Spiller, M. Hanbucken, Rep. Prog. Phys. **47**(4), 399 (1984)
95. L. Rosenthal, H. Greve, V. Zaporojtchenko, T. Strunskus, F. Faupel, M. Bonitz, J. Appl. Phys. **114**(4), 044305 (2013)
96. B.M. Smirnov, Phys.-Usp. **54**(7), 691 (2011)

Part IV
Technological Applications

Chapter 11
Microcavity and Microchannel Plasmas: General Characteristics and Emerging Applications

J. Gary Eden and Sung-Jin Park

Abstract Confining low temperature plasma in cavities or channels having mesoscopic dimensions (1–1000 μm) has opened a new avenue for plasma science and its applications. This chapter provides a brief overview of the current understanding of microplasma physics, and discusses the distinctives of microcavity plasmas with respect to conventional (macroscopic) plasmas. Notable properties of microcavity plasmas include their peak and time-averaged electron densities as well as power dissipated per unit volume. Applications of arrays of microcavity and microchannel plasmas described here include plasma printing, plasmachemical generation of commercially-valuable products such as ozone, and UHF receiving antennas comprising parallel microchannel plasmas.

11.1 Introduction

Low temperature plasma confined partially or fully by cavities having at least one mesoscopic dimension have come to be known as *microcavity plasmas*, *microchannel plasmas* or simply, *microplasmas*. As its name implies, the field of microcavity plasmas is chiefly concerned with the fundamental science and applications of non-equilibrium plasma with one or more dimensions (d) in the $1\,\mu m \leq d \leq 1\,mm$ interval. Although this new discipline in plasma physics has existed for little more than a decade, it is clear that such microplasmas have a number of intriguing properties that have no counterpart in the macroscopic plasma domain. Electron densities of 10^{14}–$10^{15}\,cm^{-3}$, for example, are achieved on a routine basis as compared to values of $10^{10} - 10^{11}\,cm^{-3}$ that are typical of conventional, low pressure plasma devices such as the fluorescent lamp. For cavity dimensions above $d \sim 30\,\mu m$, classical

J. G. Eden (✉) · S.-J. Park
Department of Electrical and Computer Engineering,
University of Illinois, Urbana, IL 61801, USA
e-mail: jgeden@illinois.edu

M. Bonitz et al. (eds.), *Complex Plasmas*, Springer Series on Atomic, Optical, and Plasma Physics 82, DOI: 10.1007/978-3-319-05437-7_11,

pd scaling (where *p* is the gas pressure) is valid, allowing stable glow discharges to be obtained on a continuous (CW) or high duty-cycle basis at pressures above one atmosphere.

The rapid growth of this field following its inception in the mid-1990s is the result of the combined effect of several factors, two of which are the unprecedented characteristics of microplasmas mentioned above and the adoption of microfabrication techniques developed extensively by the microelectronics, and microelectromechanical systems (MEMS) communities. Because of the spatial domain in which microplasmas reside, the ability to precisely and reproducibly fabricate microcavities, interconnects and other structures (such as optical gratings, waveguides, and stress-relief bridges) is of considerable value. That these techniques were available at the time the first microcavity plasmas were demonstrated (and other fabrication methods developed since) is a clear case of serendipity, and its impact on the realization of many classes of microplasma devices and arrays would be difficult to overstate. Therefore, microcavity and microchannel plasma science and applications lie at the intersection of plasma physics, materials science and engineering, physical chemistry, electronics, and photonics. Not only are the multidisciplinary dimensions to the field one of its most appealing aspects but they also reflect the disparate and far-reaching applications that are currently being pursued.

In this chapter, the general principles and characteristics of microplasmas will first be reviewed. Operating parameters and the unique opportunity that microplasmas offer to explore the interaction of a plasma with its surroundings (boundary) will be emphasized. The remainder of the chapter will discuss the design of several specific classes of microcavity and microchannel plasma devices and their applications in fundamental science, lighting, electromagnetics, and plasma chemistry. Owing to the rapid growth in this field, one chapter is insufficient to describe developments thoroughly, and the reader is encouraged to consult several excellent reviews that have appeared recently. References [1–7], in particular, provide considerable background regarding both the basic mechanisms and selected applications of microcavity plasmas.

11.2 Principles and General Properties of Microcavity Plasmas

Microcavity plasmas are best known for several characteristics, including electron density (n_e), operating pressure (p), and power loading (P_s), that are unprecedented for glow discharges. Table 11.1 summarizes several of the key parameters of microplasmas. For each parameter, values are given that reflect the approximate range reported in the literature. Each of these is addressed briefly in the discussion to follow. It should also be mentioned that Table 11.1 is intended to be representative and not exhaustive.

Operating pressure, cavity dimensions, and the specific properties of the gas, vapor, or gas mixture under study are the factors having the greatest bearing on the physics and performance of a microcavity or microchannel plasma. Most experiments

Table 11.1 Summary of the currently available range in specific parameters of microcavity plasmas. For n_e, peak and time-averaged values are not distinguished

Parameter	Values
Pressure (p)	<50 Torr → >10 bar
Electron density (n_e)	$10^{13}\,\mathrm{cm}^{-3} \rightarrow 10^{17}\,\mathrm{cm}^{-3}$
Specific power loading (P_s)	$\sim$1 kWcm^{-3} → $\sim$1 MWcm^{-3}
Characteristic cavity dimension (d)	<4 μm → 1 mm
Cavity volume	pL → nL

to date have involved cavities with a characteristic dimension (i.e., radius of a cylinder, width of a trench, or the minor axis of an ellipsoid) between 100 and 500 μm, but microplasmas have successfully been produced in inverted square pyramidal microcavities and microchannels with d as small as 10 and <4 μm, respectively [8, 9].

11.2.1 Gas and Electron Temperatures

Spectroscopic analysis, computer simulations, and electrical (voltage-current) and particle (electron and metastable) density measurements all point to the low gas temperature and nonequilibrium nature of microplasmas. Electron temperatures are typically in the range of 1–5 eV whereas gas temperatures, which are dependent upon the gas itself and the driving conditions, are generally <500 K for the rare gases but can be considerably higher in molecular gases [2, 5, 10]. For the latter, vibrational-translational (VT) relaxation in both diatomics and polyatomics can contribute significantly to heating the background gas. The method most often adopted for measuring the gas temperature in a microplasma is that of determining the rotational temperature of a diatomic molecule that is present (intentionally or not) as a minority constituent of the gas mixture. Usually observed in emission, the rotational structure of the molecular transition of interest must be at least partially resolvable (implying a molecule of low reduced mass), and the diatomic most frequently used in this regard is N_2. Its C $^3\Pi_u \rightarrow$ B $^3\Pi_g$ electronic transition, also known as the second positive system, is capable of producing intense fluorescence in several bands lying between 337.1 nm $[(v', v'') = (0, 0)]$ and 405.9 nm $[(v', v'') = (0, 3)]$. Other homonuclear diatomics that serve as temperature diagnostics internal to the plasma include N_2^+ and He_2.

Measuring or estimating the electron temperature (T_e) experimentally is a greater challenge but spectroscopic measurements are once again able to provide considerable information. A prominent example is observing optical emission from highly-excited states of neutral or ionized atoms or molecules. Because T_e and the gas temperature are instrumental in determining the relative populations of electronic states, optical emission spectra obtained while varying $|\mathbf{E}|/N$ (where $|\mathbf{E}|$ and N are the electric field strength and the background gas number density in the plasma, respectively) will yield reliable estimates of T_e. Simulations of the plasma kinetics and knowledge of the oscillator strengths for the dominant emission lines improve the estimate considerably. Early studies of fluorescence spectra from rare gas microcavity plasmas showed that emission from excited states of atomic ions and neutrals is a convenient diagnostic of the relative value of T_e and the behavior of the microplasma. At low pressures (below approximately 100 Torr) and d on the order of several hundred microns or more, the appearance of strong fluorescence from ionic excited states lying tens of eV above ground indicated that the microplasma was operating in the hollow cathode mode. As the gas pressure was increased (which reduces $|\mathbf{E}|/N$), however, the populations of the most energetic ion states observable fell rapidly, demonstrating that T_e had declined because the plasma was no longer behaving as a conventional hollow cathode discharge but rather resembled an abnormal glow.

11.2.2 Electron Densities, Scaling with d

In order to confine a plasma within a cavity, the electron density (n_e) must be sufficiently high (and, therefore, the Debye length λ_D sufficiently small) that a plasma sheath can be formed. Consequently, cavity dimensions dictate, to a degree, the minimum electron density of the plasma to be generated in the cavity. If d is reduced, for example, then n_e must increase and, because $\lambda_D \propto n_e^{1/2}$, every ten-fold reduction in d requires that n_e rise by two orders of magnitude. If $n_e \sim 10^{14}\ \text{cm}^{-3}$ for a cylindrical cavity having a radius (R) of 100 μm, then we would expect a minimum electron density of $\sim 10^{18}\ \text{cm}^{-3}$ to be necessary if a plasma is to be produced within a cavity 1 μm in radius. This simple guideline assumes that excitation of the microplasma is of sufficient duration that the sheath is fully established.

11.2.3 Plasma-Wall Interactions

Another consequence of decreasing the size of a microcavity is that the ratio of surface area-to-volume for the cavity *rises*. For a cylinder of fixed length L, for example, the surface area/volume ratio scales as R^{-1}. One would expect, therefore, that plasma-surface interactions will become of increasing importance as microcavities shrink in size. This is, indeed, the case. In larger volume plasmas and plasma devices, such as fluorescent lamps, the walls play a secondary role in the physics of the plasma, and the electrodes serve as a source (or sink) for electrons. Electric field strengths in the sheath are typically $< 10^3$ V/cm and electron production from the cathode is determined by a constant γ_{SE}, known as the secondary electron emission coefficient. For microcavity plasmas, however, the choice of cavity and electrode geometry is pivotal because they specify the spatial variation of $|\mathbf{E}|$ in the cavity. Because $|\mathbf{E}|$ in the sheath of a microplasma can reach values above 10^5 V/cm, the electronic structure of the wall material and its surface morphology are of increased importance. Such fields are sufficiently strong, in certain circumstances, to alter the electronic structure of the wall at its surface which, in turn, impacts the plasma/boundary interactions. Stated another way, the walls of a microcavity plasma need no longer be passive, but rather are able to interact actively with the plasma and, most intriguing of all, the interaction appears to be controllable in some situations. As will be discussed later in this chapter, it is now possible to vary γ_{SE} at will.

11.2.4 Breakdown of pd Scaling

As discussed by Becker and Schoenbach [5], *pd* scaling has long been known to be a reliable guide for predicting the parameters of macroscopic glow discharges when d is varied. Simply stated, *pd* scaling predicts that, if a cylindrical glow discharge (for example) has certain parameters when $d = 1$ cm and $p = 5$ Torr, similar properties are to be expected for $d = 0.5$ cm and $p = 10$ Torr (and other values of p and d for which the *pd* product is constant). Surprisingly, experiments suggest that *pd* scaling continues to be obeyed for microplasma cavities as small as $d < 50\,\mu\text{m}$. However, evidence has accumulated that below the $d \approx 30$–$50\,\mu\text{m}$ region, *pd* scaling is no longer valid [11, 12]. Go and colleagues [13] have investigated the behavior of Paschen's curve (breakdown voltage as a function of *pd*) for small gaps ($<10\,\mu\text{m}$) between electrodes exposed directly to the plasma. Because of the influence of ion-enhanced field emission [14], the discharge breakdown voltage is predicted to fall rapidly below $d \approx 7\,\mu\text{m}$. Although this effect is not observed (nor would it be expected) for dielectric barrier discharge (DBD) structures in which at least one of the electrodes is isolated electrically from the plasma by a dielectric layer [3, 9], the gas pressures at which stable glows are obtained for $d = 5$–$50\,\mu\text{m}$ microcavities [9] are considerably lower than those anticipated on the basis of *pd* scaling.

11.2.5 Electric Field Strength in the Sheath

Plasma is a term introduced by Irving Langmuir to describe the charge neutrality existing throughout virtually all of the volume of bulk, macroscopic plasmas. As microplasmas shrink in size, however, the absence of charge neutrality over an increasing portion of the plasma volume becomes apparent. Charge separation is supported by strong electric fields in the cavity and at the plasma/boundary interfaces [15–17]. This is particularly true in the sheath region in which electric field strengths above 100 kV/cm ($10\,V/\mu\text{m}$) often exist in atmospheric pressure microplasmas.

11.2.6 Gas Phase Chemistry, in situ Production of Transient Molecules

Operation of microplasmas at pressures of hundreds of Torr or more has profound implications with regard to the gas phase chemistry that is now accessible. Specifically, the door has been opened to producing, in situ, large number densities of species (such as excited molecules and radicals) for which the chemical formation mechanism requires at least one collision between heavy particles. The pressures available to conventional, low pressure glow discharges are generally too low for such reactions to play a significant role.

A group of diatomic molecules well known for the efficiency with which they are generated in high pressure nonequilibrium plasmas is the rare gas dimer family, denoted $Rg_2^\star$, where Rg represents He, Ne, Ar, Kr, or Xe and the asterisk denotes an electronic excited state of the molecule. Because these molecules are formed by the three body collision: $Rg^\star + 2Rg \xrightarrow{k_f} Rg_2^\star + Rg$, the rate of formation of these molecules ($\mathscr{R}$) varies quadratically with the Rg background gas pressure. That is: $\mathscr{R} = k_f[Rg]^2$ where the formation rate constant k_f is expressed in cm^6s^{-1} and the square brackets denote the rare gas number density, expressed in cm^{-3}. An analogous process is responsible for the formation of the rare gas dimer ions, Rg_2^+. Consequently, the comparatively high pressures at which sub-500 μm microcavity plasma devices prefer to operate is a boon for plasma chemistry. It is no surprise that, in addition to the vacuum ultraviolet (VUV) emission observed from the rare gas dimers by several laboratories, other molecules such as ArD, XeI, XeO, Xe_2I, and visible (Rydberg → Rydberg) transitions of He_2 have been observed to date from microplasmas. A wide variety of molecules can now be formed in this manner and reports of new species will undoubtedly be forthcoming. It should also be mentioned that the production efficiency for most of the molecules cited above benefits not only from the high pressure environment of a microcavity plasma but also from the specific power loadings available (tens to hundreds of $kWcm^{-3}$).

Before leaving this section, one other key aspect of microcavity plasmas should be discussed. Regardless of the gas or gas mixture in question, microcavity and microchannel plasmas are diffuse glows and quite reproducible, especially when generated in DBD systems. This contrasts with the filaments or streamers often encountered in atmospheric pressure plasmas. The source of the improved behavior of microcavity plasmas, relative to their unbounded counterparts, is the stabilization of electron loss (and production) rates that is provided by the cavity wall.

Long before low temperature plasma was intentionally confined in a cavity, free microplasmas produced at atmospheric pressure in the form of streamers or *microdischarges* were well-known and studied extensively. A prominent example is the microdischarge filaments produced in large volume DBD discharges, particularly at atmospheric pressure and in attaching gases. Kogelschatz [3] has summarized nicely the properties of such columnar microdischarges in air for a 1 mm discharge gap. Distributed throughout the plasma volume, microdischarges have electron densities of $10^{14} - 10^{15}$ cm^{-3} and, in this and other aspects, bear a striking resemblance to microcavity plasmas. A rough estimate of the expected diameter for a cylindrical microdischarge equates the rate of loss of electrons due to volumetric recombination with that arising from diffusion. Doing so for a background rare gas (Ne) yields the expression:

$$d_M = \left(\frac{D_a}{\alpha_D n_e}\right)^{\frac{1}{2}} \tag{11.1}$$

where α_D is the coefficient, expressed in $cm^3 s^{-1}$, for the dissociative recombination of Ne_2^+ (the dominant electron loss mechanism in Ne at pressures beyond a few Torr), D_a is the ambipolar diffusion coefficient, and d_M, the radius of the microdischarge, replaces the characteristic diffusion length Λ in the electron loss rate term: D_a/Λ^2. Solutions to Eq. 11.1 give values of d_M that are consistent with the observed radius of $\sim$100 µm [3]. Specifically, recalling that $D_a \simeq D_i(1 + T_e/T_i)$ and assuming the ion diffusion coefficient D_i to be $\sim 40\,cm^2 s^{-1}$ [18], $T_e \simeq 1eV$, $\alpha_D \sim 4 \cdot 10^{-8}\, T_e^{-1/2}\, cm^3 s^{-1}$ [15], and $n_e = 10^{14} cm^{-3}$, then $d_M \simeq 200$ µm. This value of d_M, calculated for Ne, is larger than the experimental value for air [3] because the latter includes an attaching gas which has the effect of accelerating localized electron loss and, thus, reducing d_M.

Microcavity plasmas are, indeed, related to the unbounded microdischarges that are commonplace in atmospheric pressure DBD plasmas. It is no coincidence that the $\sim$100 µm diameters of microdischarges [3] match well the range in cavity dimension d over which microcavity plasmas operate at 1 atmosphere. However, the introduction of the cavity wall represents a significant step forward in both stabilizing and interacting with the plasma through electron production and loss at the sheath/wall interface. Confining low temperature plasma in cavities or microchannels also has compelling implications for chemical processing, lighting and other applications that will be discussed in Sect. 11.4.

11.3 Device Structures, Fabrication, and Modes of Operation

Since the first microcavity plasmas were reported by White [19] and Schoenbach et al. [20], devices have been demonstrated in a broad range of geometries and materials. Most are designed to offer specific advantages (ease of fabrication, light extraction efficiency, current density, etc.) but all have in common two basic elements: a cavity which can also take on the form of a trench or channel, and a means to couple electrical power into the gas(es) located inside the cavity. Beyond these shared features, microcavity plasma devices diverge widely and, in this section, a sampling of several structures will be discussed.

Figure 11.1 is a cross-sectional diagram of a basic cylindrical microcavity plasma device. In this design, the microcavity is produced by machining (chemically or mechanically) through one electrode, a dielectric layer, and into the substrate or base layer which also serves as the second electrode. In some cases, the cavity will extend entirely through the substrate for the purpose of providing optical access to

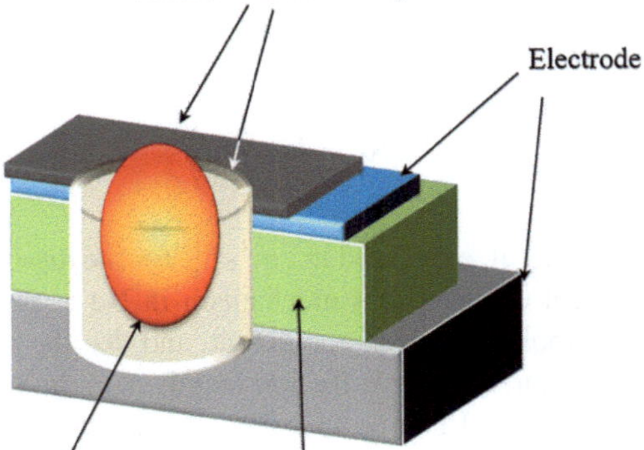

Fig. 11.1 Diagram (*not to scale*) in cross-section of a basic microcavity plasma device design. If continuous (*CW*) operation is desired, then at least a portion of the dielectric layer lining the microcavity is removed

the plasma or the extraction of plasma from the cavity with external electrodes. If continuous operation of the microplasma is desired, at least a portion of the dielectric film lining the cavity wall and separating the plasma from the electrodes must be removed. The structure of Fig. 11.1 is straightforward to fabricate and it is fair to say that it has served as the workhorse of the microcavity plasma community because of its simplicity and low cost. Early microcavity plasma devices of this type [20, 21] employed thin mica sheets as the dielectric separating the two electrodes but more recent designs have generally incorporated dielectric films. Devices similar to those of Fig. 11.1 but fabricated in a semiconductor (Si) were first reported in 1997 [22].

Cavity diameters for the original microplasma devices typically ranged from 200 to 700 μm and the cavities themselves were generally made by drilling or ultrasonic milling [20–22]. While mechanical methods are acceptable for microplasma structures of larger *d* and for a handful of devices, the adoption of micro- and nanofabrication techniques (and photolithography, in particular) is essential for realizing sub-100 μm values of *d* in devices that are fabricated precisely and uniformly from device-to-device while maintaining the surface roughness of the cavity wall below a few μm. The latter becomes of increasing importance as cavity dimensions drop, owing to field emission from the cavity walls. Even for larger cavity diameters ($d \lesssim 500\,\mu\text{m}$), drilling or other mechanical methods often leave the cavity walls coarse. Understandably, the behavior of devices incorporating such cavities is not consistently reproducible and conclusions reached with one device are not necessarily applicable to others having the same nominal dimensions. Thus, it is important that the mean (RMS) value of the cavity surface roughness be no more than a few percent of the cavity dimension *d*. Otherwise, efforts to investigate the physics of microcavity plasmas can quickly be dominated, and obscured, by surface effects.

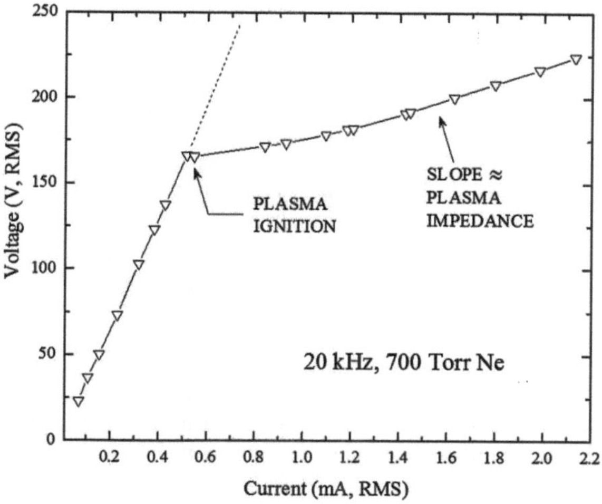

Fig. 11.2 Voltage-current (V-I) characteristics representative of those for a microcavity plasma device having a dielectric barrier design. The RMS voltage at which ignition of the plasma occurs is indicated, and the slope of the dashed line (which reflects displacement current) is the impedance $Z_C \approx (\omega C)^{-1}$ [see text for details]

When devices similar to those of Fig. 11.1 are operated continuously at rare gas pressures $\lesssim$50 Torr, they exhibit behavior typical of a traditional hollow cathode discharge. For pressures above ~100 Torr, however, the plasma characteristics transition to that of an abnormal glow for which the voltage-current (V-I) characteristic has a positive slope. V-I data representative of those observed for microcavity plasma devices having DBD structures [3, 23] are presented in Fig. 11.2. The operating point at which plasma ignition occurs is indicated on the graph, and the slope of the dashed line represents the impedance Z_C associated with the displacement current drawn by the capacitive DBD structure. For a sinusoidal driving frequency of $f = \omega/2\pi$, $Z_C \approx (\omega C_d)^{-1}$, where C_d is the capacitance of the microplasma device load. Once discharge breakdown occurs, the slope of the V-I characteristic drops abruptly owing to the reduced impedance of the microplasma itself relative to Z_C.

A planar microcavity plasma array in a DBD configuration is shown in cross-section in Fig. 11.3. This specific structure is intended for application to plasma printing [24] and the microcavities, fabricated in a dielectric sheet, typically have rectangular cross-sections and depths ranging from 30 to 200 μm. Of course, the individual microcavities and array pattern may take on any geometry desired. Thomas and co-workers [24] describe the patterning of both conducting and insulating surfaces with the system of Fig. 11.3 and gas mixtures at atmospheric pressure that include N_2 as the primary constituent but also H_2 and, depending on the specific surface to be treated, NH_3.

Several considerations, both scientific and applied, often make it desirable to have microcavity geometries other than cylindrical or rectangular. One such example is the microellipsoidal device illustrated in panel (a) of Fig. 11.4. Virtually identical

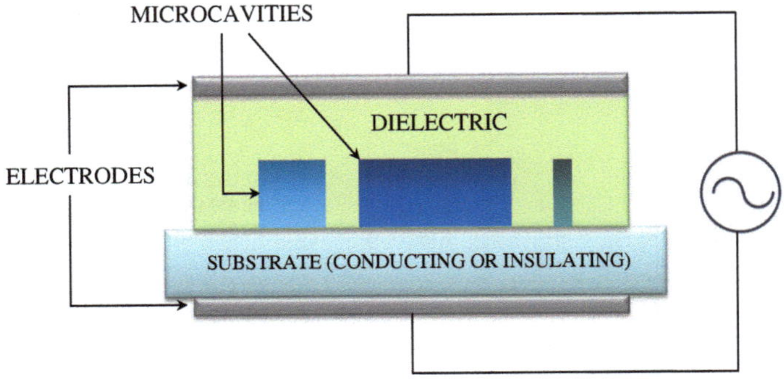

Fig. 11.3 Cross-sectional diagram (*not to scale*) of a planar microcavity plasma array having a DBD structure. This particular design has been developed for plasma printing, at atmospheric pressure, on both conducting or insulating surfaces (after M. Thomas et al. Ref. [24])

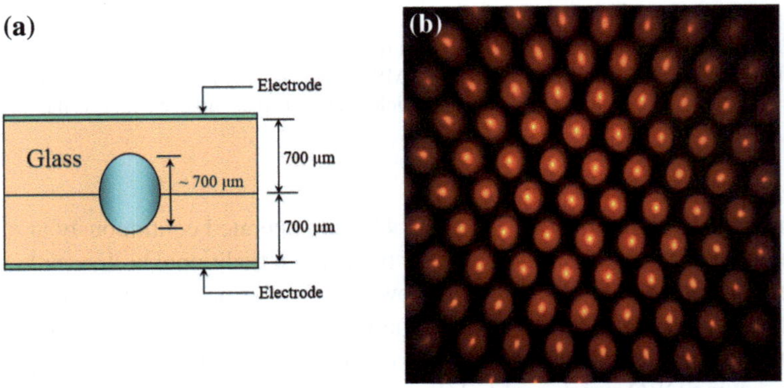

Fig. 11.4 **a** Diagram illustrating the cross-sectional design of an ellipsoidal microcavity. Formed by a micropowder ablation process, ellipsoids having major and minor axes of 600–800 and 500–600 μm, respectively, have been realized to date; **b** optical micrograph of a portion of an array of ellipsoidal microcavities having major axes 700 μm in length and operating in 740 Torr of Ne. Because of the small depth-of-field of the microscope, only the central portion of the image is in focus. Note the intensification of plasma emission at the center of each ellipsoid (photo courtesy of S.H. Sung, University of Illinois)

half-ellipsoids are fabricated into one surface of two glass sheets by the micropowder ablation process described in [25]. After matching corresponding half-ellipsoids in the two substrates, the glass sheets are hermetically sealed with the desired gas trapped in the cavities. Ellipsoids having major and minor axis lengths in the 600–800 μm and 500–600 μm intervals, respectively, have been fabricated to date at the University of Illinois. Electrodes generally consist of ~100 nm thick indium tin oxide (ITO) films deposited onto the outer faces of both glass sheets. Figure 11.4(b) is an optical micrograph of a portion of an array of ellipsoids having major and minor axis lengths of 700 and 540 μm, respectively. Operating in 740 Torr of Ne, this array clearly shows greatest power deposition at the center of the ellipsoids. Because of the difference between the dielectric constants of the cavity wall and the gas, the curvature of the microcavity wall has the effect of intensifying the electric field strength and, hence, plasma optical emission at the center of each ellipsoid. A similar effect in one dimension was demonstrated in [26] by fabricating arrays of cylindrical microcavities in co-fired ceramic tape having a relative dielectric constant of $\varepsilon_r = 40$. With electrodes straddling linear arrays of such cavities, Park et al. [26] observed the theoretically-predicted "focusing" of the electrical field strength near the center of the cavity. Panels (a) and (b) of Fig. 11.5 are time-integrated, false color images (recorded by viewing the 180 μm dia. microcavity along its axis) of the spatial distribution of the visible fluorescence intensity produced in a Ne microplasma. For a Ne pressure of 400 Torr (Fig. 11.5a), the negative glow regions on either side of the circular aperture, each produced by a specific half-cycle of the periodic driving voltage waveform, are clearly observed. When the Ne pressure is increased to 800 Torr, the intensity map of Fig. 11.5b reveals the appearance of a weak positive column. Owing to the reduced extent of the sheath and negative glow that accompanies the higher pressure, the microcavity is able to accommodate a positive column of modest length (~50 μm).

Microcavity and microchannel plasma devices fabricated in semiconductors and silicon, in particular, have the advantage of drawing upon decades of investment by the microelectronics and MEMS communities in fabrication technology. Of particular importance to microplasma researchers is the ability afforded by micro- and nanofabrication tools of producing structures that are precise and reproducible to a degree approaching atomic scale. The benefit of this capability is (as mentioned earlier) the assurance that arrays of devices, or a set of devices produced one at a time, will be essentially identical and conclusions drawn from data acquired with such arrays will not be marginalized by device-to-device variations in critical parameters such as cavity size or wall roughness. For this reason, an early microcavity design adopted at Illinois is the inverted square pyramid fabricated in Si. Formed by the intersection of four Si(111) planes, the pyramids shown in plan view by the scanning electron micrograph (SEM) of Fig. 11.6a are produced in a Si(100) surface by anisotropic etching. The emitting aperture of all cavities (base of each pyramid) is 50 × 50 μm and the apex of each pyramid lies ~37 μm below the substrate surface. With this simple design, arrays as large as 250,000 cavities (in a 500 × 500 configuration) have been fabricated and tested [27] and the fluorescence intensity has been shown to be uniform across the entire array to well within ±10 %. These pyramidal devices are generally operated as micro-DBD devices in which several thin dielectric layers coat the cavity wall, and the Si substrate (serving as one electrode) and a metal film electrode are usually driven by a 20–100 kHz sinusoidal voltage. The V-I characteristics for Si pyramidal microcavity arrays exhibit positive resistivities and, therefore, arrays of such devices require no external ballast. Rather, each of the devices is effectively self-ballasted and stable operation of large arrays is achieved. Addressable versions of pyramidal microcavity arrays have also been realized and no barriers to scaling array sizes to 1 megapixel and beyond are apparent.

A second Si-based microplasma device design is that of a trench and panel (b) of Fig. 11.6 is an SEM of a triangular channel having a width of 3.69 μm at its aperture and etched into Si(100) by a conventional, anisotropic wet etch process.

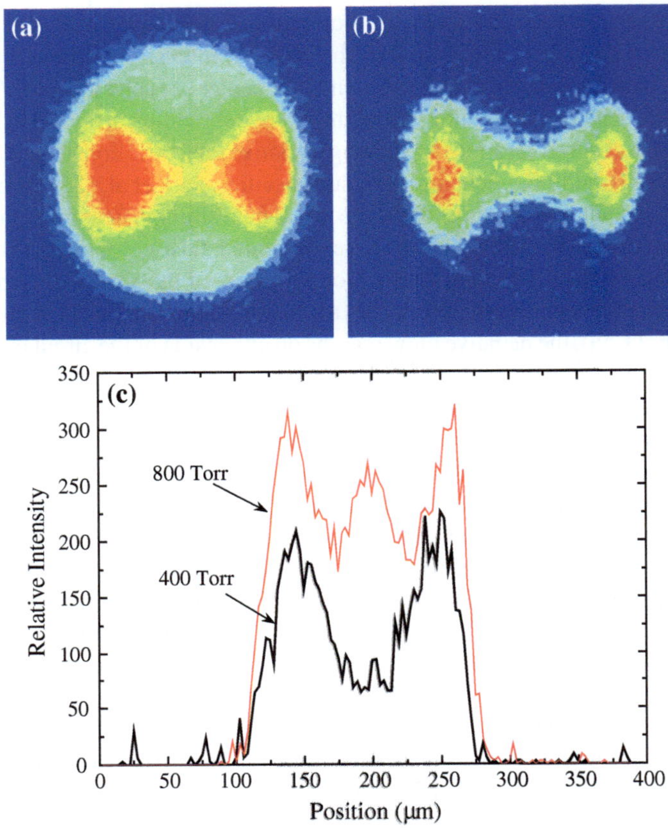

Fig. 11.5 **a**, **b** False color intensity maps of the spontaneous emission produced by a microplasma in a 180 μm dia. cylindrical cavity fabricated in ceramic. Metal electrodes are buried in the ceramic wall and are oriented vertically in the figure. Shaping of the electric field and the resulting fluorescence is a result of the dielectric constant of the ceramic ($\varepsilon_r = 40$). The false color image in (**a**) was recorded with 400 Torr Ne in the cavity whereas image (**b**) was obtained at 800 Torr. For clarity, the intensity in (**b**) was attenuated with a neutral density filter. Panel (**c**) compares intensity lineouts for the 400 and 800 Torr images. (These data are reprinted from [26] by permission of the American Institute of Physics)

The sidewalls of the trench are again Si(111) planes and, therefore, smooth at nearly the atomic level. Arrays of such trenches are fabricated into microchannel plasma devices having the structure illustrated in Fig. 11.7. Note that the upper electrode (a metal film) is flush with the rim of the microcavity along its full length. The photograph at left in Fig. 11.8 is that of an array of 50 Si microtrenches. Each has an aperture at the top of the trench of 50 μm and the length of the array is 3 cm. When operating in any of the rare gases (and other gases and gas mixtures), the array of linear microchannels generates intense and uniform fluorescence as shown by the photograph in the right half of Fig. 11.7.

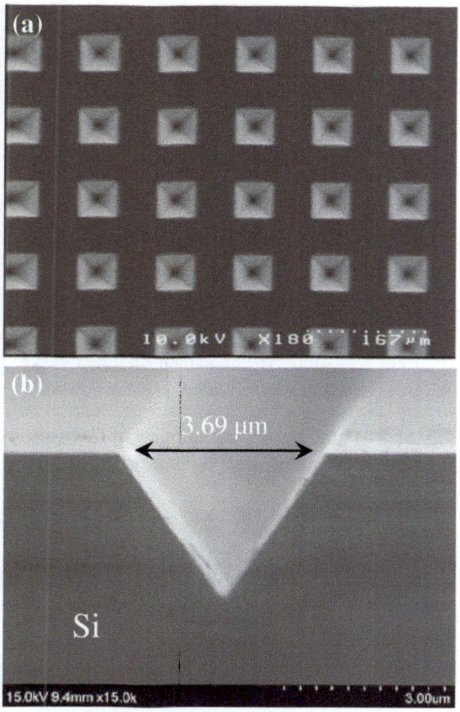

Fig. 11.6 **a** SEM in plan view (i.e., *viewed from above*) of a portion of an array of inverted square pyramid microcavities. At the base of each pyramid (*the emitting aperture*), the dimensions are 50 × 50 μm. **b** SEM of a triangular cross-section trench etched into Si(100) by an anisotropic process

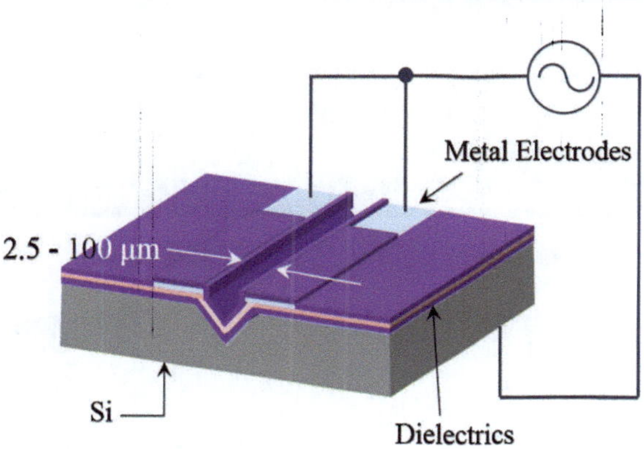

Fig. 11.7 Illustration of the construction of a Si microtrench plasma device having a triangular cross-section

Perhaps the most versatile materials system from the standpoint of microcavity fabrication and performance is that of nanoporous alumina produced electrochemically from Al. Aluminum is normally an unattractive electrode material in plasma systems (owing primarily to its melting point and vulnerability to sputtering) but, when encapsulated in alumina, it is quite robust while retaining its excellent conductivity. In 2009, Kim et al. [28] demonstrated the ability to control the cross-sectional

Fig. 11.8 Photographs in plan view of *(left)* an array of 50 Si microtrench devices, each having an aperture of 50 μm; *(Right)* Operation of the 50 channel array in 740 Torr of Ne

Fig. 11.9 (*Top*) SEM of a portion of three columns and five rows of microcavities in the form of truncated paraboloids. The diameter of each upper aperture is 164 μm; (*bottom*) magnified, cutaway view of one microcavity (with a second in the background)

profile of the wall of microcavities by combining photolithography with wet chemistry. Formed entirely from nanoporous Al_2O_3 with Al electrodes buried in, and surrounding, the microcavity wall, these devices offer the ability to simultaneously adjust the electrical and optical properties of the microcavity plasma. Figure 11.9 presents two SEMs of portions of an array of microplasma devices in which the cavities have the form of truncated paraboloids. The top panel of Fig. 11.9 shows a portion of three columns (and five rows) of paraboloidal microcavities. The three cavities in the front row are cut away so that the cavity wall profiles and surface morphology are visible. A magnified and cross-sectional view of a single cavity is shown in the lower half of Fig. 11.9. The emitting (upper) aperture of each of the microcavities in Fig. 11.9 is $\sim$164 μm. In the fabrication process described in [28], the positions of the microcavities are defined by photolithography and the specifications of the mask as well as the chemical processing conditions jointly determine the contour of the cavity walls. Tests demonstrate that these processes offer exquisite precision in defining the cavity profile while maintaining surface roughness well below 1 μm. Such a capability, coupled with numerical simulations, allows for the light extraction from the cavity (for example) and the electric field distribution in the cavity to be examined simultaneously and an optimal solution determined. Although not readily visible in Fig. 11.9, Al electrodes are buried in the nanoporous Al_2O_3 and form a ring around each microcavity [29]. In Fig. 11.10, two SEMs show a portion of an array of paraboloidal cavities that are intentionally overlapped. The upper panel of Fig. 11.10 is an SEM in plan view of a portion of an array in which the microcavities are overlapped by $\sim$5 % of the cavity aperture. When viewed at an angle to the surface normal (Fig. 11.10, lower panel), scalloping created in an array with even greater overlap of the cavities is obvious.

Arrays of microcavity devices similar to those of Figs. 11.9 and 11.10 are capable of producing intense emission in the ultraviolet (UV), visible, and infrared, depending upon the gas or gas mixture introduced to the cavities. The upper portion of Fig. 11.11 is an illustration of the cross-sectional design of a basic microcavity plasma array in which cavities are fabricated only in the upper Al/Al_2O_3 sheet. Although cylindrical cavities are shown, the process of [28] allows for the cross-sectional profile to be varied continuously from cylindrical to tapered (truncated cone) to parabolic and approaching hemispherical. The lower half of Fig. 11.11 is an optical micrograph of a section of an array of Al/Al_2O_3 devices having parabolic microcavities with an emitting aperture of 160 μm (diameter). Operating with a mixture of 20 % Xe in Ne at a total pressure of 400 Torr, this array generates VUV fluorescence that, in turn, produces green emission from a phosphor film deposited onto a window. It should be emphasized that such arrays are produced in lightweight Al foil, typically 120–250 μm in thickness. Addressable arrays have also been demonstrated and Fig. 11.12 is a diagram of such an array (viewed from above) where two identical arrays, fabricated in separate sheets, are aligned and bonded. Sets of linear electrodes, buried in the two sheets, are oriented orthogonal to each other. A simpler design, given in Fig. 11.13, relies on a trigger electrode buried in nanoporous Al_2O_3 and running beneath a column (or row) of microcavities. Fully addressable arrays

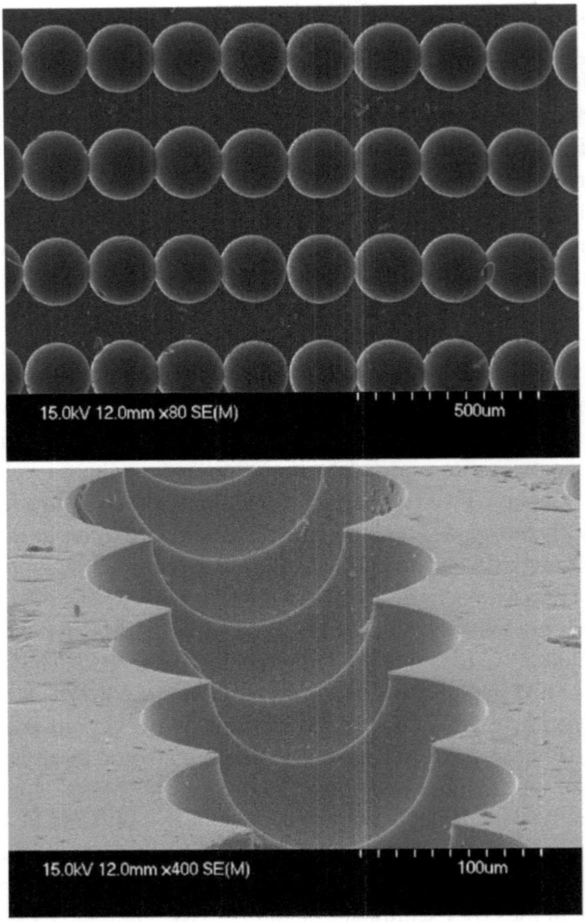

Fig. 11.10 (*Top*) SEM in plan view of a portion of several rows of partially-overlapped paraboloidal cavities fabricated in nanoporous alumina; (*bottom*) SEM of several microcavities from another array, showing the scalloped structure resulting from the intentional overlap (photo courtesy of K.S. Kim)

have been demonstrated for both the designs of Figs. 11.12 and 11.13, and further details concerning the characteristics and fabrication processes of Al/Al_2O_3 devices can be found in [28, 29].

In addition to the microcavity designs described thus far, numerous other promising materials and structures have been investigated. Among these are the fabrication of microchannel plasma devices in glass [25, 30], channels and microcavities produced in polymer by replica molding [31, 32], and cavities inherent to dielectric foams (electrets [33, 34]). Space considerations preclude a detailed description of these and other intriguing materials and structural geometries but ongoing work worldwide will undoubtedly identify new solutions for specific applications.

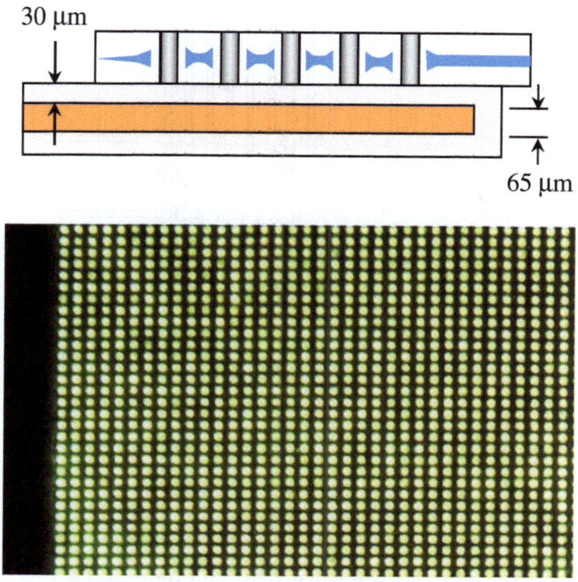

Fig. 11.11 (*Top*) Diagram in cross-section of a basic design for producing microplasmas in arrays of cylindrical (as illustrated) or parabolic cavities. The white regions represent nanoporous alumina, and microcavities are fabricated only in the *top* sheet; (*Bottom*) photograph in plan view of a section of a microplasma array in which the cavities are truncated paraboloids (cf. Fig. 11.9). A gas mixture of 20 % Xe in Ne produces green fluorescence from a phosphor film deposited onto the inner surface of a window

The discussion to this point has centered only on the first requirement for a microcavity plasma device—namely, the cavity itself. A means for coupling power into the plasma is also necessary and although most work to date has pursued DC microdischarges or low frequency (<100 kHz) sinusoidal or pulsed excitation, clever alternatives at higher frequencies have been reported. A miniaturized version of an inductively-coupled plasma source was reported by Yin et al. [35] in 1999, and two groups have generated microcavity plasmas with microstrip resonators operating in the microwave region [36–39]. Iza and Hopwood introduced the split-ring resonator in 2003 [38]. Based on a $\lambda/4$ transformer to match the power supply to a ring having a circumference of $\lambda/2$ at 900 MHz, this source efficiently produces microplasmas in 45–500 μm gaps at Ar or air pressures up to 1 atmosphere. The resonator is designed such that the maximum electric field strength occurs at the gap in the ring. A photograph of one version of the split-ring resonator is shown in Fig. 11.14 [39]. Requiring only 0.5 Wof electrical power to sustain a microplasma in air at atmospheric pressure, this system is well-suited for applications requiring portability. Recently, Hopwood and co-workers have extended the microwave stripline approach to excite arrays of microplasmas, each driven by a dedicated $\lambda/4$ transmission line. Inserting solid state switches into each microstrip line enables the microplasmas to be controlled by only a few volts.

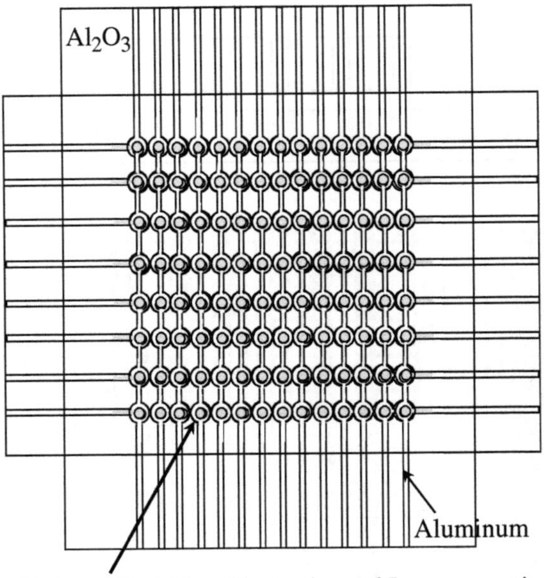

Fig. 11.12 Design of an addressable array of microcavity plasmas in which the ring electrodes surrounding each microcavity, and the linear interconnects, are formed during an electrochemical process. The addressable array comprises two thin sheets as shown

11.4 Selected Applications

The potential applications of microplasma technology are truly enormous in number and scope. Many are already being pursued and commercial products are appearing. Despite its simplicity, the microcavity plasma (and arrays thereof) is rich in potential primarily because it can be viewed from multiple perspectives. In particular, each of its three primary characteristics [(1) the emission of light (UV → infrared), (2) atmospheric pressure operation and specific power loadings >100 kW/cm^3, and (3) $n_e \gtrsim 10^{14} - 10^{15}$ cm^{-3}] suggests a set of applications. The first property, for example, points to the potential of microplasmas in displays, lighting, and disinfection, whereas the second indicates its usefulness in plasma chemistry. In this section, no attempt will be made to review the many purposes to which microplasmas are presently being devoted. Rather, a few selected applications will be discussed to illustrate the breadth of processes, systems, and functions that microcavity plasmas are able to impact.

One of the first areas in which microplasmas were applied is as a scientific tool in its own right. Becker and Schoenbach [5] have provided an excellent review of early work in applying microplasmas (operated DC or at high frequency) to analytical chemistry, and the development of vacuum ultraviolet (VUV) or ultraviolet-emitting microcavity plasmas, in particular, as radiation sources for spectroscopic analysis continues. A recent example is the work of Martin et al. [40] who demonstrated a

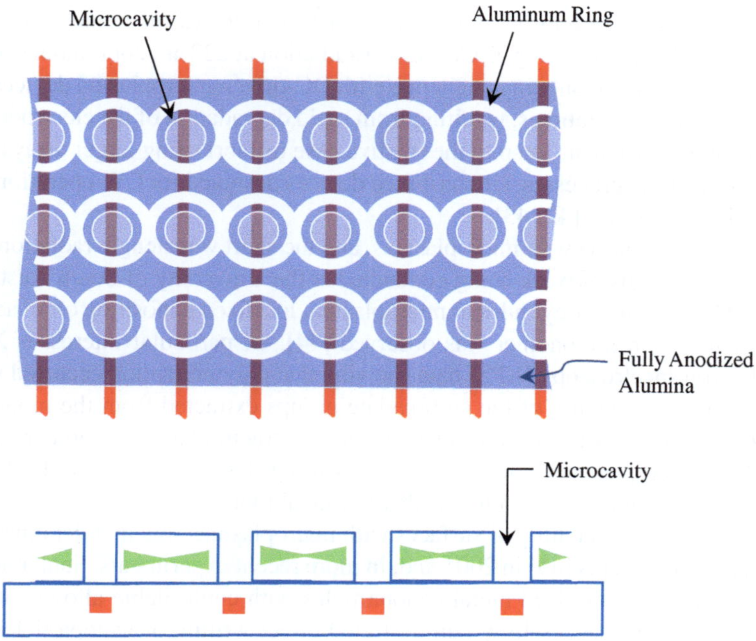

Fig. 11.13 Detailed diagrams of an alternative addressable microcavity plasma array, shown in plan view (*top*) and in cross-section (*bottom*)

Fig. 11.14 Photograph of a split-ring microwave resonator designed for producing microcavity plasmas in 500 μm wide gaps. The resonant frequency is ~900 MHz (reprinted from [39] by permission)

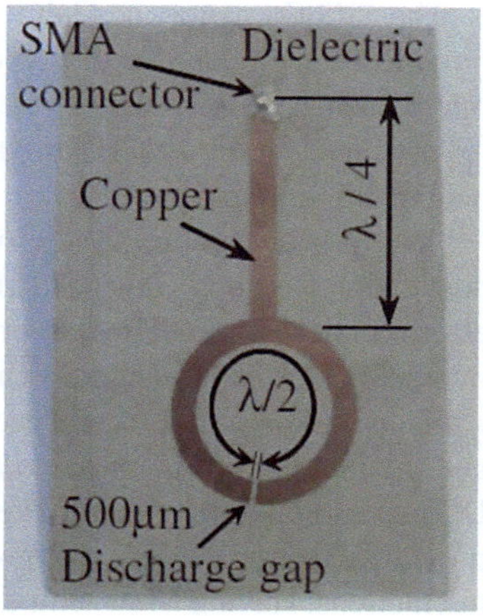

microcavity device emitting intense resonance radiation of atomic Cl in the VUV at 137.95 and 139.653 nm which correspond to the $3p^4\ {}^4P_{3/2} \rightarrow 3p^5\ {}^2P_{3/2,1/2}$ transitions of the atom. Other, weaker lines were also produced. Having measured spectral widths (FWHM) of 4.7×10^{-4} nm, these transitions are ideally-suited for the measurement of Cl ground state number densities by the absorption of resonance radiation. The same group has fabricated and tested arrays of cylindrical cavity devices, capable of generating KrCl excimer radiation at 222 nm continuously or on a pulsed basis [41]. An ongoing challenge with DC-driven microplasma devices and arrays are their short lifetimes, resulting from rapid degradation of the electrodes but the cathode, in particular. Despite the relative ease of fabricating larger arrays with CMOS-compatible processes, silicon-based devices designed for CW operation also face similar difficulties [42, 43].

Surface modification with microplasmas, and the short wavelength radiation they produce, is the focus of work being conducted at the University of South Australia. With a 400 μm diameter cylindrical microplasma, localized deposition of an acrylic acid polymer from a monomer was observed [44]. Of particular interest is X-ray photoelectron spectroscopy (XPS) data showing that polymerization occurred while retaining the functionality of the carboxylate groups extracted from the monomer. Localized plasma polymerization and other surface structural modification processes induced by microplasmas, such as plasma printing described in Sect. 11.2 [24], appear to provide materials science with a powerful tool.

Spatially-selective etching of surfaces with microplasmas was first demonstrated by Sankaran and Giapis [45] in 2001 and, in more recent experiments, Sankaran and co-workers have synthesized metal nanoparticles with controllable alloy compositions [46] and have successfully demonstrated direct writing of nanoparticles in a solid (polymer) film through the electrochemical interaction of a microplasma with metal salts in the polymer [47]. Other experiments are investigating electron transfer reactions between a plasma and a liquid [48, 49].

The Si/SiO_2 nanocones fabricated with an RF microplasma by Yang et al. [50] provide another example of the localized materials processing that is possible with microplasmas. Grown on an Fe-coated Si substrate with a 700 μm diameter plasma jet produced in an Ar/CH_4 mixture, the cones had a base diameter of 1 μm and heights up to 5 μm.

Another contribution of microplasma technology is in serving as an in situ diagnostic tool. Recently, Tai et al. [51] integrated a DC microplasma into a transmission electron microscope (TEM) to provide a route to directly observing plasma-surface interactions in real time and with a spatial resolution below 100 nm. The growth of Au islands was found to proceed according to the Stranski-Krastanov mechanism and the measured growth rate was a factor of three larger than that predicted by a DC plasma sputtering model. The plasma bipolar junction transistor (PBJT, [52, 53]) provides a unique platform with which to examine the dynamics of plasma-solid state interactions. Having a low temperature plasma serve as the collector of an otherwise all solid state transistor is a convenient way to interface an $e^- - h^+$ plasma (in a semiconductor) with an e^-–ion plasma in the gas phase. Furthermore, the influence of the collector sheath electric field on the electronic structure of the base (Si) surface and, in turn, its effect on the secondary electron emission coefficient (γ_{SE}) can be observed directly. Experiments show that γ_{SE} can be varied by a factor of $\sim$2 to 4 with a base-emitter voltage of $<$1 V [54].

A fascinating set of microplasma applications that draw upon the electromagnetic properties of arrays and unique geometric structures have been demonstrated by

Sakai, Tachibana, and colleagues [55, 56]. The scattering, absorption, and reflection of electromagnetic radiation are well-known, having been studied for decades in the context of radio and microwave propagation. By combining the dispersion relation for plasmas with specific microplasma array and geometric structures, however, microplasmas have been shown to act as metamaterials. The first experimental demonstration of a two dimensional photonic crystal [57] was reported by Sakai et al. in 2005 [58, 59]. With microplasma arrays having a pitch (center-to-center spacing) of 2.5 mm between columnar plasmas, bandgap energies corresponding to the microwave region were observed. For example, a two-dimensional array comprising ~1 mm diameter cylindrical microplasmas yields attenuations of at least 4-6 dB in the ~62 GHz region. Sakai, Tachibana, and co-workers have also successfully demonstrated plasma metamaterials constructed with a double helix wire structure in which microplasma is produced. A preliminary measurement indicates that a negative index of refraction, $-3 \leq R_e(n) \leq -1.5$, was achieved [56, 60].

With respect to displays and lighting, efforts continue to develop nanostructured cathodes as an alternative to coplanar electrodes that are common to plasma display panels. Sankaran et al. [61] recently demonstrated that crystalline diamond nanorods have characteristics attractive as microplasma cathodes. For an applied electric field strength of only 0.35 $V\mu m^{-1}$, the current density in the microplasma was measured to be $J \sim 7\,mA\,cm^{-2}$. Measurements made over a period of more than 250 h and at a fixed current density of 5 mA cm^{-2} showed a loss in current density of less than 10 %.

Of the applications of microplasmas that are being pursued at present, lighting is one of the most advanced, insofar as commercialization is concerned. Lamps that are thin (several mm in thickness), flat, and lightweight [62, 63] are currently being manufactured in sizes up to 12 × 12 sq. in. (~900 cm^2). No obstacle to scaling the lamp surface area by a factor of 2 or more is apparent. Driven (at the present time) by fluorescence from the Xe_2 molecule ($\lambda \sim 172\,nm$) that is produced efficiently in microplasmas and downconverted in a phosphor to the visible, a 12 × 12 sq. in. lamp produces more than 2500 lumens at an overall (i.e., system) efficiency above 20 lumens/W. Advanced microcavity designs at the research and development stage offer efficacies well above 30 lumens/W. These lamps do not suffer from "droop" that is often encountered with solid state lamps as the drive current is increased. The primary asset of microplasma lighting for residential and commercial interior lighting, however, is its spatial homogeneity because microplasma lighting is inherently a broad area source. In addition to producing considerable optical output (6 × 6 sq. in. lamps generate almost 600 lumens), microplasma lamps have a color rendering index (CRI) approaching 90 and the correlated color temperature (CCT) can be customer-specified throughout the 3100–5800 K range. In addition to interior lights, these lamps are finding applications in videography and cinematography, horticulture, and product inspection. The lamp "engine" itself (i.e., with the visible phosphor removed) is of considerable interest for VUV and UV disinfection, photopolymerization, and post-treatment of water.

As a final example of the applications of microplasmas, consider their potential as antennas. Experiments have shown, for example, that microchannel plasmas can be configured to serve as surprisingly sensitive antennas in the RF and microwave regions, a result that draws upon the electron densities and geometries available with microplasmas. Employing plasma columns as antennas dates back to at least 1919 when John Hettinger proposed [64] a variety of plasma-based antennas, several of which would today be classified as base-loaded verticals. Several attempts to demonstrate plasma antennas with low pressure glow discharges (such as those in fluorescent tubes) have been largely successful, but operating frequencies have been limited by the available electron densities and plasma noise and overall size were concerns.

Figure 11.15 is a schematic diagram of an electronic testing arrangement for a microchannel receiving antenna comprising 8 plasma channels, each having a width and length of 200 μm and 4 mm, respectively. Developed jointly by the University of Illinois and a Lockheed-Martin team directed by S. Macheret, this basic design generates estimated peak electron densities of $\sim 4 \times 10^{15}\,\mathrm{cm}^{-3}$ when the microchannel plasmas are driven with a 20 kHz sinusoidal voltage having an RMS value of 1.15 kV [65]. In these experiments, only the left electrode in Fig. 11.15 was powered and the current and voltage characteristics of the microplasma array were monitored by a spectrum analyzer and oscilloscope through two SMA connector output ports, labelled RF_{OUT1} and RF_{OUT2}.

As evidenced by the power spectrum presented in Fig. 11.16, plasma channels produced in 100 Torr of Ne ([Ne] $\sim 3 \times 10^{18}\,\mathrm{cm}^{-3}$) function well as a receiving antenna in the spectral region below 600 MHz. This spectrum is the spectral (Fourier) domain representation of the RF-component of the array current, and it demonstrates the detection of a number of signals, the majority of which are associated with broadcasting and governmental agency stations in the Champaign, Illinois (USA) area. At least eight of the carrier signals are more than 10 dB above the background noise floor, a remarkable result when one considers the deeply sub-wavelength dimensions of this microplasma antenna structure.

11.5 Summary and Future Challenges

The physics and applications of microplasmas have emerged together over the last 1.5 decades to form one of the most vibrant sub-fields in plasma science and technology. Any list of the most significant advances of the past few years must include: (1) Delineating in greater detail the microplasma parameter space (pushing the limits of electron density, pressure, and plasma cavity dimensions, in particular) and providing mounting evidence of the breakdown in *pd* scaling; (2) demonstrating the utility of individual microplasmas and arrays of elements for designing metamaterials in the microwave and, possibly, THz regions, and (3) interfacing microplasmas to solid and liquid materials, and demonstrating that the secondary electron emission coefficient of a surface can be controlled for at least one class of materials. Not only is plasma one of the known states of matter but it is also normally the least ordered of the four. Artificial microplasma media, such as arrays and metamaterials, will join

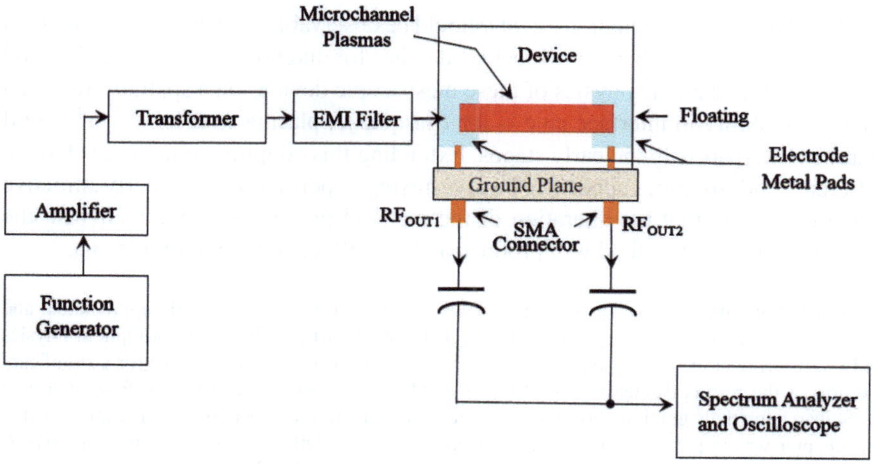

Fig. 11.15 Diagram of the electronic system employed for evaluating arrays of microchannel plasmas as receiving antennas

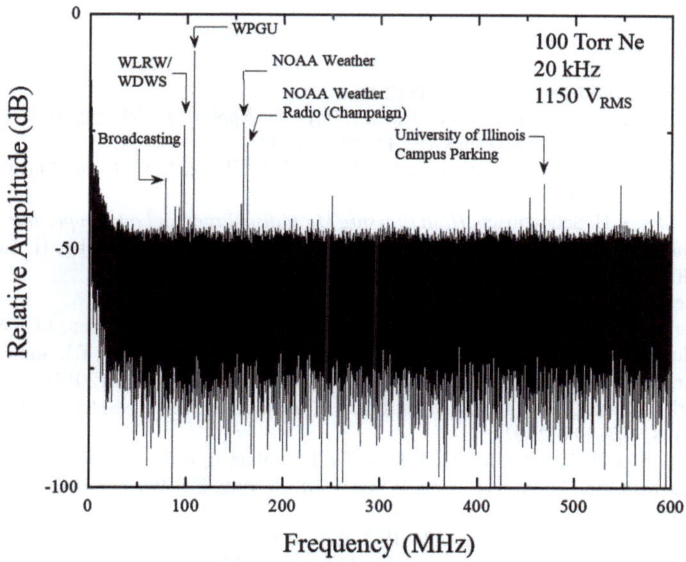

Fig. 11.16 Power spectrum of the microchannel plasma current in the region below 600 MHz. The origins of several signals in the spectrum are identified. NOAA is an acronym for the National Oceanic and Atmospheric Administration, and WLRW, WDWS, and WPGU are radio stations located in Champaign-Urbana (Illinois, USA)

other plasma states (including Coulombic crystals and plasma bullets) in beginning to reverse the reputation of plasmas for disorder.

Despite the enormous progress of the recent past, this field is young and many opportunities and challenges lie ahead. The $d = 1\,\mu m$ milestone remains to be reached, below which quantum effects should be observable, and particularly at much higher pressures (>10 bar). A critical need exists for diagnostics, both electrical and optical, to elucidate the physics of these mesoscopic devices on a spatially-resolved basis. The ability to interface an e^-- ion (gas phase) plasma with its $e^- - h^+$ (solid state) counterpart is in the early stages. Extending this coupling to materials having different band structures appears to be a promising opportunity. Particularly attractive is the likelihood that the alteration of the material properties of the wall with the plasma sheath electric field will produce a "back-effect" on the microplasma.

Acknowledgments Discussions with S. Macheret and M. Kushner are greatly appreciated, and many thanks are extended to present and former members of the Laboratory for Optical Physics and Engineering at the University of Illinois. Their tireless efforts are responsible for a significant fraction of the results discussed here. One of us (JGE) also wishes to thank G. Eisenstein and A. Shwartz for their gracious hospitality at the Israel Institute of Technology (Technion) while this chapter was in preparation. The support of the U.S. Air Office of Scientific Research (H.R. Schlossberg) under grants FA9550-10-1-0048 and FA9550-12-1-0487 is gratefully acknowledged.

References

1. K. Tachibana, IEEJ Trans. **1**, 145–155 (2006)
2. K.H. Becker, K.H. Schoenbach, J.G. Eden, J. Phys. D: Appl. Phys. **39**, R55–R70 (2006)
3. U. Kogelschatz, Contrib. Plasma Phys. **47**, 80–88 (2007)
4. F. Iza, G.J. Kim, S.M. Lee, J.K. Lee, J.L. Walsh, Y.T. Zhang, M.G. Kong, Plasma Process. Polym. **5**, 322–344 (2008)
5. K.H. Becker, K.H. Schoenbach, *High Pressure Microdischarges, in Low Temperature Plasmas: Fundamentals, Technologies, and Techniques*, 2nd edn., ed. by R. Hippler, H. Kersten, M. Schmidt and K.H. Schoenbach (Wiley-VCH, Weinheim, 2008)
6. J.G. Eden, S.J. Park, Plasma Phys. Control. Fusion **47**, B83–B92 (2005)
7. S. Samukawa, M. Hori, S. Rauf, K. Tachibana, P. Bruggeman, G. Kroesen, J.C. Whitehead, A.B. Murphy, A.F. Gutsol, S. Starikovskaia, U. Kortshagen, J.-P. Boeuf, T.J. Sommerer, M.J. Kushner, U. Czarnetzki, N. Mason, J. Phys. D: Appl. Phys. **45**, 253001 (2012)
8. S.-J. Park, J. Chen, C.J. Wagner, N.P. Ostrom, C. Liu, J.G. Eden, IEEE J. Select. Topics Quantum Electron. **8**, 387–394 (2002)
9. J.G. Eden, S.-J. Park, J.H. Cho, M.H. Kim, T.J. Houlahan Jr, B. Li, E.S. Kim, T.L. Kim, S.K. Lee, K.S. Kim, J.K. Yoon, S.H. Sung, P. Sun, C.M. Herring, C.J. Wagner, IEEE Trans. Plasma Sci. **41**, 661–675 (2013)
10. J.P. Boeuf, J. Phys. D: Appl. Phys. **36**, R53–R79 (2003)
11. S.-J. Park, J.G. Eden, Appl. Phys. Lett. **81**, 4127–4129 (2002)
12. S.-J. Park, J.G. Eden, J. Chen, C. Liu, Appl. Phys. Lett. **85**, 4869–4871 (2004)
13. R. Tirumala, D.B. Go, Appl. Phys. Lett. **97**, 151502 (2010)
14. W.S. Boyle, P. Kisliuk, Phys. Rev. **97**, 255–259 (1955)
15. M. Kushner, J. Phys. D: Appl. Phys. **95**, 846–859 (2004)
16. M. Kushner, J. Phys. D: Appl. Phys. **38**, 1633–1643 (2005)
17. H.-J. Lee, S.-J. Park, J.G. Eden, J. Phys. D: Appl. Phys. **45**, 405201 (2012)
18. J. Dutton, E.M. Williams, Proc. Phys. Soc. **84**, 171–175 (1964)
19. A.D. White, J. Appl. Phys. **30**, 711–719 (1959)
20. K.H. Schoenbach, R. Verhappen, T. Tessnow, F.E. Peterkin, W.W. Byszewski, Appl. Phys. Lett. **68**, 13–15 (1996)
21. K.H. Schoenbach, A. El-Habachi, W. Shi, M. Ciocca, Plasma Sources Sci. Technol. **6**, 468–477 (1997)

22. J.W. Frame, D.J. Wheeler, T.A. DeTemple, J.G. Eden, Appl. Phys. Lett. **71**, 1165–1167 (1997)
23. U. Kogelschatz, J. Salge, *High-Pressure Plasmas: Dielectric Barrier and Corona Discharges Properties and Technical Applications, in Low Temperature Plasmas: Fundamentals, Technologies, and Techniques* 2nd edn., ed. by R. Hippler, H. Kersten, M. Schmidt, K. H. Schoenbach (Wiley-VCH, Weinheim, 2008)
24. M. Thomas, J. Borris, A. Dohse, M. Eichler, A. Hinze, K. Lachmann, K. Nagel, C.-P. Klages, Plasma Proc. Polym. **9**, 1086–1103 (2012)
25. S.H. Sung, I.C. Hwang, S.-J. Park, J.G. Eden, Appl. Phys. Lett. **97**, 231502 (2010)
26. S.-J. Park, T.M. Spinka, J.G. Eden, Appl. Phys. Lett. **89**, 031502 (2006)
27. K.-F. Chen, N.P. Ostrom, S.-J. Park, J.G. Eden, Appl. Phys. Lett. **88**, 061121 (2006)
28. K.S. Kim, T.L. Kim, J.K. Yoon, S.-J. Park, J.G. Eden, Appl. Phys. Lett. **94**, 011503 (2009)
29. K.S. Kim, S.-J. Park, J.G. Eden, J. Phys. D: Appl. Phys. **41**, 012004 (2008)
30. A. Bass, C. Chevalier, M.W. Blades, J. Anal. Atom. Spectrom. **16**, 919–921 (2001)
31. M. Lu, S.-J. Park, B.T. Cunningham, J.G. Eden, IEEE/ASME J. Microelectromechan. Syst. **16**, 1397–1402 (2007)
32. M. Lu, B.T. Cunningham, S.-J. Park, J.G. Eden, Appl. Phys. Lett. **92**, 101928 (2008)
33. X. Qiu, R. Gerhard, A. Mellinger, IEEE Trans. Diel. Electr. Insul. **18**, 34–42 (2011)
34. A. Mellinger, O. Mellinger, IEEE Trans. Diel. Electr. Insul. **18**, 43–48 (2011)
35. Y. Yin, J. Messier, J. Hopwood, IEEE Trans. Plasma Sci. **27**, 1516–1524 (1999)
36. A.M. Bilgic, U. Engel, E. Voges, M. Kuckelheim, J.A.C. Broekaert, Plasma Sources Sci. Technol. **9**, 1–4 (2000)
37. A.M. Bilgic, E. Voges, E. Engel, J.A.C. Broekaert, J. Anal. Atom. Spectrom. **15**, 579–580 (2000)
38. F. Iza, J.A. Hopwood, IEEE Trans. Plasma Sci. **31**, 782–787 (2003)
39. F. Iza, J. Hopwood, Plasma Sources Sci. Technol. **14**, 397–406 (2005)
40. V. Martin, G. Bauville, N. Sadeghi, V. Puech, J. Phys. D: Appl. Phys. **44**, 435203 (2011)
41. V. Martin, G. Bauville, M. Fleury, V. Puech, Plasma Sources Sci. Technol. **21**, 065001 (2012)
42. R. Dussart, L.J. Overzet, P. Lefaucheux, T. Dufour, M. Kulsreshath, M.A. Mandra, T. Tillocher, O. Aubry, S. Dozias, P. Ranson, J.B. Lee, M. Goeckner, Eur. Phys. J. D. **60**, 601–608 (2010)
43. M.K. Kulsreshath, L. Schwaederle, L.J. Overzet, L. Lefaucheux, J. Ladroue, T. Tillocher, O. Aubry, M. Woytasik, G. Schelcher, R. Dussart, J. Phys. D: Appl. Phys. **45**, 285202 (2012)
44. S.A. Al-Bataineh, E.J. Szili, A. Mishra, S.-J. Park, J.G. Eden, H.J. Griesser, N.H. Voelcker, R.D. Short, D.A. Steele, Plasma Process. Polym. **8**, 695–700 (2011)
45. R.M. Sankaran, K.P. Giapis, Appl. Phys. Lett. **79**, 593–595 (2001)
46. W.-H. Chiang, R.M. Sankaran, Nat. Mater. **8**, 882–886 (2009)
47. P.A. Lin, R.M. Sankaran, Angew. Chem. Int. Ed. **50**, 10953–10956 (2011)
48. S.W. Lee, D. Liang, X.P. Gao, R.M. Sankaran, Adv. Func. Mater. **21**, 2155–2161 (2011)
49. C. Richmonds, M. Witzke, B. Bartling, S.W. Lee, J. Wainright, C.C. Liu, R.M. Sankaran, J. Am. Chem. Soc. **133**, 17582–17585 (2011)
50. Z. Yang, H. Shirai, T. Kobayashi, Y. Hasegawa, Thin Solid Films **515**, 4153–4158 (2007)
51. K. Tai, T.J. Houlahan Jr, J.G. Eden, S.J. Dillon, Sci. Rep. **3**, 1325 (2013)
52. C.J. Wagner, P.A. Tchertchian, J.G. Eden, Appl. Phys. Lett. **97**, 134102 (2010)
53. P.A. Tchertchian, C.J. Wagner, T.J. Houlahan Jr, B. Li, D.J. Sievers, J.G. Eden, Contrib. Plasma Phys. **51**, 889–905 (2011)
54. B. Li, T.J. Houlahan Jr, C.J. Wagner, J.G. Eden, Appl. Phys. Lett. **102**, 083502 (2013)
55. K. Tachibana, Pure Appl. Chem. **82**, 1189–1199 (2010)
56. O. Sakai, K. Tachibana, Plasma Sources Sci. Technol. **21**, 013001 (2012)
57. E. Yablonovitch, Science **289**, 557–559 (2000)
58. O. Sakai, T. Sakaguichi, K. Tachibana, Appl. Phys. Lett. **87**, 241505 (2005)
59. O. Sakai, Y. Kishimoto, K. Tachibana, J. Phys. D: Appl. Phys. **38**, 431–441 (2005)
60. O. Sakai, T. Naito, T. Shimomura, K. Tachibana, Thin Solid Films **518**, 3444–3448 (2010)
61. K.J. Sankaran, S. Kunuku, S.-C. Lou, J. Kurian, H.-C. Chen, C.-Y. Lee, N.-H. Tai, K.-C. Leou, C. Chen, I.-N. Lin, Nanoscale Res. Lett. **7**, 522 (2012)

62. S.-J. Park, J.D. Readle, A.J. Price, J.K. Yoon, J.G. Eden, J. Phys. D: Appl. Phys. **40**, 3907–3913 (2007)
63. J.G. Eden, S.-J. Park, C.M. Herring, J.M. Bulson, J. Phys. D: Appl. Phys. **44**, 224011 (2011)
64. J. Hettinger, *Aerial Conductor for Wireless Signaling and Other Purposes*, U.S. Patent No. 1,309,031, 8 July 1919 NULL
65. S. Macheret, S.-J. Park, J.G. Eden (unpublished data)

Chapter 12
Plasma Electrochemistry: A Novel Chemical Process for the Synthesis and Assembly of Nanomaterials

Seung Whan Lee and R. Mohan Sankaran

Abstract Electrochemical reactions are typically initiated at the interface of a solid metal electrode and ionic electrolyte. Recently, the development of a novel microplasma source has allowed non-thermal, atmospheric-pressure plasmas to be stably formed at the surface of aqueous ionic electrolytes and thin polymeric films. Thus, microplasmas can act as a gaseous electrode for electrochemical reactions which is termed "plasma electrochemistry". Plasma electrochemistry allows the synthesis and assembly of nanomaterials, and, in comparison to other chemical methods, is low cost, high purity, scalable via roll-to-roll processing, and environmentally friendly. Overall, these features should be attractive for emerging technological applications in catalysis, energy harvesting/storage, and medicine. This chapter provides technical background and motivation for this area of research, and describes recent progress and remaining scientific challenges.

12.1 Introduction to Plasma Electrochemistry

Over the last two decades, plasma processing has played a critical role in the manufacturing of materials for many industries including semiconductor fabrication where plasmas are used to etch and deposit thin films that make up integrated circuits. The unique properties of plasma-based processes such as low temperature and high purity are essential to high-fidelity pattern transfer. However, these processes are normally operated at low pressure which can entail large costs associated with operation and maintenance of vacuum equipment. Alternatively, operation near or at atmospheric pressure would be potentially cheaper, easier to implement, and highly desirable for industrial applications [1]. Therefore, one of the important challenges in plasma

S. W. Lee · R. M. Sankaran (✉)
Department of Chemical Engineering, Case Western Reserve University, Cleveland, OH 44106-7217, USA
e-mail: mohan@case.edu

M. Bonitz et al. (eds.), *Complex Plasmas*, Springer Series on Atomic, Optical, and Plasma Physics 82, DOI: 10.1007/978-3-319-05437-7_12,

processing is the development of low temperature (i.e., cold), atmospheric-pressure plasma sources. The availability of such plasmas is expected to open new applications in medicine [2–4], gas treatment [5, 6], textiles [7], surface modification [8, 9], and nanofabrication [10, 11].

An important class of low temperature, atmospheric-pressure plasmas is microplasmas. Briefly, microplasmas are defined as plasmas less than 1 mm in at least one dimension. Gas pressure, p, times electrode gap, d, is preserved at high (i.e., atmospheric) pressures in these plasmas, leading to their stability and non-thermal characteristics. Additional details of microplasmas can be found in Chaps. 11, 13, and 14.

Non-thermal operation at atmospheric pressure allows microplasmas to be stably formed with aqueous liquids. If the liquid is operated as part of the electrical circuit (e.g., current flows from the plasma to the liquid), electrochemical reactions can occur at the plasma–liquid interface [12]. For example, if the microplasma is operated cathodically, metal ions in solution are reduced to solid metal, analogous to electrodeposition. In contrast to electrodeposition where the metal coats a solid electrode, the reduction of metal ions in a plasma–liquid system leads to the nucleation and growth of metal nanoparticles (NPs). In this chapter, this concept, termed as plasma electrochemistry, is presented. First, examples of metal NP synthesis are presented. Then, model electrochemical reactions such as the reduction of ferricyanide ions [$Fe(CN)_6{}^{3-}$] to ferrocyanide ions [$Fe(CN_6)^{4-}$] to better understand the underlying chemistry are discussed. Then, plasma electrochemistry is extended from liquid systems to thin films in order to establish the following strategies for micro/nanofabrication of metal NPs: (1) 'direct writing' by scanning a microplasma across a thin film, (2) pattern transfer by combining discharge extraction and lithographic techniques, and (3) directed assembly via metallosupramolecular polymer templates. Direct writing of metal NPs is achieved by exposing spin-coated films of metal salts and polymer to a scanning microplasma. As the microplasma interacts with the film surface, the metal ions are reduced and NPs are nucleated. Thus, patterns of metal NPs close to the size of the microplasma ($\sim$150 μm) are produced in a flexible, polymeric film. In order to reduce heating and sputtering of the film and provide better control over the discharge, a two-stage discharge was developed consisting of a primary microplasma and an extracted discharge formed at a third, positively-biased electrode. Incorporating a nanoscale mask such as an alumina membrane template, the extracted discharge can be used to expose only the desired areas of the film and produce patterned metal NPs. Further reduction in the resolution of the transferred patterns is achieved by moving to a metallosupramolecular polymer template. These novel materials contain inorganic (metal) and organic (polymer) moieties that self-assemble into films with well-defined morphologies. Exposing the template to the extracted discharge results in the in situ nucleation and growth of well-dispersed, uniformly-sized metal NPs. Overall, these diverse approaches to nanofabrication based on plasma electrochemistry are low cost, scalable, and amenable to roll-to-roll processing, which is attractive for emerging technological applications in catalysis, energy, and medicine.

12.2 Plasma Electrochemical Reactions in Solution

Electrochemical reactions are typically carried out in solution at the interface of an electron conductor (a metal or a semiconductor) and an ionic conductor (an electrolyte). Plasmas are a source of charged species such as electrons and ions, making them potentially suitable for electrochemical applications. However, unlike electrochemical reactions between solid and liquid phases, the charge exchange reactions occur between the gas and liquid (or solid) phase.

Although plasmas have found numerous chemical and materials applications, only a few previous reports of electrochemical applications exist. Over 100 years ago, Gubkin studied gas discharges at the surface of an aqueous solution of silver nitrate and observed the formation of a metallic silver film [13]. Gubkin believed that electrons in a plasma could directly reduce dispersed Ag ions. Later, Hickling introduced the idea of glow discharge electrolysis (GDE) whereby electrochemical reactions such as water electrolysis were initiated by a plasma electrode [14, 15]. In contrast to Gubkin, Hickling operated the plasma as the anode and studied interactions of gaseous ions with solution species. Despite these early studies that motivated the use of plasmas in electrochemistry, plasmas have not been widely implemented in electrochemical applications mostly because non-thermal operation at atmospheric pressure is difficult.

Recently, there has been renewed interest in plasma–liquid systems for the synthesis of metal NPs [16–23]. Figure 12.1 shows a side-by-side comparison of a standard electrochemical cell versus a plasma–liquid electrochemical cell. In a standard electrochemical cell, metal cations created from anodic dissolution of metal foils or introduced in the electrolyte as a metal salt are reduced at the cathode, leading to deposition of a film. In some cases, nanoparticles are formed in solution, but isolation of the reduction products is difficult because of the presence of the solid electrode [24–27]. Replacing the metal cathode with a direct-current (dc) microplasma allows similar reduction reactions to take place, but without a solid electrode, metallic particles homogeneously nucleate in solution. A mechanism for nanoparticle formation has been hypothesized based on migration of metal ions to the plasma–liquid interface, nucleation and growth in the reaction zone, and diffusion of the nanoparticles back into the bulk of the solution [12].

The earliest report of plasma–liquid synthesis of metal NPs was the reduction of a platinum salt (H_2PtCl_6) [17]. In this case, hydrogen (H_2) gas was found necessary to reduce the Pt cations to metallic Pt NPs. More recently, argon (Ar) or oxygen (O_2) plasmas have been used to reduce metal cations in solution (no H_2) [28]. The precise mechanism for plasma reduction remains elusive. Plasmas formed with water could generate oxidizing or reducing agents in solution by electron impact dissociation or UV irradiation including atomic hydrogen and hydroxyl radicals, H_2 gas, and hydrogen peroxide which may be responsible for reduction in addition to the direct electron reduction originally proposed by Gubkin. Recently, fundamental studies of charge transfer reactions at the plasma–liquid interface have suggested that direct electron reduction is possible, detailed in the following section.

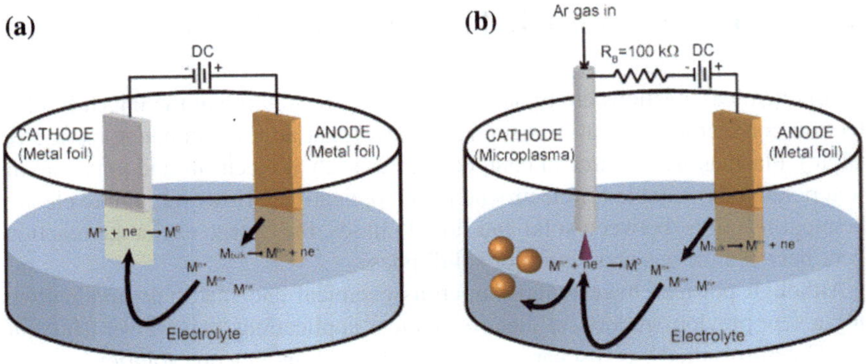

Fig. 12.1 Schematic of **a** a standard electrochemical cell with metal electrodes and **b** a plasma–liquid electrochemical cell with an atmospheric-pressure microplasma as the cathode. In the standard electrochemical cell, metal ions are plated at the cathode. In the plasma–liquid electrochemical cell, metal ions nucleate NPs

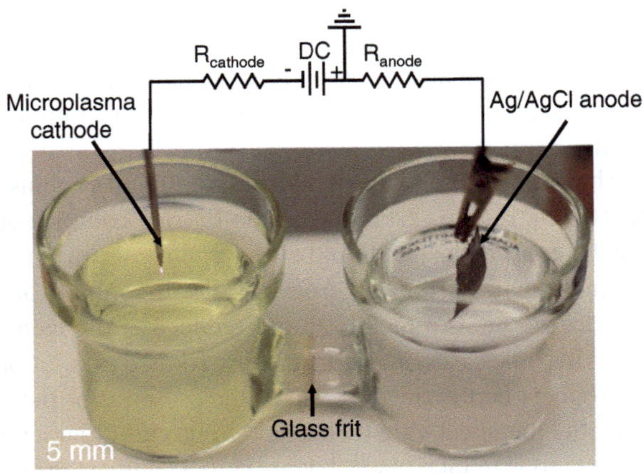

Fig. 12.2 Schematic of plasma–liquid electrochemical setup using a split glass cell for fundamental studies of electron transfer reactions. An atmospheric-pressure microplasma is formed in an Ar gas flow between a stainless-steel capillary tube and the solution surface. The counter electrode is Ag/AgCl. The catholyte bath consists of 15 mL of 0.2 mM potassium ferricyanide [$K_3Fe(CN)_6$] and 0.1 M potassium chloride (KCl) dissolved in deionized water and the anolyte bath contains only 0.1 M KCl. Modified with permission from Ref. [29], copyright 2011 American Chemical Society

12.3 Fundamental Studies of Charge Transfer Reactions at the Plasma–Liquid Interface

Plasma electrochemistry encompasses the idea that charged species in a plasma such as electrons can directly carry out electrochemical reactions in solution or at surfaces through charge transfer. While several studies of plasmas in contact with liquids have

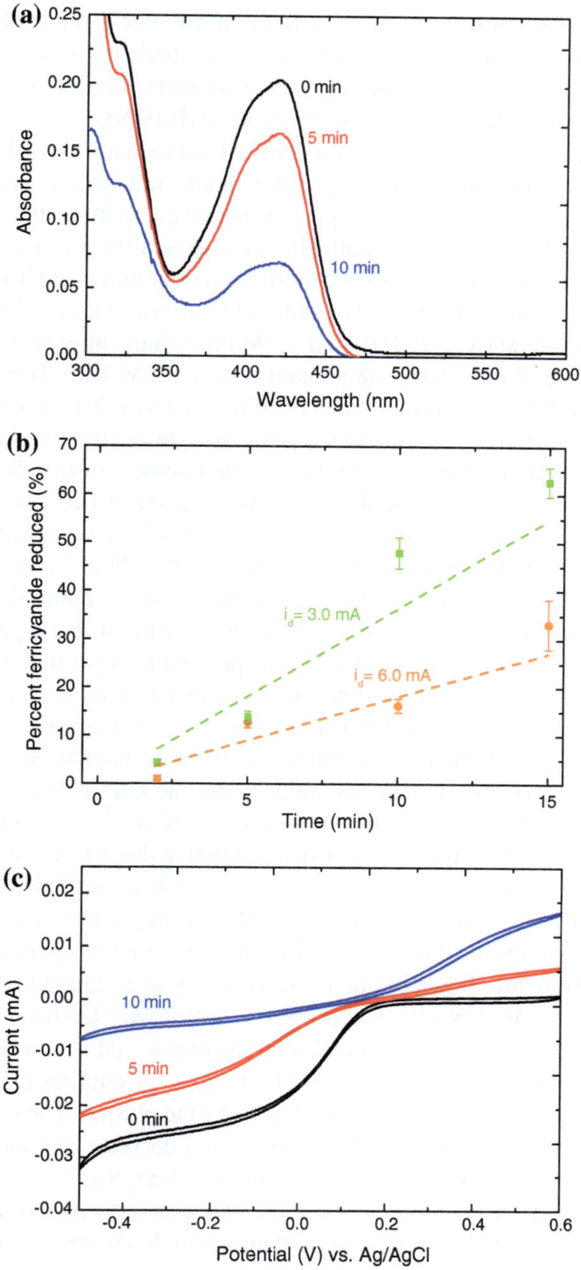

Fig. 12.3 **a** UV-vis absorbance spectra of solutions of ferricyanide after exposure to the plasma for 0, 5, and 15 min (discharge current, $i_d = 6$ mA), **b** percent of ferricyanide reduced in solution as a function of time at discharge currents (i_d) of 3 and 6 mA, and **c** cyclic voltammograms of solutions of ferricyanide after exposure to the plasma for 0, 5, and 15 min. Reprinted with permission from Ref. [29], copyright 2011 American Chemical Society

been reported in the literature, it still remains unknown if or how gas-phase electrons interact with ions in solutions in comparison to electrochemical systems involving metal electrodes. To address this question, an experiment was designed based on the well-known ferricyanide [$Fe(CN)_6^{3-}$]/ferrocyanide [$Fe(CN)_6^{4-}$] redox couple [29]. Figure 12.2 schematically shows the split glass cell setup that was used to operate the plasma–liquid system. The electrolytes in the anodic and cathodic sides were prevented from mixing by a fritted glass plug. Separate compartments for the cathode and anode ensured isolation of the half-cell reaction occurring at the plasma–liquid interface. In the anode side, an Ag/AgCl mesh was partly immersed in the electrolyte and served as the counter electrode. The catholyte bath consisted of 15 mL of 0.2 mM potassium ferricyanide [$K_3Fe(CN)_6$] and 0.1 M potassium chloride (KCl) dissolved in deionized water; the anolyte bath contained only 0.1 M KCl. The conversion of $Fe(CN)_6^{3-}$ to $Fe(CN)_6^{4-}$ was monitored by ultraviolet-visible (UV-vis) spectrophotometry. Representative UV-vis spectra of solutions exposed to a plasma ($i_d = 6$ mA) for 0, 5 and, 15 min are shown in Fig. 12.3a. The intensity of the absorbance peak at ~420 nm is directly proportional to the concentration of $Fe(CN)_6^{3-}$ in solution [30], and the decrease over time suggests that $Fe(CN)_6^{3-}$ is reduced to $Fe(CN)_6^{4-}$ by the plasma. Figure 12.3b shows the percentage of $Fe(CN)_6^{3-}$ species reduced by the plasma. The results show that $Fe(CN)_6^{3-}$ is progressively reduced over time, and the rate of reduction is proportional to the discharge current. For a given discharge current, the electron flux (number of electrons per unit area per unit time) delivered to the surface of the solution by the plasma remains constant; thus, the $Fe(CN)_6^{3-}$ contained in the solution is reduced at a relatively constant rate. When the discharge current is raised, the electron density in the plasma increases [31], a higher flux of electrons reaches the solution surface, and the rate of reduction increases. To verify that $Fe(CN)_6^{3-}$ was indeed reduced to $Fe(CN)_6^{4-}$ and not destroyed by the plasma, cyclic voltammetry was carried out with a three-electrode system comprised of a glassy carbon rotating-disk working electrode controlled by a potentiostat, an Ag/AgCl reference electrode, and a Pt mesh auxiliary electrode. Representative cyclic voltammograms of solutions of ferricyanide exposed to a plasma ($i_d = 6$ mA) for 0, 5, and 15 min are shown in Fig. 12.3c. A steady-state cathodic current plateau proportional to the $Fe(CN)_6^{3-}$ concentration is observed for the initial solution. After plasma exposure, this current plateau decreases and a new anodic current plateau appears, indicating the presence of $Fe(CN)_6^{4-}$ in solution and clearly showing that $Fe(CN)_6^{3-}$ was reduced to $Fe(CN)_6^{4-}$ by the plasma. The current plateaus for $Fe(CN)_6^{3-}$ reduction and $Fe(CN)_6^{4-}$ oxidation decrease and increase, respectively, by the same magnitude, which confirms that $Fe(CN)_6^{3-}$ was not destroyed. The current plateau for $Fe(CN)_6^{4-}$ oxidation increases with time as more and more $Fe(CN)_6^{4-}$ is generated in solution, consistent with UV-vis absorbance results.

Despite the evidence for electron transfer, a comparison of the number of electrons injected by the plasma using Faraday's law with the number of ferricyanide molecules reduced from UV-vis absorbance shows that less than 10 % of electrons are involved in ferricyanide reduction [31]. To verify that there are no current losses at the plasma–liquid interface, the current through the cell was evaluated by replacing the Ag/AgCl electrode with Ag or Cu. As shown in Fig. 12.4, the weight losses measured at the

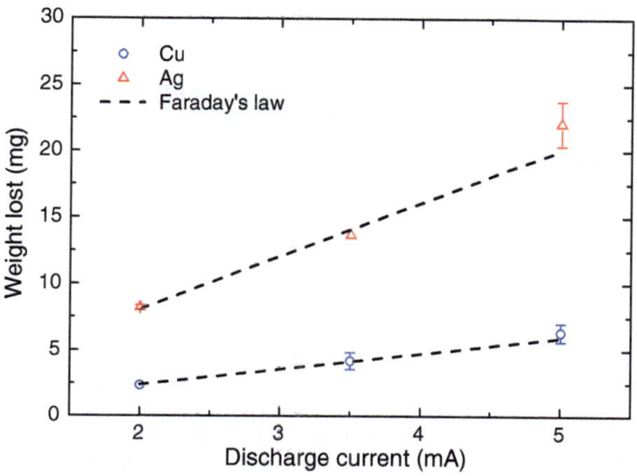

Fig. 12.4 Weight loss measured for Cu and Ag foil anodes after operating a plasma-based electrochemical cell for various times and discharge currents (i_d). The electrolytes consisted of 1 mM HCl in deionized water for Cu and 1 mM HNO_3 in deionized water for Ag. The weight losses predicted by Faraday's law are also shown *dashed line*. Reprinted with permission from Ref. [29], copyright 2011 American Chemical Society

anode are in excellent agreement with Faraday's law for both $Ag \rightarrow Ag^+ + e^-$ and $Cu \rightarrow Cu^{2+} + 2e^-$, suggesting that charge is transferred with 100 % efficiency from the plasma to the liquid. In order for reduction to occur, the donor and acceptor states in electron-transfer reactions such as this one must be of equal energy [32]. Since a plasma contains a complex distribution of electron energies, it is possible that only a fraction of the electrons with the appropriate energy reduce the $Fe(CN)_6{}^{3-}$ ions. The remaining electrons may be involved in other charge-transfer processes such as H_2 gas formation [$2H^+ + 2e^- \rightarrow H_2(g)$], leading to a lower than expected reduction of $Fe(CN)_6{}^{3-}$ molecules. In general, aqueous electrochemical systems can be characterized by water electrolysis at sufficiently high potentials, i.e., electrolytic reactions at the cathode and anode that result in the decomposition of water to H_2 and O_2 gas. The electrolysis of water under acidic conditions occurs through two primary reactions, reduction at the cathode,

$$2H^+ + 2e^- \rightarrow H_2(g) \tag{12.1}$$

and oxidation at the anode,

$$2H_2O \rightarrow O_2(g) + 4H^+ + 4e^-. \tag{12.2}$$

In order to verify that water electrolysis occurs in plasma–liquid systems via the above reactions, three independent methods of characterization were performed on a sealed glass cell [33]: (1) a pH sensitive dye containing bromothymol blue and phenol red valid over the pH range 1.0–11.0 (Hydrion One Drop Indicator 1–11) were used

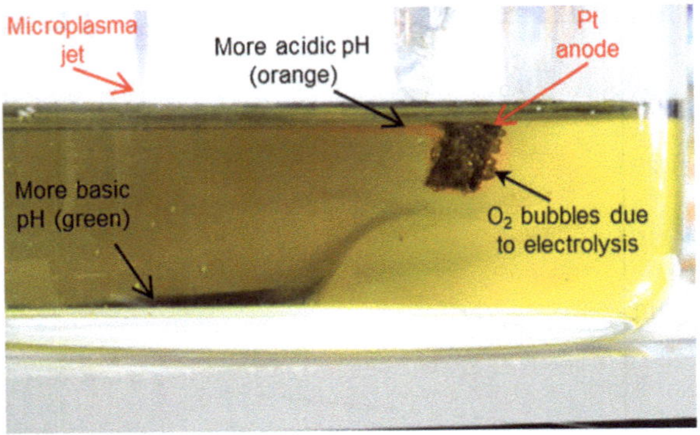

Fig. 12.5 Photograph of the microplasma formed with an aqueous solution electrode containing a pH sensitive dye. The reduction reaction at the microplasma/liquid interface results in an increase in the pH, as indicated by the *green colour* change of the dye. The oxidation reaction at the anode results in O_2 gas (bubbles) and a concomitant decrease in the pH, as indicated by the *orange colour* change. Reprinted with permission from Ref. [33], copyright 2012 IOP Society

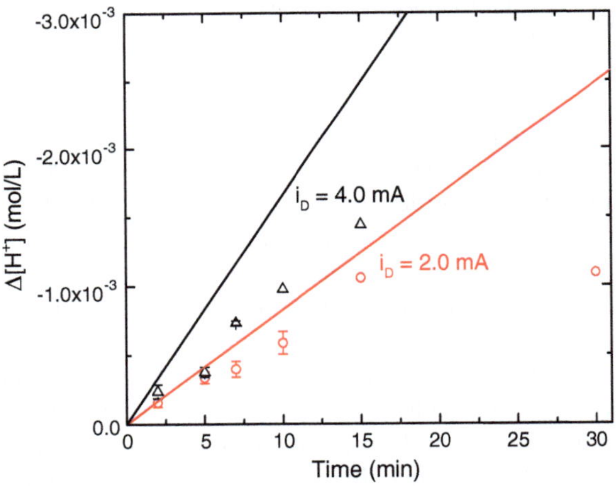

Fig. 12.6 Change in concentration of H^+ (Δ [H^+]) as a function of time in the cathode bath of a two-cell electrochemical setup operated with a microplasma cathode at a constant discharge current (i_D) of 2 and 4 mA. The concentration decreases in time ([H^+] < 0) as H^+ is reduced. *Error bars* correspond to the standard error of the measurements, and the *solid lines* indicate [H^+] as predicted by Faraday's law. Reprinted with permission from Ref. [33], copyright 2012 IOP Society

to visually observe the spatiotemporal changes in pH during cell operation; (2) the pH in the cathode bath of the split cell setup was measured after running the cell at a constant discharge current for a fixed time; and (3) the generation of H_2 gas was measured by capturing the gas product and injecting into an isotope ratio mass spectrometer.

Figure 12.5 shows a photograph of a plasma–liquid cell operating in the presence of a pH sensitive dye in tap water and sufficient HCl to buffer it to a neutral pH

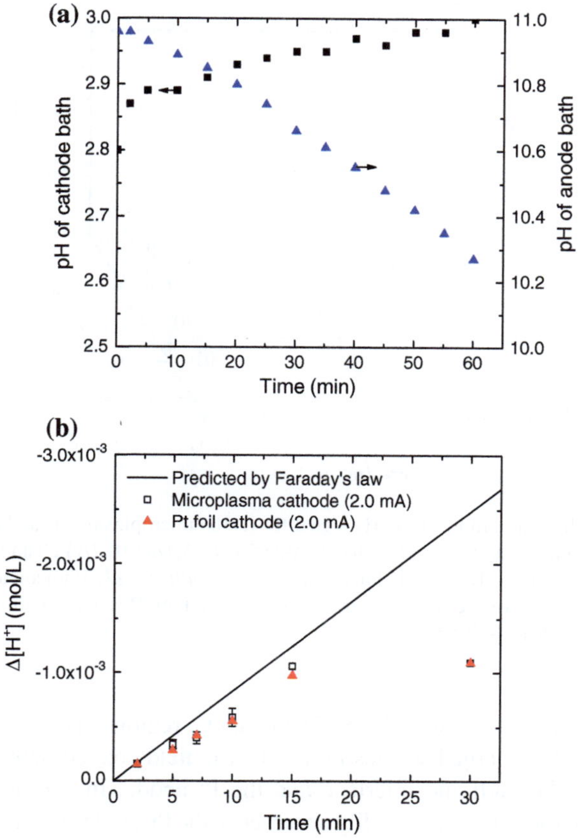

Fig. 12.7 **a** pH as a function of time for a concentration gradient setup in a two compartment glass cell. The acidic solution contained 1 mM HCl and 0.1 M KCl in deionized water and the basic solution contained 1 mM NaOH and 0.1 M KCl in deionized water. **b** Change in concentration of H^+ ($\Delta[H^+]$) as a function of time in cathode bath of a split cell electrochemical setup operated either with a microplasma cathode or a Pt foil cathode at a current of 2 mA. A Pt foil anode and catholyte and anolyte containing 1 mM HCl and 0.1 M KCl in deionized water were used in both cases. The *solid line* indicates the expected change in H^+ concentration, based on Faraday's law and assuming only the hydrogen evolution reaction ($2H^+ + 2e^- \rightarrow H_2(g)$) occurs in the catholyte. Reprinted with permission from Ref. [33], copyright 2012 IOP Society

(pH ~ 7). When the microplasma was formed on the surface, the dye immediately became green directly beneath the microplasma as a result of an increase in the pH of the solution near the plasma/liquid interface (12.1). Over a short time (seconds), the green color change extended all the way to the bottom of the cell. This may be due to advection of the reduced liquid, driven by the gas flow from the microplasma that causes a deformation of the liquid surface, or solvated electrons that induce reduction beneath the liquid surface. At the Pt foil anode, the dye turned orange when the microplasma formed, reflecting a decrease in the pH of the solution as a

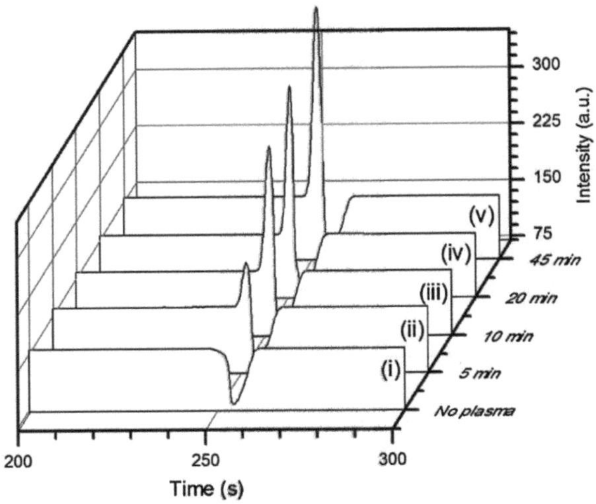

Fig. 12.8 Mass chromatograms of the H_2 signal for (*i*) the microplasma off and the microplasma operated at 1 mA with an aqueous solution electrode for (*ii*) 5, (*iii*) 10, (*iv*) 20 and (*V*) 45 min. The aqueous solution contained 1.0 mM HCl in all cases. The *negative peaks* immediately following the H_2 signal are an artifact of the mass spectrometer valve system. Reprinted with permission from Ref. [33], copyright 2012 IOP Society

result of oxidation (12.2). Both the basic and acidic regions disperse in the solution by migration of the charged species in the electric field created between the virtual cathode at the plasma/liquid interface and the Pt anode. In addition to the color changes, bubble formation was readily observed at the Pt anode. This is most probably O_2 evolution resulting from the oxidation of water (12.2).

A sealed split cell setup was used to measure pH changes (see Fig. 12.2). Solutions contained 1 mM HCl and 1.0 M KCl; the KCl was necessary to maintain conductivity as the H^+ was reduced. pH measurements were made after operating the microplasma cathode for a fixed time at constant discharge current. The pH values correspond to a H^+ concentration in solution and can be used to calculate the change in the H^+ concentration, $\Delta[H^+]$, i.e., decrease in $[H^+]$ since the pH increases as a result of H_2 gas evolution. The measurements were repeated at various times and discharge currents to obtain a statistically relevant set of data and analyze the kinetics of the hydrogen evolution reaction. Figure 12.6 shows the average $\Delta[H^+]$ obtained as a function of time for discharge currents of 2.0 and 4.0 mA. $\Delta[H^+]$ decreases with time at a relatively constant rate for times less than 15 min, and the rate at which $\Delta[H^+]$ decreases increases with discharge current. The results appear to be largely consistent with a constant flux of electrons that depends on discharge current reaching the solution surface and reducing H^+. At a given discharge current, the flux of electrons is constant over time and the H^+ in solution is reduced at a constant rate. As the current is increased, the flux of electrons increases and the rate of reduction increases. These results were compared with what would be expected assuming water electrolysis and Faraday's law based on the discharge current. The measured change in the H^+ concentration is in good agreement with Faraday's law at lower discharge current and short run times, but less than the corresponding predicted value at higher discharge current and longer run times. These discrepancies could be explained by diffusion and/or migration of H^+ between the cell compartments which would lead to a smaller than expected net change in the pH (i.e., $\Delta[H^+]$). To verify these hypotheses, two simple tests were carried out: (1) a concentration gradient was established between the two compartments and the pH was monitored to characterize the diffusion of ions across the glass frit and (2) an electrochemical cell was operated with a Pt foil cathode under the same conditions as the microplasma cathode (see Fig. 12.7). The tests showed that diffusion of ions is not significant, but under normal electrochemical operation, the H^+ concentration is lower than that predicted by Faraday's law, suggesting that migration of ions across the glass frit due to the electric field may be affecting the results at long times.

To confirm that H_2 gas was evolved as a result of the hydrogen evolution reaction, mass spectrometry was carried out. Glass or stainless-steel flasks were coupled to the outlet port of the cell to collect the gas effluent, which presumably consisted of Ar from the microplasma and the generated H_2 and O_2 gas. The captured gas was then injected into a mass spectrometer which was tuned to detect H_2 and its isotopes. Figure 12.8 shows mass chromatograms for the H_2 signal obtained from various gas samples that were collected; the time corresponds to how long it took for the sample to reach the detector after injection ($t = 0\,s$). A clear signal at approximately 250 s could be detected that confirmed the presence of H_2 gas in the injected samples. A control sample was collected with the microplasma off to ensure that this signal was real; sample (i) shows a clean baseline with no peak at 250 s. In comparison, a signal corresponding to H_2 gas was detected for all the gas samples (ii)–(v) collected with the microplasma running and the intensity of the signal increased with collection time, indicating the continuous generation of H_2 gas which is in agreement with pH measurements. Although other mechanisms for H_2 generation could not be ruled out such as, for example, dissociation of water vapor that is present in the discharge and subsequent formation of H_2 by radical recombination in the gas or liquid phase, the pH measurements suggest that the mechanism for H_2 formation is water electrolysis.

In summary, model reactions such as ferricyanide reduction and the hydrogen evolution reaction confirm that a plasma electrode is capable of transferring electrons and inducing electrochemical reactions in solution. The reaction rates are found to depend intimately on discharge parameters (e.g., discharge current) which control the electron density (or flux) and energy distribution of electrons in the plasma. The absence of a solid electrode and the ability to tune the electron density and electron energy in the plasma open a new direction for electrochemistry based on reactions between gas-phase electrons and ionic electrolytes, including NP synthesis. In addition, these reactions can be performed at the surface of thin films, the topic of the following section.

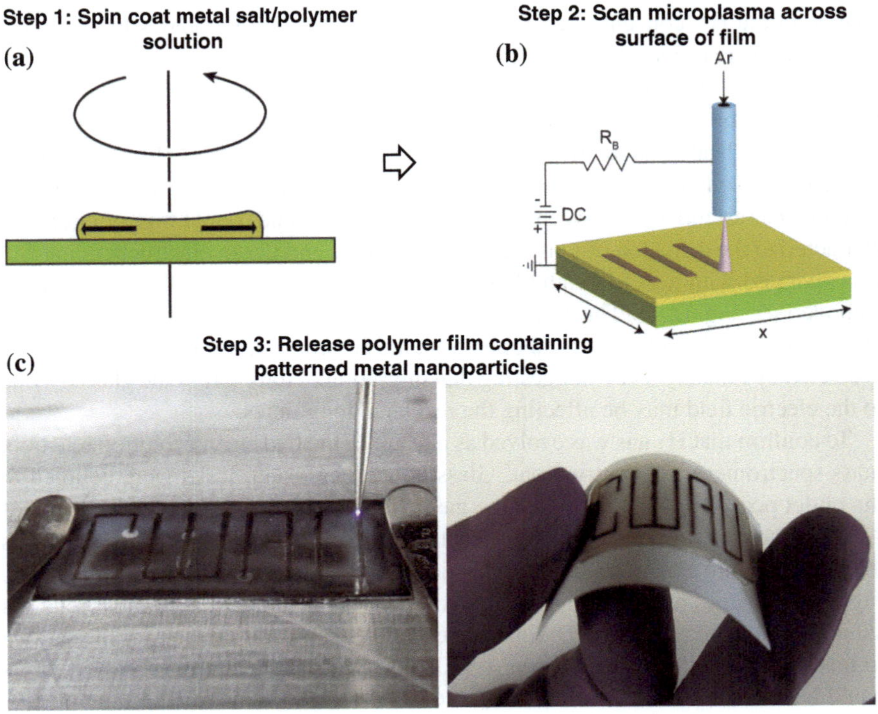

Fig. 12.9 Stepwise flow diagram of direct-write approach based on atmospheric pressure microplasmas: a solution containing metal salts and polymer is spin coated (*Step 1*) and exposed to a scanning microplasma (*Step 2*) to generate a patterned film of metal NPs. The film can be released from the substrate to obtain a free-standing film of patterning metal NPs (*Step 3*). Modified with permission from Ref. [49], copyright 2011 Wiley-VCH

12.4 Plasma Electrochemical Reactions at the Surface of Thin Films

Technological applications of metal NPs in sensors [34, 35], magnetic data storage [36], flash memory [37], and catalysis [38] to name a few, require their assembly into films and other ordered macroscopic structures. The combination of plasma electrochemistry and patterning techniques has the potential to create a new chemical approach for micro/nanofabrication that is potentially low cost and scalable. Here, three strategies for micro/nanofabrication based on plasma electrochemistry are presented: (a) direct writing based on a scanning microplasma, (b) lithographic pattern transfer using an extracted discharge, and (c) directed assembly from metallosupramolecular polymers.

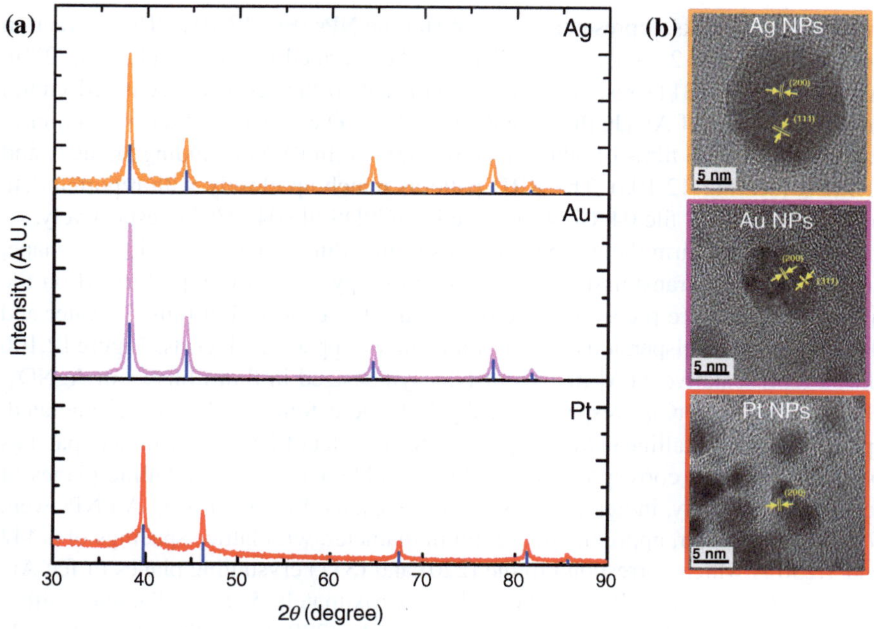

Fig. 12.10 **a** XRD spectra and **b** corresponding representative TEM images of Ag, Au, and Pt NPs synthesized by microplasma reduction of $AgNO_3$, $HAuCl_4$, and H_2PtCl_6, respectively. Reprinted with permission from Ref. [49], copyright 2011 Wiley-VCH

12.4.1 Direct Writing Based on Scanning Microplasma

The ability to initiate and control electrochemical reactions, such as the reduction of aqueous metal ions, with a gaseous (i.e., plasma) electrode enables unique applications of microplasmas. The same reactions can be carried out at a gas–solid (thin film) interface, and if the reduction is performed locally, patterns of metal NPs are generated by scanning the microplasma. Compared to forming particles in solution, the metal NPs are immobilized on a substrate for direct use in applications. We have termed this process "direct writing" since the metal NPs are produced as a pattern in a single step, analogous to other non-lithographic patterning techniques [39]. Figure 12.9 illustrates a stepwise diagram for the process. First, an aqueous solution of metal salt and polymer is spin coated onto a Si substrate. The dried film is then exposed to the microplasma to reduce the metal ions and form metal NPs. By scanning the microplasma, arbitrary patterns of metal NPs such as "**CWRU**" are fabricated. Finally, the film is released from the Si substrate to produce a free-standing, flexible polymer film containing patterned metal NPs.

The reduction of metal precursors to crystalline metal in thin films was evaluated by X-ray diffraction (XRD). A representative XRD spectrum for films of silver nitrate and polyvinyl alcohol ($AgNO_3$/PVA) after microplasma reduction shows several intense, broad peaks, representative of crystalline NPs (Fig. 12.10a). The diffraction peaks at 38.1°, 44.2°, 64.4°, and 77.4° can be ascribed to the Ag(111), Ag(200), Ag(220) and Ag(311) respectively, and indexed to the fcc (face centered cubic) crystalline phase of Ag (JCPDS file 87–0597). XRD characterization was similarly carried out for thin films of other metal precursors (in PVA) including $HAuCl_4$ and H_2PtCl_6 (see Fig. 12.10a). The XRD patterns match up closely with reported data for fcc Au (JCPDS file 04–0784) and fcc Pt (JCPDS file 04–0802), respectively.

In order to confirm the presence of NPs in the films and assess their size, shape, and morphology, transmission electron microscopy (TEM) was performed on the films. Samples were prepared by re-dissolving the reduced thin films in water and drop casting the dispersions on carbon-coated copper TEM grids. Figure 12.10b shows representative TEM images of NPs synthesized in thin films from $AgNO_3$, $HAuCl_4$, and H_2PtCl_6, respectively. Ag NPs were found to be unagglomerated, spherical, and crystalline with an approximate diameter of 20 nm. The lattice spacings observed by TEM correspond well with the (111) and (200) crystalline planes of fcc Ag, respectively, in agreement with XRD results. In comparison, Au NPs were found to be smaller, approximately 5 nm in diameter, with lattice spacings of 1.442 and 1.230 Å which correspond to the (220) and (311) crystalline planes of fcc Au, respectively. Similarly, Pt NPs were also approximately 5 nm in diameter with a lattice spacing of 1.388 Å which corresponds to the (220) crystalline plane of fcc Pt. The smaller particle sizes for Au and Pt suggests that the rate of reduction and/or particle nucleation and growth is slower for these metals than Ag.

Plasmas are characterized by a complex physicochemical environment consisting of atoms and molecules, charged species (e.g., ions and electrons), ultraviolet (UV) and vacuum UV (VUV) radiation, and hot gases that can transfer thermal energy. Metal salts in polymers have been previously reduced by many different mechanisms including electron [40–42], UV [43, 44] or VUV irradiation [45], and heating [46–48]. Although the reduction process has been successfully carried out by a microplasma, the underlying mechanism for reduction remains unclear. To address this issue, a simple control experiment was designed. As shown in Fig. 12.11, the reduction of the polymer film containing metal salt was performed with and without a grounded metal (stainless steel) mesh that could block the electrons from the microplasma. UV light and hot gases were still able to penetrate through the mesh and interact with the film. The experiments show that metal NPs are not formed when the grounded mesh is used, suggesting that electrons in the microplasma are responsible for reduction [28], which is consistent with plasma-liquid experiments.

In summary, a direct-write approach to produce patterned films of metal NPs based on plasma electrochemistry has been successfully developed. While the resolution of this process is limited by the size of the microplasma ($\sim$150 μm), ultrahigh vacuum and complex instrumentation are not required. The pattern resolution can be improved by incorporating a mask, which is the focus of the next section.

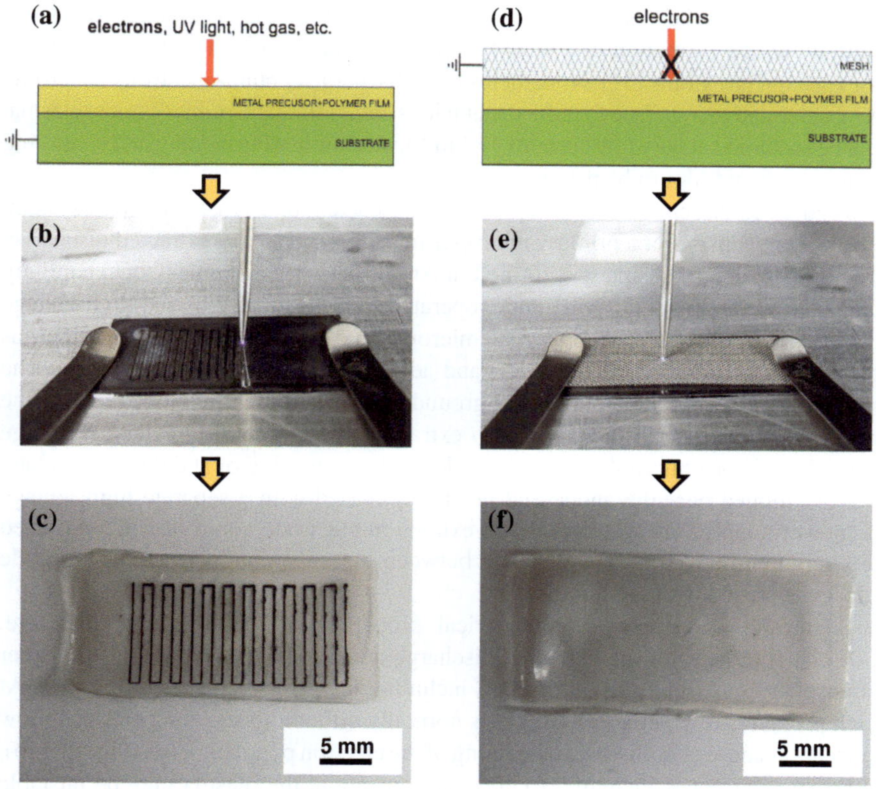

Fig. 12.11 A control experiment is demonstrated: **a–c** the microplasma is formed directly with the film to reduce the metal salt precursor to crystalline metal NPs; **d–f** a grounded mesh is placed between the microplasma and the film to block the electrons. As a result, the film shows no change (i.e., no reduction of the metal salt precursor). Reprinted with permission from Ref. [49], copyright 2011 Wiley-VCH

12.4.2 Lithographic Pattern Transfer Using Extracted Discharge

A potential drawback of the direct write approach [49] is that direct exposure of a polymer film to the microplasma could result in damage due to heating and sputtering. In addition, the resolution of direct writing is currently limited to ~150 μm, because of the size of the microplasma source which is limited by the size of the electrode. Additional confinement of the microplasma could reduce its footprint and allow finer patterns to be fabricated. Alternatively, the microplasma process could be combined with lithographic techniques to block and expose a smaller region of the film. However, heating and sputtering by the microplasma could again damage the mask. To address this issue, an extracted discharge has been developed. The idea is based on the generation of electron beams from hollow cathode discharges where a positively-biased electrode is used to extract and accelerate electrons [50]. A key difference is that the electrons are extracted at higher pressures, which leads to ionization and the formation of a weakly ionized discharge due to collisions with the background gas atoms [31]. Nonetheless, the discharge contains electrons which can be accelerated to a surface to reduce metal ions and, unlike the primary microplasma, can be sustained at very low currents (<1 mA), thus mitigating heating and sputtering of the mask and film to be patterned.

Figure 12.12a schematically depicts the low-current discharge extraction system. The system consists of a primary microplasma and a secondary extracted discharge. The primary microplasma was formed in Ar gas between a stainless steel capillary tube cathode and a metal grid anode, operated by a negatively-biased high voltage direct current (dc) power supply. The microplasma current was controlled after gas breakdown by a ballast resistor (R_1) and adjusting the power supply voltage. The resistor (R_2) between the anode and ground was used to continuously monitor the microplasma current ($I_{microplasma}$). To extract a low-current, secondary discharge from the primary microplasma, a third electrode, consisting of stainless steel plate, was positioned near the anode and positively biased with a separate high voltage dc power supply. The voltage on the extraction electrode controlled the extracted discharge current (I_{ed}). The distance between the anode and extraction electrode (d_{ee}) was varied with a micrometer.

Figure 12.12b–d shows the electrical properties of the extracted discharge. I–V characteristics of the extracted discharge are significantly different than other atmospheric-pressure plasma sources, including the primary microplasma [51]. At high pressures (e.g., atmospheric), it is normally difficult to sustain a plasma at low currents because of collisional quenching of the electron population (see Fig. 12.12b). When the current is increased to enhance ionization, the plasma may be unstable because of the glow-to-arc transition [31], and gas heating and sputtering become problematic. In comparison, the extracted discharge is sustainable at atmospheric pressure and low currents, as confirmed by the uninterrupted, nonlinear region of the I–V curves in Fig. 12.12c. At higher discharge currents, the I–V curve indicates complete gas breakdown, resulting in a discharge with similar properties to other atmospheric-pressure plasmas. The extracted discharge system is particularly easier to control than the microplasma in the low current regime which is characterized by weak ionization of the background gas, and has the potential to reduce heating and sputtering of an exposed thin film.

To demonstrate lithographic pattern transfer with the extracted discharge system, porous alumina templates were placed on $AgNO_3$/PVA thin films. After exposing the masked film to the extracted discharge, the film was rinsed in deionized water to remove the unexposed polymer (Fig. 12.13a) [33]. Scanning electron microscopy (SEM) images of patterned films are shown in Fig. 12.13b–d. For the 150 nm template, patterned Ag with dimensions of 150 nm are visible after 10 min of exposure (Fig. 12.13b). A longer exposure time (30 min) leads to a higher density of patterned Ag and a small increase in the pattern resolution (Fig. 12.13c). Smaller scale patterns were achieved by using lower currents; patterns with sizes as small as 150 nm were successfully transferred (Fig. 12.13d). However, the density of patterned Ag was significantly lower. This may be caused by the inability of electrons to pass through the

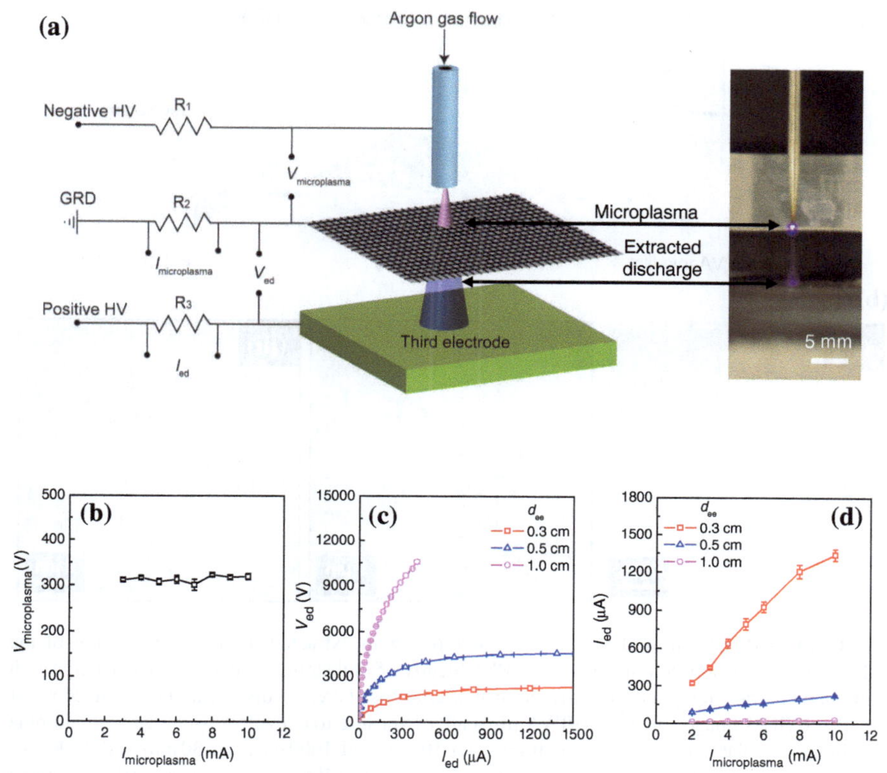

Fig. 12.12 **a** Schematic diagram of the extracted discharge system consisting of the microplasma and a positively-biased third electrode. The resistor network was used to measure the I–V characteristics of the microplasma ($I_{\text{microplasma}}$, $V_{\text{microplasma}}$) and the extracted discharge (I_{ed}, V_{ed}). Photo corresponds to an extracted discharge current of 800 μA and an extraction electrode gap distance (d_{ee}) of 0.5 cm. **b** I–V characteristics of the microplasma. **c** I–V characteristics of the extracted discharge at various extraction electrode gaps (d_{ee}) [microplasma current, $I_{\text{microplasma}} = 4$ mA]. **d** Extracted discharge current as a function of microplasma current at various extraction electrode gaps (d_{ee}) [extraction voltage, $V_{\text{ed}} = 3.1$ kV]. Reprinted with permission from Ref. [52], copyright 2012 American Vacuum Society

pores at low current. Future studies are planned to relate the discharge properties and mask geometry to pattern transfer.

In summary, the generation of a low-current, weakly ionized, extracted discharge allows lithographic pattern transfer via plasma electrochemistry. Compared to direct writing, direct interaction of the "soft" polymeric samples with the plasma is avoided, and masking reduces the dimension of the transferred patterns by approximately 3 orders of magnitude (150 μm vs. 150 nm). Further reductions in the size of the transferred pattern may be possible by self/directed assembly approaches, discussed in the next section.

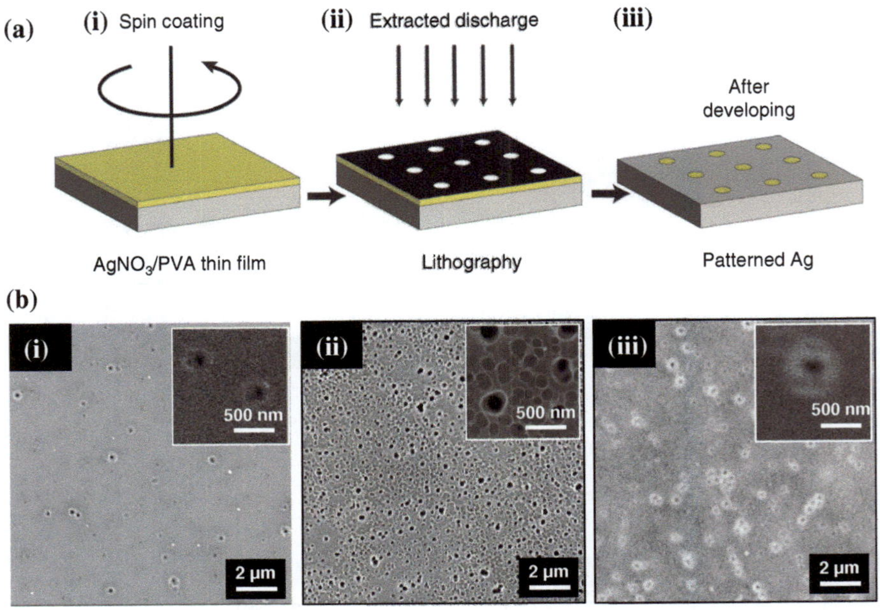

Fig. 12.13 **a** Process flow diagram for lithography with extracted discharge consisting of the following steps: (*i*) spin coating of $AgNO_3$/PVA blend on Si substrate, (*ii*) masking of discharge with alumina membrane template, and (iii) removal of unexposed area by dissolving in deionized water for 30 s. **b** SEM images of $AgNO_3$/PVA films after exposure to extracted discharge with alumina membrane template at following conditions: (*i*) 10 min and 1000 µA, (*ii*) 30 min and 1000 µA, and (*iii*) 30 min and 200 µA. Reprinted with permission from Ref. [52], copyright 2012 American Vacuum Society

12.4.3 Directed Assembly from Metallosupramolecular Polymers

Lithographic approaches have allowed a reduction in the dimension of patterns produced by plasma electrochemistry [52]. However, the pattern size remains limited by the mask. Recently, alumina pores [53] or block copolymers [54, 55] have been fabricated by chemical methods and characterized by features as small as 1 nm. To produce patterned metal nanostrucututres, the nanoscale domains in these templates can be filled with a metal precursor such as $HAuCl_4$ and reduced to form metal NPs by plasma exposure [56], chemical reduction [57], or thermal treatment [58].

Alternatively, metal-containing polymers [59–62] where the metal is an intricate component of the polymer backbone are interesting candidates for templating since the metal is already present and does not need to be separately loaded into the polymer template. These hybrid inorganic/organic films are characterized by the processability of polymers and the functionalities of the metal which allows pre-assembled metal/polymer films to be obtained by phase separation of the inorganic and organic constituents. A subclass of this field is metallosupramolecular polymers [63, 64] which can be prepared using a wide variety of metal-ions and polymers [65–68].

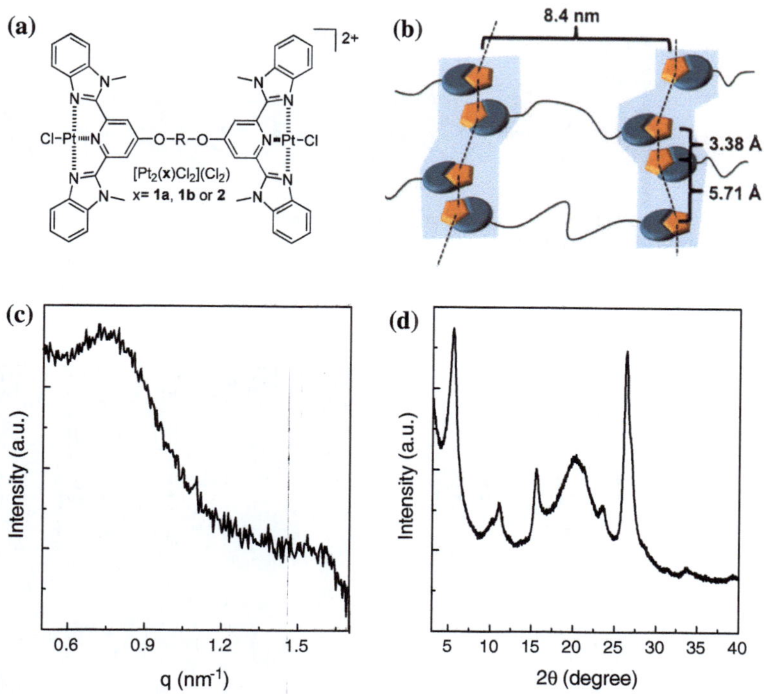

Fig. 12.14 Metallosupramolecular polymers as templates for NP formation. **a** Chemical structure of Pt-containing supramolecular polymers $[Pt_2(\mathbf{1a})Cl_2](Cl_2)$. Average molecular weight (Mn) is 4200 g/mol (**1a**). **b** Schematic of the proposed morphology of $[Pt2(\mathbf{1a})Cl_2](Cl_2)$ suggested by **c** small-angle X-ray scattering (SAXS) diffractogram and **d** wide-angle X-ray scattering (WAXS) diffractogram. Reprinted with permission from Ref. [69], copyright 2012 American Chemical Society

For example, Pt-containing supramolecular polymer adopt lamellar morphologies [69] in the solid state as a result of phase separation of the hard ionic, metal–ligand complexes and the soft polymer core (Fig. 12.14).

Here, metal NP formation from metallosupramolecular polymer films is carried out by exposing to an extracted discharge (Fig. 12.15a). Experiments were carried out with carbon-coated Cu TEM grids as the substrate in order to image the particles immediately after they were formed. Solutions of Pt-containing supramolecular polymer were spin-coated onto the TEM grids at 2000 rpm for 60 s and dried in vacuum oven at 40 °C overnight. The TEM grids were placed on the third electrode of the extracted discharge setup, and treated at an extracted discharge current of 200 μA for different times (30, 60, and 180 min). The solutions and films were kept covered with aluminum foil at all times, before and after plasma reduction, in a desiccator at −10 °C to avoid reaction with light or water. Figure 12.15b shows TEM images of the as-deposited (before plasma reduction) Pt-containing supramolecular film and the same film after plasma reduction at 0 μA for 1 h (inset). No NPs are observed confirming that particle reduction does not occur from our preparation and storage

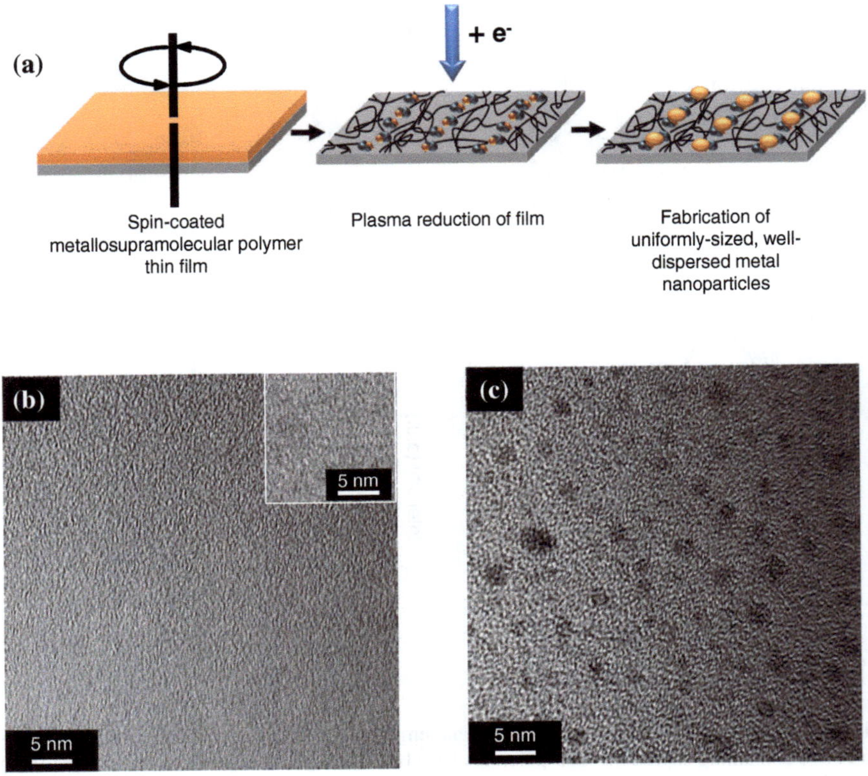

Fig. 12.15 **a** Process flowchart for fabrication of metal NPs from metallosupramolecular polymer templete. TEM images of $[Pt_2(\mathbf{1a})Cl_2](Cl_2)$ before and after plasma reduction for the following exposure times **b** as-deposited and **c** 0.5 h. The extracted discharge was operated at a constant current of 200 μA. *Inset* of **a** shows a high-resolution TEM image of the film after exposing to 0 μA for 1 h. Reprinted with permission from Ref. [69], copyright 2012 American Chemical Society

of the material. In comparison, NPs could be clearly observed after plasma reduction at 200 μA for 30 min (Fig. 12.15c). The NPs are unagglomerated, spherical, and crystalline, with highly uniform size.

In order to further understand the mechanism for NP formation by plasma reduction of Pt-containing supramolecular polymers, two perturbations were explored: (1) the molecular weight of the polymer backbone, which impacts the Pt concentration or loading in the resulting films, was varied from $M_n = 4200$ g/mol ("**1a**") to $M_n = 9600$ g/mol ("**1b**"), and (2) the nature of the polymeric backbone in the macromonomer was varied from a relatively polar poly(THF) backbone ("**1a**" and "**1b**") to a non-polar hydrocarbon backbone ("**2**"). Representative TEM images of various Pt-containing supramolecular polymer films after plasma reduction for 30 min and 3 h are shown in Fig. 12.16. Interestingly, both an increase in the molecular weight of the backbone or switching to a less polar backbone lead to an observable decrease in the size of the formed Pt particles.

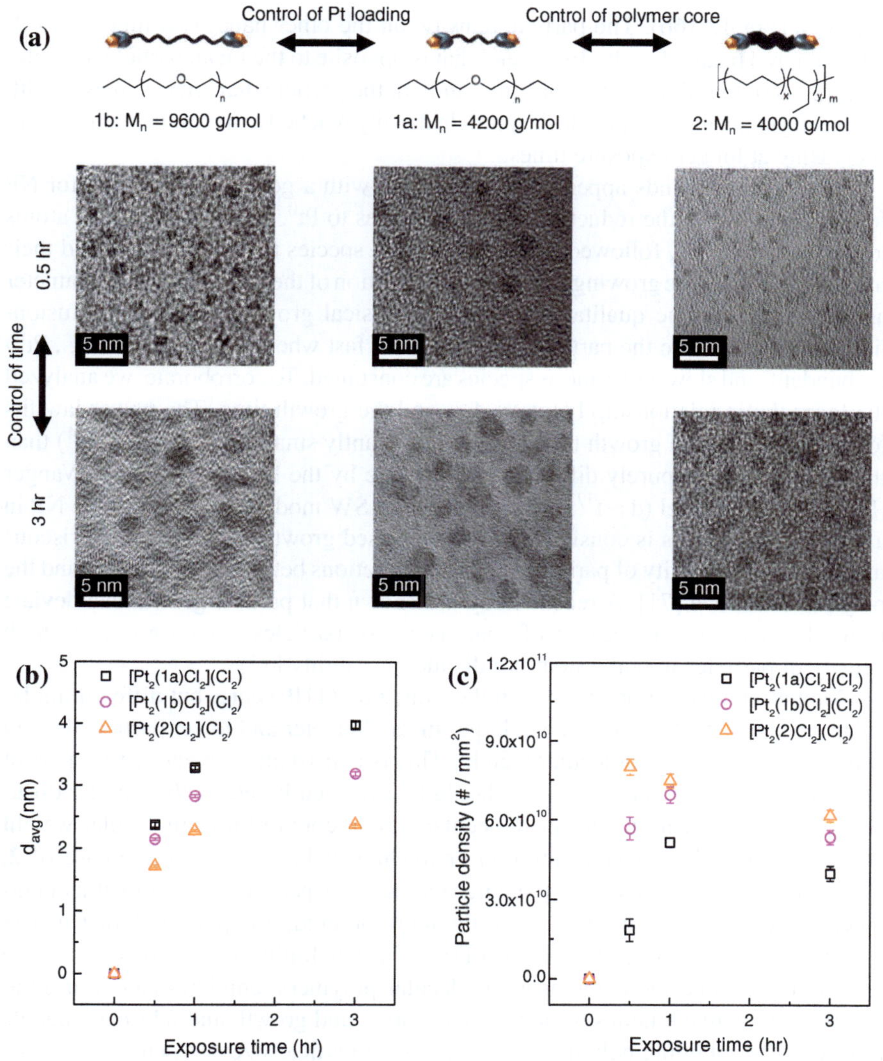

Fig. 12.16 **a** Schematic of potential approaches to control nucleation and growth via macromonomer weight (i.e., loading of Pt) or macromonomer core, and TEM images of the respective films after plasma reduction. **b** Average particle diameter, d_{avg}, as a function of time. **c** Particle density as a function of time. d_{avg} and particle density were obtained from a statistical analysis of TEM measurements of the various samples. *Error bars* correspond to the standard error. Reprinted with permission from Ref. [69], copyright 2012 American Chemical Society

From measurements of particles in the TEM images, a statistical analysis of the particle size and particle density were carried out for the various metallosupramolecular polymers (Fig. 12.16b and c). At any given exposure time, Pt NPs synthesized from **1a** were found to have the largest mean diameter, followed by **1b**, and, finally, **2** (Fig. 12.16b). The particle density, on the other hand, was highest for **2**, followed by **1b**, and, finally, **1a**, a trend that is opposite to the mean particle diameter (Fig. 12.16c). For all the polymer films studied, the particle diameter increased with increasing exposure time, while the particle density reached a maximum value before decreasing at longer exposure times.

Overall, these trends appear to be consistent with a general mechanism for NP formation based on the reduction of Pt complexes to Pt^0, clustering of these atoms to nucleate particles, followed by diffusion of Pt species through the film and their incorporation into the growing particle. The evolution of the average particle diameter as a function of time qualitatively follows classical growth models of diffusion-limited growth where the particles initially grow fast when the precursor (e.g., Pt^0) is abundant, and slow with time as species are consumed. To corroborate, we analyzed the logarithmic relationship between d_{avg} and the growth time. The power law fits to our data indicated growth trajectories significantly smaller (ca. $d_{avg} < t^{1/4}$) than that predicted for a purely diffusion-limited case by the Lifshitz–Slyozov–Wanger (LSW) growth model ($d \sim t^{1/3}$) [70]. Since the LSW model assumes a single NP in a solution phase, this is consistent with suppressed growth due to a highly viscous medium, a high density of particles, and/or interactions between the particles and the polymer backbone [71]. A recent study has shown that particle growth can deviate from classical models because of coalescence of particles, which given the high density of particles in our system is a distinct possibility [72].

Comparing macromonomers with the same poly(THF) core, but different molecular weights, i.e., **1a** versus **1b**, a larger mean diameter and lower particle density was obtained at lower molecular weight. This is consistent with easier transport of the Pt cations or Pt^0 species which should be expected based on the smaller diffusion distances present in a lower molecular weight core (a lower molecular weight also results in a higher concentration or loading of Pt^{2+} ions). In the case of **2**, the reason a higher density and smaller diameter of particles are formed as compared to **1a** or **1b** may be because of the more ordered, less polar polymer matrix [poly(ethylene-co-butylene) vs. poly(THF)] which limits atom diffusion. These results suggest that the metallosupramolecular polymers templates can be used to control the relative kinetics of particle nucleation and growth and fabricate tunable films of NP-containing polymer films in terms of particle size and density.

In summary, exposure of metallosupramolecular polymer films to the extracted discharge results in reduction of the metal component and the formation of highly-uniform, crystalline, well-dispersed metal NPs. Particle size and density are found to be controlled by exposure time, molecular weight of the polymer core, and polarity of the polymer core. The molecular level control over the formation of metal NPs in a polymer matrix is potentially attractive for applications where the properties of the nanocomposites materials must be tuned, such as flexible electronics.

12.5 Applications of Plasma Electrochemistry to H_2O_2 Detection

Iridium (Ir) NPs have attracted considerable interest for H_2O_2 detection due to their high conductivity and high electrocatalytic activity [73]. Despite this promise, synthesis of Ir NPs remains a challenge. Ir NPs have been typically prepared by chemical reduction of an Ir salt, such as $IrCl_3$, in solution by a strong reducing agent such as sodium borohydride ($NaBH_4$), hydrazine, or dimethyl formamide, that pose serious environmental and biological risks, the latter of which could be a problem for in vivo applications [74, 75]. Plasma electrochemistry offers an alternative and green synthetic route for the synthesis of Ir NPs. Here, we study the synthesis of Ir NPs by plasma reduction and show that it compares favorably with Ir NPs synthesized by chemical reduction for electrocatalytic detection of H_2O_2.

Ir NPs were synthesized by chemical reduction using a standard procedure involving $NaBH_4$ [76, 77]. Briefly, a solution containing 1.8 mM $IrCl_3$ and 1 mM citric acid was titrated to pH = 7.0 by adding 100 mM NaOH dropwise. Then, 10 mM $NaBH_4$ was slowly added dropwise until the amount of electrons released by the $NaBH_4$ reaction with water was ~1.4 times the theoretical value required to completely reduce Ir^{3+}. To synthesize Ir NPs in solution by plasma reduction, the solution containing 1 mM $IrCl_3$ in deionized water and 1 mM of citric acid as a stabilizer was prepared and a microplasma was operated at a constant current of 4 for 1 h.

The chemical composition of the NPs was assessed by X-ray photoelectron spectroscopy (XPS). Figure 12.17a and b show survey spectra for Ir NPs synthesized by chemical reduction and plasma reduction, respectively. In addition to peaks corresponding to Ir, C, and O, Na is observed in the spectrum collected from Ir NPs synthesized by chemical reduction. The Na is most probably residual contamination from the reducing agent, $NaBH_4$, or the NaOH that is used to adjust the pH of the solution during synthesis. In comparison, the spectrum for Ir NPs synthesized by plasma reduction does not show any metallic impurities. Additional XPS analysis was performed by high-resolution scans from 58 to 70 eV of the samples which includes the Ir 4f and Na 2s regions. All spectral peaks were de-convoluted, background subtracted using the Shirley method, and fitted by Gaussian–Lorentzian functions. The Ir 4f region consists of the $4f_{7/2}$ and $4f_{5/2}$ peaks at 61.7 and 64.7 eV, respectively, which can be assigned to metallic Ir (Ir^0), and the $4f_{7/2}$ and $4f_{5/2}$ peaks at 62.7 and 65.6 eV, respectively, which can be assigned to IrO_2 (Ir^{4+}) [78, 79]. The Na 2s peak is located at 63.75 eV [80]. Peaks corresponding to IrO_2 are observed for the Ir NPs synthesized by both chemical reduction and plasma reduction; the oxidation supports the O 1s peaks observed in the survey spectra. A key difference between the samples is the large peak corresponding to Na in the spectrum for the chemically-reduced sample. The Na was present despite washing and centrifuging of the sample, suggesting that Na is adsorbed on the surface of or incorporated in the bulk of the Ir NPs synthesized by chemical reduction.

Previously, plasma electrochemistry has been carried on thin films to directly reduce dispersed metal salts to supported metal NPs [52, 69]. As proof-of-concept, an Ir NP-based electrode was fabricated in a single step by this method and its electro-

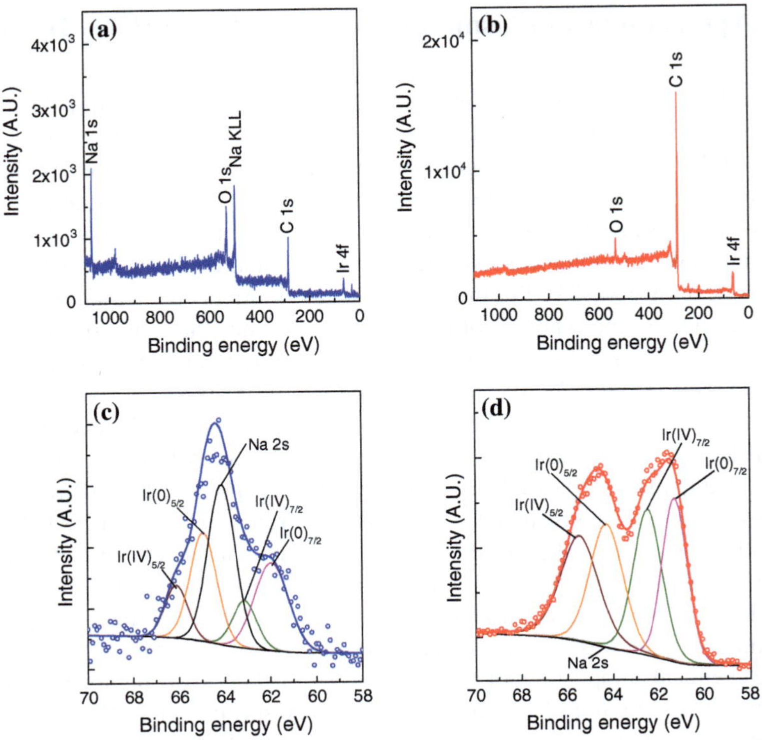

Fig. 12.17 XPS survey spectra of Ir NPs synthesized by **a** chemical reduction and **b** plasma electrochemical reduction, and high resolution spectra of the Ir 4f regions of Ir NPs synthesized by **c** chemical reduction and **d** plasma reduction. Reprinted with permission from Ref. [77], copyright 2013 Elsevier B. V

catalytic activity toward H_2O_2 oxidation was evaluated. Figure 12.18a schematically illustrates the following steps that were carried out for fabrication and testing: (1) an $IrCl_3$/C ink was prepared and drop cast on a PET substrate, (2) the thin film was exposed to a remote atmospheric-pressure plasma, and (3) the amperometric response of the free-standing, flexible electrode was evaluated. A photo of the flexible Ir NP-based electrode is shown in Fig. 12.18b. Figure 12.18c shows a clear difference in the real-time amperometric response to 2 mM H_2O_2 measured at 0.6 V before and after plasma reduction, confirming the presence of electrocatalytically active Ir NPs.

In summary, plasma electrochemistry has the potential for the synthesis of NPs with uniform size, crystallinity, and high purity. In addition, the ability to fabricate thin films of NPs in a single step is attractive for roll-to-roll manufacturing of flexible sensors and electronics, including electrocatalytic materials for the detection of H_2O_2.

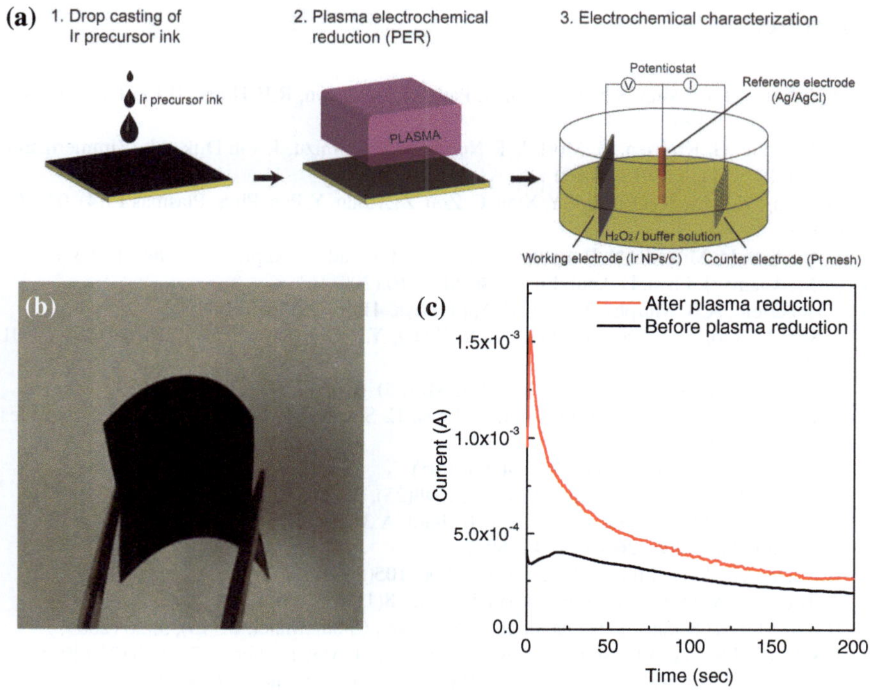

Fig. 12.18 **a** Schematic illustration of steps for single-step fabrication of Ir NP-based electrode for H_2O_2 sensing via exposure of drop cast $IrCl_3$/C thin film to remote atmospheric-pressure plasma setup. **b** Photo of the free-standing, flexible Ir NP-based electrode after plasma reduction, confirming the fabrication of electrocatalytically active Ir NPs for H_2O_2 detection. **c** Real-time amperometric response of the Ir NP-based electrode fabricated in a single step to H_2O_2(2 mM). The applied potential was +0.6 V (vs. Ag/AgCl). Reprinted with permission from Ref. [77], copyright 2013 Elsevier B. V

12.6 Summary and Conclusions

Microplasmas operate stably and non-thermally at atmospheric pressure and are a source of ions, electrons, and other electronically excited states at ambient conditions. Overall, these features allow microplasmas to be formed at the surface of aqueous solutions and polymeric thin films to serve as a gaseous electrode for electrochemical synthesis. In particular, the synthesis of metal NPs by electrochemical reduction of metal salts dispersed in solution or in a polymer film opens up a new direction for the fabrication of colloidal and patterned nanomaterials. The advantages of plasma electrochemistry can be summarized as the following: (1) the process is atmospheric pressure and room temperature, (2) the process is high purity and environmentally friendly (no H_2 gas or other chemicals are required to reduce the metal ions), and (3) the process has potential for low-cost, large-scale (roll-to-roll) micro/nanofabrication.

References

1. A. Schutze, J.Y. Jeong, S.E. Babayan, J. Park, G.S. Selwyn, R.F. Hicks, IEEE T Plasma Sci. **26**(6), 1685 (1998)
2. M.G. Kong, G. Kroesen, G. Morfill, T. Nosenko, T. Shimizu, J. van Dijk, J.L. Zimmermann, New J. Phys. **11**, 115012 (2009)
3. Q. Xiong, X.P. Lu, K. Ostrikov, Y. Xian, C. Zou, Z. Xiong, Y. Pan, Phys. Plasmas **17**(4), 043506 (2010)
4. U. Cvelbar, M. Mozetic, N. Hauptman, M. Klanjsek-Gunde, J. Appl. Phys. **106**(103303) (2009)
5. R. McAdams, J. Phys. D Appl. Phys. **34**(18), 2810 (2001)
6. J. Heberlein, A.B. Murphy, J. Phys. D Appl. Phys. **41**(5), 053001 (2008)
7. J. Sun, L. Yao, Z.Q. Gao, S.J. Peng, C.X. Wang, Y.P. Qiu, Surf. Coat. Tech. **204**(24), 4101 (2010)
8. L. Bardos, H. Barankova, Thin Solid Films **518**(23), 6705 (2010)
9. T. Desmet, R. Morent, N. De Geyter, C. Leys, E. Schacht, P. Dubruel, Biomacromolecules **10**(9), 2351 (2009)
10. K. Ostrikov, Rev. Mod. Phys. **77**(2), 489 (2005)
11. I. Denysenko, K. Ostrikov, Appl. Phys. Lett. **90**(25), 251501 (2007).
12. R. Akolkar, R.M. Sankaran, J. Vac. Sci. Technol. A **31**(5), 050811 (2013)
13. J. Gubkin, Ann. Phys. **268**(9), 114 (1887)
14. A.R. Denaro, A. Hickling, J. Electrochem. Soc. **105**(5), 265 (1958)
15. A. Hickling, M.D. Ingram, J. Electroanal. Chem. **8**(1), 65 (1964)
16. H. Furusho, K. Kitano, S. Hamaguchi, Y. Nagasaki, Chem. Mater. **21**(15), 3526 (2009)
17. I.G. Koo, M.S. Lee, J.H. Shim, J.H. Ahn, W.M. Lee, J. Mater. Chem. **15**(38), 4125 (2005)
18. K. Furuya, Y. Hirowatari, T. Ishioka, A. Harata, Chem. Lett. **36**(9), 1088 (2007)
19. J. Hieda, N. Saito, O. Takai, J. Vac. Sci. Technol. A **26**(4), 854 (2008)
20. Y.B. Xie, C.J. Liu, Plasma Process. Polym. **5**(3), 239 (2008)
21. Y. Shimizu, K. Kawaguchi, T. Sasaki, N. Koshizaki, Appl. Phys. Lett. **94**(19), 191504 (2009)
22. S. Bhattacharyya, D. Staack, E.A. Vitol, R. Singhal, A. Fridman, G. Friedman, Y. Gogotsi, Adv. Mater. **21**(40), 4039 (2009)
23. T. Kaneko, K. Baba, R. Hatakeyama, Plasma Phys. Contr. Fus. **51**(12), 124011 (2009).
24. M.T. Reetz, W. Helbig, S.A. Quaiser, U. Stimming, N. Breuer, R. Vogel, Science **267**(5196), 367 (1995)
25. M.T. Reetz, W. Helbig, J. Am. Chem. Soc. **116**(16), 7401 (1994)
26. L. Rodriguez-Sanchez, M.C. Blanco, M.A. Lopez-Quintela, J. Phys. Chem. B **104**(41), 9683 (2000)
27. C.J. Huang, P.H. Chiu, Y.H. Wang, K.L. Chen, J.J. Linn, C.F. Yang, J. Electrochem. Soc. **153**(12), D193 (2006)
28. J.J. Zou, Y.P. Zhang, C.J. Liu, Langmuir **22**(26), 11388 (2006)
29. C. Richmonds, M. Witzke, B. Bartling, S.W. Lee, J. Wainright, C.C. Liu, R.M. Sankaran, J. Am. Chem. Soc. **133**(44), 17582 (2011)
30. T.H. Huang, G. Salter, S.L. Kahn, Y.M. Gindt, J. Chem. Educ. **84**(9), 1461 (2007)
31. R.H. Stark, K.H. Schoenbach, J. Appl. Phys. **85**(4), 2075 (1999)
32. H. Tributsch, L. Pohlmann, Science **279**(5358), 1891 (1998)
33. M. Witzke, P. Rumbach, D.B. Go, R.M. Sankaran, J. Phys. D Appl. Phys. **45**(44), 442001 (2012)
34. D.O. Shin, J.R. Jeong, T.H. Han, C.M. Koo, H.J. Park, Y.T. Lim, S.O. Kim, J. Mater. Chem. **20**(34), 7241 (2010)
35. A.J. Haes, W.P. Hall, L. Chang, W.L. Klein, R.P. Van Duyne, Nano Lett. **4**(6), 1029 (2004)
36. K. Naito, H. Hieda, M. Sakurai, Y. Kamata, K. Asakawa, IEEE T Magn. **38**(5), 1949 (2002)
37. J.S. Lee, J. Cho, C. Lee, I. Kim, J. Park, Y.M. Kim, H. Shin, J. Lee, F. Caruso, Nat. Nanotechnol. **2**(12), 790 (2007)
38. J.X. Huang, A.R. Tao, S. Connor, R.R. He, P.D. Yang, Nano Lett. **6**(3), 524 (2006)

39. S.W. Lee, R.M. Sankaran, Mater. Today **16**(4), 117 (2013)
40. R. Abargues, J. Marques-Hueso, J. Canet-Ferrer, E. Pedrueza, J.L. Valdes, E. Jimenez, J.P. Martinez-Pastor, Nanotechnology **19**(35), 355308 (2008)
41. E.U. Donev, J.T. Hastings, Nano Lett. **9**(7), 2715 (2009)
42. K.A. Bogle, S.D. Dhole, V.N. Bhoraskar, Nanotechnology **17**(13), 3204 (2006)
43. M.K. Abyaneh, D. Paramanik, S. Varma, S.W. Gosavi, S.K. Kulkarni, J. Phys. D Appl. Phys. **40**(12), 3771 (2007)
44. Y. Zhou, C.Y. Wang, Y.R. Zhu, Z.Y. Chen, Chem. Mater. **11**(9), 2310 (1999)
45. J.Z. Li, K. Kamata, S. Watanabe, T. Iyoda, Adv. Mater. **19**(9), 1267 (2007)
46. J.H. Dai, M.L. Bruening, Nano Lett. **2**(5), 497 (2002)
47. X.P. Sun, Y.L. Luo, Mater Lett. **59**(29–30), 3847 (2005)
48. S. Porel, S. Singh, S.S. Harsha, D.N. Rao, T.P. Radhakrishnan, Chem. Mater. **17**(1), 9 (2005)
49. S.W. Lee, D. Liang, X.P.A. Gao, R.M. Sankaran, Adv. Funct. Mater. **21**(11), 2155 (2011)
50. J.R. Bayless, R.C. Knechtli, G.N. Mercer, IEEE J. Quantum Electron. **10**(2), 213 (1974)
51. R.M. Sankaran, K.P. Giapis, J. Appl. Phys. **92**(5), 2406 (2002)
52. S.W. Lee, H. Zamani, P.X.L. Feng, R.M. Sankaran, J. Vac. Sci. Technol. B **30**(1), 010603 (2012)
53. M.S. Sander, L.S. Tan, Adv. Funct. Mater. **13**(5), 393 (2003)
54. Q.S. Wei, Y. Lin, E.R. Anderson, A.L. Briseno, S.P. Gido, J.J. Watkins, ACS Nano **6**(2), 1188 (2012)
55. J.G. Son, W.K. Bae, H.M. Kang, P.F. Nealey, K. Char, ACS Nano **3**(12), 3927 (2009)
56. R. Glass, M. Moller, J.P. Spatz, Nanotechnology **14**(10), 1153 (2003)
57. J.P. Spatz, S. Mossmer, M. Moller, Chem.-Eur. J. **2**(12), 1552 (1996)
58. J.U. Kim, S.H. Cha, K. Shin, J.Y. Jho, J.C. Lee, Adv. Mater. **16**(5), 459 (2004)
59. G.R. Whittell, M.D. Hager, U.S. Schubert, I. Manners, Nat. Mater. **10**(3), 176 (2011)
60. M.L. Mejia, K. Agapiou, X.P. Yang, B.J. Holliday, J. Am. Chem. Soc. **131**(51), 18196 (2009)
61. J.Y. Cheng, C.A. Ross, V.Z.H. Chan, E.L. Thomas, R.G.H. Lammertink, G.J. Vancso, Adv. Mater. **13**(15), 1174 (2001)
62. R.B. Grubbs, J. Polym. Sci. Polym. Chem. **43**(19), 4323 (2005)
63. J.D. Fox, S.J. Rowan, Macromolecules **42**(18), 6823 (2009)
64. V.A. Friese, D.G. Kurth, Coord. Chem. Rev. **252**(1–2), 199 (2008)
65. R. Dobrawa, F. Wurthner, J. Polym. Sci. Polym. Chem. **43**(21), 4981 (2005)
66. U.S. Schubert, C. Eschbaumer, Angew. Chem. Int. Ed. **41**(16), 2893 (2002)
67. G.R. Whittell, I. Manners, Adv. Mater. **19**(21), 3439 (2007)
68. D.G. Kurth, M. Higuchi, Soft Matter **2**(11), 915 (2006)
69. S.W. Lee, J.R. Kumpfer, P.A. Lin, G.D. Li, X.P.A. Gao, S.J. Rowan, R.M. Sankaran, Macromolecules **45**(20), 8201 (2012)
70. I.M. Lifshitz, V.V. Slyozov, J. Phys. Chem. Solids. **19**(1–2), 35 (1961)
71. T.J. Woehl, J.E. Evans, L. Arslan, W.D. Ristenpart, N.D. Browning, ACS Nano **6**(10), 8599 (2012)
72. H.M. Zheng, R.K. Smith, Y.W. Jun, C. Kisielowski, U. Dahmen, A.P. Alivisatos, Science **324**(5932), 1309 (2009)
73. J. Shen, L. Dudik, C.C. Liu, Sens. Actuators, B-Chem. **125**(1), 106 (2007)
74. P. Raveendran, J. Fu, S.L. Wallen, J. Am. Chem. Soc. **125**(46), 13940 (2003)
75. J. Virkutyte, R.S. Varma, Chem. Sci. **2**(5), 837 (2011)
76. Y. Shimazaki, Y. Kobayashi, S. Yamada, T. Miwa, M. Konno, J. Colloid Interf. Sci. **292**(1), 122 (2005)
77. S.W. Lee, M. Janyasupab, C.C. Liu, R.M. Sankaran, Catal Today **211**, 137 (2013)
78. R.H. Horng, D.S. Wuu, L.H. Wu, M.K. Lee, Thin Solid Films **373**(1–2), 231 (2000)
79. K. Kuribayashi, Y. Fujita, H. Isige, T. Iwanuma, Mater. Sci. Eng. B-Solid **109**(1–3), 188 (2004)
80. A. Barrie, F.J. Street, J. Electron Spectrosc. **7**(1), 1 (1975)

Chapter 13
Progress in Large-Scale Ozone Generation Using Microplasmas

Jose Lopez

Abstract Large-scale ozone generation for municipal and industrial purposes has been entirely based on the creation of microplasmas or microdischarges which are a type of non-thermal or cold plasma created using a reactor known as a dielectric barrier discharge (DBD). Even though microplasma generated ozone has been in continuous use for over a hundred years especially in water treatment, recent shifts in environmental awareness and sustainability have lead to a surge of ozone generating facilities throughout the world. Along with the global expansion of this environmental cleaning application various new discoveries have emerged in the science and technology of ozone generation. This chapter will briefly describe some of the historical and recent developments of this plasma application. A more comprehensive overview of the discharge physics, plasma chemistry, and the biological oxidation capabilities of ozone will be addressed along with a review of some of the most recent breakthrough developments and challenges in large-scale ozone generation technology.

13.1 Introduction

Ever since people settled in one location for longer than a single day, sources of water have been fundamental to their choice of settlement. From the moment that these settlements began to grow beyond a few families, provisions had to be made for water supplies from springs, streams, wells, or reservoirs of rainwater. Once cities emerged, supplying water to their residents became an overarching priority. However, the prerequisite of an abundant enough water source for a civilization is only part of the necessity. The water has to be somehow treated, effectively delivered, and divided among the dependent population. As the global human population has grown, so have

J. Lopez (✉)
Department of Physics, Seton Hall University, South Orange, NJ 07079, USA
e-mail: jose.lopez1@shu.edu

M. Bonitz et al. (eds.), *Complex Plasmas*, Springer Series on Atomic, Optical, and Plasma Physics 82, DOI: 10.1007/978-3-319-05437-7_13,

the demands on water quality. The concern for water quality is strongly linked to the overall health and well-being of the local and global population which has increased the demand for water treatment processes [1]. In its broadest context water treatment can be defined as the handling of a water supply with the intent of transforming the water quality to meet specific criteria and standards [2]. In the public sector, the main intent is water purification to provide clean water of desirable drinkable quality that is further free of any water borne diseases. In the industrial sector, the primary goal is obtaining water in sufficient quantity and quality to allow for the development and continued growth of various industrial processes.

One of the well established methods for large-scale water treatment for public utilities and industrial applications has been the treatment of water supplies with ozone (O_3). Large-scale ozone generation for water treatment was the first major breakthrough application of non-equilibrium or cold plasmas. As early as 1857, Werner von Siemens discovered an efficient method for generating ozone by passing air or oxygen through an alternating current discharge tube where one of the electrode surfaces was covered with an insulating or dielectric barrier material [3]. Siemens' discovery came at the same time when the foundations of bacteriology had been recently established through the work of the French microbiologist Louis Pasteur and the investigative work of the German medical doctor Robert Koch and British medical doctor John Snow [4, 5]. It was discovered during the mid-1800s that infectious diseases like cholera and typhoid fever were caused by living microorganisms which in certain situations were spread by contaminated water.

Ozone is an extremely effective oxidant which is only surpassed in its oxidizing power by fluorine or radicals like OH or O atoms. Siemens succeeded in persuading Ohlmüller, professor at the Imperial Prussian Department of Health, to test the effect of ozone exposure on cholera, typhus, and coli bacteria. The result was complete sterility after ozone treatment [6]. Soon after the first official documentation of these bactericidal properties emerged and large quantity ozone production started in small water treatment plants in Oudshoorn, Holland (1893). Within the following years various major drinking water plants using ozone disinfection were constructed in Russia (St. Petersburg 1905), in France (Nice 1907, Chartres 1908, Paris 1909) and in Spain (Madrid 1910). The water works at St. Petersburg had the capability to treat 50,000 m^3 of drinking water per day with ozone, while those of Paris reached up to 90,000 m^3 daily production. Thus, historically ozonation was the first successful attempt of disinfecting drinking water on a large scale [7].

Ever since Siemens' discovery, ozone generating technology has been closely linked with the development of water purification processes. In many countries ozonation in water treatment was later replaced by the more cost-effective processes using chlorine (Cl) or chlorine compounds, which are not only cheaper but also more soluble in water than ozone. However, recent concerns about potentially harmful chlorine induced carcinogenic byproducts have spurred on the enhanced use of ozone once again. Many European cities and some Canadian cities have fully abandoned chlorination in favor of ozone technology to disinfect water. In Europe, currently more than 3000 cities use ozone to disinfect their municipal water supplies. Waterworks in the United States, Japan, and China along with emerging global powers like

Brazil are all increasingly turning to ozone, in order to be able to meet more stringent legislative demands about disinfection byproducts like trihalomethanes (THMs) and haloacetic acids. These compounds can be formed when chlorine is added to the raw water containing organic water pollutants or humic materials. Some THMs are suspected to cause cancer. For this reason many leading experts in water treatment processes consider ozone treatment the technology of choice for potable water [8, 9]. Even with the current growing global demand for large-scale ozone production for water disinfection and many other modern industrial processes, the main generation method of O_3 remains microplasmas or microdischarges a type of atmospheric or higher pressure non-equilibrium or cold plasmas created with device known as a dielectric barrier discharge (DBD) [10].

13.2 Ozone and Its Various Applications

Ozone (O_3) is a triangular shaped molecule with a bond angle of 117° and equal bond lengths of 0.128 nm [11]. Ozone is a practically colorless gas with a characteristic pungent odor [12, 13]. Due to its oxidizing power, ozone finds applications as a potent germicide and viricide as well as a bleaching agent. In many applications ozone is increasingly used to replace other oxidants such as fluorine and chlorine that present environmental problems and safety hazards. Since strong oxidants are chemically active species, their storage, handling, and transportation involve substantial hazards. An important issue is further residues and side reactions. In all respects ozone is a superior choice due to its eventual side product degradation to oxygen. As a consequence of its inherent instability ozone is neither stored nor shipped. It is always generated on site at a generation rate controlled by its consumption needs for the specific process.

The most prevalent and important application of ozone remains mostly for the treatment of water. It is capable of oxidizing many organic and inorganic compounds in water. Ozone chemistry in water involves the generation of hydroxyl free radicals, very reactive species approaching diffusion controlled reaction rates for many solutes such as aromatic hydrocarbons, unsaturated compounds, aliphatic alcohols, and formic acid [14]. Ozone breaks double carbon bonds of DNA in living organisms. Microorganisms are killed by ozone as it causes cell lysis or lysing which causes the cell walls to rupture. Figure 13.1 shows the approximate necessary ozone concentration amounts in milligrams per liter of water (mg/l) that effectively limit or completely suppress microbial growth. O_3 further attacks all bacteria, virus, cysts, and spores in varying degrees.

Aside from applications in drinking water purification, ozone disinfection is used to treat water in swimming pools, cooling towers, and other water reservoirs. The effectiveness of ozone for water treatment depends as previously mentioned on the ozone concentration in water coupled with the ozone exposure or contact time. Ozone treatment further finds applications in municipal wastewater treatment plants. The environmental remediation of soil and groundwater uses ozone. The food and

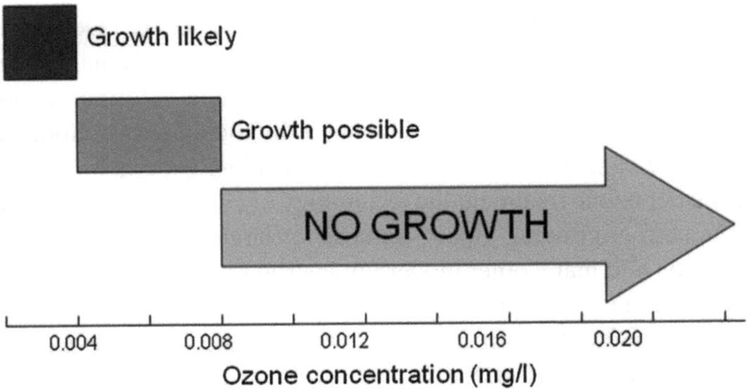

Fig. 13.1 Approximate ozone concentration (mg/l) ranges where water borne microbial growth can be completely suppressed

beverage manufacturers are heavy users of ozone especially the multibillion dollar industry of bottled water. Many zoos and especially aquariums are heavy users of ozone treatments to maintain the health of their resident animals and fish. Various industrial processes such as wood pulp bleaching for paper production use ozone treatment. Ozone is extensively used by the pharmaceutical and semiconductor industry to produce ultra-pure water for their chemical synthesis processes and microchip fabrication. Table 13.1 presents some examples of the various water treatment applications of ozone listing the ozone concentration ranges and ozone treatment duration or contact times.

13.3 Ozone Formation with Microplasmas

Ozone can be generated from pure oxygen gas or directly from air through electrical gas discharges in which the electron, atomic, or molecular collision energies are high enough to dissociate O_2 molecules. The overall gas temperature has to be kept low enough for the O_3 molecules to survive without undergoing thermal decomposition. Low temperature is absolutely necessary because the ozone molecules decay rapidly at high temperatures. For this reason, effective O_3 synthesis has to be performed at low temperatures which are typically at or near room temperature. Further, a relatively high pressure is required because ozone formation evolves from a three body reaction involving an oxygen atom, an O_2 molecule, and a third collision partner of either O_2 or N_2. This specific low-temperature and high pressure environment is specifically found in microplasmas or sometimes referred to as a microdischarges which are nonequilibrium or cold (non-thermal) plasma discharges. Microplasmas are directly created by a dielectric barrier discharge (DBD). Each one of the individual microplasmas created in the DBD acts like a miniature ozone producing factory.

Table 13.1 Examples of the various water treatment applications of ozone

Application	O_3 concentration (mg/l)	Contact time
Ultra-pure water	0.05–0.25	sec.–min.
Water bottling	0.4–1.0	5–10 min.
Swimming pools and spa	0.1–0.75	4 min.
Potable water, taste and odor disinfection	1.5–5.0	5–10 min.
Microfloculation	1.0–3.0	5–10 min.
Lignin removal	3.0–10.0	10–20 min.
Municipal wastewater	5.0–15.0	15–30 min.

Table 13.2 Characteristic parameters of the dielectric barrier discharge

Voltage, V_{pp}	0.5–20 kV
Repetition frequency, f	50 Hz–500 kHz
Pressure, p	1–3 bar
Gap distance, g	0.2–5 mm
Dielectric material thickness, d	0.5–2 mm
Relative dielectric permittivity, ε_r	5–10 (glass)
Electric field strength E of first breakdown	$\approx$ 150 Td($p = 1$bar, $T = 300$ K)
Electric field	0.1–100 kV/cm
Electron energy	1–10 eV
Electron density	10^{14} cm^{-3}

All these aforementioned requirements make microplasmas, these specific type of non-thermal atmospheric or higher pressure discharges, ideally suited for ozone generation.

13.3.1 Role of the Dielectric Barrier Discharge

The ideal plasma discharge type for large-scale ozone generation is the dielectric barrier discharge (DBD) which produces many microplasmas or microdischarges. The dielectric barrier discharge is inherently well-suited for applications involving volume plasma chemical synthesis such as high-volume ozone generation. It combines the large volume excitation manifested in a glow discharge with the possibility to operate at high pressure.

The main structural elements of a DBD configuration are shown in Fig. 13.2. The primary characteristic of the DBD is that a dielectric layer covers one or both of the metal electrodes. The dielectric surface is the principal factor in the proper functioning of the created discharge [15]. Once ionization occurs at a location in the discharge gap, the transported charge accumulates on the dielectric surfaces. The field due to this charge reduces the overall electric field in the gap and interrupts the current flow after a few nanoseconds. The duration of the current pulse depends

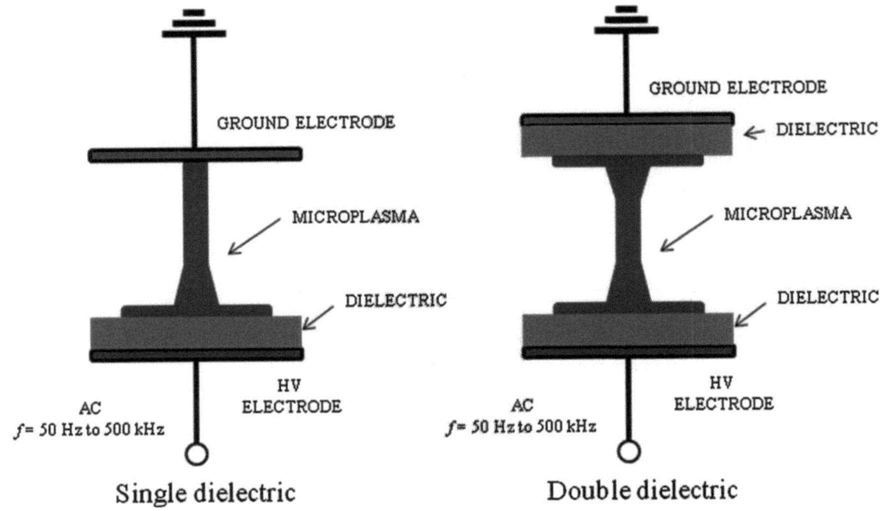

Fig. 13.2 A characteristic dielectric barrier discharge is covered with either a single dielectric on one electrode or a double dielectric on each of the electrodes. The microplasmas occur between the electrode gaps

on the pressure and the effective ionization characteristics of the gas as well as on the dielectric properties [16].

By applying an AC sinusoidal voltage of sufficient amplitude which can range from 0.5 to 20 kV depending on the gap separation, a large number of induced filamentary microplasmas (thus named due to their micron dimensional sizes). These microplasmas are randomly distributed in space and time. At the maximum and minimum of the applied voltage the displacement current is zero ($dV/dt = 0$) and the microplasma activity stops, only to start again when the breakdown field is reached in the gap during the next halfwave. The dielectric serves two primary functions: it limits the amount of charge transported by a single microplasma and distributes the microplasmas over the entire electrode area. By limiting the charge transported, the dielectric barrier creates a situation where there is a large distribution of filaments that contain energetic electrons. The electrical energy is easily channeled into the electrons and not into the ions or neutral gas particles. It leads to nonequilibrium or cold plasmas which depend on the pressure and the effective ionization characteristics of the gas as well as on the dielectric properties [16]. In most situations the reduced electric field at breakdown corresponds to electron energies of about 1–10 eV. This is an appropriate energy range for the excitation of atomic and molecular species and for the breaking of chemical bonds, which are all ideal scenarios for ozone synthesis. The typical operational parameters and observed physical phenomena of the dielectric barrier discharge are listed in Table 13.2 [17].

13.3.2 Kinetics of Ozone Formation

The electric discharge produced by a dielectric barrier discharge causes air or oxygen to undergo chemical reactions induced by the initial electron impact that causes ion formation and dissociation of the various discharge species. There exists a rather extensive body of literature of these various reactions [17]. However, as it turns out, the ion reactions play only a minor role in ozone formation while the main ozone synthesis path can be described by tracing the reactions of the oxygen (O) atoms generated by electron impact dissociation of O_2 along with other O atoms produced in reactions with N and N_2 and those of a few excited molecular states especially in air discharges. However, as previously emphasized the predominant initial key step toward ozone formation is the dissociation of O_2 molecules by electron impact. Commencing from ground-state O_2 molecules, two reaction paths toward dissociation occur [15, 18]. One reaction (13.1) via excitation of the $A^3\Sigma_\mu^-$ level with a threshold energy of 6 eV and another reaction (13.2) via excitation of the $B^3\Sigma_\mu^-$ level with a threshold of 8.4 eV.

$$e + O_2 \rightarrow e + O_2(A^3\Sigma_\mu^-) \rightarrow e + O(^3P) + O(^3P) \quad (13.1)$$

$$e + O_2 \rightarrow e + O_2(B^3\Sigma_\mu^-) \rightarrow e + O(^1D) + O(^3P) \quad (13.2)$$

In the subsequent step with only pure oxygen feed gas, which is used in most commercial ozonizers, O_3 formation is a relatively straightforward process. O_3 always originates from a three-body reaction of an oxygen atom reacting with two O_2 molecules

$$O + O_2 + O_2 \rightarrow O_3^* + O_2 \rightarrow O_3 + O_2 \quad (13.3)$$

where O_3^* stands for a transient excited state in which the O_3 molecule is initially formed. The time scale for O_3 formation in atmospheric-pressure oxygen is a few microseconds.

Fast side reactions (reactions 13.4–13.7 below) converting O atoms or O_3 into O_2 compete with the ozone formation and pose an upper limit on the O atom concentration or the degree of O_2 dissociation tolerable in the DBD

$$O + O + O_2 \rightarrow 2O_2 \quad (13.4)$$

$$O + O_3 + O_2 \rightarrow 3O_2 \quad (13.5)$$

$$O(^1D) + O_3 \rightarrow 2O_2 \quad (13.6)$$

$$O + O_3^* + O_2 \rightarrow 3O_2 \quad (13.7)$$

With air feed gas the discharge physics becomes more complicated. The presence of nitrogen atoms and excited atomic and molecular species, as well as the nitrogen ions N^+ or N_2^+ , adds to the complexity of the reaction system. Once again, the ions are of minor importance for ozone formation. Excitation and dissociation of nitrogen molecules, however, does lead to a number of additional reaction paths involving nitrogen atoms and the excited molecular states $N_2(A^3\Sigma_u^+)$ and $N_2(B^3\Pi g)$, that can produce additional O atoms for ozone generation [9, 10].

$$N + O_2 \rightarrow NO + O \tag{13.8}$$

$$N + NO \rightarrow N + O \tag{13.9}$$

$$N + NO_2 \rightarrow N_2O + O \tag{13.10}$$

$$N_2(A, B) + O_2 \rightarrow N_2 + 2O \tag{13.11}$$

$$N_2(A) + O_2 \rightarrow N_2O + O \tag{13.12}$$

These oxygen atoms, generated in addition to those obtained from direct electron impact dissociation of O_2, contribute about 50 % of the ozone formed in air, which now takes longer, roughly about 100 μs. The result is that a substantial fraction of the electron energy, initially lost in collisions with nitrogen molecules is recovered and utilized for ozone generation through reactions (13.8–13.12). In addition to ozone a variety of nitrogen oxide species are generated: NO, N_2O, NO_2, NO_3, and N_2O_5. All these species have been measured at realistic ozone generating conditions [17]. In the presence of ozone only the highest oxidation stage N_2O_5 is detected in addition to the rather stable molecule N_2O (nitrous oxide, laughing gas). Due to these additional reactions and, as expected, the dissociation of N_2 requiring higher electron energies, the maximum generated ozone concentration in air is less than that in pure oxygen.

13.4 Standard Ozone Generator Configurations

Aside from the original ozone discharge reactor proposed by Siemens various different electrode configurations have been proposed where at least one dielectric barrier (insulator) is used to limit the discharge current between the metal electrode(s). Figure 13.3 shows two different dielectric barrier discharge configurations.

The presence of the dielectric barrier prevents DC operation because the insulating material cannot pass a DC current. AC or pulsed operation is possible because any voltage variation *dV/dt* will result in a displacement current in the dielectric barrier(s). DBDs are operated in air or oxygen, at pressures between 1 and 3 bar, with electrode separations between 0.1 mm to several cm, frequency ranges from line frequency to microwave frequencies, and at voltages ranging from about a few hundred volts to several tens of kilovolts. DBDs in different gases and gas mixtures have been studied at various pressure levels. Better understanding of the discharge properties and of the ozone formation process has benefited greatly from large-scale ozone generation

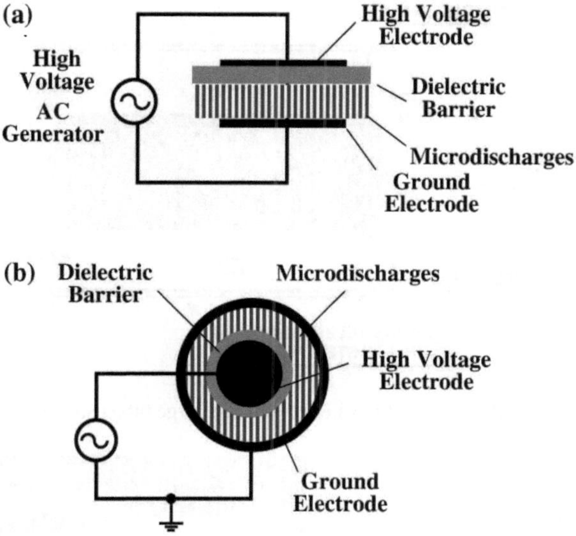

Fig. 13.3 **a** Planar and **b** cylindrical dielectric barrier discharge configurations

research [14–16]. Major improvements have recently been obtained by tailoring microdischarge properties in air or in oxygen where recombination of oxygen atoms is minimized and ozone formation is optimized. This can be achieved by adjusting the width of the discharge space, the operating pressure, the properties of the dielectric barrier, and the temperature of the cooling medium.

Planar and cylindrical electrode configurations are used in the construction of ozonizers. However, the majority of commercial ozone generators use cylindrical electrodes forming narrow discharge spaces of 0.5–1 mm in radial width. The outer electrode is normally a stainless steel tube which has a ground potential and is water cooled to allow for proper heat dissipation as demonstrated in Fig. 13.4. These cylindrical tubes range in length from 1 to 4 m. A traditional contemporary ozonizer (Fig. 13.5) has as the coaxial inner electrode a glass tube which is closed at one side and has an inner metal coating used as a high voltage electrode. However the most modern and efficient ozone generator designs have as an inner electrode made with a closed steel cylinder which is completely covered by a dielectric layer of ceramic or enamel coating (Fig. 13.6). The general advantages of these enamel coated dielectric electrodes is due to the reduced number of dielectric covered electrodes necessary and lower operating voltages which allow for lower power consumption leading to more efficient ozone generation. However, in both these designs the feed gas is streaming in the axial direction through the discharge region between the inner and outer tubes. Each volume element of the flowing gas is exposed to the action of many microplasmas which leads to the generation of ozone.

Classical ozone generators were operated at line frequencies and step-up transformers were used to reach the desired voltage levels necessary for the gas

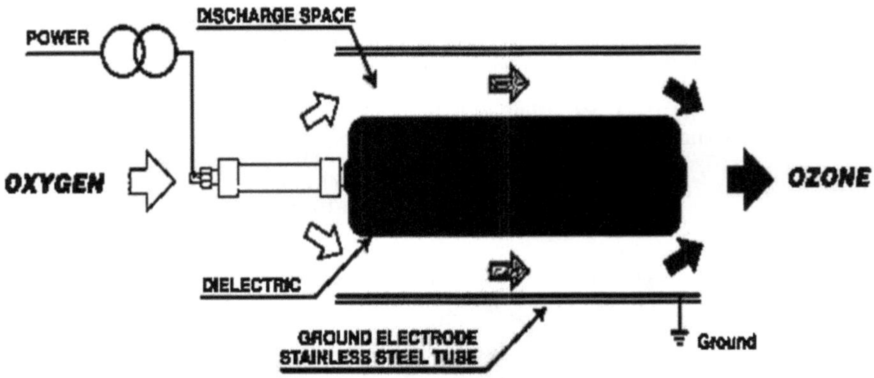

Fig. 13.4 The typical configuration of a water cooled discharge tube used in an ozone generator

Fig. 13.5 A traditional large-scale ozone generator with the grounded, water cooled outside electrode and glass dielectrics with metal electrode coating inside the glass tube. (*Picture courtesy of Degrémont Technologies—Ozonia*)

breakdown and the microplasma formation. However, dielectric failure was a common issue since all the dielectric electrodes were connected in parallel. To circumvent some of these issues, high voltage fuses were used to disconnect faulty elements. Modern large-scale ozone generators take advantage of solid state power semiconductors [19–22]. They utilize thysitor insulated gate bipolar transistors (IGBT) and other high power semiconductor electronics technology as frequency converters and inverter components to create square-wave currents or special pulse trains in the frequency range between 500 Hz and 5 kHz. Using this power technology, applied voltages can be reduced to the range of about 5 kV which effectively reduces the possibility of dielectric failure. Figure 13.7 demonstrates a typical modern day power supply unit (PSU) used in large-scale ozone generators. With larger ozone generators power factor compensation has become an important issue. In modern PSUs the

Fig. 13.6 A modern large-scale ozone generator with enamel coated dielectrics (*Picture courtesy of Degrémont Technologies—Ozonia*)

Fig. 13.7 An advanced power supply unit (PSU) used in modern ozone generator (*Picture courtesy of Degrémont Technologies—Ozonia*)

power densities now reach 1–10 kW/m^2 of electrode area. As previously emphasized improved dielectric material coupled with using high voltage semiconductor technology at higher frequencies, has brought several advantages of increased power at lower voltage, fast shut off, and improved process control.

As previously discussed, large ozone generators use several hundred discharge tubes and can produce up to 100 kg ozone per hour. The feed gas for most ozone generators is either air or pure oxygen. Depending on the feed gas, ozone concentration of up to 8 weight percent (wt%) can be obtained with air and up to 18 wt% with pure oxygen (where 1 wt% equals 10,000 ppm O_3). Advanced water treatment processes utilize ozone at typical concentrations up to 12 wt%. Depending on the desired ozone concentration the energy required to produce 1 kg O_3 ranges from 7.5–10 kWh in oxygen and from about 15–20 kWh in air. Most large ozone generators use pure oxygen or oxygen blended with about 1–3 % nitrogen (N_2) as a feed gas. The addition, nitrogen has a pronounced beneficial effect on ozone generation: a demonstrated longer ozone production efficiency maintenance and avoidance of ozone generation efficiency decline with time, as seen with pure oxygen operation [23, 24].

It is essential that the feed gas whether air, pure oxygen, or O_2/N_2 mixture be processed to remove humidity. Humidity has a strong influence on the surface conductivity of the dielectric and on the properties of the microdischarges. Further, trace impurities such as H_2, NO_x, and hydrocarbons have an adverse effect on ozone formation [24]. Some of these impurities lead to detrimental atomic oxygen recombination to O_2, NO_x or OH, while other impurities cause catalytic ozone destruction. For this reason many large ozone generation installations in polluted areas take precautions to prevent hydrocarbons accumulation in their liquid oxygen supply.

Ozone generator reactors are essentially built like heat exchangers since the ozone formation efficiency and the stability of the O_3 molecule deteriorate at elevated temperature. Since nonequilibrium discharges are suited for ozone generation, an efficient cooling of the discharge gap is mandatory. These configuration requirements for a large-scale ozone generator are of course all contingent upon the amount of ozone that must be produced for the specific application.

All large-scale ozone generators are essentially designed depending on the particular requirement imposed by the application. However, in the particular application for water treatment, the facility design tends to follow a typical engineering construction scheme as seen in Fig. 13.8. The feed gas whether air or oxygen is usually stored in a storage tank in the vicinity of the processing facility. Many facilities use liquid oxygen (LOX) due to its larger storage capability which requires the oxygen or admixture of O_2/N_2 to pass through a vaporizer before being fed into the ozone generator. The generated ozone is usually bubble diffused into a storage tank of the water supply. The unused ozone after the water treatment process is sent through an ozone destructor and vented out to the atmosphere.

13.5 Recent Developments in Large-Scale Ozone Generators

There are a variety of factors that influence the design of modern ozone generators. Low electrical power consumption, ease of operation, reduced down-time, and low maintenance are of crucial importance in ozonizers for large-scale public utilities or industrial applications. On the other hand, smaller-scale applications place a premium on ease of operation and reduced cost per ozonizer unit over power efficiency. The

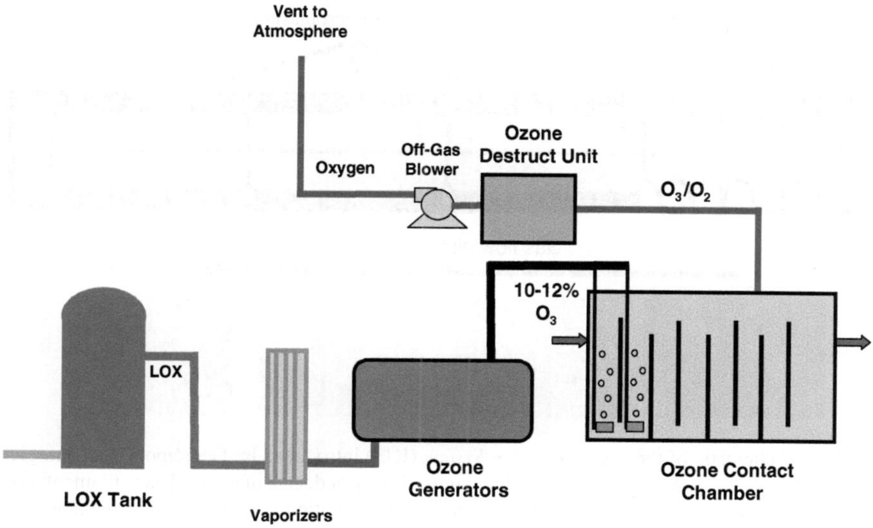

Fig. 13.8 A typical schematic design of a water treatment plan that uses ozone

scale of the ozonizer required for a specific application has led to a large range of different ozone generator designs that are commercially available today.

In large-scale industrial applications, ozone concentrations ranging from 9 to 13 wt%, in an almost pure oxygen feed gas, can be achieved. Quantities, ranging from 20 kg to several hundred kilograms of ozone per hour, can be provided to the application user. As would be expected, the ozone concentration rises with the amount of energy deposited in the gas, at least at low ozone concentrations. Ultimately, there is a saturation concentration, strongly depended on the gas temperature in the discharge gap, which in turn can be regulated with an effective water cooling heat exchanger. Cooling water temperatures around 30 °C are found at most ozone production sites [25].

A critical aspect of next generation large-scale ozonizer technology with increased ozone generation efficiency has come from investigating the correlation of the electrode arrangements and the power induction. Predictions from a finite-element model, that simulated the power induction into the gas along the length of the discharge gap, and further took into account the local background conditions, have been correlated with experimental data using a single adjustable fitting parameter. Additionally, several electrode arrangements of different design were selected, manufactured, simulated with the model, and tested in a series of experiments. These results of the modeling study and the experimentation had the objective of proposing a new method to improve the ozone generation efficiency, and are exclusively applicable to large-scale ozonizers [26–28].

It was found from this basic research that the electrode arrangement of an ozone generator may be optimized to the needs of the application and/or to the necessary operational conditions. A complete optimization scheme for large-scale systems

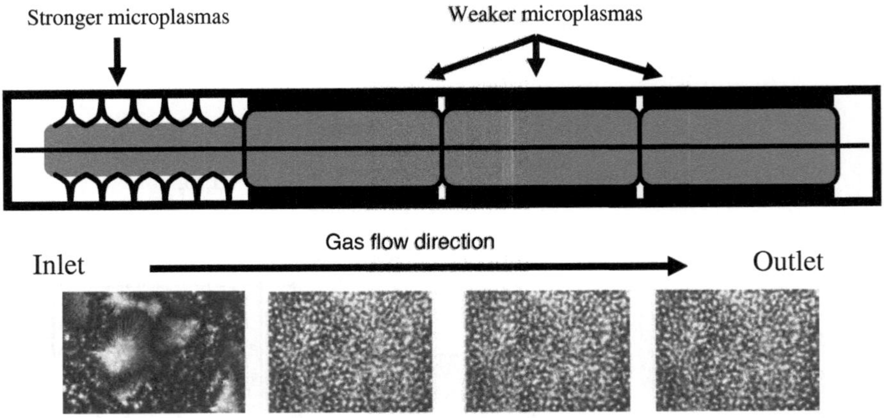

Fig. 13.9 Schematic of the *Intelligent Gap System* (IGS) introduced by Degrémont Technologies (Ozonia) Ltd. in 2007. A short high-load inlet region with lower degree of microplasma filamentation is followed by a longer region with a high degree of filamentation. The degree of filamentation is also illustrated by the Lichtenberg pictures [28] situated at the bottom of the electrode centers

would include adjusting the power induction pattern to increase power efficiency and tailoring the discharge filaments or miroplasmas to maximize robustness for all conditions of operation. Figure 13.9 illustrates the conceptual design needed for the realization of the discontinuous feed gas processing. This ozone generator was manufactured and successfully operated on a large scale [28].

The ability to optimize the microplasmas for the ozone formation process means essentially finding a compromise between having excessive energy losses due to ions in weak microplasmas and avoiding undesired arcing and chemical side effects, if the microplasmas become too strong. The "strength" of a microplasma can be characterized by the amount of charge transported by the microdischarge or by the energy density deposited in the discharge channel after termination of the current flow [28]. After the termination of a microplasma, the generated ozone diffuses out of the volume of the current filament. Since the electric field is reduced at this particular location, further microplasmas will occur at other positions in the gap. The increase of ozone concentration essentially results from the action of a large number of microplasmas. The ozone that is formed will typically be exposed to the action of several hundred or thousand microplasmas before it reaches the end of the discharge tube. The optimum ozonizer design provides the appropriate power induction to raise the amount of energy deposited in the feed gas to tailor the necessary number of microplasmas that will lead to higher ozone generation concentrations.

The exact implementation of these criteria leads to an ozonizer with vastly improved ozone generation properties. As illustrated in Fig 13.10, the classical shape of an efficiency curve is shown as the dashed line, which has a pronounced concentration limit. The measured data can be compared to the classical curve in the low ozone concentration region only. An almost linear trend, which deviates from zero- and first-order kinetics suggested by Hirth et al. [29, 30], is found in the high ozone

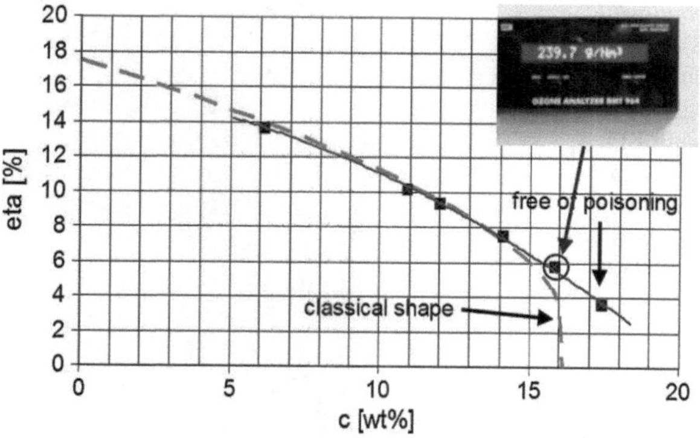

Fig. 13.10 Ozone concentration versus efficiency at constant power density, cooling water temperature, pressure, and frequency. The classical curve fits only in the low ozone concentration region. The high ozone concentration region shows a linear trend

concentration region. An efficient and robust system is obtained through an optimized discharge arrangement, i.e. from a tailored degree of discharge filamentation or density of microplasmas.

The proof of concept was obtained on a large scale on a duplex vessel with a total active ozone generation area of 112.7 m^2 and a production capacity of approximately 60 kg/h at 10 wt% of ozone concentration. An ozone concentration of approximately 16 wt% (see Fig. 13.10) was achieved and maintained, even at increased cooling water temperatures, and at high power densities.

In summary, the ability to tailor the degree of filamentation, or the capability to control the number of microplasmas at low and high ozone concentrations, benefits the ozone generation process in terms of reduced power consumption and an increased regime of useful ozone concentration generation. The conditions for large-scale ozone synthesis in a dielectric barrier discharge require highly filamented discharges where the preferred characteristic of the microplasmas is Townsend-like at reduced electric fields E/N around 190 Td. Furthermore, smaller discharge gaps than 0.3 mm require thick dielectric layers to maintain a good degree of filamentation at reasonable efficiency. This technology is best suited for large-scale ozone production with capacities of up to several hundred kilograms per hour, at concentrations between 6 and 13 wt%. Corresponding electrode designs have a length of up to 3 m and are bundled to several hundred m^2 of active DBD area. This novel approach to operate a dielectric barrier discharge for ozone generation, the so-called 'Intelligent Gap System' (IGS) was introduced by Degrémont Technologies—Ozonia in late 2007 as a new ozonizer technology for large-scale ozone generation [27, 28]. The IGS ozonizer seen in a final production version in Fig. 13.11 is currently the most efficient large-scale ozone generation technology available commercially for public and industrial water treatment.

Fig. 13.11 An *Intelligent Gap System* (IGS) ozone generator in production in early 2008 at Degrémont Technologies-Ozonia Ltd. for final installation at a municipal water treatment facility. (*Picture courtesy of Degrémont Technologies—Ozonia*)

13.6 Challenges in Large-Scale Ozone Generators

The increasing problem of feed gases of lower grade liquid oxygen (LOX) containing higher levels of trace hydrocarbon impurities lead to detrimental by-product formations in dielectric barrier discharge (DBD) for large-scale ozone generation. Studies of hydrocarbon impurities using the kinetics of methane (CH_4) conversion in DBDs as a typical contaminant species and the subsequent HNO_3 formation are presented in this section. The role of methane was investigated by means of gas-phase plasma diagnostics, supported by detailed process modeling, and in-situ and ex-situ by-product analysis is explained. The by-products formation in the plasma with the presence of CH_4, were found to differ significantly in oxygen-fed generators as compared to generators fed with oxygen/nitrogen mixtures. The amount of HNO_3 formed depends on the concentration of NO_x formed in the plasma and the amount of CH_4 that is converted, but not on the O_3 concentration. In this section, an investigations of CH_4 concentrations of up to 1.95 wt% of the feed gas are used to illustrate some of the challenges encountered in large-scale ozone generation. The rate of deterioration of the overall ozone generator performance was found to be affected by the concentration of nitrogen in the oxygen/nitrogen mixture.

13.6.1 By-Product Formation in DBDs

The kinetics of by-product formation in a dielectric barrier discharge (DBD) plasma depends on three major factors: gas mixture, discharge operating conditions (gas pressure, gas temperature, mode and frequency of applied voltage, electrode arrangement, power density, etc.), and the electrode and dielectric materials.

The most comprehensive account of experimental investigations into the kinetics of NO_x formation in the air-fed, DBD-based ozone generators with discharge gaps around 1 mm can be found in publications by Kogelschatz and co-workers dating back to the mid-1980s [31, 32]. Extensive modeling efforts [33] helped elucidate the basic plasma chemical reaction pathways for a gas composition close to 80 % N_2 and 20 % O_2. However, the issue of possible deleterious effects of hydrocarbon impurities in the feed gas on the operation of the ozone generator was not addressed. There were previous attempts to investigate the effect of various impurities in the feed gas on the ozone efficiency, such as by Cromwell and Manley [34], Inoue and Sagino [35], and lately by Salvermoser et al. [36] in ozone generation by xenon excimer lamps. Cromwell and Manley investigated the effect of various impurities and their concentrations: H_2, moisture, Freon 12 (CCl_2F_2), Ar, carbon oxides and N_2. Their results were important for the development of appropriate feed gas to optimize commercial ozonizers. Inoue and Sagino studied the effect of hydrocarbons, specifically that of the cyclohexane, n-hexane and n-heptane, on the rate of ozone formation. They showed, at least in laboratory conditions, a uniformly large effect of these hydrocarbons on ozone efficiency. Salvermoser et al. tackled the effect of moisture on the photochemical ozone generation and found significant effect of moisture on the achievable ozone concentration of excimer lamps. The investigation explained qualitatively the fast catalytic ozone destruction cycle by OH and HO_2 radicals.

In this section we review results of studies carried out to elucidate the effects of hydrocarbon contamination (up to concentrations of close to 2 wt% in the feed gas) on the performance of a single-tube commercial DBD ozone generator. The key issue is whether trace hydrocarbon contaminants significantly affect or change the plasma chemistry or whether, in sufficiently low concentrations, simply act as a "spectator" without impacting the plasma chemical reaction pathways that govern the ozone generation.

Figure 13.12 shows the potential pathways of NO_x formation in dry air [37]. As reported by Kogelschatz [31, 32], only nitrous oxide (N_2O) and dinitrogen pentoxide (N_2O_5) were found at the outlet of the air-fed DBD ozone generator. The kinetics of NO_x formation was first modeled by Eliasson [38]. Equivalent results were obtained from improved simulations carried out by Pitchford [39].

Kogelschatz and Müller [31] suggested that of the two possible reactions leading to the formation of HNO_3:

$$N_2O_5 + H_2O \rightarrow 2HNO_3 \tag{13.13}$$

$$NO_2 +_* OH \rightarrow HNO_3 \tag{13.14}$$

the fast radical reaction (13.14) as the main route to HNO_3 formation, a conclusion based in part on the key role of NO_2 formed in the plasma. As a consequence, one might expect a strong correlation between the measured NO_x and OH emission inten-

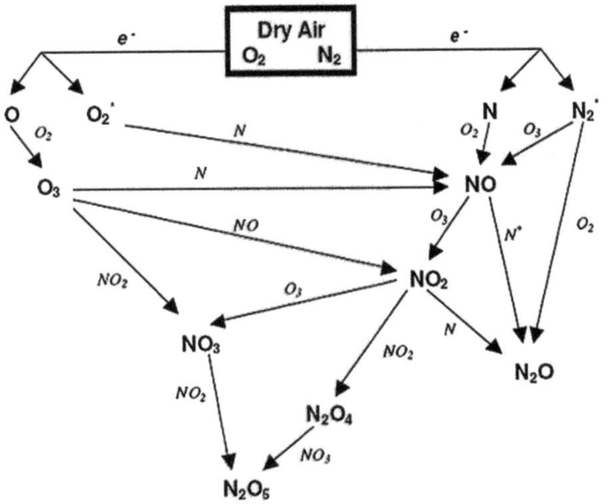

Fig. 13.12 Schematic diagram of the primary chemical reactions in a dry air plasma following electron impact on N_2 and O_2. Only the formation reactions up to the formation of N_2O_5 are shown [37]

sities obtained by optical emission spectroscopy. Since methane reactions producing HNO_3 in the discharge also involve NO_x compounds, a simultaneous investigation of NO_x and CH_4 appears desirable.

13.6.2 Hydrocarbon Contamination in DBDs

Recent experimental research carried out using electrical and optical analysis demonstrates that the main by-products measured in the gas stream emanating from an oxygen/nitrogen-fed DBD ozone generator were N_2O_5 and N_2O [40]. The dependence of their formation on the ozone concentration, at 3 wt% nitrogen admixture, is shown in Fig. 13.13. The measurements were repeated for several electrodes and for different discharge gap geometries in the IGS (Intelligent Gap System) [26–28]. A comparison of the AT95 (discharge gap of 0.4 mm) and the LG (discharge gap of 0.32 mm) curves indicates a marginal decrease in the NO_x formation for the smaller discharge gap. We note that AT95 and LG electrode designs and materials are proprietary products of Degrémont Technologies. These proprietary electrodes differ from each other by capacitances of the dielectric.

The Intelligent Gap System (IGS) technology tailors the microplasmas from the generator inlet to the generator outlet by means of a varying discharge gaps over the entire generator length, combined with appropriate dielectric properties of the electrodes. Strong discharges are created at the generator inlet while softer discharges are utilized to better synthesize the ozone towards the generator outlet [28].

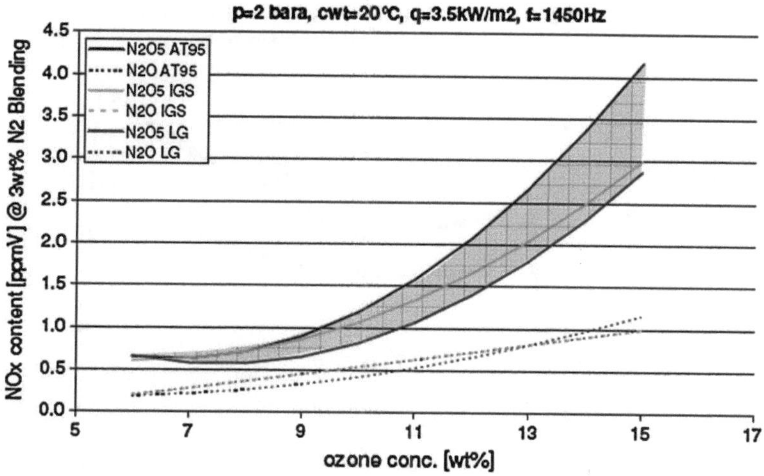

Fig. 13.13 Amount of N_2O_5 and N_2O formed in the ozone generator as a function of ozone concentration at 3 wt% nitrogen admixture for various electrode arrangements

The following results were conclusively observed:

- The amount of NO_x formed scales linearly with the concentration of the nitrogen admixture
- The determining factor in the NO_x formation is the ozone concentration
- The N_2O_5 formation is a non-linear function of the ozone concentration; a second-order regression formula is needed to fit the data.
- The N_2O concentration is essentially a linear function of the ozone concentration.

13.6.3 The Role of Hydrocarbons

The key question here is whether methane or hydrocarbons in general, are taking part in the plasma chemical reaction pathways or play the role of a mere "spectator" and do not significantly change the reactions. Figure 13.14 shows emissions in the 390–465 nm region for methane concentrations from 0.23 to 1.95 wt%. The emission feature at 436 nm, attributable to the CH radical, is visible even at the lower methane concentrations. The presence of the CH radical emission is indicative of the fragmentation of the parent CH_4 molecule in the plasma and the formation of reactive CH_x ($x = 1 - 3$) species and thus provides indirect evidence that methane is more than a mere "spectator" because of the chemically reactive nature of its dissociation fragments.

Furthermore, high concentrations of methane admixtures in the range of 1–2 wt% resulted in almost immediate formation of droplet in the discharge tube, visible to the naked eye, in conjunction with the rapid deposition of a "dirt" layer on the ceramic

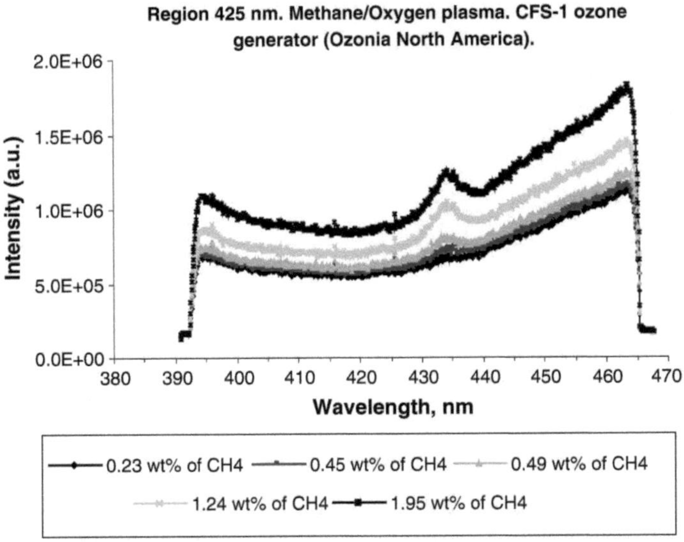

Fig. 13.14 CH peak in the discharge process as a function of methane concentration in the feed gas

coated electrode, a change in the operating characteristics of the ozone generator, and arcing inside the discharge tube. Exact location of arcing could not be determined.

The exact location of the deposited "dirt" layer and its temporal growth pattern after ignition of the ozone generator yielded useful information about the kinetics of by-product formation. Analysis of the deposits in conjunction with spectroscopic data led to the proposal of two scenarios concerning the methane reactions in the discharge process, one for the case of feed gas consisting of pure O_2 and a CH_4 admixture (Fig. 13.15) and one for the case of a feed gas consisting of O_2 admixed with N_2 and CH_4 (Fig. 13.16). There is a very important difference between the two cases.

Admixing small amounts of methane to pure oxygen does not result in droplet formation and deposits in the first 1/3 of the length of the discharge tube. For the remaining 2/3 of the length of the tube, the deposition of the "dirt" layer is balanced by vaporization (via sputtering) resulting in a thin and rather uniform layer of deposits, mostly oxides from the stainless steel discharge tube. The conversion of CH_4 proceeds via collisional processes and ultimately leads to the formation of CO_2 and H_2O.

In the case of oxygen with N_2 and CH_4 admixtures, it takes only a small amount of N_2 (approximately 200 ppmV) before deposits appear on the enamel coated electrode, very close to the entrance to the discharge tube (the "clean" part extends for only about 1/8 of the length of the electrode). As before, methane is being converted and H_2O and CO_2 are formed, but in the presence of nitrogen, NO_x compounds are formed, and, subsequently, nitric acid (HNO_3) is being produced. Vaporization

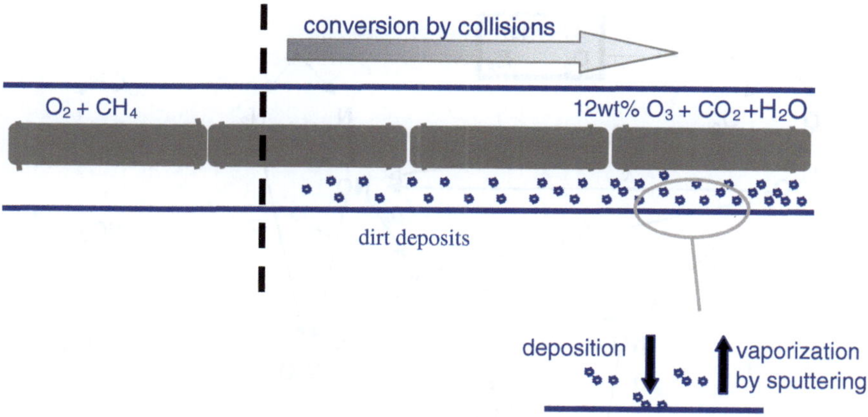

Fig. 13.15 Illustration of the methane conversion along the discharge tube for pure oxygen with a methane admixture (around 1–2 wt%). Droplets are visible at about 1/3 of the total length of the discharge tube

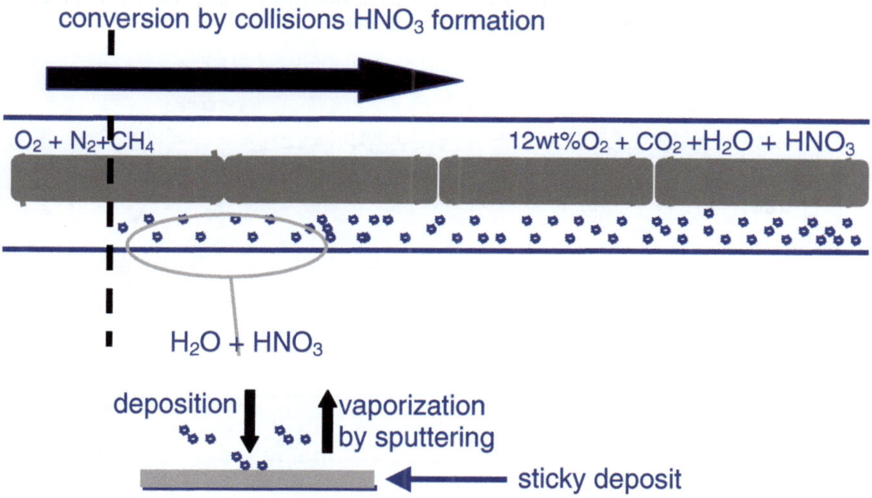

Fig. 13.16 Illustration of the deposition and built-up of HNO_3 and H_2O. Minute N_2 admixtures are sufficient to change the by-product formation processes drastically

via sputtering is insufficient to remove all the deposits and a permanent "sticky" layer containing HNO_3 begins to form and grow on the ceramic coated electrode. Its thickness increases along the length of the electrode.

The general picture of the relevant air plasma chemistry, presented in Fig. 13.12, changes when impurities such as methane are introduced. Reactions, resulting in HNO_3 formation, are shown in Fig. 13.17.

Results from emission spectroscopic studies confirm the above model of the effect of methane on the plasma chemistry inside the discharge tube. Figure 13.18 shows

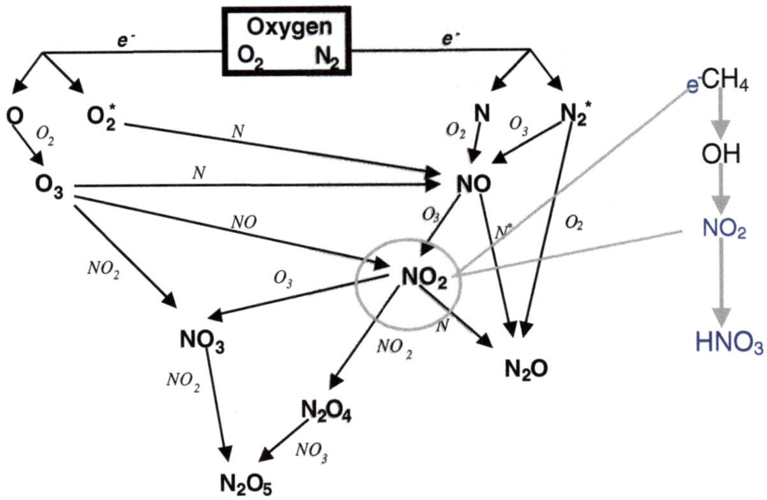

Fig. 13.17 Schematic diagram of the primary chemical reactions in a dry air plasma in the presence of CH_4. Formation path of HNO_3 is schematically indicated

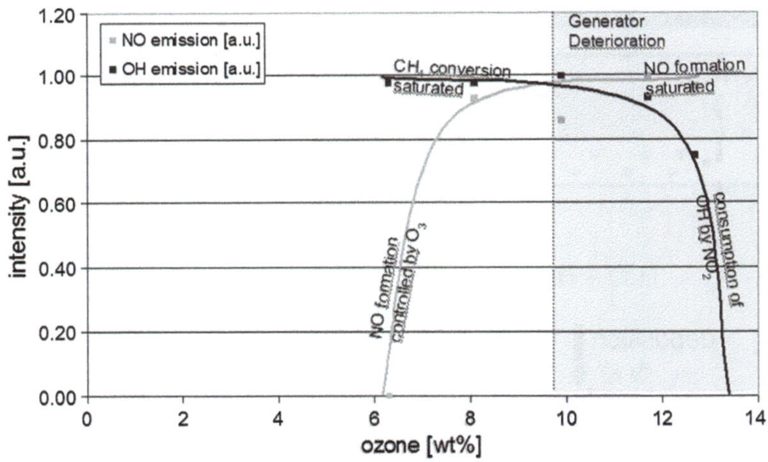

Fig. 13.18 Relative OH (280 nm band) and NO (248 nm band) emissions (200–400 nm) versus ozone concentration (approx. 5 wt% N_2, 1000 ppmV of CH_4, p = 2.2bar, cooling water temperature of 11° C, excitation frequency of 1450 Hz)

the NO and OH optical emissions from a mixture of oxygen with 5 wt% of nitrogen and a 1000 ppmV admixture of methane in the 200–400 nm region as a function of ozone concentration.

Assuming the OH signal to be a valid indicator of the methane conversion and comparing the OH signal with the NO signal, the following observations can be made:

1. For low ozone concentrations, the NO signal is low, whereas the OH signal is saturated.
2. With increasing ozone concentration, the NO signal rises and saturates above 10 wt% of ozone concentration. Conversely, the OH signal decreases and vanishes for ozone concentration higher than 10 wt%.

Therefore, the methane conversion is not driven by the ozone concentration, since in that case a linear increase of OH signal with increasing ozone concentration would be expected. The disappearance of the OH signal can be explained by the formation of HNO_3, since the OH radicals in the plasma react with the NO_x by-products. This conclusion is supported by the fact that the NO signal at higher ozone concentration reaches a plateau, indicating saturation.

In summary, the OH emission intensity is not correlated with the ozone concentration as long as the ozone concentration remains below about 12 wt%; the signal vanishes rapidly for higher ozone concentrations and becomes undetectable above about 13.5 wt% ozone. The NO_x by-product formation starts between 6 and 8 wt% ozone concentration and quickly saturates. The NO_x by-product formation does not appear to be affected by the presence of OH radicals. The evaluation of by-products from IR spectroscopy indicated that two HNO_3 molecules are present for each N_2O_5 molecule. This leads to the following overall conclusions:

- The HNO_3 formation is governed by the amount of OH formed in the plasma
- The availability of NO limits the HNO_3 formation
- NO_2 is a prerequisite intermediate product needed for the formation of N_2O_5, which, in turn, is governed by the formed in the discharge NO.
- The formation of HNO_3 follows the fast radical reaction route 13.14.

These results indicate that methane is most probably converted/dissociated by electron impact and is not affected by the ozone concentration. The H_2O formation is limited by the availability of OH, NO, and NO_2 in pure O_2 with methane traces in low concentrations. Therefore, the deposits on the electrodes are reversible. On the other hand, if the HNO_3 is produced in significant amounts, it is more likely to deposit on electrodes and stick permanently, thus leading to an irreversible deterioration of the electrode surfaces.

13.6.4 Hazards of Hydrocarbons

The detrimental effect of hydrocarbons such as methane on the ozone generator plasma chemistry is contingent upon the formation of the so-called "sticky deposit" on the surface of the covered electrode, which is a liquid, dust, and HNO_3 coating. Once the deposit dries, there are no other means to restore the electrode than by physical cleaning. Figures 13.19 and 13.20 show pictures of deposits formed at the exits of the discharge tube, which caused device failures. In both cases, insufficient control of the nitrogen admixture leading to formation of HNO_3 in the presence of moisture and/or hydrocarbons was the root cause of the problem.

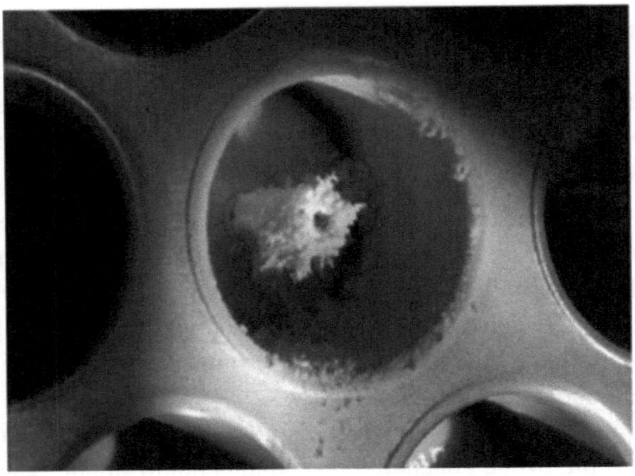

Fig. 13.19 Picture of an amorphous/crystalline deposit formed in the ozone generator. (*Picture courtesy of Degrémont Technologies—Ozonia*)

Fig. 13.20 Picture of an initially liquid deposit, which subsequently dried out and then was found to be "baked" on the electrode surface. The resulting sand-like crust plugs the tube and causes failure of the device (*Picture courtesy of Degrémont Technologies—Ozonia*)

The deterioration of the active discharge area is generally a slow process, extending from a few months to a few years. However, the following reactions describe mechanisms that affect the power efficiency immediately by destroying ozone and thus lowering the maximum achievable ozone generation:

$$O_3 + 2^{*}OH \rightarrow 2O_2 + H_2O \tag{13.15}$$

$$O_3 + CO \rightarrow O_2 + CO_2 \tag{13.16}$$

A model accounting for these reactions effects on the power efficiency was already described in the literature and was validated for large scale ozone generator units during the performance measurements reported by Lang [25] and the large-scale validation of Degrémont Technologies—Ozonia's proprietary IGS technology [26–28].

In conclusion, the kinetics of NO_x by-products in DBD-based ozone generators follow similar principles in the air-and oxygen-fed generators (even in oxygen/nitrogen mixtures with only traces of nitrogen). The amount of N_2O_5 formed is found to depend on only two parameters, the ozone concentration and the amount of nitrogen admixture. In the presence of hydrocarbons such as CH_4, the byproducts of the discharge were found to differ significantly in oxygen-fed generators as compared to the byproducts of oxygen/nitrogen mix-fed generators. The first case produces just humidity, the second HNO_3. Humidity is detrimental to ozone efficiency during the process. However, the produced HNO_3 affects very strongly electrode's surface conductivity

13.7 Conclusions and Future Progress

Although a rather mature application, large-scale ozone generation using dielectric barrier discharges has continued to progress and improve as a technology throughout the years. Several reasons for this progress and renaissance are the emergent global demand for environmentally safe and sustainable water treatment technologies spurred on by the increasing global population. The continued drive to better understand the fundamental physics of microplasmas and the plasma chemical reactions brought on by current modeling work and experimental research will continue to improve this technology. The development of cost and energy efficient power electronics coupled with the electrical engineering of matching power supply units to tailor DBD properties will progress this application. Future progress in next generation large-scale ozone generators is expected from ultra-short nanosecond high voltage pulses, advanced dielectric materials, and the continued better understanding and control of the integral ozone producing microplasmas.

References

1. American Water Works Association, J. Edzwald, Water Quality & Treatment Handbook: A Handbook on Drinking Water, 6th edn. (McGraw-Hill Professional, New York, 2010).
2. M. Kavanaugh, in Chapter 1–"Water Treatment". Process Technologies for Water Treatment, ed. by S. Stucki (Plenum, New York, 1988), pp. 1–24.
3. W. Siemens, Poggendorfs Ann. Phys. Chem. **102**, 66–122 (1857)
4. R.D. Morris, *The Blue Death: Disease, Disaster, and the Water We Drink* (Harper, New York, 2007)

5. S. Johnson, *The Ghost Map: The Story of London's Most Terrifying Epidemic and How It Changed Science, Cities, and the Modern World* (Riverhead Trade, New York, 2007)
6. W. Ohlmüller, *Ueber die Einwirkung des Ozons auf Bakterien* (Springer Verlag, Berlin, 1891)
7. U. Kogelschatz, in Proceedings IOA 17th Ozone World Congress. Industrial Ozone Generation: Historical Perspective, Current Status and Future Prospects, vol. II.1 (Strasbourg, 2005), pp. 1–10.
8. K.L. Rakness, *Ozone in Drinking Water Treatment: Process Design, Operation, and Optimization* (American Water Works Association, Denver, 2005)
9. J.L. Lopez, in Chapter 2–"Ozone Generation with Cold Plasmas for Water Treatment Processes, ed. by Biological and Environmental Applications of Gas Discharge Plasmas. G. Brelles-Mariño (Nova Science Publisher Inc, New York, 2010), pp. 33–48.
10. K.H. Becker, U. Kogelschatz, K.H. Schoenbach, R.J. Barker (eds.), History of Non-Equilibrium Air Discharges - Corona Discharges. Series in Plasma Physics: Non-Equilibrium Air Plasmas at Atmospheric Pressure. (Institute of Physics Publishing Ltd, Bristol, 2005), pp. 47–68.
11. A. Kalemos, A. Mavridis, Electronic structure and bonding of ozone. J. Chem. Phys. **129**, 054312 (2008)
12. M. Horváth, L. Bilitzky, J. Hüttner, *Ozone* (Elsevier Science Publishing, New York, 1985)
13. J. Wojtowicz, *"Ozone" in Kirk-Othmer Encyclopedia of Chemical Technology*, 4th edn. (Wiley, New York, 1996)
14. J. Hoigné, in Handbook of Environmental Chemistry, Part C: Quality and Treatment of Drinking Water II, ed. by J. Hrubec. Chemistry of Aqueous Ozone and Transformation of Pollutants by Ozonation and Advanced Oxidation Processes, vol 5 (Springer Verlag, Berlin, 1998), pp. 83–141.
15. B. Eliasson, U. Kogelschatz, Nonequilibrium volume plasma chemical processing. IEEE Trans. Plasma Sci. **19**(6), 1063–1077 (1991)
16. B. Eliasson, M. Hirth, U. Kogelschatz, Ozone synthesis from oxygen in dielectric barrier discharges. J. Phys. D Appl. Phys. **20**, 1421–1437 (1987)
17. H.E. Wagner, R. Brandenburg et al., The barrier discharge: basic properties and applications to surface treatment. Vacuum **71**, 417–436 (2003)
18. U. Kogelschatz, B. Eliasson, W. Egli, Dielectric-Barrier discharges. Principle and applications. J. Phys Iv Fr. 7 C4, 47–66 (1997).
19. U. Kogelschatz, in Advanced Ozone Generation, in *Process Technologies for Water Treatment*, ed. by S. Stucki (Plenum Press, New York, 1988), pp. 87–120
20. U. Kogelschatz, B. Eliasson, in Ozone Generation and Applications in Handbook of Electrostatic Processes, eds. by J.S. Chang, A.J. Kelly, J.M. Crowley (Marcel Dekker, New York, 1955), pp. 581–605.
21. U. Kogelschatz. in Proceedings of the International Ozone Symposium. (Basel, Switzerland, 1999), pp. 253–265.
22. U. Kogelschatz, Dielectric-barrier discharges: their history, discharge physics, and industrial applications. Plasma Chem. Plasma Process. **23**(1), 1–46 (2003)
23. M. Taguchi, K. Yamaschiro, T. Takano, H. Itoh, Extreme decrease of ozone product using high pure oxygen. Plasma Process. Polym. **4**, 719–727 (2007)
24. G. Vezzù, A. Freilich, J.L. Lopez, Fundamental Effects of N_2 and CH_4 in Ozone Generators. Technical Report - Degremont Technologies HKH 108717 (2007).
25. H.V. Lang, G. Vezzù, M. Suter, in *Proceedings IOA 17th Ozone World Congress. Science and Experience Based Improvements of industrial Ozone Generators*, vol II.3 (Strasbourg, 2005), pp. 1–10
26. G. Vezzù. Intelligent Gap System. Technical Report, Dübendorf, Switzerland: Ozonia Ltd: HKH-107000 (2005)
27. G. Vezzù, R. Merz, R. Gisler, M. Müller, B. Paolini, J.L. Lopez, A. Freilich, K.H. Becker. in *Proceedings of the 2007 IOA IUVA World Congress*. Evolution of Industrial Ozone Generation, (Los Angeles, 2007)
28. G. Vezzù, J.L. Lopez, A. Freilich, K.H. Becker, Optimization of large-scale ozone generators. IEEE Trans. Plasma Sci. **37**(6), 890–896 (2009)

29. M. Hirth, U. Kogelschatz, Zusammenhang zwischen Entladungsstruktur und Wirkungsgrad im Siemens-Ozonisator. Baden, Switzerland: Brown Boveri Corporation Technical Report: KLR 81–72 B (1981)
30. M. Hirth, U. Kogelschatz, J. Müller. Der Einfluss von geometrischen Toleranzen auf den Wirkungsgrad von Ozonisatoren. Baden, Switzerland: Brown Boveri Corporation. Technical Report: KLR 83–104 B (1983)
31. U. Kogelschatz, J. Müller. Messungen zur Stickoxidbildung in Luftozonisatoren. Baden: Brown Boveri Corporate Research Technical Report: KLR 85–50 B (1985)
32. U. Kogelschatz, P. Baessler, Determination of nitrous oxide and dinitrogen pentoxide concentration in the output of air-fed ozone generators of high power density. Ozone Sci. Eng. **9**, 195–206 (1987)
33. U. Kogelschatz, B. Eliasson, M. Hirth, Ozone generation from oxygen and air: discharge physics and reaction mechanisms. Ozone Sci. Eng. **10**, 367–378 (1988)
34. W.E. Cromwell, T.C. Manley, Effect of gaseous diluents on energy yield of ozone generation from oxygen. Ozone Chem. Technol. Adv. Chem. Ser. **21**, 304–312 (1959)
35. E. Inoue, K. Sugino, Inhibiting action of hydrocarbons on ozone formation by silent electrical discharge. Ozone Chem. Technol. Adv. Chem. Ser. **21**, 313–312 (1959)
36. M. Salvermoser, D.E. Murnick, U. Kogelschatz, Influence of water vapor on photochemical ozone generation with 172 nm Xenon excimer lamps. Ozone Sci. Eng. **30**, 228–237 (2008). Also in Eur. Phys. J. Appl. Phys. **47**(2), 22812 (2009)
37. M. Figus, Rotational and vibrational temperature measurements in capillary plasma electrode (CPE) discharges. Master's Thesis, Stevens Institute of Technology, QC718.5.S6 F54 (2004)
38. B. Eliasson, J. Müller, Electrical Discharge in Oxygen, Efficiency and Concentration of the Ozone Production. Baden Brown Boveri Corporate Research; Technical Report: KLR 79-31 B (1979)
39. L.C. Pitchford, E. Sciamma, Calculated excited and radical populations in a DBD discharge in N2/O2 mixtures. Ozonia Ltd, Dübendorf, Switzerland. Technical Report HKH108665 (2005)
40. J. Lopez, G. Vezzù, A. Freilich, B. Paolini, Effects of hydrocarbon contamination on ozone generation with dielectric barrier discharges. Eur. Phys. J. D **67**, 180 (2013)

Chapter 14
Dental Applications of Atmospheric-Pressure Non-Thermal Plasmas

WeiDong Zhu, Kurt Becker, Jie Pan, Jue Zhang and Jing Fang

Abstract This chapter presents a summary of selected recent research efforts devoted to the use of low-temperature (or non-thermal) atmospheric-pressure plasmas in various dental applications. Areas of application that will be discussed here in some depth include the surface modification and functionalization of selected dental surfaces and materials, plasma-induced and plasma-assisted in-vitro teeth bleaching, and dental bacteria and bacterial biofilm inactivation in root canals. This chapter is not intended to provide a comprehensive overview of all dental applications of atmospheric-pressure, non-thermal plasmas, nor to review the diverse research efforts carried out by many groups in this field, but rather focus on a few applications, which have been the recent research focus of the authors and their collaborators.

14.1 Introduction

Non-thermal plasmas have been used quite commonly for various biological applications such as inactivation of microorganisms or the sterilization of medical devices since the mid 1990s [1, 2]. By contrast, their use in biomedical and

W. Zhu (✉)
Department of Applied Science and Technology, Saint Peter's University, Jersey City, NJ 07306, USA
e-mail: wzhu@saintpeters.edu

K. Becker
Department of Applied Physics, Department of Mechanical and Aerospace Engineering, New York University Polytechnic School of Engineering, Brooklyn, NY, USA

J. Pan
Department of General Dentistry, Peking University School of Stomatology, Beijing, China

J. Zhang · J. Fang
College of Engineering, Peking University, Beijing, China

Academy for Advanced Interdisciplinary Studies, Peking University, Beijing, China

M. Bonitz et al. (eds.), *Complex Plasmas*, Springer Series on Atomic, Optical, and Plasma Physics 82, DOI: 10.1007/978-3-319-05437-7_14,

medical (incl. dental) applications is a fairly recent endeavor, with a series of pertinent publications starting to appear in the literature around 2005. Plasma physicists across the world have been teaming up with biologists, medical doctors and medical and dental researchers to explore areas that plasmas had never been used in before. Since 2009, special sessions dedicated to the medical applications of plasmas have been organized as part of many international plasma physics conferences such as the International Conference on Plasma Science (ICOPS), the Gaseous Electronics Conference (GEC), and the International Workshop on Microplasmas (IWM). An international conference entirely devoted to medical application of plasmas, the International Conference on Plasma Medicine, was started in 2007 and has since been held approximately every 18 months. A few special issues on this topic have also been published by journals such as the New Journal of Physics, the IEEE Transaction on Plasma Science, and Plasma Processes and Polymers. For a recent in-depth review of the field of plasma medicine, we refer the reader to reference [3]. One of the first reported examples of the use of a non-thermal plasma in dental applications is the work of Stoffels and co-workers [4], who developed a "plasma needle" that they used to treat dental cavities [5]. Many more dental applications of non-thermal plasmas have been reported in the literature in recent years, in fact, far too many to provide a comprehensive review of the field and of the multi-faceted research efforts in this field in a single book chapter of a limited scope. Thus, we want to emphasize that the current chapter does not attempt to review all aspects of dental applications of non-thermal plasma nor review comprehensively the work of the large number of groups engaged in this endeavor, but rather has the purpose to give a snapshot of a limited number of selected efforts in the areas of plasma surface modification and functionalization, plasma-assisted and plasma-induced teeth bleaching, and dental bacteria and bacterial biofilm inactivation in root canals, with an emphasis on the work carried out by the authors and their collaborators.

14.2 Atmospheric-Pressure Plasmas and Plasma Sources

14.2.1 Overview

A brief review of the historical roots of atmospheric-pressure discharge plasmas can be found in reference [6], from where we take a few pertinent facts for the brief summary given below. The observation that air can pass electrical charges was first reported by Coulomb in 1785 [7], who verified that two oppositely charged metal spheres gradually lose their charges due to leakage through the surrounding air. Systematic studies of atmospheric-pressure arc discharges in air date back to about 1800 [8]. In addition to these hot arc discharges, cold glow discharges were also investigated. Faraday was probably the first to realize that an ionized gas had unique properties [9–11]. Many experiments were carried out at reduced pressure, which had the advantage that only moderate voltages were required to start the discharge and that the discharge plasma could fill the entire discharge vessel (generation of large-volume plasmas). The downside was that vacuum pumps and vacuum chambers were required and progress depended to some extent on the advancement of vacuum technology (and the availability of adequate voltage sources). More refined experiments with rarefied gases started at the beginning of the twentieth century. From the very beginning it was obvious that cold glow discharge plasmas had different properties than the hot arc discharges. For a long time it was believed that glow discharges which are characterized by hot electrons and essentially cold heavy particles (atoms, molecules, ion at or near room temperature) could exist only at low pressure.

Systematic studies of atmospheric-pressure plasmas in air, rare gases, and a variety of gas mixtures started in the 1980s showing that non-equilibrium plasma conditions with electron energies substantially higher than those of heavy particles, and properties resembling those of low pressure glow discharges, can exist also at much higher pressure, for example in atmospheric pressure air [12–17]. Moreover, single collision conditions no longer prevail in high-pressure plasmas. Step-wise processes, i.e. electron-impact excitation of excited states and collisions of excited species with other particles resulting in new energy transfer routes and three-body collisions leading e.g., to the formation of excimers become prominent [13]. This allows for many applications of high-pressure plasmas such as high power lasers, opening switches, novel plasma processing applications and sputtering, electromagnetic (EM) absorbers and reflectors, remediation of gaseous pollutants, medical sterilization and biological decontamination, and excimer lamps and other non-coherent vacuum-ultraviolet (VUV) light sources [14–16]. However, self-sustained diffuse discharges tend to be unstable at high pressure due to their susceptibility to filamentation and the transition to an arc (see references [12, 17] and references therein to earlier work), which limits their practical utility. Many attempts have been made to resolve this problem. In addition to the well-known atmospheric-pressure dielectric barrier discharge, in which one or both electrodes are covered by a dielectric [18], there are two other frequently employed approaches to generate and maintain stable glow-type discharge plasmas at elevated pressures, (1) pulsed excitation, which essentially extinguishes the glow phase before the glow undergoes the transition into an arc and (2) confining the plasma in a cavity, where at least one dimension is at or below about 1 mm, so-called microplasmas, in which interactions of the plasma with the surrounding wall/boundary contribute to the stabilization.

Numerous publications describing and characterizing a variety of atmospheric-pressure plasma sources have appeared in the literature in the last 15 years. We refer the reader to the 2007 review by Laroussi and Akan [3] for a review and extensive references to original papers. Biocompatible plasma sources, i.e. plasma sources for the generation and maintenance of plasmas that will be in direct contact with cells and living tissue, must satisfy stringent criteria in terms of the electrical, chemical, and thermal properties of the plasma so as not to harm the cells and tissue. Two recent reviews [19, 20] provide extensive summaries of plasma sources used in various (bio) medical applications. Here we will give brief description of three biocompatible plasma sources used in the work reported in this chapter.

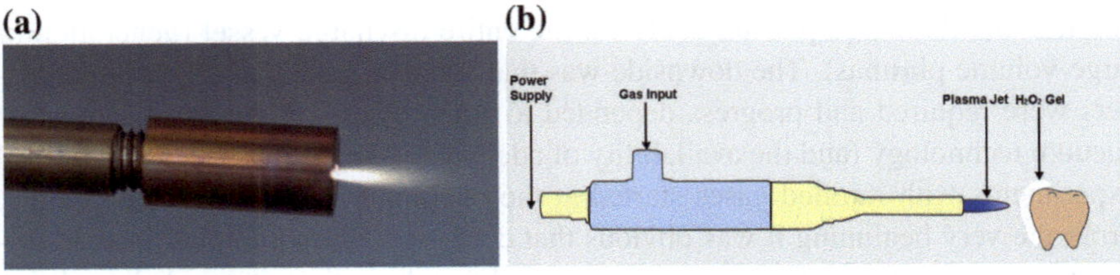

Fig. 14.1 **a** Picture of the atmospheric-pressure cold PMJ. **b** Schematic diagram of the PMJ treatment of an extracted tooth [23]

14.2.2 The Plasma MicroJet (PMJ) Source

The Plasma MicroJet (PMJ) device is modeled after the source first described by Kolb and co-workers [21] with a few minor modifications. It is comprised of two copper tubings separated by an alumina tubing. The critical distance between the ends of the copper tubings is set to ∼0.5 mm. An end cap with a circular opening (∼0.8 mm in diameter, ∼1 mm in depth) is attached to the outer copper tubing to match the diameter of the pore on the inner copper tubing for optimal operation. The inner copper tubing is completely embedded in the device and powered by a dc power supply (negatively biased). The outer electrode is grounded for safety reasons. Compressed air or a rare gas (such as helium and argon) with admixtures of a molecular gas (typically oxygen) is used as the working gas at a flow rate of 4–5 standard liters per minute (slm). The discharge sustaining voltage is in the range of 400–600 V with an operating current of 20–35 mA. For a detailed device description, we refer the reader to reference [22]. The generated PMJ is usually between 8–10 mm in length. Figure 14.1a shows a picture of an air PMJ and Fig. 14.1b depicts schematically the interaction of the PMJ with a tooth in a teeth bleaching application.

14.2.3 The INP kINPen Source

The kINPen 09 source shown in Fig. 14.2 was developed by researchers at the INP Greifswald [24, 25]. It consists of a hand-held unit for the generation of a plasma jet at atmospheric pressure, a rf power supply, and a gas supply unit and works with all rare gases (especially argon), but also with nitrogen and air and admixtures of molecular gases to the rare gas. The plasma is ignited at the tip of a centered electrode inside a capillary and blown out by the gas flow to form the plasma jet. Because of the contracted plasma resulting from this type of generation, the kINPen 09 is especially suited for focused treatments of small spots. Large-scale treatments can be realized by moving the jet over a selected area. These features make it interesting for the treatment of parts of complex geometries and cavities, e.g. in operative dentistry or dermatology. If needed, plasma jets can also be arranged in arrays to adapt to specialized geometries.

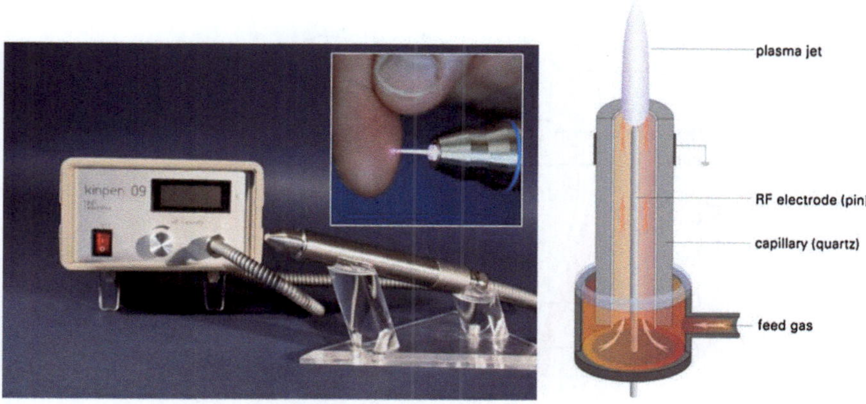

Fig. 14.2 A photograph of INP kINPen 09 and a schematic diagram of the device [24]

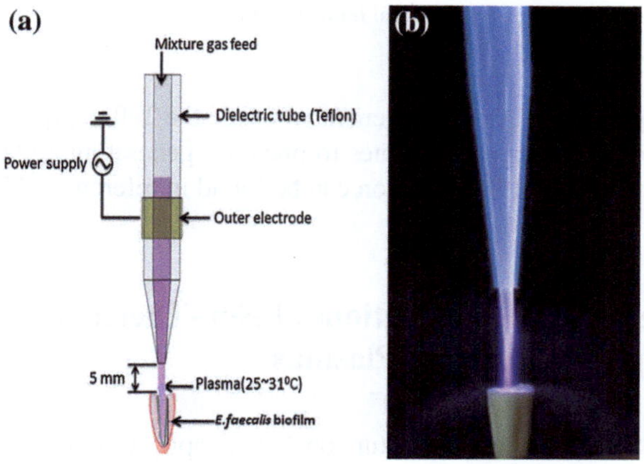

Fig. 14.3 **a** Schematic diagram of the plasma pipette (PP) device and **b** photograph of the PP during an in vitro root canal treatment [27]

14.2.4 The Plasma Pipette (PP) Source

The Plasma Pipette (PP) device resembles to some extend the Plasma Pencil that was introduced by Laroussi and Lu first in 2006 [26]. Unlike the double electrode arrangement of the Plasma Pencil, a single annular copper electrode (floating at high electrical potential) is wrapped around a Teflon pipette tube with inner and out diameters of 7 and 10 mm, respectively. The inner diameter of the pipette tube tapers down to about 1.5 mm at the exit. Figure 14.3 shows a schematic diagram of the PP device (Fig. 14.4a) and a photograph of the in vitro plasma treatment of a single root canal tooth (Fig. 14.3b). The annular electrode is connected to a low frequency (10 kHz) ac power supply (sinusoidal wave with a peak voltage of 18 kV) via a 1 $M\Omega$ ballast resistor. Various noble gases and noble gas with molecular gas admixture are

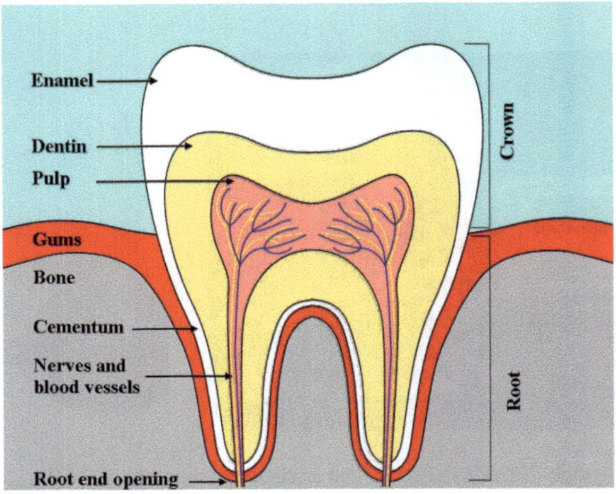

Fig. 14.4 A schematic structure of a normal human tooth

used as working gas. The plasma is generated inside the Teflon pipette tube near the powered electrode, and then propagates to open air, generating a plasma jet of ~5 cm length. Details of the PP generation can be found in reference [27].

14.3 Selected Dental Applications of Non-Thermal Atmospheric-Pressure Plasmas

There is a significant amount of literature on dental applications of non-thermal plasmas (at both low and atmospheric pressures) reflecting the work done by many groups active in this field [23, 28, 29]. Instead of a comprehensive review of all the dental applications of non-thermal plasmas, we focus on the work carried out by the authors and their collaborators with the atmospheric-pressure non-thermal plasma sources described in Sect. 14.2, in the following three areas: (1) the interactions of non-thermal atmospheric-pressure plasmas with dental materials; (2) plasma-induced and plasma-assisted *in vitro* teeth bleaching; and (3) plasma root canal treatment.

14.3.1 Human Tooth

Before we discuss the dental applications of non-thermal atmospheric-pressure plasmas, it is beneficial to briefly describe the structure of the human tooth. Figure 14.4 shows schematically the structure of a normal human tooth, which consists of the crown, and the root and is comprised of enamel, dentine, cementum and pulp. Enamel, which is primarily composed of minerals (96 %, mainly hydroxyapatite), water and organic material, is the hardest and highly mineralized structure in the human body, yet it is also very brittle. Normally, it is the only part of the tooth exposed to the environment [30]. The enamel is supported by its underlying dentine, a protective layer that supports the crown and protects the pulp. It consists of mineralized connective tissue with bundles of collagen filaments surrounded by mineral crystals. Dentine is porous and made up of inorganic material (70 %), organic material (20 %), and water (10 %) [31]. Dentine is softer than enamel and, as a result, dentin decays faster than enamel. This decay leads to cavities, if not treated properly. Cementum, which is softer than either enamel or dentin, is found in the root covering the dentine. Its composition is almost similar to that of the bones, but lacks vascularization. Cementum consists of inorganic material (45 %) (mainly hydroxyapatite), organic material (33 %) (mainly collagen) and water (22 %) [31]. The pulp contains blood vessels and nerves. The blood and nerve cells are supplied from the tissues of the jaw via the root.

14.3.2 Interactions of Non-Thermal Atmospheric-Pressure Plasmas with Dental Materials

14.3.2.1 Background

The interaction of non-thermal plasmas with surfaces has been a subject of intense studies for many decades. Both atmospheric-pressure plasmas and low-pressure plasmas have been widely used to modify and functionalize a large variety of surfaces including polymers and biomaterials. We refer the reader to a few recent reviews on the topic [32–35]. A particularly illustrative summary was given by Schmidt [34], who categorized the interaction of plasmas with surfaces as follows:

1. *Etching and sputtering*. Etching means the removal of bulk material, a process which includes chemical reactions producing volatile compounds that contain atoms of the bulk material. Sputtering is a physical process which removes bulk atoms from the surface by collisions of energetic ions. These processes lead to a weighable loss of bulk substance.
2. *Deposition*. Deposition of films of non-substrate material changes the mechanical (tribology), chemical (corrosion protection), and optical (reflecting and decorative) properties of materials.
3. *Cleaning*. Cleaning refers to the removal of material on the surface, a process which may or may not also result in the removal of bulk material.
4. *Functionalization and grafting*. Functionalization creates functional groups and/or cross links on the surface by chemical reactions between gas-phase species and surface species/reactive sites and/or surface species. Grafting refers specifically to surface reactions between gas phase species and polymeric materials. The mass yield or loss in these processes is typically very small. Functionalization changes

the surface wettability (typically enhances the wettability), the adhesion, lamination to other films, the printability, and other coating applications. Biological properties may also be influenced.

5. *Interstitial modifications*. Interstitial modifications occur e.g. by ion implantation and lead to the hardening of metal surfaces.

The depth up to which the different processes affect the surface typically range from 1 nm (functionalization) to 10–100 nm (etching), and up to 1 μm for coatings. We note that in reality the different processes are often not separated, but occur simultaneously, e.g. cleaning may include sputtering or functionalization or etching may occur along with deposition and/or sputtering.

In the context of the plasma-assisted and/or plasma-initiated surface treatment processes of interest in dentistry, we mention inactivation of bacteria and bacterial biofilms inside the oral cavity as well as inactivation/decontamination of dental materials, implants, and biomaterials relevant to the bone tissue/implant interface as well as the functionalization of materials as topics of current research. For an overview of the vast literature on these topics, we refer the reader to a few reviews, which contain numerous references to the original literature. In addition to Ref. [33], we note the 2012 article by Kona et al. [36], the 2011 review of Yoshari et al. [37], which addresses the functionalization of dental impact surfaces by non-thermal plasmas (NTP) and the 2010 reviews by Smitha and Chaitanya [38] and Ritts et al. [39] on dentin surface treatment by a non-thermal plasma.

Bonding issues such as the stable adhesion to ceramic and alloy surfaces aimed at enhancing the bond strength and bond stability for long-term clinical performance continues to be an important goal for reliable clinical performance of restorative dental materials. Homogeneous and stable bonds depend on the optimal wettability of the adherent surface by the liquid adhesive. Wetting, the first stage in adhesion, determines the actual contact area between the adhesive and the solid, thus affecting the adhesion strength and it determines the energetics of the detachment of the adhesive from the solid. Optimal wettability is achieved when the surface energy (SE) of a substrate is maximized relative to the liquid adhesive to lower the liquid contact angle on the substrate. Adequate bonding depends on both micromechanical interlocking and physico-chemical treatments to assure wetting by the adhesive. The physical treatment consists mainly in the creation of a rough and tribochemically altered surface and chemical treatment entails the application of ceramic and/or alloy primers to the surface. Alloys have been treated physically for bonding purposes e.g. by electrolytic etching, sandblasting, deposition of a silica layer and laser-sintered cobalt chrome (see e.g. [40] and references therein), occasionally combined with the application of alloy primers. However, some methods are less effective for reactive metals such as titanium, which, along with Ti alloys, is a preferred material for dental implants [41].

Similar efforts have been made for ceramics, initially using glass ceramic systems etched with hydrofluoric acid followed by silane application. However, the subsequent utilization of structural ceramics of high chemical inertness and surface homogeneity such as high strength yttrium stabilized tetragonal zirconia polycrystals (Y-TZP) required modified resin bonding cementation concepts. Chemical and mechanical procedures to promote adequate bonds to Y-TZP included the combination of airborne particle abrasion with phosphate ester monomers, in particular, 10-methacryloyloxydecyl dihydrogenphosphate (MDP). However, airborne abrasion in conjunction with MDP has a tendency to induce micro-cracks on Y-TZP substrates at the intergrain level and may also introduce microgaps and structural defects that might be detrimental for the longevity of the restoration. Thus, alternative approaches to improve the bonding mechanisms for Y-TZP-based materials (as well as Ti) are desirable.

14.3.2.2 Atmospheric Pressure Non-Thermal Plasma Treatment of Dental Material

Low-temperature plasmas have been used in many industrial applications to change (mostly enhance) the wettability of surfaces through the surface energy (SE) changes of substrates without changing the nature of the bulk of the substrate [42]. We report the results of a series of experiments aimed at elucidating the effect of a nonthermal plasma interacting with Ti and Y-TZP surfaces in terms of changes in the SE, changes in the surface properties as determined from optical interferometry and via X-ray photoelectron spectroscopy (XPS), and changes in the contact angle using a commercial MDP primer dispensed on Y-TZP and Ti surfaces [40]. We used the kINPen 09 plasma jet described earlier and exposed Ti and Y-TZP discs to an Ar/O_2 (1–5 %) plasma for 5, 10, and 20 s. The effects of the plasma treatment were benchmarked against discs in control groups that were not exposed to the plasma.

The contact angle of liquid on a solid surface can serve as a direct, yet only qualitative indication of surface energy: the smaller the contact angle, the higher the surface energy value. Images of the contact angle of water before and after plasma treatment for both Ti and Y-TZP are shown in Fig. 14.5. Significant increases in the SE values were observed for all plasma treated groups relative to their untreated counterpart. More specifically, the increase in SE values can be attributed mainly to an increase in the polar component of the SE between plasma-treated and non-treated samples, while comparatively smaller variations occurred for the disperse component of the SE between plasma-treated and non-treated samples [40]. The plasma exposure time (5, 10, or 20 s) appeared to have no discernible effect on the SE values within plasma-treated groups for both substrates. The plasma treatment resulted in a uniform reduction in the disperse values of the SE for the plasma-treated Ti discs compared with baseline. The disperse component showed a slight reduction for the Y-TZP (5 and 10 s exposure) groups and a statistically insignificant small increase for the Y-TZP (20 s exposure) group.

The surface chemistry evaluation of Y-TZP and Ti surfaces by XPS showed that the plasma exposure strongly influenced the concentration of atomic O and C on the substrate surface for both materials, with a 20–25 % increase of O and a roughly 40 % decrease of C (as shown in Fig. 14.6). The presence of O_2 in the operating gas of the plasma device promotes the formation of active peroxide radicals

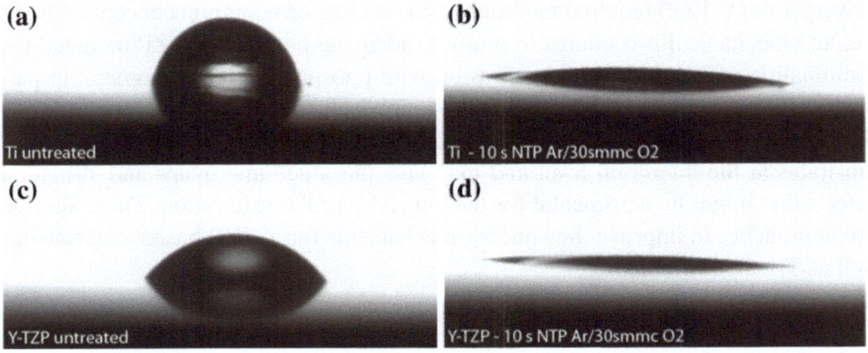

Fig. 14.5 Contact angle of a water droplet dispersed on untreated, **a** and **c**, and plasma-treated (10 s), **b** and **d**, Ti and Y-TZP. The contact angles were 82° and 57° for Ti and Y-TZP, respectively, before plasma exposure and essentially 0° after plasma treatment [40]

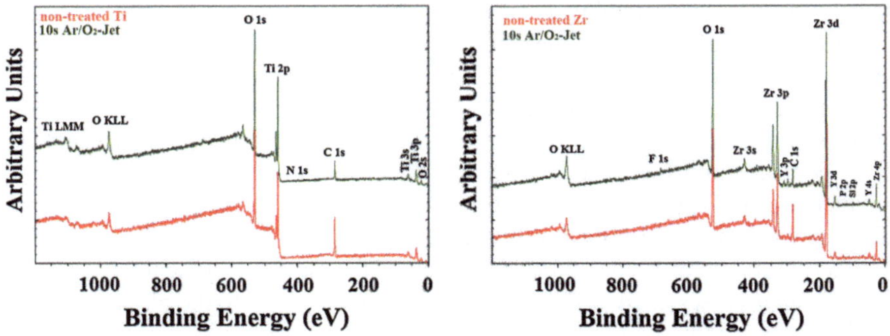

Fig. 14.6 XPS spectra of Ti (*left*) and Y-TZP (*right*) showing the changes in the concentrations of O and C on the substrate surface prior and after plasma treatment [40]

(R-O-O·, R: free organic radical), which initiate chemical surface changes even on inert materials such as dense crystalline sintered ceramics. The use of O_2 also promotes the incorporation of additional functional groups (C–O, C–OH) into the top layer of the treated surface.

Excited particles, atoms, and UV radiation also stimulate chemical changes at the surfaces such as the breaking of R-H and C-C bonds and initiate secondary reactions of atomic hydrogen. The presence of oxygen in the plasma also has a cleaning effect by removing organic contamination through the breaking of C-H and C-C bonds, thus reducing the hydrocarbons adsorbed on the surface. Surface cleaning of metals can result in surface energies that approach 70 mN/m, which ensures excellent wettability [42]. Since plasma treatment increased the SE values to more than 60 mN/m for both materials compared with values of 35–40 mN/m for the untreated samples, it was not surprising that higher wettability was observed relative to non-treated substrates. This is in agreement with results reported elsewhere [35, 43]. However, our results show that the increase in SE values is primarily due to an increase in the polar component on plasma-treated samples of Ti and Y-TZP. In fact, the disperse component, perhaps related to adsorbed -C-H moieties, was found to decrease after plasma treatment for Ti, but remained essentially unchanged for Y-TZP. Although the surface modification is expected to depend, in addition to the operating gas and the pressure, on the plasma exposure time, we found no significant difference in the SE values for Y-TZP and Ti for 5, 10, or 20 s of plasma exposure.

The contact angle obtained for the MDP primer dispensed on the Y-TZP surface was reduced to 0° after plasma treatment, indicating that total wetting of the surface was achieved. Chemical bonding occurs between the MDP monomer and metal ions present on restorative dental ceramics. As Y-TZP contains metal oxides, the plasma treatment most likely changed the chemical bonds involving these metal oxides via secondary forces such as Van der Waals forces or hydrogen bonds and the MDP primer. A more specific mechanism may be the generation of -OH groups on the Y-TZP surface by the plasma, thus increasing the acid base reaction between the surface and the acidic phosphate of the MDP molecule. This is supported by the XPS findings of increased O on the surface after treatment.

Unlike the observation for the non-treated Y-TZP, a 0° contact angle with MDP was already observed for the non-treated Ti surface. Titanium is a highly reactive element, and various chemical elements such as oxygen have high affinity for titanium, spontaneously forming an oxidized surface, which induces a high wettability on the Ti surface and thereby is beneficial for the bonding mechanisms as the phosphate ester group of MDP can bond to the metal oxides.

Overall, all the proposed NTP regimes applied may be regarded as an attractive alternative for increasing the SE and wettability of two widely utilized biomaterials.

14.3.3 Plasma-Induced and Plasma-Assisted in vitro Teeth Bleaching

14.3.3.1 Background

Nowadays, dental bleaching (also known as tooth whitening) is of great popularity and importance in cosmetic dentistry. The distinction between bleaching and whitening, according to the Food and Drug Administration (FDA), is that whitening restores the natural tooth color and bleaching whitens the teeth beyond that. For reasons of convenience, we will not distinguish bleaching from whitening in this chapter and use the terms interchangeably. The color of teeth is determined by the combined effects of the intrinsic color of teeth and the presence of extrinsic stains formed on the tooth surface by chromogenic agents, such as medications (in particular tetracycline), beverages (coffee, tea and red wine), and tobacco (smoking or chewing). Tooth stains are mainly composed of single-bond or double-bond organic compound chromophores. The whitening of teeth can probably be traced back to the ancient Romans, who used a tooth paste made from human urine. Dentists later attributed the cleaning of teeth to the ammonia in the urine. Modern tooth-bleaching treatments are based primarily on the bleaching effect induced by hydrogen peroxide (H_2O_2) and include (1) in-office or power bleaching, (2) dentist-prescribed home bleaching done by the patient, and (3) over-the-counter (OTC) bleaching kits. Home bleaching uses low levels of the bleaching agent on the teeth via a mouth guard and treatment duration is about 2–4 weeks. OTC kits are self-applied to the teeth and contain low concentrations of H_2O_2 or carbamide peroxide and require usually more time for optimum whitening compared to processes using prescription whitening agents. In-office bleaching generally uses relatively high doses of whitening agents (25–35 % H_2O_2) on the teeth for a fast effect. Some recent approaches use light sources at various wavelengths to accelerate the bleaching process. However, it is not clear that the application of light significantly improves the efficacy of the bleaching materials [44–46].

14.3.3.2 Plasma Induced and Plasma-Assisted Teeth Bleaching

The application of atmospheric pressure non-thermal plasmas in teeth bleaching was first reported by Lee and coworkers [47]. They used an ac driven hybrid helium plasma jet and 28 % H_2O_2. It was not clear whether the H_2O_2 used in their study was introduced in liquid form or embedded in a whitening gel. Nevertheless, the results were astounding. When the plasma was used in combination with H_2O_2, the bleaching efficacy was improved by a factor of 3 compared to using H_2O_2 alone. Around the same time, Sun et al. reported the effective bleaching of extracted teeth with dental whitening gel (35 % H_2O_2) assisted by a PMJ source operated in compressed air as the working gas [23].

In the work of Sun et al. [23], forty intact caries-free human teeth, chosen from premolars extracted for orthodontic reasons, were cleaned thoroughly and evenly divided into two groups: Group A was treated only with clinical dental whitening gel (Beyond Technology Corp.), containing 35 % H_2O_2, for 20 min at room temperature; Group B was treated with both the H_2O_2 gel and the PMJ, in which the tooth was placed at 1 cm from the exit nozzle of the plasma device and the H_2O_2 gel (1–2 mm thick) was replenished roughly every 30 s, since the air flow inevitably pushed the gel off the tooth in about 30 s. All teeth were stored in a 0.1 % Thymol solution at 4 °C before use.

Photographs of teeth before and after the treatment are shown in Fig. 14.7a and b, respectively. The teeth in group B showed clearly a better bleaching effect than those in group A. In order to quantify the results, the images of the tooth specimens were analyzed with a spectrophotometer and transformed into the CIE (L^*, a^*, b^*) color classification scheme, a standard color classification scheme used for evaluating tooth whitening in a clinical setting. This system allows color specification in a 3-D space. The L^* axis represent the degree of brightness within a sample, ranging from 0 (black) to 100 (white). The a^* plane denotes the degree of green/red, while the b^* plane represents the degree of blue/yellow in the sample. The overall change of colorimetric value of the specimen before and after the treatment was calculated using

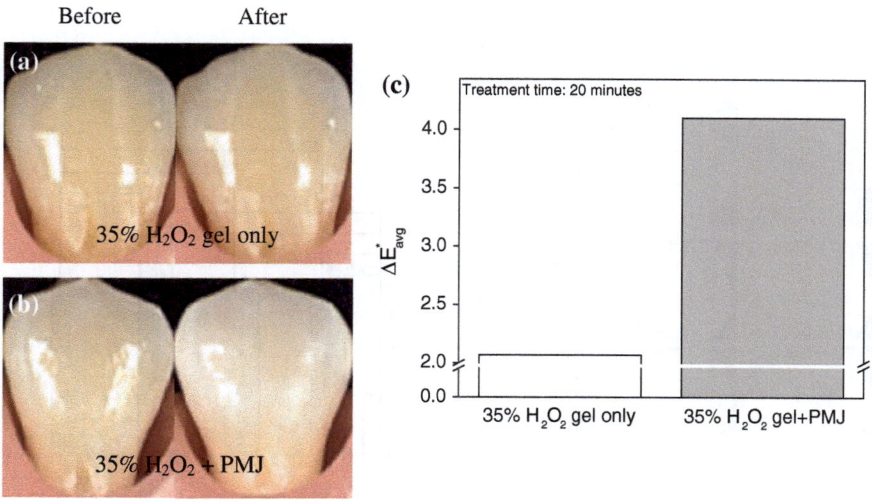

Fig. 14.7 Photographs of typical teeth before (*left*) and after (*right*) 20 min treatment by **a** dental gel (35 % H_2O_2) only (Group A) and **b** dental gel (35 % H_2O_2) and PMJ (Group B) [23]; **c** Average colorimetric value ΔE^*_{avg} of the teeth in the two groups

$$\Delta E^* = \sqrt{(\Delta L^*)^2 + (\Delta a^*)^2 + (\Delta b^*)^2}$$

The average colorimetric value ΔE^*_{avg} of teeth in groups A and B are shown in Fig. 14.7c. The value for the teeth in groups A was slightly above 2, whereas the value for teeth in group B reached a value of approximately 4.4.

In a follow-up study, plasma-assisted teeth bleaching by dental whitening gel with other concentrations of H_2O_2 (6, 15 and 25 %) was also investigated [48]. Here the sample size was 10 teeth per group and the same experimental procedures were followed. Photographs of teeth from these groups are shown in Fig. 14.8 together with those from the 35 % H_2O_2 + PMJ group and the control group (in which only 35 % H_2O_2 was applied). Also shown in Fig. 14.8 are the average colorimetric value ΔE^*_{avg}. The ΔE^*_{avg} value increased by a factor of about 2.5–3.0 in the plasma treated groups, regardless of the concentration of H_2O_2. Statistical analysis (ANOVA) showed a significant difference between the plasma treated groups and the control group. However, no significant difference of the ΔE^* values was observed among the various plasma treated groups.

There had been attempts to apply the PMJ directly on teeth for bleaching purposes. However, enamel surface dehydration due to the continuous flow of air (at a temperature slightly above 37 °C) may also lead to a temporary whitening effect (demineralization). Therefore, saline solution (0.9 % NaCl) was applied to the tooth surface and replenished every 30 s in the PMJ treatment process, to keep the tooth surface moist. Teeth exposed to an air flow and the saline solution for 20 min were used as negative control group [49]. The bleaching efficacy is compared to that obtained

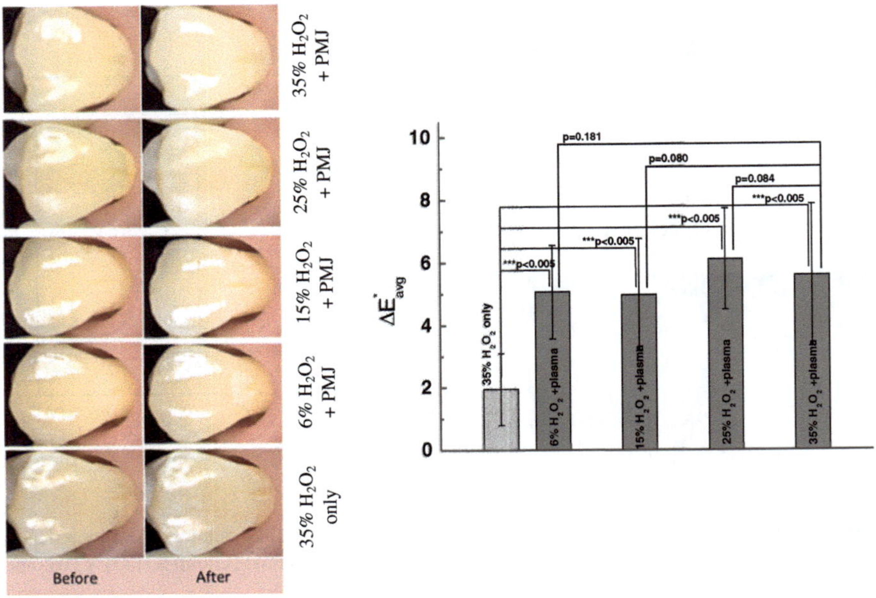

Fig. 14.8 Photographs of teeth before and after the plasma treatment and average colorimetric values of teeth from various treatment groups [48]

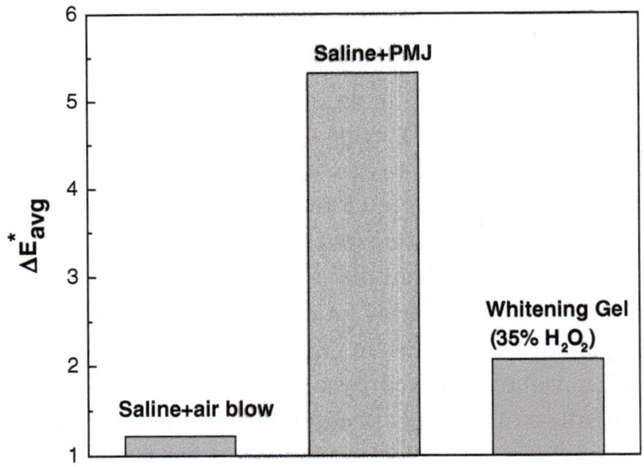

Fig. 14.9 Average colorimetric value ΔE^*_{avg} of the teeth treated with saline solution and air blow at room temperature, with saline solution and PMJ and with whitening gel (35 % H_2O_2) at room temperature. Teeth were treated for 20 min in all groups (10 teeth per group) [49]

from the whitening gel (35 % H_2O_2) alone and shown in Fig. 14.9 (sampling size: 10 teeth in each case). It is interesting to note that using the PMJ in conjunction with a saline solution can achieve the same level of bleaching efficacy as the whitening gel with the PMJ.

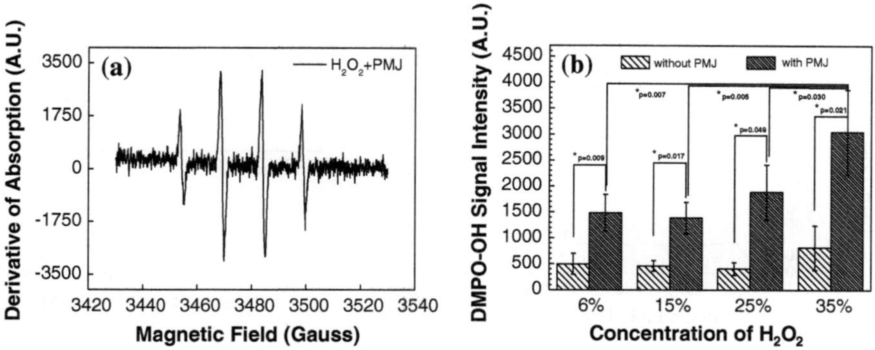

Fig. 14.10 **a** ESR spectrum of DMPO-OH (with a peak intensity ratio of 1:2:2:1) [23]; **b** Intensity of DMPO-OH signals at different H_2O_2 concentrations before and after employing the plasma [48]

14.3.3.3 Bleaching Mechanism

It is believed that the hydroxyl radical (·OH) reacts with the chromophores and breaks macromolecular stains into smaller molecules that are colorless [50, 51]. ·OH generated by different concentrations of H_2O_2 before and after the plasma treatment were measured by using an electron spin resonance (ESR) spin-trapping method [52]. DMPO (5,5-dimethyl-1-pyrroline-1-oxide) was used to spin trap ·OH at each concentration of H_2O_2. For each plasma treatment group, the plasma was applied at a fixed distance of 10 mm (from the exit nozzle of the device to the surface of the liquid) for 20 s to obtain the best signal-to-noise ratio in the ESR spectrum. The direct spin trapping reaction between DMPO and ·OH produces the spin adduct DMPO-OH that is characterized by a quartet ESR spectrum with a peak intensity ratio of 1:2:2:1, as illustrated in Fig. 14.10a. Half heights of the second peak of the quartet spectra were chosen to represent DMPO-OH intensities and are plotted in Fig. 14.10b.

The DMPO-OH signal intensity due to the plasma treatment increases by a factor of 2–3 compared to the untreated cases. Meanwhile, a significant difference in the DMPO-OH signal intensities was found between the 6, 15, 25 % H_2O_2 plasma treated groups and the 35 % H_2O_2 plasma treated group.

The fact that the increase of the ·OH concentration in the 35% group did not result in a better whitening efficacy indicates that, while hydroxyl radicals are important, they are not the only radicals that contribute to the tooth bleaching. Other species such as atomic oxygen (O), the superoxide anion ($O_2^{-\cdot}$) and singlet molecular oxygen ($^1O_2$) must also play a role in the tooth bleaching process. O is readily generated though electron impact dissociation of molecular oxygen, and serves as the precursor of other reactive oxygen species. $O_2^{-\cdot}$, or rather HOO$^\cdot$ (the conjugate acid of $O_2^{-\cdot}$) can in principle be spin trapped by DMPO too, forming spin adduct DMPO-OOH, which typically shows 12 consecutive peaks in ESR spectrum. However, the direct detection of $O_2^{-\cdot}$ through spin trapping with DMPO is rather difficult since $O_2^{-\cdot}$ reacts

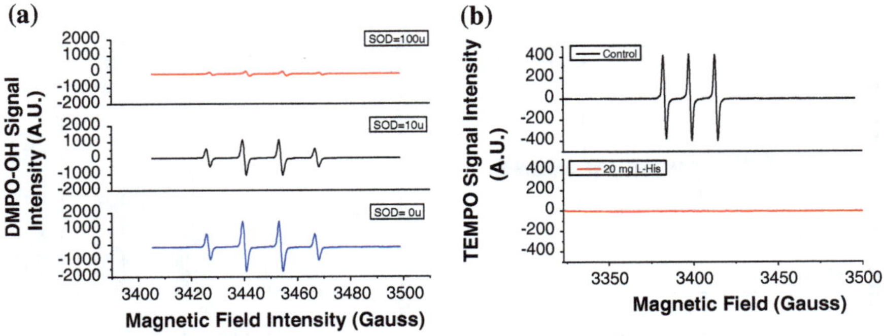

Fig. 14.11 **a** DEPO-OH signal and the influence of SOD and **b** TEMPO signal and the influence of L-His [49]

with H_2O_2, and forms ·OH via a so-called Haber-Weiss reaction ($O_2^{-\cdot} + H_2O_2 \rightarrow \cdot OH + OH^- + O_2$). This reaction is enormously accelerated by the catalysis of transition metals such as Fe^{2+} and Cu^+ (Fenton reaction) [53]. Therefore, Superoxide Dismutase (SOD) was added into the system to indirectly evaluate the existence of $O_2^{-\cdot}$. When an increased amount of SOD is added into the system, DMPO-OH signal decreases accordingly (as shown in Fig. 14.11a). This indicates that $O_2^{-\cdot}$ not only exist in the system, but also very likely serves as one of the precursors of ·OH.

$^1O_2$is believed to be generated directly in the plasma and subsequently delivered to the tooth surface. The lifetime of $^1O_2$ (in particular $O_2(^1\Delta_g)$) was calculated to be 72 min [54]. This number is considerably reduced at higher concentration of oxygen and in the presence of other molecular species to the range of milliseconds in the gaseous phase. TEMP (2,2,6,6-Tetramethylpiperidine) is used to spin trap $^1O_2$. The spin trap TEMPO, a triplet pattern in ESR spectrum with intensity ratio of 1:1:1 [55] is detected directly in PMJ treated distilled water. To further verify the existence of $^1O_2$, L-histidine (L-His, a $^1O_2$ quencher) was added into the system before PMJ treatment. TEMPO signal decreases with the increase of L-His and completely disappears when 20 mg L-His is added into the system (as shown in Fig. 14.11b). There must be an interplay among the various radicals ·OH, O, $O_2^{-\cdot}$ and $O_2(^1\Delta_g)$ during the bleaching process, which is currently still under investigation.

14.3.3.4 Evaluation of the Enamel Surface

A few tooth characterization studies were carried out for the 35 % H_2O_2 alone group, the H_2O_2 (various concentrations) + PMJ groups and the saline + PMJ group, including (1) the elevation of tooth surface temperature; (2) the enamel surface morphological and elemental change; and (3) the enamel surface microhardness change. We discuss these studies below focusing on the H_2O_2 only and various H_2O_2 + PMJ cases.

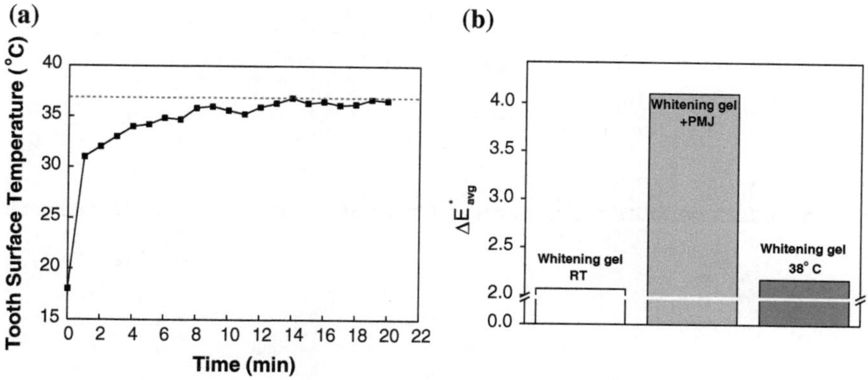

Fig. 14.12 **a** Tooth surface temperature as a function of PMJ treatment time (no whitening gel applied). **b** Average colorimetric value ΔE^*_{avg} of the teeth treated with whitening gel (35 % H_2O_2) at room temperature, with whitening gel and PMJ, and with whitening gel at 38 °C [23]

(1) Evaluation of tooth surface temperature

Prolonged exposure of teeth to temperatures higher than 42 °C (critical temperature) may cause thermal injury to the bone [56]. Therefore, the tooth surface temperature (without the presence of whitening gel) was monitored at the standard treatment distance when the PMJ was applied. The results are shown in Fig. 14.12a.

The tooth surface temperature stabilized at 37 °C after 8 min of plasma treatment. This is well below the critical temperature. In recent studies, we measured the tooth pulp chamber temperature (which really matters most) with a K-type thermocouple buried in the pulp chamber filled with a heat conducting compound. Starting with an initial temperature of 24 °C, the temperature dropped 2 °C in the first minute and rose 8 °C in the following 3 min. Subsequently, the temperature increased at a much slower pace (~0.2 °C/min). After a treatment time of 20 min, the tooth pulp chamber temperature did not increase to the critical temperature.

While thermal damage is not a concern, activation of H_2O_2 due to the increase in the temperature might still be possible. A group of teeth receiving whitening gel was placed in an incubator at 38 °C for 20 min. The average colorimetric value is compared to that from the group receiving whitening gel at room temperature. The results are shown in Fig. 14.12b. No significant difference was found between the two groups. Therefore, simply elevating the temperature (to 38 °C) does not contribute significantly in the bleaching process.

(2) Enamel surface morphological and elemental change

The enamel surface morphology and the elemental composition were determined before and after different treatments (for 15 min) via scanning electron microscopy (SEM) and energy-dispersive X-ray spectroscopy (EDX), respectively. Details of the procedures of these evaluations can be found in reference [57].

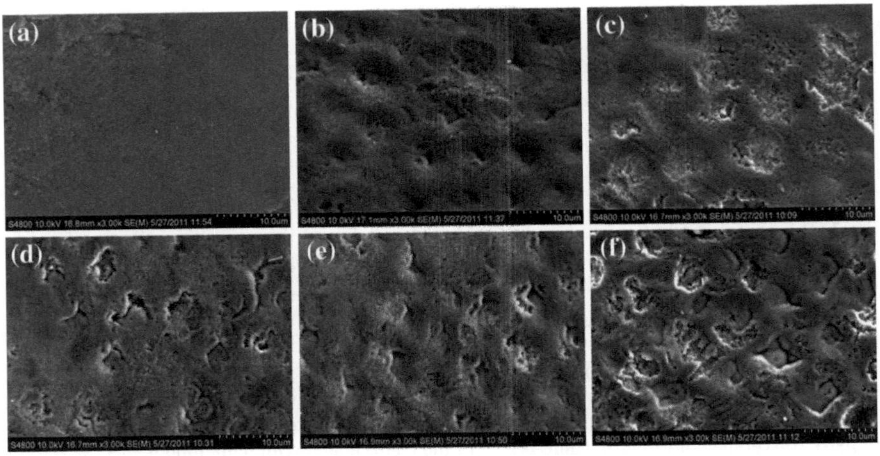

Fig. 14.13 SEM photographs (×3000) of the enamel surface after various treatments: **a** no treatment; **b** 35 % H_2O_2, no plasma; and **c–f** plasma treatment with H_2O_2 at concentrations of 6, 15, 25, and 35 %, respectively [57]

Figure 14.13 shows the enamel surface morphology of a representative tooth from each group at a magnification of 3000.

The morphology of any enamel surface covered with the bleaching agent (see Fig. 14.13b–f) shows a variety of etching patterns not found for the teeth in the blank control group (see Fig. 14.13a). These changes are attributed to demineralization during the treatment process. Whereas the teeth in the plasma-treated groups with a comparatively low concentration of hydrogen peroxide (see Fig. 14.13c, d) show etching and a porosity similar to the teeth in the negative control group (Fig. 14.13b), we find a higher degree of porosity and a higher density of etched structures in teeth that were plasma treated in the presence of higher H_2O_2 concentrations (Fig. 14.13e, f).

The demineralization (erosion) patterns observed in Fig. 14.13 are related to the distribution of various enamel crystals in the enamel rods. We refer interested readers to reference [57] for a detailed discussion and references therein. The highest morphological change observed in the 35 % H_2O_2 and PMJ group is similar to that required in dental bonding procedures [58] and is therefore considered acceptable. Nevertheless, the degree of change can somewhat be controlled by the concentration of H_2O_2 used in the process.

The EDX data in terms of the atomic percentages of Ca, O, and P (primary elemental compounds of tooth enamel) are displayed in Fig. 14.14 for teeth from all six groups mentioned in Fig. 14.13. The changes induced by the treatment are generally minimal, except perhaps for the group with 35 % H_2O_2 and PMJ, which shows an enhanced O concentration and a decline in the P and Ca concentrations. The largely unchanged Ca concentration suggests that the resistance to acid dissolution is acceptable under these experimental conditions. No significant difference of the O/Ca/P atom concentrations between the blank control group and each treated group was observed.

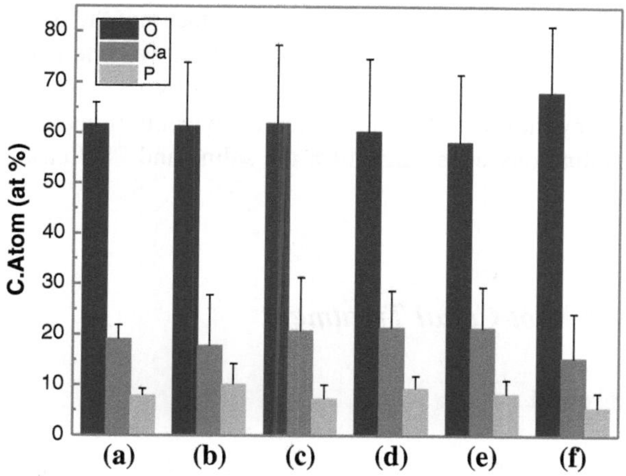

Fig. 14.14 The mean value and the standard deviations of O/Ca/P atom amount on the enamel surface for three teeth in **a** no treatment; **b** 35 % H_2O_2, no plasma; and **c–f** plasma treatment with H_2O_2 at concentrations of 6, 15, 25, and 35 %, respectively. The elemental composition shown for the blank control group is used to represent that of teeth in all groups prior to any treatment. O/Ca/P atom amount between the blank group and each treated group has no significant difference [57]

Table 14.1 The mean values and standard deviations of enamel microhardness of samples in various groups before and after 15 min treatments

Group	Before	After	Change
No treatment	358.87 ± 15.23	319.67 ± 22.17	39.20 ± 19.88
35 % H_2O_2	302.83 ± 55.65	250.22 ± 63.35	44.20 ± 26.96
6 % H_2O_2+PMJ	335.78 ± 23.30	261.56 ± 30.72	64.20 ± 16.89
15 % H_2O_2+PMJ	312.00 ± 28.61	267.22 ± 35.76	39.40 ± 12.90
25 % H_2O_2+PMJ	336.67 ± 31.66	283.22 ± 9.35	41.46 ± 20.84
35 % H_2O_2+PMJ	314.29 ± 51.16	287.29 ± 55.25	46.40 ± 20.47

Mean values of the changes of microhardness (and standard deviations) within each group are also included as a separate column [57]

(3) Enamel surface microhardness evaluation

The microhardness of the enamel surface was measured before and after treatment (for 15 min) in each experiment group using a Shimadzu HMV tester with a Knoop indenter at a load of 0.9807 N for 15 s. Detailed procedures can again be found in reference [57]. Table 14.1 shows the mean and standard deviation of the microhardness values.

In all groups, the microhardness was reduced after a 15 min treatment. However, the change was barely outside the combined uncertainties of the measured values before and after the treatment. An ANOVA analysis of the mean shows no significant differences in the change of the microhardness between the teeth in the blank control group and those in the treated groups. The reduction in the microhardness of the teeth in the blank control group may be attributed to the fact that the teeth were stored in liquid and dried before each microhardness test. Thus, we conclude that the change in the microhardness was not significantly affected by the use of 35 % hydrogen peroxide alone or by the use of H_2O_2 at different concentrations in conjunction with PMJ. Similar evaluations were carried for the saline and PMJ case and the results were similar.

14.3.4 Plasma Root Canal Treatment

14.3.4.1 Background

Over 700 oral bacterial species are known to exist to date [59]. They usually live in harmony with the host until either the balance of the composition of the microflora is altered or they enter cites that are usually not accessible to them [60]. Oral bacteria (and bacterial biofilm) can grow on several distinct surfaces in the oral cavity: teeth, mucosal surface and gingival crevices, each of which has its unique physical and biological properties. Teeth are the only hard non-shedding surface of the body and have their own complex structure (as described in Sect. 14.3.1). At the center of the tooth lies a hollow area called pulp chamber that contains nerve, blood vessel and lymph tissue. The blood vessels receive nutrition and the nerves send signals to the brain. This hollow area narrows down and extends to the root of the tooth where it often splits into more intricate anatomical branches. Oral bacteria such as *Enterococcus Faecalis (E. faecalis)* may enter the pulp chamber through cracks or cavities, causing tooth pulp and consequently root canal infection.

Human root canal infection is normally treated with methods such as mechanical cleaning, irrigation, laser irradiation, ultrasound, and the application of hypochlorite and other antibacterial compounds [61]. The purpose of root canal therapy is to eliminate the infection of the root canal system and prevent re-infection. However, these traditional treatment methods cannot completely eliminate bacteria (or bacterial biofilm) from endodontic sites [62], with remaining cultures ranging from 40–60 % [63, 64]. This is attributed to the complex anatomy of root canal systems, the diffused localization of endodontic infections, the limited diffusibility of antimicrobials and the relative antimicrobial resistance of infecting bacteria [63]. The majority of these re-infection cases are attributed to *Enterococcus Faecalis (E. faecalis)*, which can grow within dentinal tubules and re-infect the obturated root canal [65, 66]. *E. faecalis* can be easily destroyed in an open environment, but it becomes more resistant when growing in the root canal system [67]. According to previous studies, proteases, such as serine protease, gelatinase, and collagen-binding protein (Ace) are secreted and easily bind to dentin firmly [68]. Because of these biological properties, *E. faecalis* has the capacity to live in dentinal tubules and endure prolonged times of starvation [69, 70] and *E. faecalis* biofilms are orders of magnitude more resistant to phagocytosis, antibodies, or antimicrobials than planktonic bacteria [71, 72]. This section will focus on the in vitro treatment of root canals in extracted human teeth that are innoculated with planktonic *E. faecalis* and *E. faecalis* bacterial biofilm with the PMJ and the PP, respectively. Re-infection prevention will also be discussed biefly.

14.3.4.2 Inactivation of Planktonic *E. faecalis* in Root Canal

Premixed research grade argon with 2 vol % O_2 is used as the working gas in the PMJ device. The other operating parameters are similar to those described in Sect. 14.2. The length of the PMJ reaches approximately 5 mm, which is where the inoculated teeth are placed. All extracted human teeth were decoronated leaving a root length of 15 mm. The root canals were prepared with up to 40# Ni–Ti hand instruments. Each apical foramen was sealed with composite resin. All specimens were sterilized by an autoclave before further treatment. *E. faecalis* (ATCC29212) was cultured in a Luria-Bertani (LB) medium for 18 h until it reached the logarithmic growth phase. Then 10 μl fresh diluted suspension of 10^6 CFU/ml (colony forming units per milliliter) was injected into the root canals. This concentration is chosen since it is close to the clinical situation. Some teeth received gas flow only for 2, 4, 6 and 8 min, serving as the negative control; the others received PMJ treatment for the same time periods. Each condition was repeated for at least three times for statistical evaluation purposes. After each treatment, the root canals were rinsed with Luria-Bertani (LB) medium for five times. The collected suspensions was further mixed the LB medium and plated on a Petri dish. CFU evaluation was carried out after standard culturing procedure. The inactivation rate is defined as $\left(1 - \mathrm{CFU}_{\mathrm{treated}}/\mathrm{CFU}_{\mathrm{control}}\right) \times 100\,\%$. A detailed experimental description can be found in reference [73].

Figure 14.15 shows the inactivation rates of *E. faecalis* in a root canal after various PMJ treatment times. The inactivation rate gradually increases with the treatment time and reaches 98.8 % after an 8-min PMJ treatment.

In traditional root canal therapy, other than biomechanical preparation and root canal shaping, intracanal medicament is usually applied to the infected root canal before obturation for the purpose of re-infection prevention. The re-infection of root canal after 8, 10, 20, 30, 40 min of PMJ treatment was evaluated after the treated teeth were incubation for 7 days at 37 °C. The results are compared with traditional intracanal medicament treatments with formocresol (FC), camphor phenol (CP) and calcium hydroxide ($Ca(OH)_2$) and plotted in Fig. 14.16. The re-infection rate is defined as: $\mathrm{CFU}_{\mathrm{treated}}/\mathrm{CFU}_{\mathrm{control}} \times 100\,\%$ [73]. Root canals treated with a PMJ for 8 min showed a re-infection rate about 62 % after 7 days. The re-infection rate decreased gradually with the increase of the PMJ treatment time, reaching 0.8 % for teeth treated with the PMJ for 30 min. The samples treated with the PMJ for 40 min showed no detectable re-infection, while the samples treated with intracanal medicaments for 7 consecutive days showed a re-infection rate of 0 % (FC), 2.8 % (CP) and 4.2 % ($Ca(OH)_2$), respectively.

Teeth from the control group and the groups treated with the PMJ for 8 and 30 min were further examined with scanning electron microscopy. SEM pictures of the whole root canal, the middle third and the apical third of the root canal were taken with magnifications of 100, 5000 and 5000, respectively and are shown in Fig. 14.17.

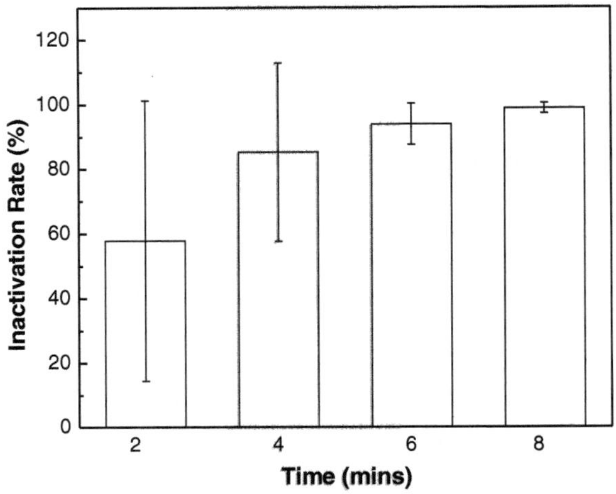

Fig. 14.15 Inactivation of *E. faecalis* in human tooth canal at different PMJ treatment time. Inactivation rate is defined as $\left(1\text{-CFU}_{\text{treated}}/\text{CFU}_{\text{control}}\right) \times 100\,\%$ [73]

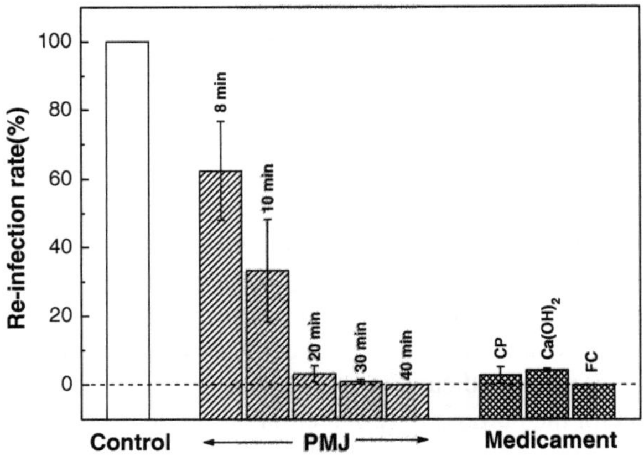

Fig. 14.16 Re-infection evaluation of *E. faecalis* infected root canal treated by PMJ (for 8, 10, 20, 30 and 40 min) and traditional intracanal medicaments: formocresol (FC), camphor phenol (CP) and calcium hydroxide ($Ca(OH)_2$) in 7 days [73]

In the control group, bacteria cover the entire root canal surface from the top third to the apical third (Fig. 14.17a–c). The root canal is partially cleaned after 8 min PMJ treatment with a clear contrast line separating the zones with and without bacteria (arrow in Fig. 14.17d), indicating the active species generated by plasma have a limited penetration depth in 8 min. The middle third of the tooth root canal was cleaned predominantly without bacteria observed around the opening of the dentinal tubules (Fig. 14.17e). A considerable fraction of bacteria still survived in the apical third of the root canal (Fig. 14.17f). This could explain the high re-infection rate in 7 days with an 8 min PMJ treatment. When the treatment time was extended

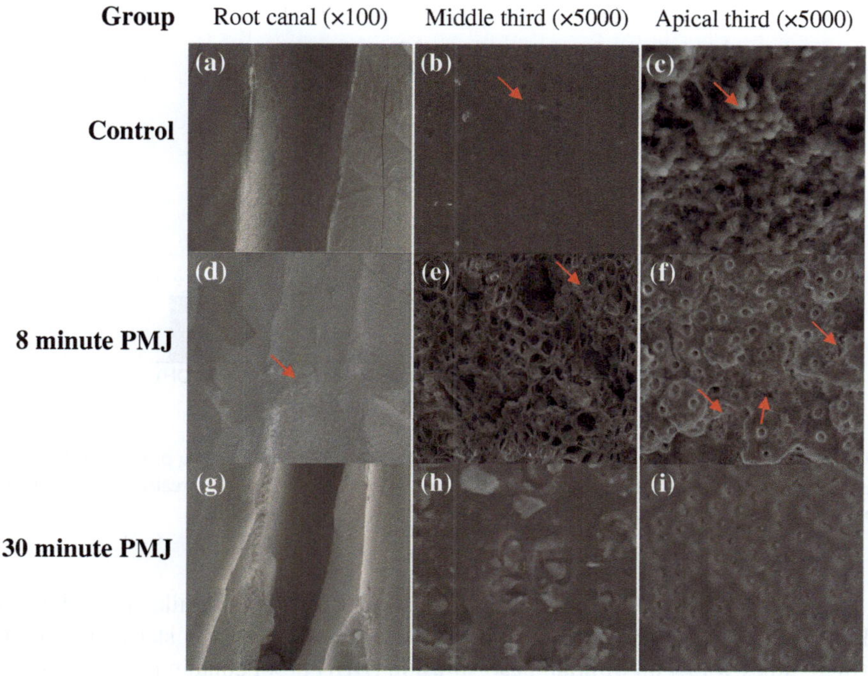

Fig. 14.17 SEM images of *E. faecalis* infected tooth root canal from the control group: **a–c**, 8 min PMJ treatment group: (**d–f**), and 30 min PMJ treatment group: **g–i**. *Red arrows* in the pictures indicate where *E. faecalis* can be seen, including inside the dentinal tubule [73]

to 30 min, the entire root canal was thoroughly cleaned without bacteria observed in the SEM pictures (Fig. 14.17g–i), suggesting appropriate PMJ treatment could effectively prevent re-infection.

14.3.4.3 Inactivation of *E. faecalis* Biofilms in Root Canal

Biofilms are structured microbial communities and are much more resistant to common treatment methods than the planktonic bacteria. $Ca(OH)_2$ is reported to have limited effectiveness in eliminating bacterial biofilms from human root canals [74, 75]. In this situation, more effective disinfection methods are needed to eradicate *E. faecalis* biofilm from persistent root canal infections. The plasma pipette (PP) was used to inactivate *E. faecalis* biofilm in root canals in vitro. Its effectiveness is compared to the results from the traditional $Ca(OH)_2$ intracanal treatment. During each treatment, the distance between the tip of the PP source and the sample was approximately 5 mm and the gas temperature at the top of the root canal ranged from 25 to 31 °C.

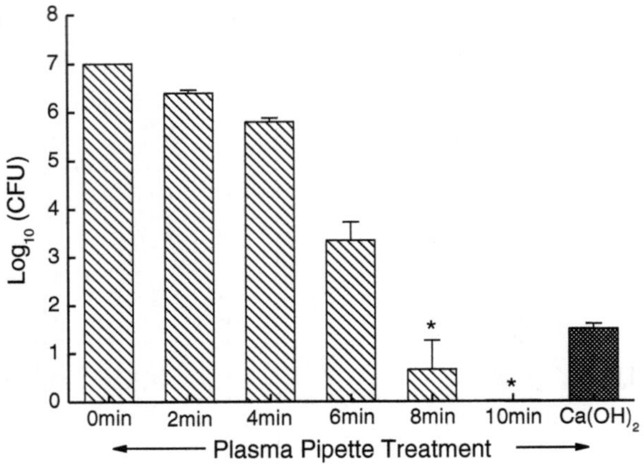

Fig. 14.18 The inactivation of *E. faecalis* biofilms by an Ar/O_2 (2 %) plasma pipette for 2, 4, 6, 8 and 10 min. 0 min treatment is the negative control and the 7 day $Ca(OH)_2$ treatment group is the positive control [27]

The preparation of the single root canal teeth follows a similar procedure as described in the previous section. The specimens were placed in sterile microcentrifuge tubes with 1 ml of brain heart infusion (BHI) broth containing 10^8 CFU/ml of *E. faecalis*. Each specimen was incubated anaerobically for 7 days and 1 ml sterile BHI broth was refreshed every 2 days to ensure viability of the bacteria. *E. faecalis* biofilm formed in root canal after 7 days of culturing, and are treated with PP for 2, 4, 6, 8 and 10 min. Infected teeth without PP treatment and treated with $Ca(OH)_2$ for 7 days at 37 °C are used as negative and positive controls, respectively. The detailed procedure can be found in reference [27]. After PP treatment of the teeth, 15 μl ultra-pure water was injected into the root canal. A smooth broach was used to mill in the canal for 1 min, and then the two # 15 sterile paper points were inserted into the canal to draw the bacteria out. The paper points contaminated with the bacteria were put into a centrifuge tube with 1 ml ultra-pure water and the process was repeated three times. The centrifuge tube was then shaken for 1 min and 100 μl of the liquid was removed to be cultured on a petri dish for a count of the CFUs. The same procedure was also performed with the positive and negative controls. After 7 days incubation, the initial total CFU count in a specimen reached 10^7. The decrease of the logarithm of the CFU is shown in Fig. 14.18. The degree of inactivation of *E. faecalis* biofilm by PP treatment is proportional to the plasma exposure time. A 2, 4 or 6 min PP treatment is less effective than the conventional 7 day $Ca(OH)_2$ treatment. However, longer plasma exposure times of 8 and 10 min showed statistically significantly better results than the $Ca(OH)_2$ group. No residual CFUs were detected in the 10 min PP treated samples.

Successful growth and destruction of the *E. faecalis* biofilm was verified via SEM and Confocal Scanning Laser Microscopy (CSLM) analysis. Freshly prepared clean roots (group A), *E. faecalis* biofilm inoculated root canals without (group B) and with

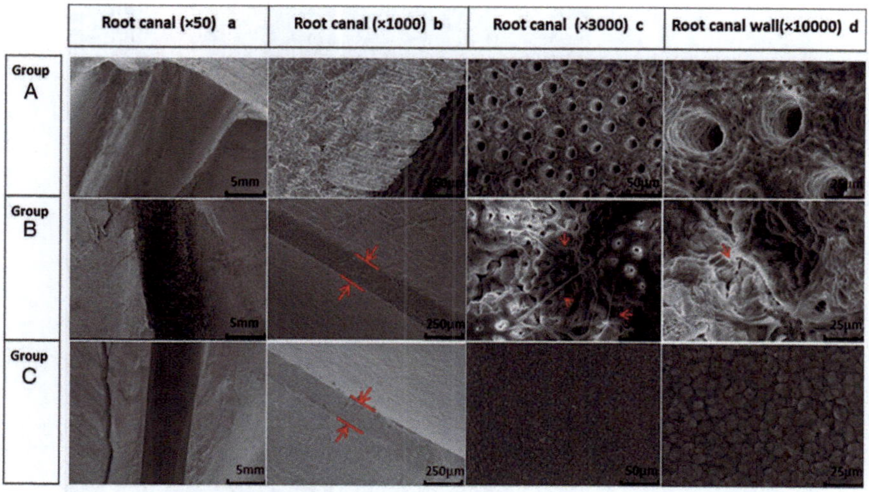

Fig. 14.19 SEM micrographs of *Group A* (prepared root canal), *Group B* (*E. faecalis* biofilm infected root canal) and *Group C* (*E. faecalis* biofilms infected root canal after 10 min plasma treatment) at magnifications of 50× (a), 1000× (b), 3000× (c) and 10000× (**d**). The distance between two arrows (*Group B-b*, *Group C-b*) refers to the thickness of the *E. faecalis* biofilm. The arrows in *Group B-c,d* indicate mycelia in the biofilm [27]

(group C) a 10 min PP treatment were included in this analysis. The SEM pictures were taken at the apical third of the root canals.

As shown in Fig. 14.19, the smear layer of the root canal was entirely removed after preparation (Group A). After 7 days of culturing, the *E. faecalis* biofilm successfully formed in the root canals, with a thickness of about 100 μm (Group B-b). Many layers of bacteria accumulated around the opening of the dentinal tubules with numerous mycelia cross-linking (red arrows in Group B-c, d). After 10 min of cold plasma treatment, we observed that the regular structure of the biofilm was destroyed and replaced with ruptured bacteria. The mycelia and bacteria spheres in the *E. faecalis* biofilm on the surfaces of dentinal tubules disappeared (Group C-c,d). It is noteworthy to state that the debris of the biofilm remained on the canal surfaces (Group C-b).

CSLM images were obtained from two halves of one root specimen (Fig. 14.20a), one half treated with plasma. As shown in Fig. 14.20b, in the untreated half of the sample, most areas of the biofilm are green, indicating that the bacteria are alive. In addition, live bacteria are also clearly observed in the dentinal tubules (lighter green in Fig. 14.20b. The PP treated part shows the biofilm with a pronounced red area indicating that the bacteria are dead (Fig. 14.20c). Bacteria inside the dentinal tubules in the PP treated part is likely dead too since no green stains were observed. The thickness of treated *E. faecalis* biofilms was similar to that of the control specimen (indicated by the arrows in Fig. 14.20).

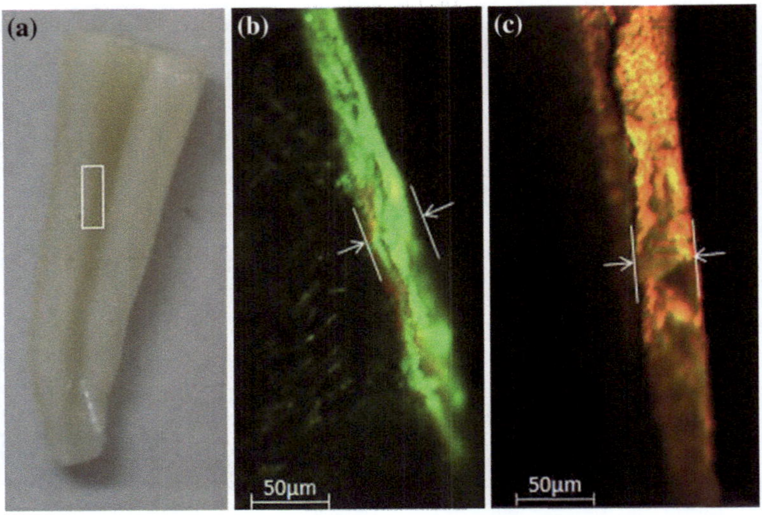

Fig. 14.20 Image of a dissected teeth showing the whole root canal **a** and CSLM images obtained from the mid-area of root canals (box in **a**) showing bacteria in the untreated *E. faecalis* biofilms (**b**), which are all alive (*green areas*) and bacteria in *E. faecalis* biofilm treated by PP source for 10 min (**c**), which are all dead (*red areas*). The distance between two arrows represents the thickness of biofilm [27]

14.3.4.4 Inactivation Mechanism

Optical emission spectroscopy in the spectral range from the ultraviolet to the near infrared region was used to evaluate the reactive species produced in the PMJ. Details of the instrumentation and the experimental setup can be found in reference [73]. Figure 14.21 shows the end-on emission spectrum when the PMJ was operated in air at a discharge current of 30 mA and a gas flow rate of 5 slm. Similar results were also obtained for the PP device (with O emissions less pronounced).

The spectrum is dominated by atomic argon emissions, mostly attributed to direct electron impact excitation of Ar. Strong emissions from excited atomic oxygen at 777 and 844 nm are also observed in the spectrum. The oxygen emission at 777 nm is due to dissociative excitation by electronic collisions of the ground state oxygen molecules [76]. The oxygen emission at 844 nm, however, is mostly due to direct electron impact with ground state oxygen atoms. The excited state of oxygen can also be populated by dissociative excitation of O_2 by the excited argon metastables: $\mathbf{Ar^* + O_2 \rightarrow O + O^* + Ar}$ [77]. The inset in Fig. 14.21 shows emissions from the 2nd positive system of $N_2 \left(\mathbf{C_3 \prod_u \rightarrow B^3 \prod_g}\right)$ with the most intense band head at 337 nm. OH band spectra between 306 and 310 nm, two atomic copper lines at 325 and 327 nm are also observable in this region. OH is likely generated from H_2O (from ambient air) via dissociative excitation by argon metastables. Copper emissions are due to the electrode material. Aside from the direct interaction with bacteria, atomic oxygen can also serve as the precursor for the production of ozone via three-body

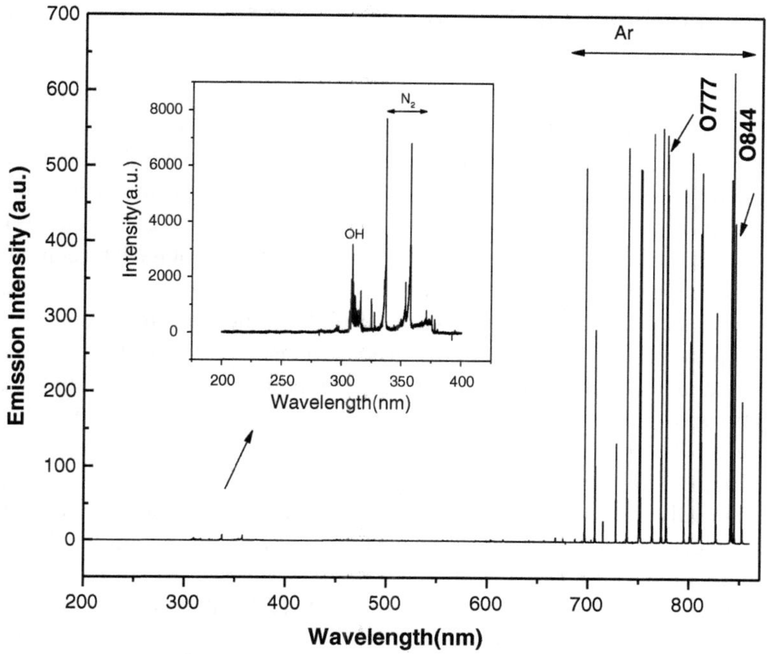

Fig. 14.21 End on optical emission spectrum of Ar/O_2 PMJ operated in air. The inset shows emission spectrum in 200–400 nm where OH and N_2 emissions are observed [73]

reaction: $\mathbf{O + O_2 + M \rightarrow O_3 + M}$. Ozone concentrations 1 cm away from the exit nozzle of the PMJ were evaluated with an Ozone Analyzer and found to be about 40 ppm. This is lower than the so-called "high ozone concentration" threshold of 50 ppm [78]. Nevertheless, one cannot neglect the antimicrobial effects of ozone when it is prescribed in sufficient concentration, and applied for an adequate period of time [79].

ESR spin trapping experiments were also carried out to evaluate the existence of ·OH, $O_2^{-\cdot}$ and $^1O_2$ in distilled water, following the procedures described in Sect. 14.3.3.3 and in reference [73]. Results similar to those shown in Fig. 14.11 were obtained. The reactive species in the gaseous plasma stream can reach places that are hard or even impossible to reach by traditional medicaments. These reactive species can interact directly with the bacteria via, for example, direct oxidation of fatty acids [80]. Charge accumulation on cell membranes from ions produced in the downstream of the plasma can cause the rupture of membrane via coulombic forces [81]. Some reactive species may be the result of the reaction of the reactive species with the surface water on the tooth. Some longer lived reactive species from the plasma can dissolve in the surface water on the tooth and interact with the bacteria indirectly. Whether dissolved or created in water, these reactive species may be transported to the cytoplasm through a series of mechanisms [82], and lead to chain oxidation and thus destruction of DNA molecules as well as cellular membranes and other cell components [83].

14.4 Conclusions and Outlook

This chapter describes a few selected dental applications of plasmas, namely the interaction of plasmas with dental materials, plasma-induced and plasma assisted in vitro teeth bleaching, and plasma root canal treatment, using three non-thermal atmospheric pressure plasma sources. It has been shown that plasmas can modify the surface property of biocompatible materials such as Ti and Y-TZP for better bonding purposes; plasmas can also significantly improve the efficacy of teeth bleaching using dental whitening gel (even at reduced concentrations of H_2O_2 down to 6 %); and can effectively inactivate bacteria and bacterial biofilms in tooth root canal, with re-infection prevention. This chapter only gives a snapshot of the research done by the authors and their collaborators, rather than a comprehensive review of all the dental applications of non-thermal plasmas carried out by many groups. The take-away message is: Plasma Really Works, at least in vitro. For the plasma-dental material interaction, in vitro studies are perfectly sufficient. But the latter two applications must eventually involve in vivo experiments. The roadmap from in vitro to in vivo is not yet clear. Much has yet to be learned before any plasma source can be used on a human tooth in a clinical setting.

One aspect that deserves further attention is the fact that there is multitude of different atmospheric pressure non-thermal plasma sources, utilizing different working gases and power sources. Each has its own advantage, but there is no benchmark to which these sources can be compared to or commonly accepted standards to which they are held, i.e., it is often hard to assess whether the device is safe enough to be directly applied to a living organism. A systematic toxicity study of the different plasma sources is also still lacking. The ideal plasma device should be designed with the high voltage electrode completely shielded and in a fashion that prevents the build-up of excessive static charges on the surfaces of the device to avoid any electric shock to the patient. The working gas also has to be chosen properly, so that the plasma works with sufficient efficacy in a reasonable amount of time, without the generation of unsafe levels of toxic by-products (such as ozone and NO_x). This is of consequence not only for the patient, but also the device operator, who may be exposed to the plasma and their by-products for a prolonged period of time.

As a final remark, most papers describing dental (and (bio) medical) applications of plasmas appear in traditional physics and engineering journals which rarely attract the attention of clinicians, medical researchers, dentists, and dental researchers. Efforts must be made to create a common platform that all parties involved in this multidisciplinary research area can embrace and can use to communicate. Dedicated conference sessions and special journal issues that focus on (bio) medical (including dental) applications of plasmas should become part of various medical conferences and medical journals.

Acknowledgments This work was supported in part by Bioelectrics Inc. (U.S.), by the Peking University Biomed-X Foundation, by the Ministry of Science and Technology (MOST) Program of the China International Science and Technology Cooperation under Grant 2009DFB30370, by the China National Basic Research Program under Grant 2007CB935602, and by the Fundamental

Research Funds for the Central Universities under Grant lzujbky-2012-159. W. Zhu would like to acknowledge partial support from the Electro-Energetic Physics Program of the U.S. Air Force Office of Scientific Research under Grant Number FA9550-08-1-0332 for the fundamental studies of the PMJ and the initial development of PP. The authors would like to thank Ms. Ruixue Wang for her assistance in the preparation of this manuscript at its early stage.

References

1. M. Laroussi, I. Alexeff, W. Kang, IEEE Trans. Plasma Sci. **28**, 184–188 (2000)
2. M. Laroussi, IEEE Trans. Plasma Sci. **24**, 1188–1191 (1996)
3. M. Laroussi, T. Akan, Plasma Proc. Polym. **4**, 777 (2007)
4. E. Stoffels, A.J. Flikweert, W.W. Stoffels, G.M.W. Kroesen, Plasma Sources Sci. Technol. **11**, 383 (2002)
5. E. Stoffles, R.E.J. Sladek, I.E. Kieft, H. kersten, R. Wiese, PPCF 46, B167 (2004).
6. K. Becker, U. Kogelschatz, K. Schoenbach, R. Barker (eds.), *Non-Equilibrium Air Plasmas at Atmospheric Pressure, History of Non-Equilibrium Air Discharges* Chap. 2 (U. Kogelschatz, Y.S. Akishev, and A.P Mapartovich) (IOP Publishing, Bristol, 2005)
7. M. Coulomb, Me'm. Acad. R. des Sci. de Paris 612–638 (1785)
8. W. Bridgewater and S. Kurtz (eds.), *The Columbia Encyclopedia*, 3rd edn. (Columbia University Press, New York, 1963)
9. M. Faraday, *Experimental Researches in Electricity*, vol. I (Taylor and Francis, London, 1839)
10. M. Faraday, *Experimental Researches in Electricity*, vol. II (Taylor and Francis, London, 1844)
11. M. Faraday, *Experimental Researches in Electricity*, vol. III (Taylor and Francis, London, 1855)
12. E.E. Kunhardt, IEEE Trans. Plasma Sci. **28**, 1 (2000)
13. K.H. Becker, N.M. Masoud, K.E. Martus, K.H. Schoenbach, Eur. Phys. J. D **35**, 279–297 (2005)
14. K. Becker, U. Kogelschatz, K. Schoenbach, R. Barker (eds.) Non-Equilibrium Air Plasmas at Atmospheric Pressure, (IOP Publication, Bristol, 2005)
15. K. Becker, J.G. Eden, K.H. Schoenbach, J. Phys. D **39**, R55–R70 (2006)
16. R. Foest, M. Schmidt, K. Becker, Int. J. Mass Spectrom. 248, 87–102 (2005); Invited Topical Review.
17. Y. Raizer, *Gas Discharge Physics* (Springer, Heidelberg, 1991)
18. U. Kogelschatz, Plasma Proc. Polym. **23**, 1–46 (2003)
19. K.-D. Weltmann, E. Kindel, T. von Woedtke, M. Hähnel, M. Stieber, R. Brandenburg, Pure Appl. Chem. **82**, 1223 (2010)
20. G.Y. Park, S.J. Park, M.Y. Choi, I.G. Koo, J.H. Byun, J.W. Hong, J.Y. Sim, G.J. Collins, J.K. Lee, Plasma Sources Sci. Technol. **21**, 043001 (2012)
21. J.K. Kolb, A.-A.H. Mohamed, R.O. Price, R.J. Swanson, A. Bowman, R.L. Chiavarini, M. Stacey, K.H. Schoenbach, Appl. Phys. Lett. **92**, 241501–241503 (2008)
22. H. Feng, P. Sun, Y. Chai, G. Tong, J. Zhang, W. Zhu, J. Fang, IEEE Trans. Plasma Sci. **37**, 121–127 (2009)
23. P. Sun, J. Pan, Y. Tian, N. Bai, H. Wu, L. Wang, C. Yu, J. Zhang, W. Zhu, K.H. Becker, J. Fang, IEEE Trans. Plasma Sci. **38**, 1892 (2010)
24. K.-D. Weltmann, E. Kindel, R. Brandenburg, C. Meyer, R. Bussiahn, Ch. Wilke, Th von Woedtke. Contrib. Plasma Phys. **49**, 631 (2009)
25. S. Bornholdt, M. Wolter, H. Kersten, Eur. Phys. J. D **60**, 653 (2010)
26. X. Lu, M. Laroussi, J. Appl. Phys. **100**, 063302 (2006)
27. J. Pan, K. Sun, Y. Liang, P. Sun, J. Zhang, W. Zhu, J. Fang, K.H. Becker, J. Endod. **39**, 105–110 (2013)
28. C. Jiang, M.-T. Chen, C. Schaudinn, A. Gorur, P.T. Vernier, J.W. Costerton, D. Jaramillo, P. Sedghizadeh, M.A. Gundersen, Plasma Proc. Polym. **6**, 479–483 (2009)

29. X. Lu, Y. Cao, P. Yang, Q. Xiong, Z. Xiong, Y. Xian, Y. Pan, IEEE Trans. Plasma Sci. **37**, 668–673 (2009)
30. P.D. Marsh, M.V. Martin, *Oral Microbiology*, 4th edn. (Wright, Oxford, 1999)
31. A.R. Ten Cate, *Oral Histology: Development, Structure, and Function*, 5th edn. (Mosby-Year Book, Saint Louis, 1998)
32. J. Meichsner, M. Schmidt, R. Schneider, H.-E. Wagner (eds.), *Nonthermal Plasma Chemistry and Physics*. Plasma Surface Chemistry (CRC Press, Boca Raton, 2013), pp. 312–387
33. Q. Song, Y.H. Li, A.C. Ritts, B. Yang, M. Chen, L. Hong, C. Xu, X. Yao, Y. Wang, Nonthermal atmospheric plasma treatment for deactivation of oral bacteria and improvement of dental restoration, in *Plasma for Bio-Decontamination, Medicine, and Food Safety, NATO Science for Peace and Security Series A: Chemistry and Biology*, ed. by Z. Machala, K. Hensel, Y. Akishev (Springer, Dordecht, 2012), pp. 215–230
34. M. Schmidt, Surface treatment, in *Non-Equilibrium Air Plasmas at Atmospheric Pressure*, ed. by K. Becker, U. Kogelschatz, K. Schoenbach, R. Barker (IOP Publication, Bristol, 2005), pp. 597–621
35. A. Fridman, *Plasma Chemistry* (Cambridge University Press, New York, 2008)
36. I. Kona, L. Jablonski, A. Kramer, K.-D. Weltmann, T. Kocher, Medical Plasma, in *Dentistry: a Future Therapy for Peri-implantitis, Plasma for Bio-Decontamination, Medicine, and Food Safety, NATO Science for Peace and Security Series A: Chemistry and Biology*, ed. by Z. Machala, K. Hensel, Y. Akishev (Springer, Dordecht, 2012), pp. 191–200
37. M. Yoshari, K. Matsuzaka, T. Inoue, Jap. Dental Sci. Rev. **47**, 89 (2011)
38. T. Smitha, N.B. Chaitanya, Ind. J. Dent. Adv. **2**, 210 (2010)
39. A.C. Ritts, H. Li, Q. Yu, C. Xu, X. Yao, L. Hong, Y. wang, Europ. J. Oral Sci. **118**, 510 (2010)
40. N.R.F.A. Silva, P.G. Coelho, G.B. Valverde, K. Becker, R. Ihrke, A. Quade, V.P. Thompson, J. Biomed. Mat. Res. **99B**, 199 (2011)
41. M.A.M. Silva, A.E. Martinelli, C. Alves Jr, R.M. Nascimento, M.P. Tavora, C.D. Vilar, Surf. Coat. Techn. **200**, 2618 (2006)
42. H.Y. Choi, S.H. Kim, J. Jang, Adv. Mater. **16**, 732 (2004)
43. D.K. Owens, R.C. Wendt, J. Appl. Polym. Sci. **13**, 1741 (1969)
44. N.U. Wetter, M.C. Barroso, J.E.P. Pelino, Lasers Surg. Med. **35**, 254 (2004)
45. M. Tavares, J. Stultz, M. Newman, V. Smith, R. Kent, E. Carpino, J.M. Goodson, J. Am. Dent. Assoc. **134**, 167 (2003)
46. M. Sulieman, Dent. Update **32**, 101 (2005)
47. H.W. Lee, G.J. Kim, J.M. Kim, J.K. Park, J.K. Lee, G.C. Kim, J. Endod. **35**, 587–591 (2009)
48. J. Pan, X. Yang, K. Sun, J. Wang, P. Sun, H. Wu, K.H. Becker, W. Zhu, J. Zhang, J. Fang, IEEE Trans. Plasma Sci. **41**, 325 (2013)
49. J. Pan, P. Sun, Y. Tian, H. Zhou, H. Wu, N. Bai, F. Liu, W. Zhu, J. Zhang, K. Becker, J. Fang, IEEE Trans. Plasma Sci. **38**, 3143 (2010)
50. D.K. Hein, B.J. Ploeger, J.K. Hartup, R.S. Wagstaff, T.M. Palmer, L.D. Hansen, Compendium Continuing Educ. Dent. **24**, 340 (2003)
51. M. Kashima-Tanaka, Y. Tsujimoto, K. Kawamoto, N. Senda, K. Ito, M. Yamazaki, J. Endod. **29**, 141 (2003)
52. B. Halliwell, J. Gutteridge, *Free Radicals in Biology and Medicine*, 3th edn. (Oxford University Press, Cambridge, 1999), p. 351
53. H. Wu, P. Sun, H. Feng, H. Zhou, R. Wang, Y. Liang, J. Lu, W. Zhu, J. Zhang, J. Fang, Plasma Plasma Proc. Polym. **9**, 417–424 (2012)
54. C. Schweitzer, R. Schmidt, Chem. Rev. **103**, 1685 (2003)
55. É. Hideg, C. Spectea, I. Vass, Photosynth. Res. **39**, 2 (1994)
56. A. Eriksson, T. Albrektsson, B. Grane, D. McQueen, Int. J. Oral. Surg. **11**, 115–121 (1982)
57. J. Wang, X. Yang, K. Sun, P. Sun, J. Pan, W. Zhu, K.H. Becker, J. Zhang, J. Fang, IEEE Trans. Plasma Sci. **40**, 2157 (2012)
58. L. Emma, Rodrı'guez-Vilchis, R. Contreras-Bulnes, I. Sa'nchez-Flores, E.C. Samano, Photomed. Laser Surg. **28**, 207–212 (2011)
59. E. Pennisi, Science **307**, 1899–1901 (2005)

60. R.E.J. Sladek, Plasma needle: non-thermal atmospheric plasmas in dentistry, PhD thesis, Technische Universiteit Eindhoven, 2006
61. M. Colak, S. Evcil, Y.Z. Bayindir, N.J. Yigit, J. Contemp. Dent. Pract. **6**, 94 (2005)
62. T. Matsuo, T. Shirakami, K. Ozaki, T. Nakanishi, H. Yumoto, S. Ebisu, J. Endod. **29**, 194–200 (2003)
63. J.A. Soares, M.R. Leonardo, L.A. Silva, M.T. Filho, I.Y. Ito, J. Appl. Oral. Sci. **14**, 355–64 (2006)
64. J.F. Siqueira Jr, I.N. Rôças, S.R. Santos, K.C. Lima, F.A. Magalhães, M. de Uzeda, J. Endod. **28**, 181–184 (2002)
65. G. Sundqvist, D. Figdor, S. Persson, U. Sjögren, Oral Surg. Oral Med. Oral Pathol. Oral Radiol. Endod. **85**, 86 (1998)
66. G. Dahlèn, W. Samuelsson, A. Molander, C. Reit, Oral Microbiol. Immun. **15**, 309 (2000)
67. R.P. Roach, J.F. Hatton, M.J. Gillespie, J. Endod. **27**, 657–660 (2001)
68. T.S. Hubble, J.F. Hatton, S.R. Nallapareddy, B.E. Murray, M.J. Gillespie, Oral Microbiol. Immunol. **18**, 121–126 (2003)
69. R.M. Love, Int. Endod. J. **34**, 399–405 (2001)
70. D. Figdor, J.K. Davies, G. Sundqvist, Oral Microbiol. Immunol. **18**, 234–239 (2003)
71. J.W. Distel, J.F. Hatton, M.J. Gillespie, J. Endod. **28**, 689–693 (2002)
72. J.W. Costerton, P.S. Stewart, E.P. Greenberg, Science **284**, 1318–1322 (1999)
73. R. Wang, H. Zhou, P. Sun, H. Wu, J. Pan, W. Zhu, J. Zhang, J. Fang, Plasma Med. **1**, 143–155 (2011)
74. L.B. Peters, A.-J. van Winkelhoff, J.F. Buijs, P.R. Wesselink, J. Endod. **35**, 13–21 (2002)
75. C. Sathorn, P. Parashos, H. Messer, Int. Endod. J. **40**, 91–99 (2007)
76. J. Jask, P. Macko, V. Martisovits, P. Luckac, P. Veis, Czechoslov. J. Phys. **54**, 1297–1303 (2004)
77. S. Saloum, M. Naddaf, J. Appl. Phys. **82**, 742–747 (2008)
78. V. Bocci, *Oxygen-Ozone therapy, a critical evaluation* (Kluwer Academic Publishers, Dordrecht, 2002)
79. W.J. Kowalski, W.P. Bahnfleth, B.A. Striebig, T.S. Whittam, AIHA J. **64**, 222–227 (2003)
80. M. Laroussi, IEEE Trans. Plasma Sci. **37**, 714–725 (2009)
81. M. Laroussi, IEEE Trans. Plasma Sci. **30**, 1409–1415 (2002)
82. X. Deng, J. Shi, M.G. Kong, IEEE Trans. Plasma Sci. **34**, 1310–1315 (2006)
83. D. Dobrynin, G. Fridman, G. Friedman, A. Fridman, New J. Phy. **11**, 115020 (2009)

Index

M. Bonitz et al. (eds.), *Complex Plasmas*, Springer Series on Atomic, Optical, and Plasma Physics 82, DOI: 10.1007/978-3-319-05437-7,

MIX
Papier aus verantwortungsvollen Quellen
Paper from responsible sources
FSC® C105338

If you have any concerns about our products,
you can contact us on
ProductSafety@springernature.com

In case Publisher is established outside the EU,
the EU authorized representative is:
Springer Nature Customer Service Center GmbH
Europaplatz 3, 69115 Heidelberg, Germany

Printed by Libri Plureos GmbH
in Hamburg, Germany